Discovering Biological Science

PRENTICE-HALL INTERMEDIATE SCIENCE SERIES

SERIES EDITOR:

William A. Andrews
Professor and Chairman of Science Education
Faculty of Education, University of Toronto

Understanding Science 1
Understanding Science 2
Physical Science: An Introductory Study
Biological Science: An Introductory Study
Discovering Physical Science
Discovering Biological Science
  — Student Text
  — Teacher's Guide

William A. Andrews, Faculty of Education, University of Toronto
Brenda J. Andrews, Department of Medical Genetics, University of Toronto
David A. Balconi, Leaside High School
Nancy J. Purcell, Dr. F. J. Donevan Collegiate Institute

Discovering Biological Science

Prentice-Hall Canada Inc., Scarborough, Ontario

CANADIAN CATALOGUING IN PUBLICATION DATA

Main entry under title:
   Discovering biological science

For use in secondary schools.
Includes index.
ISBN 0-13-215616-4

1. Biology.   I. Andrews, William A., 1930-

QH308.7.D57   574   C82-095185-4

© 1983 by Prentice-Hall Canada Inc., Scarborough, Ontario

All Rights Reserved.
No part of this book may be reproduced in any form or by any means without permission in writing from the publisher.

Prentice-Hall Inc., Englewood Cliffs, New Jersey
Prentice-Hall International, Inc., London
Prentice-Hall of Australia, Pty., Ltd., Sydney
Prentice-Hall of India Pvt., Ltd., New Delhi
Prentice-Hall of Japan, Inc., Tokyo
Prentice-Hall of Southeast Asia (PTE) Ltd., Singapore
Editora Prentice-Hall do Brasil Ltda., Rio de Janeiro

ISBN 0-13-215616-4

Metric Commission Canada has granted permission for use of the National Symbol for Metric conversion. The metric usage in this text has been reviewed by the Metric Screening Office of the Canadian General Standards Board.

Production Editors: Rebecca Vogan, Iris Skeoch
Art Director: Joe Chin
Illustrator: James Loates Illustrating
Production: Alan Terakawa
Compositor: Canadian Composition Limited

Printed and bound in Canada by The Bryant Press Limited

1 2 3 4 5 6 7 BP 89 88 87 86 85 84 83

Contents

Unit One: Life and Its Many Forms
Chapter 1: What is Biology? *4*
Chapter 2: Characteristics of Living Things *13*
Chapter 3: Classification of Living Things *21*

Unit Two: Cells: The Building Blocks of Life
Chapter 4: The Microscope *42*
Chapter 5: The Cell and Its Parts *52*
Chapter 6: Cells at Work: How Substances Enter and Leave Cells *67*
Chapter 7: Cells at Work: Photosynthesis and Respiration *80*
Chapter 8: Cells at Work: How Living Things Grow *92*

Unit Three: Monerans, Protists, and Fungi: Three Kingdoms of Living Things
Chapter 9: Monerans and Viruses *102*
Chapter 10: Bacteria and Us *117*
Chapter 11: Protists *132*
Chapter 12: Fungi *147*

Unit Four: The Plant Kingdom
Chapter 13: A Survey of the Plant Kingdom *164*
Chapter 14: Reproduction in Flowering Plants *180*
Chapter 15: Roots, Stems, and Leaves *196*
Chapter 16: Growing Plants *215*

Unit Five: The Animal Kingdom
Chapter 17: A Survey of the Animal Kingdom *234*
Chapter 18: Feeding and Digestion *244*
Chapter 19: Breathing Systems *262*
Chapter 20: Internal Transport Systems *274*
Chapter 21: Reproductive Systems *289*

Unit Six: Human Biology
Chapter 22: Food and Nutrition *308*
Chapter 23: The Digestive System *327*
Chapter 24: The Breathing System *342*
Chapter 25: The Circulatory System *355*
Chapter 26: The Excretory and Reproductive Systems *377*
Chapter 27: Structural Systems *391*
Chapter 28: Control Systems *405*

Unit Seven: Continuity
Chapter 29: Basic Genetics *422*
Chapter 30: Genes, Chromosomes, and DNA *435*
Chapter 31: Importance of Genetics *452*
Chapter 32: Evolution *468*

Unit Eight: Ecology
Chapter 33: How Living Things Interact *482*
Chapter 34: Matter and Energy in Ecosystems *499*
Chapter 35: Aquatic Ecosystems *511*
Chapter 36: Terrestrial Ecosystems *528*

Unit Nine: Building a Better Future
Chapter 37: Tobacco, Alcohol, and Other Drugs *554*
Chapter 38: Human Population Growth: Problems and Solutions *568*
Chapter 39: Conserving Natural Resources *580*

Acknowledgments

The authors wish to acknowledge the competent professional help received from the staff of Prentice-Hall Canada Inc. in the production of this text. In particular, we extend our thanks to Rebecca Vogan and Iris Skeoch for their skillful editorial work and to Rand Paterson and Alan Terakawa for their coordination of the production aspect of this text. We also thank Steve Lane for his untiring and valuable assistance in the planning and development of this text.

We wish, further, to thank the many teachers who reviewed the manuscript and offered their constructive criticisms. Many students volunteered to serve as models in the photographs. In particular, we wish to acknowledge the assistance of Janice Palmer and her Environmental Studies students at North Toronto Collegiate Institute and the assistance of Donna Moore and her Environmental Science students at David and Mary Thomson Collegiate Institute. We are grateful for the photographs that these people and those mentioned in the photo credits below provided in order to make our book more appealing and useful.

We would be remiss if we did not express our appreciation of the imaginative, attractive, and accurate art work of James Loates. Finally, we extend a special word of appreciation to Lois Andrews for her skillful and dedicated preparation of the final manuscript and the index.

W. A. Andrews
Editor and Senior Author

Photo Credits

Cover photo: W. Griebeling/Miller Services. Figs. 1-1; 1-3; 1-4; 2-8; 4-0; 4-6; 5-3; 7-2; 18-1; 22-0; 27-7; 27-11; 28-9; 30-7; 31-8; 35-3; 37-0; 39-15: Janice Palmer. Fig. 2-3: courtesy of USDA Photo. Fig. 2-4: courtesy of Manfred C. Schmid. Figs. 2-5; 12-3,D; 12-3,E; 12-3,F; 12-14; 29-0; 31-12,B; 31-12,C; 31-16; 39-2; 39-6; 39-7: courtesy of the Ontario Ministry of Agriculture and Food. Figs. 5-8; 5-9; 5-12: courtesy of Dr. K. Kovacs and Dr. E. Howarth, Department of Pathology, St. Michael's Hospital, Toronto and Ena Ilse, Sandra Briggs, Donna McComb and Sandra Cohen of the Electron Microscopy Unit of St. Michael's Hospital, Toronto. Fig. 9-0: courtesy of Eric L. Vogan. Figs. 6-1; 9-2,A; 9-2,B; 11-1,B; 13-1,A; 13-1,B: courtesy of Dr. Pamela Stokes, Botany Department, University of Toronto. Figs. 9-2,C; 9-2,D: courtesy of Ontario Ministry of Health, Laboratory Services. Figs. 9-10; 9-11; 9-13: courtesy of Professor Frances Doane, Department of Microbiology and Parasitology, University of Toronto. Fig. 10-1: courtesy of Dr. P. C. Fitz-James and K. Ebisuzaki. Fig. 11-1,A: Gilbert L. Twiest, excerpted from Fresh-water Plants and Animals, sound filmstrip series, Prentice-Hall Media. Figs. 12-3,A; 12-3,B; 12-3,C; 13-0; 13-1,C;

30-14; 32-3,A; 32-3,B; 32-10,A; 32-10,B; 33-2,A; 33-2,B; 33-8; 33-14; 36-6; 36-25; 36-30; 39-1; 39-4; 39-11: courtesy of the Ontario Ministry of Natural Resources. Fig. 22-7: courtesy of World Vision Canada. Fig. 24-10: courtesy of the Ontario Ministry of Municipal Affairs and Housing. Fig. 26-3: courtesy of Health Sciences Media Services, Sunnybrook Medical Centre, Toronto. Figs. 30-10,A; 30-10,B; 31-6: courtesy of Kim Mellick, Cytogenetics Laboratory, Hospital for Sick Children, Toronto. Fig. 30-15: courtesy of Dr. V. Markovic, Surrey Place Centre, Toronto. Figs. 31-2,A; 31-2,B; 31-4,C: courtesy of Dr. J. Crookston, Department of Hematology, Toronto General Hospital, Toronto. Fig. 31-5: courtesy of Cystic Fibrosis Foundation, Toronto. Fig. 32-5: Tony Evangelista. Fig. 33-2,E: courtesy of NASA. Fig. 36-11: courtesy of J. C. Ritchie. Fig. 36-12: Duncan M. Cameron, Jr. Figs. 36-13,A; 36-13,B: T. C. Dauphiné, Jr. Fig. 36-15: Eric S. Grace. Figs. 36-16; 36-33: Donna K. Moore. Fig. 36-19: courtesy of the Ontario Ministry of Natural Resources and Information Canada. Fig. 38-5: courtesy of The United Church Observer. Fig. 38-6: courtesy of The Toronto Star. All other photos by W. A. Andrews.

To the Student

Welcome to DISCOVERING BIOLOGICAL SCIENCE! This book deals with a science called biology. BIOLOGY *is the study of living things*: what they are made of, what they do, and how their parts work. It's important to know about the living things that share the earth with us. We hope you will enjoy this course and learn many interesting and useful things.

This book is divided into 9 main parts called units.

Unit 1 Life and Its Many Forms
Unit 2 Cells: The Building Blocks of Life
Unit 3 Monerans, Protists, and Fungi:
       Three Kingdoms of Living Things
Unit 4 The Plant Kingdom
Unit 5 The Animal Kingdom
Unit 6 Human Biology
Unit 7 Continuity
Unit 8 Ecology
Unit 9 Building a Better Future

In Unit 1 you will study the nature of life itself. You will also look at the 5 kingdoms of living things: monerans, protists, fungi, plants, and animals. In Units 2 to 5 you will study living things from the smallest cell to the largest animal.

In Unit 6 you will learn how the human body works. You will also find answers to important questions like these: How does my heart work? Why is regular exercise important? Am I in good shape? What happens to the food I eat? Do I eat the right foods? Will smoking give me lung cancer? How did I get here in the first place?

In Unit 7 you will learn about genetics: How are traits inherited? How are new breeds of plants and animals developed? What is a clone? In Unit 8 you will study ecology: How do all living things depend on one another? Finally, in Unit 9 you will look at solutions to some problems: pollution; overpopulation; shortages of resources; shortages of energy.

How To Use This Book

This book was written to make the learning of biology interesting and clear. The following points are some of the features that help make learning easier. To get the most out of this text, you should keep these points in mind.

1. Reading Sections

Each chapter has some reading sections. Read them carefully. They provide the basic information that you need to do the activities. They also summarize and explain activities. As well, they provide interesting examples from the world around us.

Always do the questions in the *Section Review* at the end of a section. They aren't difficult. You can find the answers to all the questions in the section. Finding these answers will help you understand what you have read.

2. Laboratory Activities

Always read the procedure before you begin the activity. Then follow the steps carefully. Finally, try all the *Discussion* questions. They are included to help you find out if you understand the activity. Some questions are more difficult than others. Don't be discouraged if you can't do all of them. But make sure you understand the answers when they are discussed in class.

3. Chapter Overview

Each chapter begins with an overview of the reading and activity sections in it. Read this overview before you start the chapter. It will give you a feeling for what the chapter is about.

4. Chapter Summary

Every chapter ends with a list of *Main Ideas*. Always read them. If you don't understand any of them, ask your teacher for help. A *Glossary* is also included at the end of each chapter. It gives the meanings of key terms. When necessary, it also gives their pronunciations. Use the glossary to review the key terms.

5. Chapter Questions

The end of each chapter includes five kinds of questions:

a) *True or False Items.* These items are pure recall. They test what you have remembered from the chapter. You can find answers to them by looking back through the chapter. However, don't look back until you have tried to answer the questions.

b) *Completion Items.* These items serve the same purpose as the true or false items.

c) *Multiple Choice Items.* Some of these are recall questions. Like the two previous types, they test only how well you remember the material. However, other items will help you find out how well you *understand* the material.

d) *Using Your Knowledge.* These questions give you a chance to use the knowledge you have gained in the chapter. Usually they deal with real world applications that are interesting and useful.

e) *Investigations.* These include further classroom activities, home projects, library projects, debates, and so on. The investigations give you a chance to work on your own.

W. A. Andrews
Editor and Senior Author

To the Teacher

Prentice-Hall Canada Inc., Educational Book Division, and the authors of DISCOVERING BIOLOGICAL SCIENCE are committed to the publication of instructional materials that are as bias-free as possible. This text was evaluated for bias prior to publication.

The authors and publisher of this book also recognize the importance of appropriate reading levels and have therefore made every effort to ensure the highest possible degree of readability in the text. The content has been selected, organized, and written at a level suitable to the intended audience. Standard readability tests have been applied at several stages in the text preparation to ensure an appropriate reading level.

Readability tests, however, can only provide a rough indication of a book's reading level. Research indicates that readability is affected by much more than word or sentence length; factors such as presentation, format and design, none of which are considered in the usual readability tests, also greatly influence the ease with which students read a book.

One other important factor affecting readability is the extent to which the text is motivational for students. Thus the following features were incorporated into this book to increase reader comprehension further. Page references are given to provide examples of most features.

Real World Examples:
Wherever possible, the text relates the theory to the everyday world. This motivates students and makes the content more meaningful by showing that applications of biological science are all around us. In addition, some of the applications of biological science are developed in major sections or entire chapters. Examples include: disease defence and the prevention of food spoilage (Ch. 10), how to grow plants (Ch. 16), how the human body works and its proper care (Unit 6), artificial respiration (Ch. 24), the effect of sewage on lakes (p. 508), water pollution (Ch. 35), tobacco, alcohol, and other drugs (Ch. 37), human population growth (Ch. 38), conservation (Ch. 39).

Student Involvement:
The text recognizes that students learn best if they have concrete, hands-on experiences. Hence the book features over 100 **laboratory activities** integrated with the narratives. Instructions for each lab activity guide students to understand exactly what they are to do and observe in order to complete the activity successfully. Examples of typical laboratory activities can be found on pp. 154 and 519. Each

activity consists of a short **paragraph** spelling out the purpose of the activity, a statement of the **problem** to be investigated, a list of required **materials**, the **procedure** carefully and thoroughly laid out in a step-by-step fashion, and **questions** at the conclusion of the activity. In addition, safety in the laboratory is stressed at all times. The word **"CAUTION"** in green ink draws the student's attention to specific precautions whenever necessary.

Questions:
The text features a wealth of questions. **Review questions** after each narrative section (pp. 44, 235, 570) are at the recall level and assist students with their reading and understanding of the narratives. These questions also help students to prepare notes that summarize the material of the narratives. **Discussion** questions after each lab activity (pp. 119, 366) are at the recall and comprehension level. In addition, some items require application of knowledge and skills from the activity to real-life situations. Furthermore, **study questions** at the end of each chapter help students recall, apply and extend important content (pp. 65, 433). These consist of: *true or false* (recall); *sentence completion* (recall); *multiple choice* (recall and comprehension); *using your knowledge* (some application questions); and *investigations* (a few lab activities, home projects, debates, library projects, etc.).

Attractive Format:
The book employs an attractive format to appeal to students and encourage the use of the text. It features an uncluttered, single-column format. In addition, generous use is made of headings and subheadings as well as second colour.

Numerous Illustrations:
Over 500 line drawings and photographs provide visual reinforcement of the printed word.

Topic Development:
Recognizing that most students using the text have short attention spans, the book is broken up into units, each consisting of several short chapters. Each chapter contains some narrative and some activity sections.

New Terms:
The terms are printed in second colour and are clearly explained in context when first introduced. Many of the terms are phonetically pronounced (p. 342). The most important of these terms are then listed (with their meanings) at the end of the chapter (p. 352).

Chapter Openings:
Each chapter opens with a brief outline of content to enable students to see at a glance where the chapter is headed. These introductions

consist of an overview of both the narrative and the activity sections (p. 215).

Chapter Conclusions:
Each chapter concludes with a short listing of the main ideas of the chapter and a glossary of the major terms of the chapter, including their pronunciations (when required) and meanings (p. 590).

End of Text:
The book concludes with a comprehensive index to help students use and find important biological science terms and information (p. 593).

Life and Its Many Forms

CHAPTER 1
What Is Biology?

CHAPTER 2
Characteristics of Living Things

CHAPTER 3
Classification of Living Things

A potato looks much more like a rock than a horse. Yet, in most ways, the potato is more like the horse than the rock. Why is this so? It's because the potato and horse are living and the rock is nonliving.

What do we mean by "living"? What do living things like a potato and a horse have in common? How do they differ from nonliving things? And, how do living things differ from one another? In this unit you will find answers to questions such as these.

Fig. 1-0 How many kinds of living things do you see? What do they have in common?

1 What Is Biology?

1.1 Why Study Biology?
1.2 How Do We Study Biology?
1.3 Activity: What Makes Radish Seeds Germinate?

What is biology? What good is biology to me? Why should I study biology? What will I be doing in this course?

You are probably asking yourself questions like these as you begin this course. Let's answer them before we go further.

1.1 Why Study Biology?

It's fun.... It's useful.... It's important. These are three good reasons for studying biology.

Biology Is Fun

Biology *is the study of life.* It's the study of living things — how they live; how they grow; how they reproduce; how they interact with one another; how they interact with their surroundings.

All of us are curious about the living things that share the earth with us. And it's fun satisfying that curiosity. In this course you will spend many enjoyable hours at the microscope (Fig. 1-1). You will do interesting experiments with plants and animals. And you will study exciting topics like ecology and genetics. We don't promise that every day will be exciting. But most will.

What you learn in this course can increase your enjoyment for many years to come. Do you like outdoor activities (Fig. 1-2)? Do you grow plants? Do you have a garden? Do you have a pet? Do you like watching birds and other animals? Do you run, play hockey, or swim? All of these hobbies are fun. But they are more fun when you understand the living things involved, whether the living thing is a plant, an animal, or yourself. Yes, biology is fun. And biology is for everyone.

Biology Is Useful

Biology is useful because it makes hobbies more fun. It is also useful because it gives us information that helps us every day. Here are some examples of questions you will be able to answer after studying biology.
- What foods should I eat if I want to be a better athlete?
- How important is regular exercise?

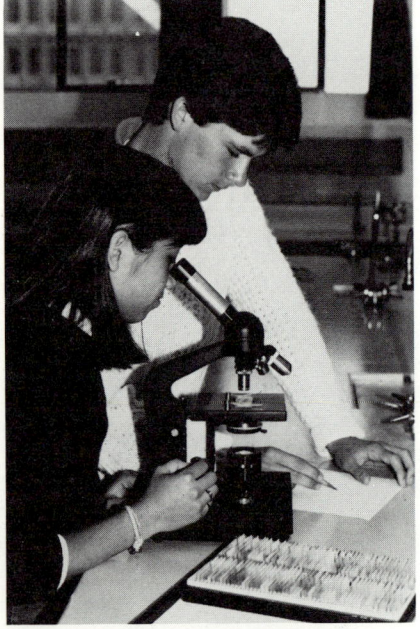

Fig. 1-1 You will use a microscope often in this course.

Fig. 1-2 Being outdoors is more fun when you know about the living things around you.

- Will smoking affect my health?
- What causes food to spoil?
- When should a lawn be fertilized?
- How can I grow better plants in the house?
- What's the best thing to do for a cold?

Biology Is Important

You already know two reasons why biology is important to all of us. It gives us hobbies. It also gives us useful information about ourselves and other living things. But there are many other reasons why biology is important. Here are a few examples.

- A knowledge of biology helps doctors do kidney transplants to save lives.
- As biologists learn more and more, cures for cancer, diabetes, and other serious diseases may be found.
- Soil erosion is destroying some of our best farmland. Biologists can tell farmers what to plant to control erosion.
- Every year the world needs more food. Biologists are developing grains and other foods that grow faster and produce larger crops.
- Biologists are developing trees that can produce fuel to help with the energy crisis.
- Biologists study endangered plants and animals so that we can stop them from becoming extinct.

Section Review

1. How can a knowledge of biology make some hobbies more fun?
2. In what ways is biology useful to us?
3. Give five examples which show that biology is important.

1.2 How Do We Study Biology?

All three people in Figure 1-3 are studying biology. That is, they are studying living things. Also, they are using the same method for studying living things. It is called the **scientific method**.

There is no one scientific method that all scientists must use in their studies. However, most scientists tend to follow six steps. We will use those six steps in this book. We hope that you will learn how to use them in your daily life. The steps are:

1. Recognizing a problem
2. Collecting information
3. Making a hypothesis
4. Doing an experiment
5. Observing and recording results
6. Making a conclusion

Now, let us take a look at each of these steps.

A B C

Fig. 1-3 All three people are studying biology. What do you think each one is doing? Why do we say all four are studying biology?

1. Recognizing a Problem

All scientific studies begin with the **recognition of a problem** Someone must recognize that there is a problem to be solved. That is, someone must see something worth studying. Here are some examples of problems.
- Why do plants need light?
- Why do leaves change colour in the fall?
- How do birds find their way during migration?
- How do fish get oxygen out of water?
- What causes lung cancer?

Usually curiosity helps us recognize a problem. Most of us wonder about the hows and whys of things that happen around us. Scientists wonder too. They are always asking questions about things that happen in their experiments. That's their job.

2. Collecting Information on the Problem

Suppose you wanted to find out why plants need light. What should you do? You should first see if anyone else has studied this problem. Perhaps you can find out in a library (Fig. 1-4). Or perhaps you could visit a greenhouse or a florist. In any case, you should begin your study by **collecting information** on the problem.

You should always do some reading before you do experiments. Before you do any experiments in this book, read the related material.

Scientists begin their studies the same way. They collect all the related information they can. They get it from books, journals, and other sources. Then they can build on the work of others. If they did not do so, science would never make any progress. We would be finding out the same things over and over again.

Fig. 1-4 A library has information that will help you solve many problems.

3. Making a Hypothesis

Sometimes the information you collect solves the problem. If this happens, the scientific method goes no further. For example, many

6 Chapter 1

A

B

Fig. 1-5 Will plants die if they get no light?

Fig. 1-6 Always record the data as you do the experiment.

biologists have studied why plants need light. If you accept their conclusions, the problem is solved. If you do not accept them, you should go on with the scientific method. That is, you should do experiments to see why plants need light.

Before doing an experiment, a scientist usually makes a **hypothesis** [hy-POTH-uh-sis]. A hypothesis is a prediction of the results of the experiment. Sometimes we call it an educated guess. What good is a hypothesis? It gives the scientist something to work toward. It directs the scientist's work. Thus a scientist studying the effects of light on plants might make this hypothesis. *Plants must have light or they will die.*

This hypothesis may make sense to you. But that does not make it true. No hypothesis is true until experiments show it to be so.

4. Doing the Experiment

The next step, then, is to make up an **experiment** [ex-PER-i-ment] to test the hypothesis. Suppose you wanted to test the hypothesis: *Plants must have light or they will die.* What could you do? You could put some plants in bright light and some in no light (Fig. 1-5). Suppose the plants in the bright light lived and the rest died. Could you conclude that the hypothesis was correct? Not necessarily. Perhaps the plants in the dark died because you gave them too much or too little water. Or perhaps it was too hot or too cold in the dark for plants to live. Or perhaps the plants in the dark had the wrong kind of soil. Or perhaps a combination of factors killed the plants in the dark.

What could you do, then, to prove the hypothesis? You must do the experiment under **controlled conditions**. All factors that can affect the results (**variables**) must be controlled (kept the same) except the one being studied. Thus both sets of plants must be planted in similar soil. They must be kept at the same temperature. And they must be watered the same amount. Now if the plants in the dark die, you can be more certain that lack of light was the cause.

Light is just one **variable** [VAI-ri-a-bul] that can make plants live or die. Soil, temperature, and water are others. These three variables must be controlled if you want to study the effect of light on plants. How could you control these variables? Can you think of any other variables that must be controlled?

5. Observing and Recording Results

A scientist always records all **observations** [ob-sur-VAY-shuns] made during an experiment. You should do the same (Fig. 1-6). This may be done in a table, graph, diagram, or written paragraph. Such information is called the **data** [DA-tuh] of the experiment. Table 1-1 shows a data table for an experiment to see if plants need light to live. *You must always be careful not to confuse observations and conclusions.* For example, if you say "the plants in the dark died" you have made a conclusion. If you simply state what happened in the

Section 1.2   7

Table 1-1   Effect of Light on Plants

Type of Plant	Appearance at start		Appearance in 1 week		Appearance in 2 weeks	
	Light	Dark	Light	Dark	Light	Dark
Geranium						
African Violet						
Coleus						
Jade plant						

experiment, you are giving your observations. *An observation is what you see, smell, feel, taste, or hear.* Therefore, when making observations, stick to what you see, smell, feel, taste, or hear. Do not draw conclusions until all the evidence is in. You can *observe* that plants in the dark turn brown and fall down. Then, on the basis of that and other observations, you can conclude that the plants died.

6. Making a Conclusion

Once the observations have been made, you draw them together and make a **conclusion** [con-CLUE-zhun]. If the hypothesis was correct, the conclusion will be the same as the hypothesis.

Usually more experiments are done to check the conclusion. If many experiments support the conclusion, it may be called a **theory** [THEE-oh-ree]. And, if the theory stands the test of time, it may become a **law**.

If the experiments do not support the conclusion, the original hypothesis must be changed. It may even have to be discarded. Then the whole process must start over again.

Summary of Scientific Method

Figure 1-7 shows the six steps in the scientific method. The scientific method begins with the **recognition of a problem**. This usually happens when a person's curiosity is aroused. The next step is the **collection of information**. This is done by reading and talking to other people. This is followed by the formation of a **hypothesis**, or prediction. You try to guess the answer to the problem using the information you have collected. Next comes an **experiment** to test the hypothesis. The experiment must have controls. Then **observations** can be made and recorded. They may lead to a **conclusion** that solves the problem.

Section Review

1. **a)** How do all scientific studies begin?
   b) Why is curiosity important to scientists?
2. Why should you collect information on the problem before you

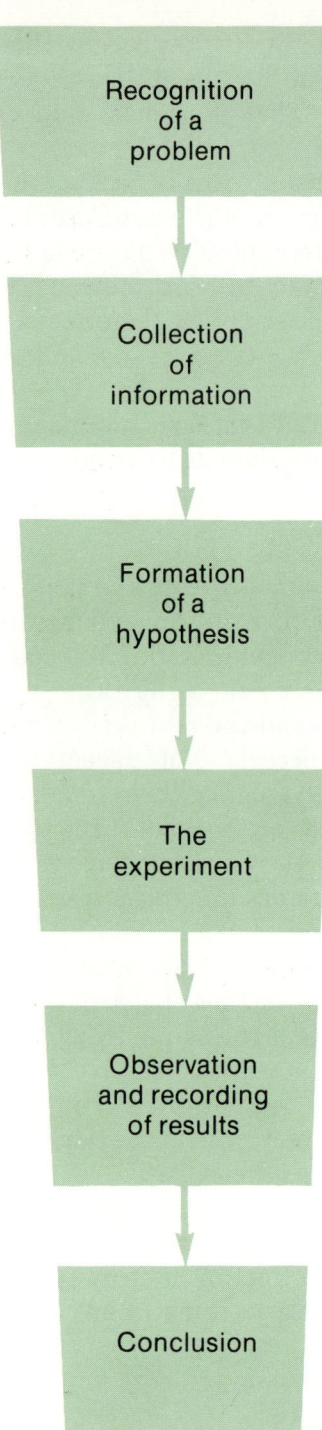

Fig. 1-7 The six main steps in the scientific method

start doing experiments?
3. a) What is a hypothesis?
   b) Of what use is a hypothesis to us?
4. What is the difference between a hypothesis and a theory? between a theory and a law?
5. a) What is a variable?
   b) What does "controlled conditions" mean?
6. a) What are data?
   b) What is an observation?
   c) How does a conclusion differ from an observation?
7. List, in order, the six main steps in the scientific method.

1.3 ACTIVITY What Makes Radish Seeds Germinate?

The purpose of this activity is to let you try the scientific method. Radish seeds won't germinate (begin to grow) while they are in the package. They will, however, germinate when they are planted. What causes germination? Is it the soil? the air? moisture? heat in the soil? Or is it a combination of factors?

Problem

What causes germination of radish seeds?

Materials

shallow dish (2)   water   potting soil
paper towel   radish seeds

Procedure

Since this is the first time you have used the scientific method in this course, we will give you some help. Remember that the scientific method has six steps: problem, collecting of information, hypothesis, experimentation, observing and recording results, conclusion. Let's look at these steps one at a time.

1. Problem

Put this heading in your notebook. Then write the problem under it.

2. Collecting of Information

Put this heading in your notebook. Now use the index of this book to see if you can find any information on the problem. A visit to the library or reading the information on the seed package may also be helpful. Make a summary in your notebook of your findings.

3. Hypothesis

Put this heading in your notebook. Under it write a hypothesis. Several hypotheses are possible. Here is an example: *Water is needed for the germination of radish seeds.*

4. The Experiment

Put this heading in your notebook. Now, make up an experiment to test your hypothesis. Write the steps in your notebook. Include a diagram. Figure 1-8 shows how you could test the hypothesis we used as an example. Think about these questions: What variable is being tested? What other variables could affect the results? Are those variables controlled in this experiment? Why is the control dish needed?

5. Observations

Put this heading in your notebook. Record all your data under this heading. Continue to observe and record results until you feel you have enough data to support or disprove the hypothesis.

6. Conclusion

Put this heading in your notebook. Then use your observations to draw a conclusion. Write the conclusion in your notebook.

7. Further Studies

Most experiments raise more questions than they answer. As you did your experiment, did you think of any more problems that you could study? If so, write them in your notebook.

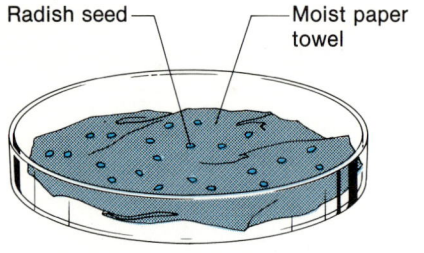

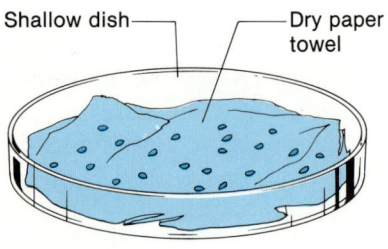

Fig. 1-8 Is water needed for the germination of radish seeds?

Main Ideas

1. We study biology because it's fun, useful, and important.
2. The scientific method has six main steps: recognition of a problem, collection of information, making a hypothesis, doing experiments, making and recording observations, and drawing conclusions.
3. The scientific method can be used to find out what causes radish seeds to germinate.

Glossary

biology		the study of life
conclusion	con-CLUE-zhun	a general statement made by drawing observations together

data	DA-tuh	the facts, figures, and other information gathered during experiments
experiment	ex-PER-i-ment	the action taken to test a hypothesis or gather data
hypothesis	hy-POTH-uh-sis	a prediction or educated guess that helps direct the course of an experiment
law		a theory that has stood the test of time
observation	ob-sur-VAY-shun	what a person discovers with one of the five senses during an experiment
theory	THEE-oh-ree	an explanation based on data from many experiments
variable	VAI-ri-a-bul	a factor that can affect the results

Study Questions

A. True or False

Decide whether each of the following statements is true or false. If the sentence is false, rewrite it to make it true. (Do not write in this book.)

1. Biology deals with living things.
2. A variable is any factor that can affect the results of an experiment.
3. A theory has been tested more than a law.

B. Completion

Complete each of the following sentences with a word or phrase that will make the sentence correct. (Do not write in this book.)

1. Before doing an experiment you should make up a _____ that predicts the results.
2. An experiment must always be done under _____ conditions.
3. If the hypothesis is correct, the _____ will be the same as the hypothesis.

C. Multiple Choice

Each of the following statements or questions is followed by four responses. Choose the correct response in each case. (Do not write in this book.)

1. George noticed that geranium plants in a window turned their leaves toward the light. He wrote this statement in his notebook: *Geranium plants need bright light.* This statement is best called
   a) a guess   b) an observation   c) a conclusion   d) a problem

2. A scientific theory
   a) is a conclusion that is supported by many experiments
   b) can never be wrong
   c) can never be changed
   d) is just an educated guess
3. A hypothesis is
   a) always true
   b) a prediction of the results of an experiment
   c) true until experiments prove otherwise
   d) based only on observations

D. Using Your Knowledge

1. Even if your hypothesis turns out to be wrong, you have not wasted your time doing an experiment. Why?
2. Suppose you want to find out if cats like one cat food better than other brands. Describe the controlled experiment you would do.
3. Suppose you are shopping for a box of cereal. You want the best value for your money. What variables should you consider before making your selection from the shelves?

E. Investigations

1. Radish seeds do not require light for germination. Perhaps someone in your class discovered this fact in Activity 1.3. Is this fact true for all types of seeds? Design and conduct an experiment to test your hypothesis. If possible, include corn, bean, and endive seeds in your testing.
2. Radish seeds do not require light for germination. But, do the seedlings need light in order to grow? Design and conduct an experiment to test your hypothesis.
3. Interview a person working in a biology-related job. What does this person do? What education does this person have?
4. Go to the guidance counsellor's office. Find five jobs that require a knowledge of biology. Describe them.

2 Characteristics of Living Things

2.1 Activity: Living or Non-living?
2.2 Living Things Move
2.3 Living Things Respond to Stimuli
2.4 Living Things Grow
2.5 Living Things Need Energy
2.6 Living Things Carry Out Metabolism
2.7 Living Things Are Made of Cells
2.8 Living Things Reproduce

Look at Figure 2-1. Which things are living and which are not? You will certainly agree that the person, dog, and plants are alive. Each is a living thing. We call a living thing an **organism**. You will probably agree that the rock, water, and air are non-living. But think, for a moment, about the apple, the seeds on the plants, and a bird's egg in a tree (Fig. 2-2). Are they living or non-living?

Will movement tell us if something is alive? Not always. Rivers and cars move, but they are not alive. What, then, is life? What is the difference between a *living* and a *non-living* thing? What are the characteristics of living things?

Fig. 2-1 Can you name the living and non-living things in this photograph? (*left*)

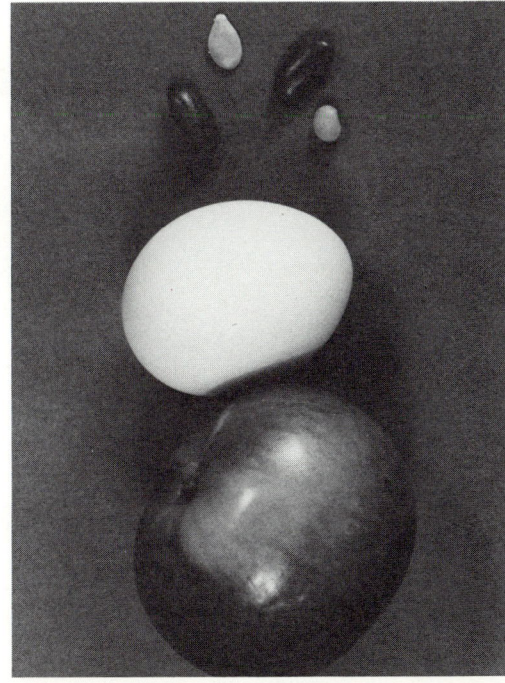

Fig. 2-2 Are the apple, seeds, and egg living or non-living? How do you know? (*right*)

Introduction 13

This chapter begins with an activity. Its purpose is to let you discover as many of the characteristics of living things as you can. The rest of the chapter discusses those characteristics.

2.1 ACTIVITY Living or Non-living?

Your teacher has provided a collection of several things. You are to decide which are living and which are non-living. After you have done this, you should know some of the characteristics of living things.

Problem

Which things are living and which are non-living?

Materials

a variety of things, such as houseplants, animals, rocks, soil, potatoes, carrots, apples, bean and corn seeds, peanuts, water, salt, and eggs

Procedure

a. Copy Table 2-1 into your notebook.
b. Study one of the things closely. Write its name in Column A. Write whether or not you think it is living or non-living in Column B. This is your hypothesis.
c. Write in Column C the reasons for your hypothesis.
d. Write in Column D further experiments you would like to do to test your hypothesis.

Fig. 2-3 The Venus' flytrap. It traps insects and other prey by quickly moving the halves of its leaves together.

Table 2-1 Living or Non-living?

Column A Thing	Column B Hypothesis (living or non-living)	Column C Reason for your hypothesis	Column D Further tests that should be done
Plant #1	Living	It grows	Does it move?

Discussion

1. Which things are living?
2. Which things are non-living?
3. Which things were hard to classify as living or non-living?
4. Make a list of the characteristics of living things.

2.2 Living Things Move

Before

After

Fig. 2-4 The sensitive plant. In just seconds a normal leaf folds up if touched.

Fig. 2-5 The leaves in this field of sunflowers have turned to face the sun.

Animals Can Move

Humans can walk. Birds can fly. Fish can swim. In fact, most animals can move from place to place. Such movement is called **locomotion** [loh-ko-MOH-shun].

The parts of animals also move. Insects can move their antennas. Rabbits can move their ears. Humans can blink their eyes. Animals cannot do these things after they are dead. Therefore movement is a characteristic of living animals.

Plants Can Move

Plants can move, too. Some, like the Venus' flytrap, move quickly (Fig. 2-3). Its leaves are in two halves. The halves move together to trap insects and other prey. The leaves of the sensitive plant also move quickly. If you touch them, they fold up (Fig. 2-4).

Other plants move, too, but more slowly. For example, many plants turn their leaves toward the sun (Fig. 2-5). Also, the opening of flower buds involves motion.

Biological Movement

Of course, non-living things can also move. A rock can roll. But something has to push it. A car can move. But it needs a motor to move it. A sailboat can move. But it needs the wind to blow it along.

Living things are self-movers. They do not need motors, the wind, or a push. The movement is caused by the living thing itself. We call

Section 2.2   15

this movement **biological movement**. Such movement is a characteristic of living things.

Section Review

1. What is locomotion?
2. Give four examples of biological motion in plants.
3. On a windy day the branches of a tree move. Is this biological motion? Why?

2.3 Living Things Respond to Stimuli

Fig. 2-6 A geranium plant in a window turns its leaves toward the light. What is the stimulus? What is the response?

A **stimulus** [STIM-yuh-lus] is a change in an organism's environment (surroundings) that causes activity in the organism. Such activity is called a **response**. Thus we say that organisms respond to stimuli. For example, when you whistle, your dog comes. The whistle is the stimulus. The coming of the dog is the response.

All living things respond to stimuli. When a bright light is shone in your face, your eyes respond by closing. If you touch something hot, you respond by quickly pulling back. If you try to swat a fly, it may respond by flying away. If a geranium plant is put in a window, it responds by turning its leaves toward the light (Fig. 2-6).

Even single-celled organisms like an amoeba respond to stimuli. You will study later how such organisms respond to heat, light, touch, and other stimuli.

Section Review

1. What is a stimulus?
2. What is a response?
3. Name three stimuli that can affect you. State your response to each stimulus.

2.4 Living Things Grow

All living things grow at some time during their lives. However, many non-living things can grow also. You have probably seen an icicle grow. How, then, can we say that growth is a characteristic of living things?

An icicle grows by adding more ice to its surface. But the growth of living things is quite different from this. A dog does not grow by adding more dog to its surface. Nor does a tree grow by adding more tree to its surface. Organisms cannot grow by adding more of their own material to their surfaces. They must combine food, water, minerals, and other substances to make new cells. In other words, the new material is added from within the organisms.

16 Chapter 2

Section Review

1. Explain the difference between the growth of a living and a non-living thing.

2.5 Living Things Need Energy

All living things need energy. It is needed to support movement, growth, and other life processes.

Most of the energy used by living things comes, in the first place, from the sun (Fig. 2-7). Plants carry out **photosynthesis** [foh-toh-SIN-thih-sis]. This process stores some of the sun's energy in foods such as glucose (a sugar). The plants use some of the glucose during a process called **respiration** [res-puh-RAY-shun]. This releases the energy the plants need to carry out life processes. Excess food is stored for future use.

Some animals get their supply of energy by eating plants. Others get their supply of energy by eating other animals. But, whatever the source, all living things need energy.

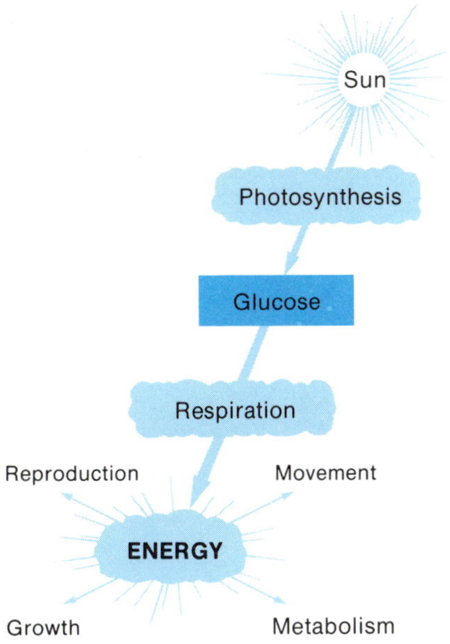

Fig. 2-7 Energy for movement, growth, and other life processes comes first from the sun.

Section Review

1. Why do organisms need energy?
2. Explain how photosynthesis and respiration provide a plant with its energy needs.
3. How do animals get the energy they need?

2.6 Living Things Carry Out Metabolism

Fig. 2-8 You need a well-balanced metabolism to be healthy.

Metabolism [meh-TAH-bol-iz-im] is all the chemical changes that occur within an organism. There are two types of metabolism. In one type, complex molecules are broken down into simpler molecules. **Respiration** is an example of this type. During respiration, foods (complex molecules) are broken down into water, carbon dioxide, and other simple molecules. **Digestion** is another example. This process breaks down foods into smaller molecules that can be used by organisms.

The other type of metabolism puts simple molecules together to make complex molecules. After foods have been digested, the parts are put together by organisms to make more of themselves. **Photosynthesis** is another example of this type of metabolism. During photosynthesis, carbon dioxide and water (simple molecules) are put together to make foods (complex molecules).

Metabolism includes all chemical changes that take place in organisms (Fig. 2-8). Metabolism is a characteristic of living things.

Section 2.6   17

Section Review

1. What is metabolism?
2. Why is metabolism necessary?
3. What role does digestion play in metabolism?
4. Why are most organisms warmer than their environments?

2.7 Living Things Are Made of Cells

All living things are made of **cells**. Some have only one cell. Amoebas and bacteria are examples. Others have millions upon millions of cells. You, for example, have about sixty million million cells (60 000 000 000 000).

All cells contain **protoplasm** [PROT-to-plaz-em]. This is the living material of cells. Only living things can make cells and their protoplasm. They do this by *organizing* (bringing together) other materials such as digested foods. This is why living things are called *organisms*.

Section Review

1. What is protoplasm?
2. What is a cell?
3. Why are living things called organisms?

2.8 Living Things Reproduce

Only living things can produce offspring like themselves. Trout lay eggs that become young trout. Robins lay eggs that become young robins. Horses give birth to horses. Corn seeds grow into corn. Only life can produce life. And life produces life like itself.

Sooner or later most living things grow old and die. Some insects live only a few weeks. An average person in Canada lives about 70 a. A horse lives about 30 a. Some Douglas fir trees in British Columbia live over 1000 a (Fig. 2-9). But, in the end, most living things wear out and die. **Reproduction**, however, ensures that each species survives, even though individuals die.

Some simple organisms such as bacteria and amoebas seem to live forever. They reproduce by splitting in two. Clearly such organisms never die of old age! Why, then, is it necessary for them to reproduce?

Section Review

1. Why is reproduction a characteristic of living things?
2. Why is reproduction necessary?

Main Ideas

1. Living things move under their own power.
2. Living things respond to stimuli from their environments.
3. Living things grow by adding new material from within themselves.
4. Living things need energy for life processes.
5. Living things carry out metabolism.
6. Living things are made of cells.
7. Living things reproduce similar living things.

Glossary

cell		the basic unit of life
locomotion	loh-ko-MOH-shun	the movement of an organism from place to place
metabolism	meh-TAH-bol-iz-im	all the chemical changes that occur in an organism
organism		a living thing
protoplasm	PROH-to-plaz-em	the living material of cells
response		activity in an organism caused by a stimulus
stimulus	STIM-yuh-lus	a change in an organism's environment that causes a response

Fig. 2-9 This Douglas fir tree is over 1000 a old. It will die one day. But reproduction ensures that more Douglas firs will take its place.

Study Questions

A. True or False

Decide whether each of the following statements is true or false. If the sentence is false, rewrite it to make it true. (Do not write in this book.)
1. Plants can only move slowly.
2. A stimulus causes a response in an organism.
3. Only living things can grow.

B. Completion

Complete each of the following sentences with a word or phrase that will make the sentence correct. (Do not write in this book.)
1. The movement of an animal from place to place is called ▬▬▬ .
2. If your mouth "waters" when you smell food, you have responded to a ▬▬▬ .
3. Energy for life processes is released by a process called ▬▬▬ .

Study Questions 19

C. Multiple Choice

Each of the following statements or questions is followed by four responses. Choose the correct response in each case. (Do not write in this book.)

1. Biological movement
    a) always involves locomotion
    b) occurs only in plants
    c) is caused by the organism itself
    d) occurs only in animals
2. When we get cold, we often shiver. The stimulus to which we are responding is
    a) the forming of "goose bumps"
    b) exercise to help warm us up
    c) the wearing of more clothes
    d) a decrease in temperature
3. All the chemical changes that take place in an organism are called
    a) respiration   b) metabolism   c) excretion   d) digestion

D. Using Your Knowledge

1. Name an animal that has movement but not locomotion.
2. Give an example of movement in a plant that is not described in this chapter.
3. Make a table in your notebook with two columns. Call one column Stimulus and the other Response. Now look at the following list. It contains a mixture of stimuli and responses. Pick out those that are stimuli. Write them in the Stimulus column. Then opposite each stimulus write a response that might occur.
    Now, write the responses from the list in the Response column. Then opposite each response write the stimulus that might have caused it.
    a) the petting of a dog
    b) a baby stops crying
    c) the squeaking of chalk on a chalkboard
    d) the drooping of the leaves on a plant
    e) the downward growth of a plant's roots
    f) a loud noise behind your back
    g) a skunk's odour
4. Is an egg alive? Give reasons to support your opinion.
5. A horse, a grasshopper, and a tulip don't look much alike. In what ways are they alike?

E. Investigations

1. Conduct an experiment to show that plants respond to light. Write a report of your findings.
2. Go into a dark room with a partner and a flashlight. Find out how the eye responds to a sudden bright light.
3. Find out the differences between sexual and asexual reproduction.

3 Classification of Living Things

3.1 What Is Classification?
3.2 Why Do We Classify Organisms?
3.3 Scientific Names for Organisms
3.4 The Main Classification Groups
3.5 The Five Kingdoms of Living Things
3.6 Activity: How a Classification Key Works
3.7 Activity: Using a Key to Identify Coniferous Trees

What kind of bug is this? What kind of snake is that? What's the name of this big tree? Look at that bird. What kind is it? Look at all the living things in the water! What are they (Fig. 3-1)?

You have probably asked such questions many times. And you probably didn't always get an answer. But that's not surprising. Even a biologist can name from memory only a small fraction of the living things on earth.

In this chapter you will learn how organisms are named. You will also learn how to use keys to identify organisms.

Fig. 3-1 This picture was taken in a tidal pool on the Gaspé peninsula. How many kinds of organisms do you see? How many can you name?

Introduction 21

3.1 What Is Classification?

Some people collect stamps. Others collect coins. And others collect hockey cards. What do you collect?

No matter what you collect, you probably do what most collectors do. You sort your collection into groups that contain similar things. For example, the stamp collector puts all Canadian stamps in one book and all American stamps in another. When you sort things in this way, you are *classifying* them. There is an important reason for classifying things: Classification makes it easier to keep track of what we have. **Classification** *is the grouping of similiar things for a specific reason.*

Section Review

1. What is classification?
2. What is the main reason for classifying things?

3.2 Why Do We Classify Organisms?

People have been classifying organisms since 300 B.C. At that time only about 1000 kinds of organisms were known. Therefore Aristotle used the simple method shown in Figure 3-2. However, this method won't work today. Organisms have been discovered that won't fit into this scheme. Where, for example, would a bacterium be placed?

Biologists think that there may be as many as 4 500 000 kinds of organisms on earth. Of these, only 1 500 000 have been identified and named. That leaves 3 000 000 to be discovered! Biologists believe that most of these will be found in the tropics (Fig. 3-3). The tropical forests are rapidly being cut. Biologists are moving through the forests with the foresters. Every day new kinds of organisms are

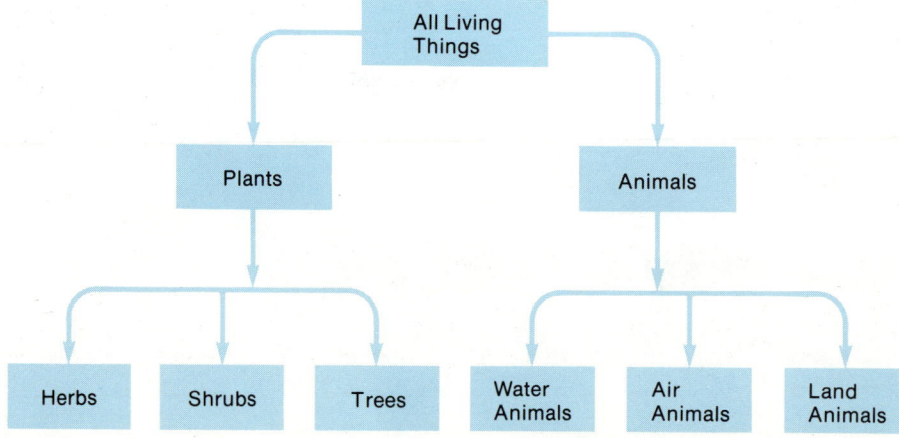

Fig. 3-2 Aristotle classified animals according to where they live. He classified plants according to the structure of their stems.

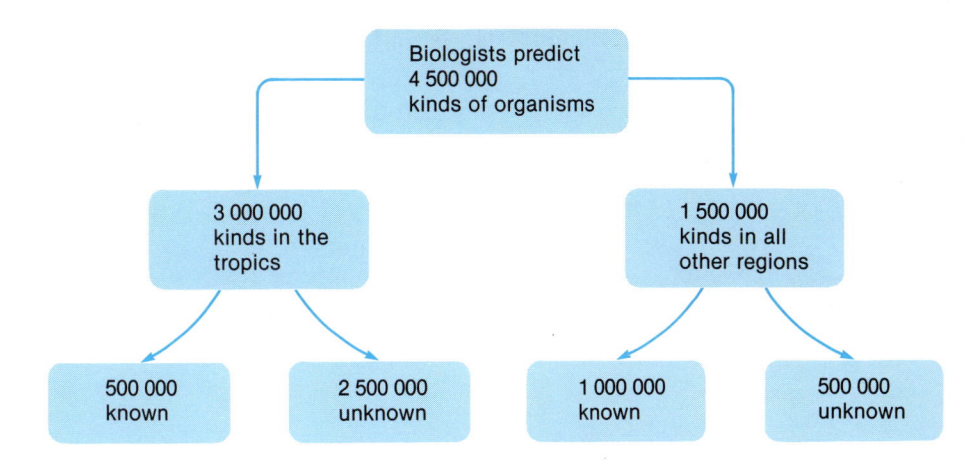

Fig. 3-3 Many organisms still must be found and named.

[Diagram: Biologists predict 4 500 000 kinds of organisms → 3 000 000 kinds in the tropics (500 000 known, 2 500 000 unknown) and 1 500 000 kinds in all other regions (1 000 000 known, 500 000 unknown)]

found. But this story has a sad ending. Biologists think that the loss of the forests may make over 1 000 000 kinds of organisms extinct during the next 10-20 a. And another 1 000 000 will become extinct in the next few decades.

As new organisms are found, biologists classify them. Why do they do this? They classify organisms for the same reason that people classify stamps. Classification makes it easier to keep track of the organisms. No one can remember details about 1 500 000 organisms. Therefore the organisms are classified, or put into groups. Each group contains organisms that are similar in many ways. It is easier to remember information about the groups since there are fewer of them. The rest of this chapter shows you how organisms are classified.

Section Review

1. Why won't Aristotle's classification method work today?
2. Why do biologists classify organisms?

3.3 Scientific Names for Organisms

In order to classify organisms, they must have names. How can we name 1 500 000 kinds of organisms? No language on earth has that many words.

By 1753, Carolus Linnaeus [lih-NAY-us], a Swedish biologist had developed a way of naming organisms. He also began our modern classification system.

Homo *sapiens*
GENUS SPECIES

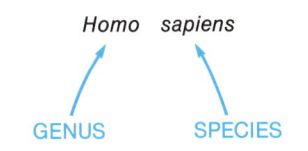

Fig. 3-4 A biological name has two words.

Scientific Names

Linnaeus used structural features as the basis of his classification system. That is, he grouped organisms that were alike in structure.

Section 3.3

He called each group a **species** (plural also **species**). Thus all house cats belong to one species; all horses belong to one species; and all modern-day humans belong to one species.

Linnaeus' naming system is simple. He gave each species a name that has two words. He used Latin words because all scientists wrote in Latin at that time. Table 3-1 shows the scientific names of some common species.

Table 3-1   Scientific Names

Common name	Scientific name
House cat	*Felis domesticus*
Horse	*Equus caballus*
Human	*Homo sapiens*
Dog	*Canis familiaris*
Sugar maple tree	*Acer saccharum*
White pine tree	*Pinus strobus*

The first word of each name is called the **genus** (plural: **genera**). The second word is called the **species** (Fig. 3-4). Note that the genus begins with a capital letter. The species does not. Also notice that they are printed in italics. (In your notes they should be underlined.)

Meaning of Species

Linnaeus defined a species as a group of organisms that were alike in structure. This definition raises a question: How alike? The dogs in Figure 3-5 have the same species name *familiaris*. Yet they are different in many ways. A more precise definition of species is needed.

Today we use the following to define a species:
1. The organisms must be alike in most respects (just as Linnaeus had said).
2. The organisms must interbreed under natural conditions.
3. The offspring must be fertile (able to produce offspring).

Thus all breeds of dogs are the same species. They can interbreed and the offspring are fertile.

Meaning of Genus

A genus is a group of similar species. Figure 3-6 shows three species in the *Canis* (dog) genus. The animals are alike. But they are different enough that they are not the same species.

A genus name can be used only once. Thus *Canis* is only used for a

Collie
Canis familiaris

Irish terrier
Canis familiaris

Whippet
Canis familiaris

Boston bull terrier
Canis familiaris

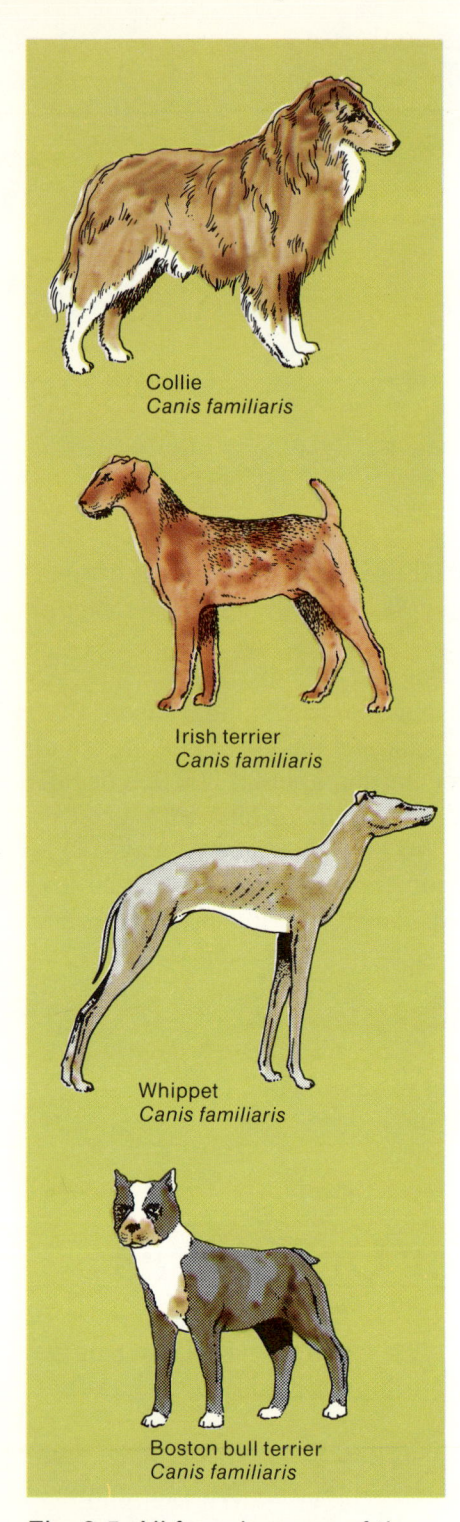

Fig. 3-5 All four dogs are of the same species. What would be a good definition of species?

certain group of dog-like animals. A species name can be used often. Table 3-2 lists some animals with the same species name, *canadensis*.

Table 3-2 Animals with the Species Name *canadensis*

Common name	Scientific name
Beaver	*Castor canadensis*
Elk	*Cervus canadensis*
Lynx	*Lynx canadensis*
Canada goose	*Branta canadensis*

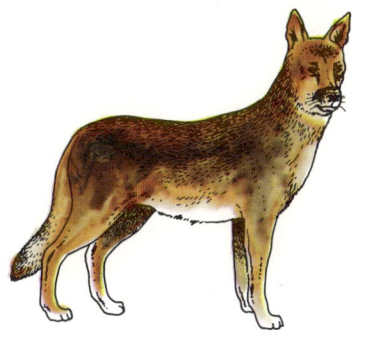

German shepherd
Canis familiaris

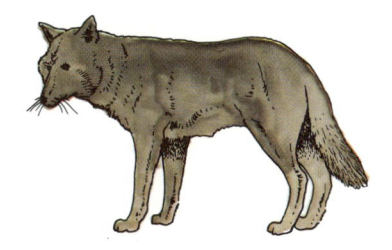

Coyote
Canis latrans

Gray wolf
Canis lupus

Fig. 3-6 Different species in the same genus have much in common.

Why Use Scientific Names?

You are probably saying by now: Why do we need these big Latin names? Why can't we just call a beaver a beaver and an elk an elk? You can, if you are sure everyone knows which animal you mean. But that is not always the case. Figure 3-7 shows a well-known member of the cat family. It has many common names. Among them are cougar, mountain lion, puma, panther, and painter. But it has only one scientific name, *Felis concolor*. Therefore, if you want to be sure that people everywhere know which cat you mean, you would call it *Felis concolor*.

Section Review

1. How did Linnaeus name organisms?
2. What do we mean by "species" today?
3. What is a genus? Give an example to illustrate your answer.
4. When is it important to use scientific names?

3.4 The Main Classification Groups

Species and Genus

These are the first two of the main classification groups. Remember that similar **species** are grouped to form a **genus**. For example, the three dog-like animals in Figure 3-6 are in the same genus, *Canis*. The three cat-like animals in Figure 3-7 are in the same genus, *Felis*.

Family

Similar genera (singular: genus) are grouped to form a **family**. For example, the red fox is in the genus *Vulpes* (Fig. 3-8). Dogs, wolves,

Section 3.4 25

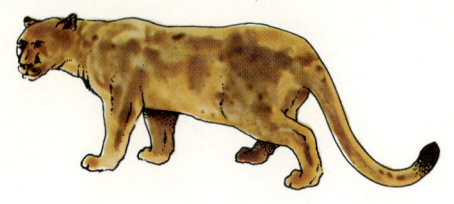

Cougar
Felis concolor

Jaguar
Felis onca

Tiger
Felis tigris

Fig. 3-7 The top cat has many common names, but only one scientific name, *Felis concolor*. All three cats are in the *Felis* genus.

and coyotes are in the genus *Canis*. The *Vulpes* genus and *Canis* genus are grouped in the Canidae (dog) family.

Order

Similar families are grouped to form an **order**. For example, the dog family (Canidae) and the cat family (Felidae) are in the order Carnivora (flesh-eaters).

Class

Similar orders are grouped to form a **class**. For example, rodents (order Rodentia) and carnivores (order Carnivora) are in the class Mammalia.

Phylum

Similar classes are grouped to form a **phylum** [FIE-lum]. For example, the phylum Chordata includes all classes in which the animals have a spinal cord. This includes the class Mammalia (mammals) and class Aves (birds). In the phyla Fungi and Plantae, phyla are often called **divisions**.

Kingdom

Finally, similar phyla are grouped to form a **kingdom**. For example, all animal phyla make up the kingdom Animalia. And all plant phyla make up the kingdom Plantae.

Summary

There are seven main classification groups:

Kingdom
    Phylum
        Class
            Order
                Family
                    Genus
                        Species

This "saying" should help you remember the order: **K**ing **P**hillip **C**annot **O**perate on **F**rogs' **G**uts **S**uccessfully!

Table 3-3 gives the classification of a human, a dog and a cat. Study it carefully. No one expects you to remember the scientific names. But you should understand how the system works.

Fig. 3-8 The red fox, *Vulpes fulva*, is in the dog family, Canidae.

Table 3-3 Classification of Three Animals

Group	Human	Dog	Cat
Kingdom	Animalia	Animalia	Animalia
Phylum	Chordata	Chordata	Chordata
Class	Mammalia	Mammalia	Mammalia
Order	Primates	Carnivora	Carnivora
Family	Hominidae	Canidae	Felidae
Genus	*Homo*	*Canis*	*Felis*
Species	*sapiens*	*familiaris*	*domesticus*

Section Review

1. List, in order, the seven main classification groups, beginning with kingdom.
2. Look at Table 3-3. Are dogs more closely related to humans or cats? How do you know?

3.5 The Five Kingdoms of Living Things

Most biologists group living things into five kingdoms. Three of these kingdoms should be quite familiar to you. The kingdom **Animalia** includes all animals. The kingdom **Plantae** includes all plants. And the kingdom **Fungi** includes all fungi such as mushrooms.

The other two kingdoms contain tiny organisms. The kingdom **Monera** includes bacteria and blue-green algae. The kingdom **Protista** includes many one-celled algae and protozoans.

The following six pages show some organisms in the main phyla of each kingdom. Where possible, your teacher will show you specimens of these organisms. Keep in mind that this section is here just to give you an overview of life. You are not expected to memorize the names of the organisms. You will study many of them in later chapters.

Section Review

1. Name the five kingdoms of living things. Arrange the kingdoms in order from simplest to most complex.
2. State the kingdom in which each of the following organisms is found: human, oak tree, earthworm, bread mould, chicken, trout, fern, clam, amoeba, grasshopper, spider, cholera germ, frog, rose, mushroom.

1. Kingdom Monera

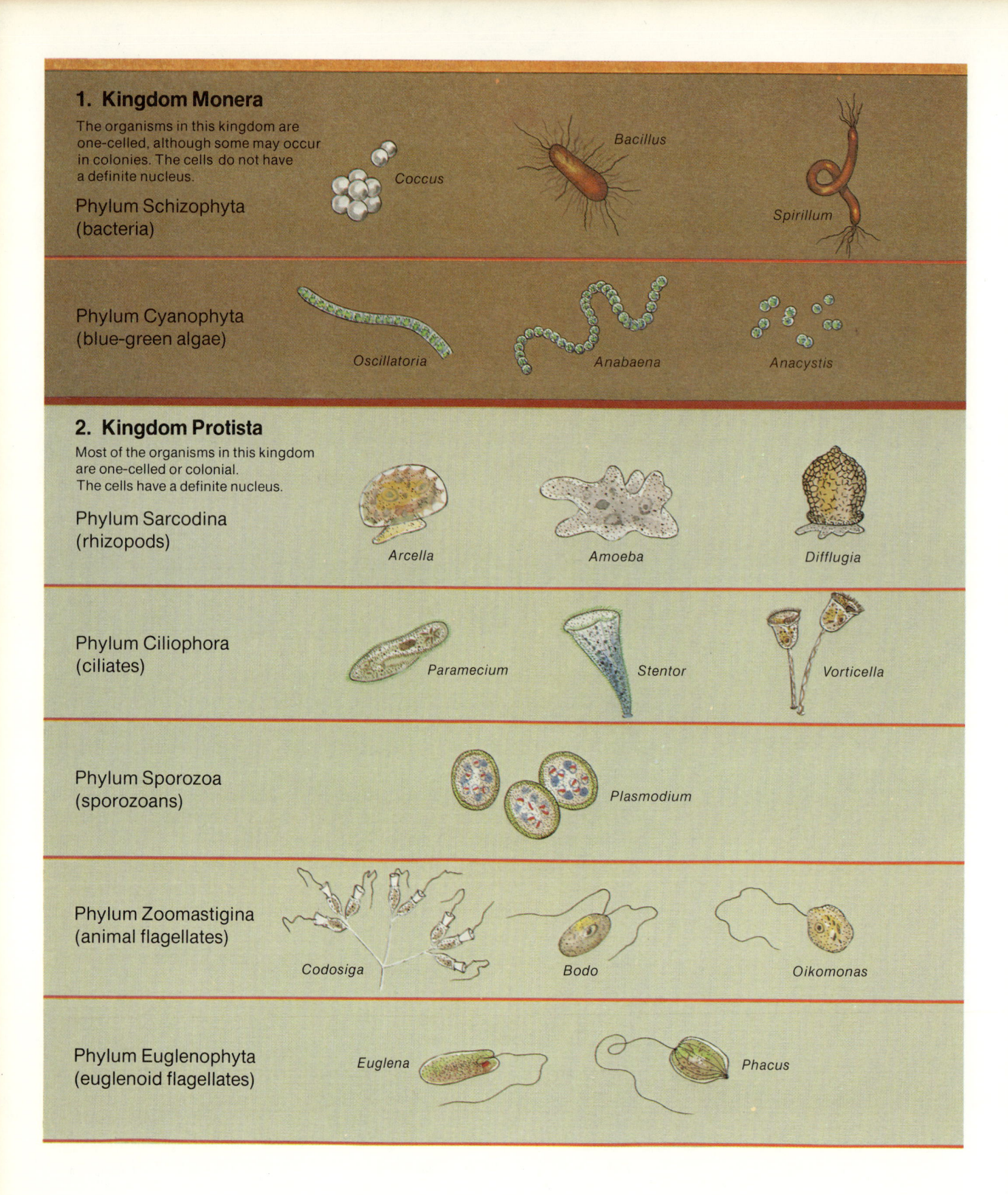

The organisms in this kingdom are one-celled, although some may occur in colonies. The cells do not have a definite nucleus.

Phylum Schizophyta (bacteria)

Coccus Bacillus Spirillum

Phylum Cyanophyta (blue-green algae)

Oscillatoria Anabaena Anacystis

2. Kingdom Protista

Most of the organisms in this kingdom are one-celled or colonial. The cells have a definite nucleus.

Phylum Sarcodina (rhizopods)

Arcella Amoeba Difflugia

Phylum Ciliophora (ciliates)

Paramecium Stentor Vorticella

Phylum Sporozoa (sporozoans)

Plasmodium

Phylum Zoomastigina (animal flagellates)

Codosiga Bodo Oikomonas

Phylum Euglenophyta (euglenoid flagellates)

Euglena Phacus

28 Chapter 3

Phylum Chrysophyta (golden algae—includes diatoms)

Diatoma *Dinobryon* *Tabellaria* *Asterionella*

Phylum Pyrrhophyta (includes dinoflagellates)

Ceratium *Peridinium*

3. Kingdom Fungi

This kingdom includes fungi and slime moulds.

Division Myxomycota (slime moulds)

Physarum *Lycogala*

Division Zygomycota (conjugation fungi)

Rhizopus (bread mould)

Division Ascomycota (sac fungi)

Saccharomyces (yeast) *Penicillium* *Morchella* (morel)

Division Basidiomycota (club fungi)

Amanita *Agaricus*

Division Deuteromycota (imperfect fungi)

Fusarium

Section 3.5

4. Kingdom Plantae

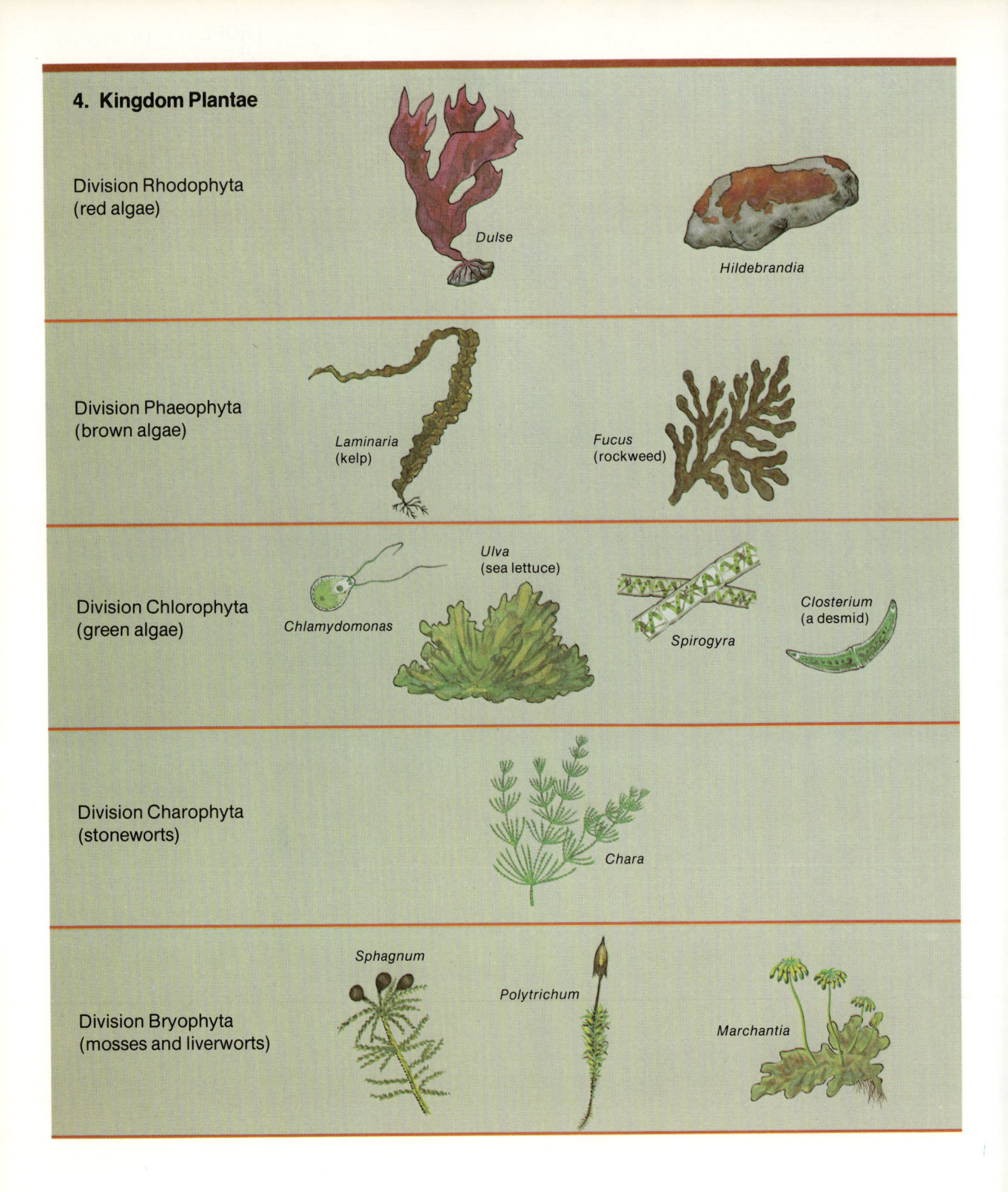

Division Rhodophyta (red algae)

Dulse

Hildebrandia

Division Phaeophyta (brown algae)

Laminaria (kelp)

Fucus (rockweed)

Division Chlorophyta (green algae)

Chlamydomonas

Ulva (sea lettuce)

Spirogyra

Closterium (a desmid)

Division Charophyta (stoneworts)

Chara

Division Bryophyta (mosses and liverworts)

Sphagnum

Polytrichum

Marchantia

30   Chapter 3

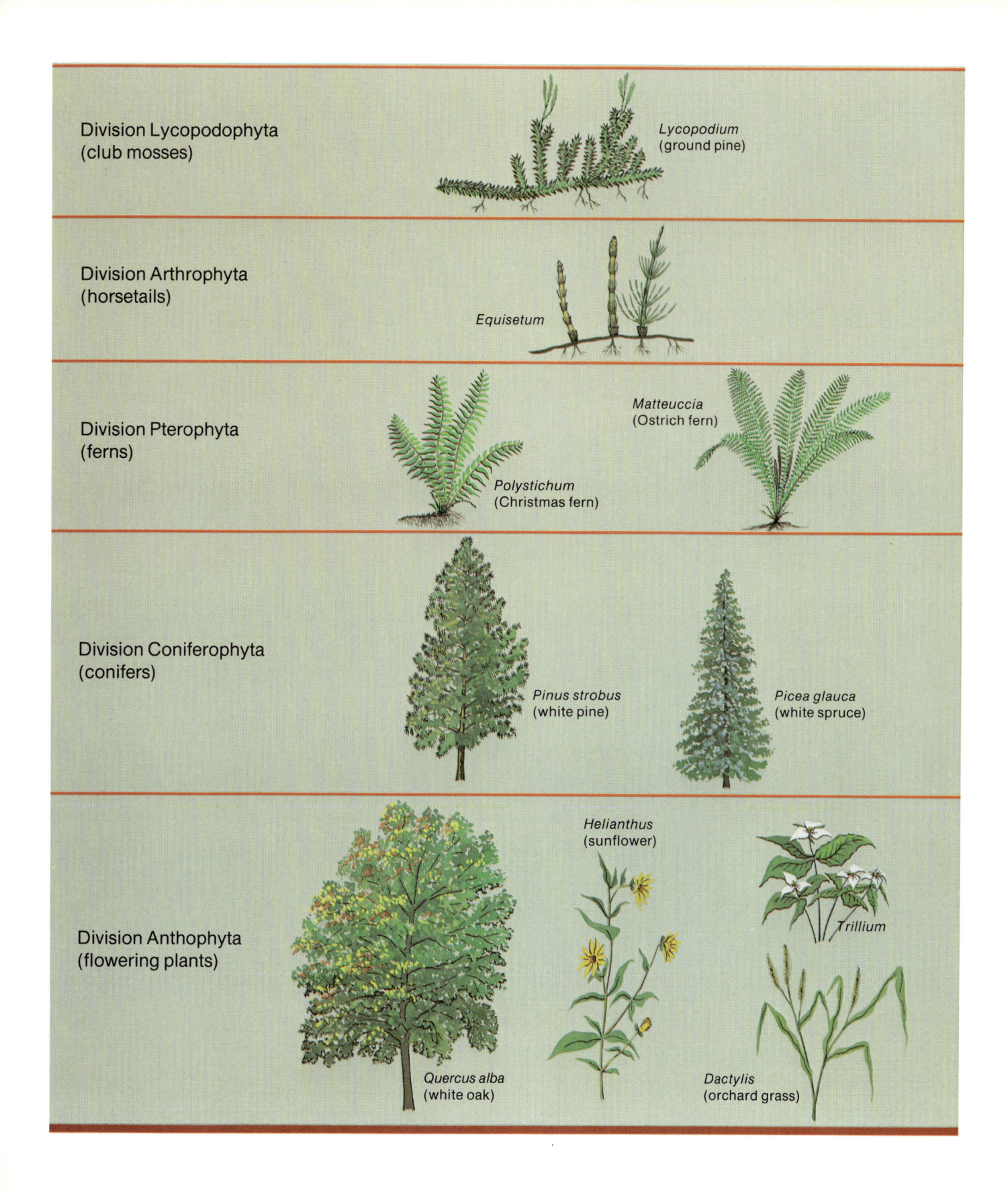

Division Lycopodophyta (club mosses) — *Lycopodium* (ground pine)

Division Arthrophyta (horsetails) — *Equisetum*

Division Pterophyta (ferns) — *Polystichum* (Christmas fern); *Matteuccia* (Ostrich fern)

Division Coniferophyta (conifers) — *Pinus strobus* (white pine); *Picea glauca* (white spruce)

Division Anthophyta (flowering plants) — *Quercus alba* (white oak); *Helianthus* (sunflower); *Trillium*; *Dactylis* (orchard grass)

Section 3.5

5. Kingdom Animalia

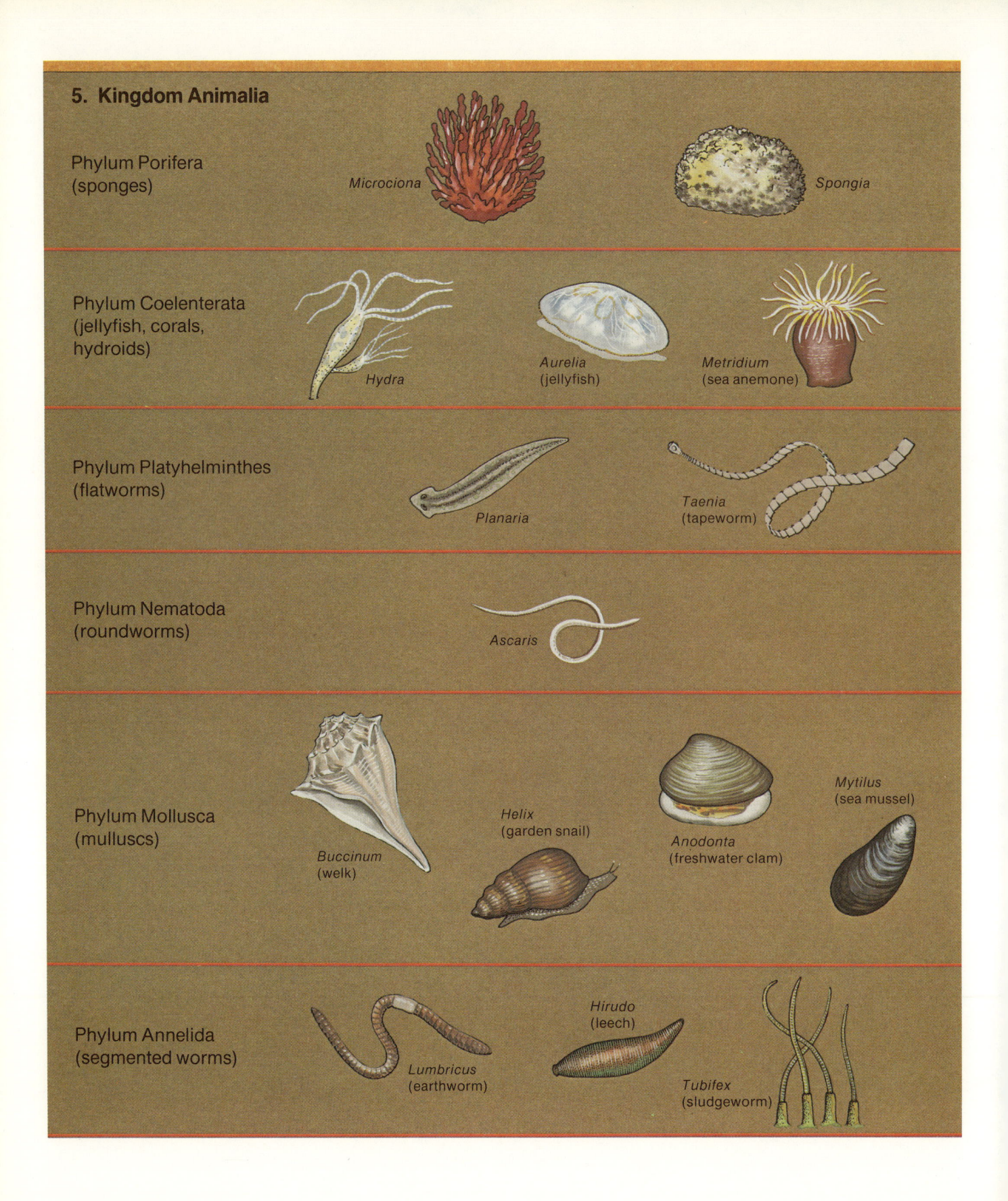

Phylum Porifera (sponges)

Microciona

Spongia

Phylum Coelenterata (jellyfish, corals, hydroids)

Hydra

Aurelia (jellyfish)

Metridium (sea anemone)

Phylum Platyhelminthes (flatworms)

Planaria

Taenia (tapeworm)

Phylum Nematoda (roundworms)

Ascaris

Phylum Mollusca (mulluscs)

Buccinum (welk)

Helix (garden snail)

Anodonta (freshwater clam)

Mytilus (sea mussel)

Phylum Annelida (segmented worms)

Lumbricus (earthworm)

Hirudo (leech)

Tubifex (sludgeworm)

32 Chapter 3

Phylum Arthropoda
(jointed-foot animals)

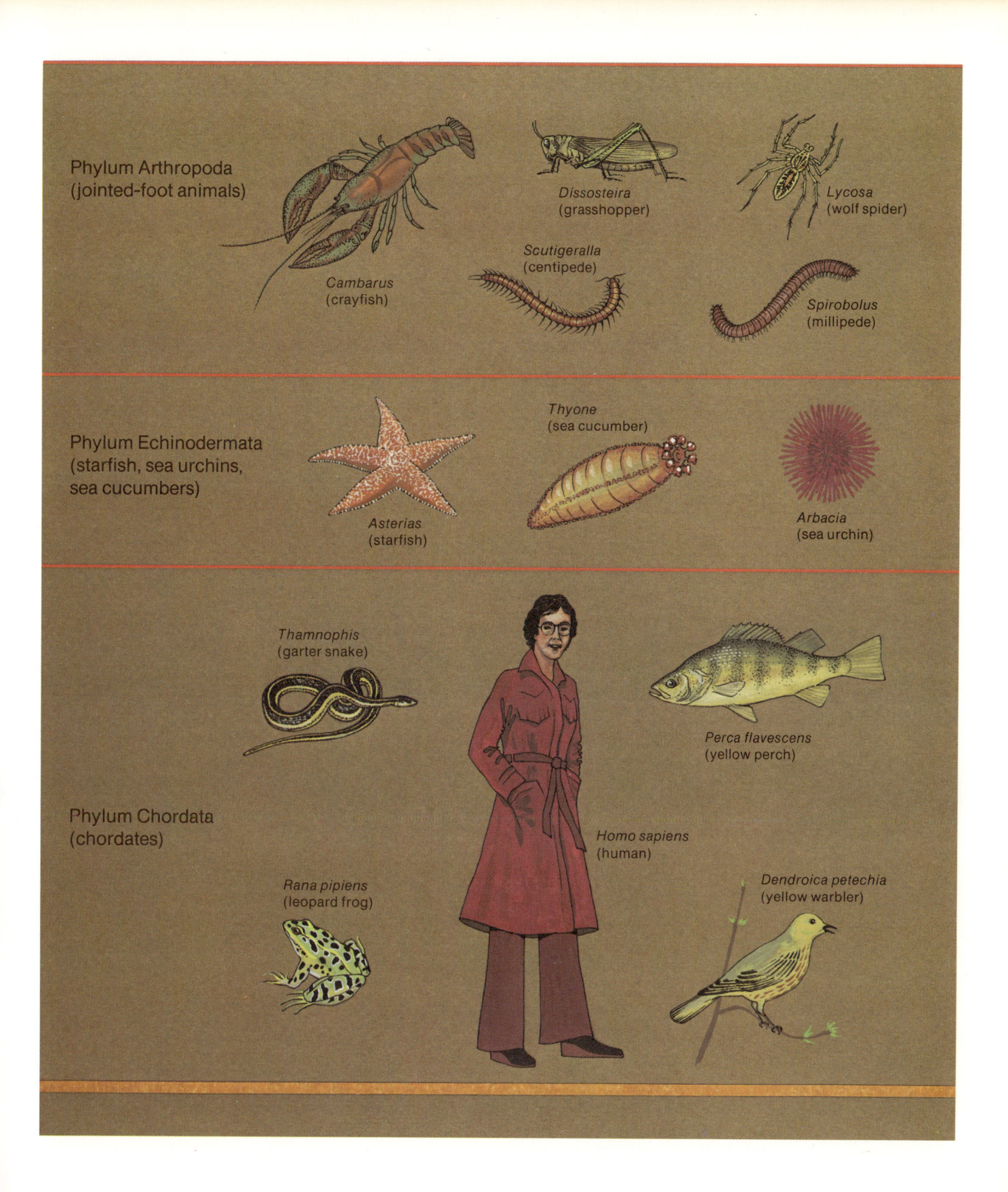

Cambarus
(crayfish)

Dissosteira
(grasshopper)

Scutigeralla
(centipede)

Lycosa
(wolf spider)

Spirobolus
(millipede)

Phylum Echinodermata
(starfish, sea urchins, sea cucumbers)

Asterias
(starfish)

Thyone
(sea cucumber)

Arbacia
(sea urchin)

Phylum Chordata
(chordates)

Thamnophis
(garter snake)

Homo sapiens
(human)

Perca flavescens
(yellow perch)

Rana pipiens
(leopard frog)

Dendroica petechia
(yellow warbler)

Section 3.5

3.6 ACTIVITY How a Classification Key Works

Suppose you have found an animal. You don't know what it is, but you would like to know. There are about 1 200 000 known species of animal. Therefore, to look through books hoping to spot a picture of the animal would be a hopeless task. However, the animal has likely been classified. And biologists have made up **classification keys** to help you find out what the animal is.

In this activity you will use two keys to identify organisms. You will use drawings instead of live organisms.

Classification Key A

This simple key is for five animals that you likely know. Its purpose is to let you see how a key works. Start with the first animal in Figure 3-9. Work down the key until you have identified it. Then do the same for each of the other animals.

1a. Has wings	butterfly
1b. Has no wings	go to 2
2a. Has 8 or fewer legs	spider
2b. Has more than 8 legs	go to 3
3a. Has 10 or fewer legs	crayfish
3b. Has more than 10 legs	go to 4
4a. Body flattened; 1 pair of legs per body segment. . . .	centipede
4b. Body rounded; 2 pairs of legs per body segment . . .	millipede

Classification Key B

Figure 3-10 shows 10 aquatic animals. This key will let you identify some of them only as far as phylum. Most of them are identified to class and some as far as order. To identify them to family, genus, or species, you would need a much more complex key. Your teacher will work through the key once with you to show you how it works.

1a. Legs present.	go to 2
1b. Legs absent	go to 5
2a. 3 pairs of legs	class Insecta (insects) (except Diptera)
2b. More than 3 pairs of legs	go to 3
3a. 4 pairs of legs	class Arachnida (spiders and mites)
3b. More than 4 pairs of legs	class Crustacea: go to 4

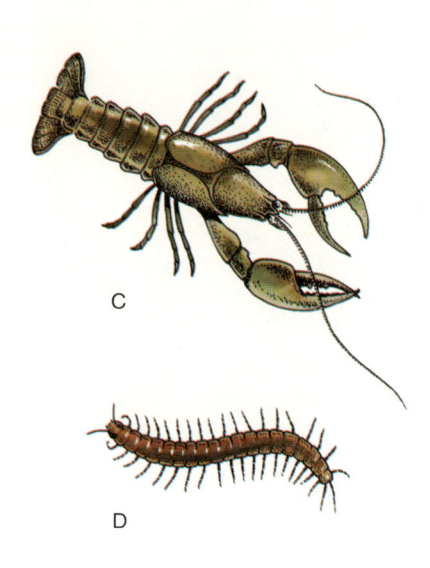

Fig. 3-9 Can you identify these animals using classification key A?

34 Chapter 3

Fig. 3-10 Some common aquatic animals. Can you find out their names?

4a.	Body flattened top to bottom	order Isopoda (sow bugs)
4b.	Body flattened sideways; swims on its side	order Ampipoda (scuds)
5a.	Has a head	order Diptera (flies)
5b.	Does not have a head	go to 6
6a.	Does not have body segments	go to 7
6b.	Has body segments............	go to 8
7a.	Body round....................	phylum Nematoda (roundworms)
7b.	Body flat.....................	phylum Platyhelminthes (flatworms)
8a.	Has suckers at both ends	class Hirudinea (leeches)
8b.	Has no suckers	class Oligochaeta (earthworms, sludgeworms)

Section 3.6

3.7 ACTIVITY Using a Key to Identify Coniferous Trees

Coniferous trees [ko-NIF-eh-rus] are trees with cones. They are called **conifers**. Most, but not all, are **evergreens**. That is, they hold their leaves (needles) all year. Two in this area are **deciduous** (drop their leaves in winter).

Some conifers, like white pine and white spruce, are used as wood pulp to make paper. Still others are used to landscape homes and parks. Most conifers are excellent habitat for wildlife.

Conifers are very important in Canada. Can you identify them? This key will help you identify the native species and common non-native species. Try it on the samples your teacher provides. Then try it on trees in your area.

Fig. 3-11 Some conifers have needle-like leaves.

Fig. 3-12 Some conifers have scale-like leaves that hide the twigs.

Key to Conifers of Ontario

A	Leaves needle-like, not hiding twig (see Fig. 3-11)	Leaves in bundles	→ B
		Leaves single	→ BB
AA	Leaves scale-like, hiding twig (see Fig. 3-12)	Twigs leaf-covered and flat	→ White cedar
		Twigs leaf-covered and rounded	→ Eastern red cedar
B	Leaves (needles) in bundles	Deciduous; 10-50 needles per bundle	→ C
		Evergreen; 2-5 needles per bundle	→ D (pine)
C	Deciduous	Cones 1 cm long	→ Tamarack (American larch)
		Cones 2-5 cm long	→ European larch
D	Pine	5 leaves per bundle	→ White pine
		2-3 leaves per bundle	→ E
E	2-3 leaves per bundle	3 leaves per bundle	→ Pitch pine
		2 leaves per bundle	→ F
F	2 leaves per bundle	Leaves usually under 5 cm long	→ G
		Leaves usually over 5 cm long	→ Red pine
G	Leaves usually under 5 cm long	Leaves 2-5 cm; cones curved; mature cones often closed	→ Jack pine
		Leaves 4-5 cm; bluish green and twisted; young bark orangish-red; mature cones usually open	→ Scots pine

36 Chapter 3

BB	Leaves single	Leaves 2-sided (flat)	→ H
		Leaves 4-sided	→ CC (spruce)
H	Leaves 2-sided (flat)	Leaves with stalk	→ I
		Leaves stalkless	→ Balsam fir
I	Leaves with stalk	A shrub; leaves pointed at top, yellow-green above & pale green below	→ Canada yew
		A tree; leaves rounded at tip; many lengths of leaves on same twig	→ Hemlock
CC	Leaves 4-sided (spruce)	Leaves green	→ J
		Leaves silvery-blue, sharp & very stiff	→ Blue spruce
J	Leaves green	Leaves roll easily between fingers; cones 2-5 cm long	→ K
		Leaves slightly flattened; do not roll easily; cones 10-15 cm long	→ Norway spruce
K	Leaves roll easily; cones 2-5 cm long	Cones 2-4 cm long; twigs with dense short hairs; not common in Southern Ontario	→ Black spruce
		Cones 4-5 cm long; twigs usually hairless	→ White spruce

Main Ideas

1. Classification is the grouping of similar things for a specific reason.
2. Biologists classify organisms so that they can keep track of them.
3. Scientific names consist of two words, a genus and a species.
4. There are seven main classification groups: kingdom, phylum, class, order, family, genus, species.
5. There are five kingdoms of organisms: Monera, Protista, Fungi, Plantae, Animalia.
6. Organisms can be identified using classification keys.

Glossary

class	a group of similar families
classification	the grouping of similar things for a specific reason
family	a group of similar genera

genus	JE-nus	a group of similar species
kingdom		a group of similar phyla
order		a group of similar families
phylum	FIE-lum	a group of similar classes
species		similar organisms that interbreed and produce fertile offspring

Study Questions

A. True or False

Decide whether each of the following statements is true or false. If the sentence is false, rewrite it to make it true. (Do not write in this book.)

1. Classification is the grouping of similar things for a specific reason.
2. Biologists have discovered 4 500 000 species of organisms.
3. A family is made up of similar orders.

B. Completion

Complete each of the following sentences with a word or phrase that will make the sentence correct. (Do not write in this book.)

1. A class is made up of similar ▓▓▓ .
2. The simplest organisms are in the kingdom ▓▓▓ .
3. Humans, dogs, and cats are in the same kingdom, phylum, and ▓▓▓ .

C. Multiple Choice

Each of the following statements or questions is followed by four responses. Choose the correct response in each case. (Do not write in this book.)

1. The scientific names of all organisms have two words. Those words are
    a) the genus and species
    b) the family and genus
    c) the family and species
    d) the family and order
2. Most of the organisms in the kingdom Protista are
    a) one-celled without a nucleus
    b) one-celled or colonial with a nucleus
    c) tiny animals, often with hair-like tails
    d) tiny animals of all shapes
3. Three animals have the scientific names *Canis familiaris*, *Canis latrans*, and *Canis lupus*. These animals are
    a) dog-like animals of the same species
    b) in the same genus but not the same family

c) in the same family but not the same genus
   d) in the same genus and family

D. Using Your Knowledge

1. How do librarians classify books? Why do they do this?
2. Describe how things are classified in a supermarket.
3. a) Why will cutting the tropical forests make species of organisms go extinct?
   b) What do you think should be done about this matter?
4. Four members of the dog family are the red fox (*Vulpes fulva*), the Arctic fox (*Alopex lagopus*), the coyote (*Canis latrans*), and the wolf (*Canis lupus*).
   a) Which are more closely related, the red fox and Arctic fox, or the coyote and wolf? How do you know?
   b) Which are most closely related to domestic dogs?

E. Investigations

1. Make a list of all the examples of classification you can find in your home.
2. Make a classification key that someone else could use to identify about ten makes of cars.
3. A horse (*Equus caballus*) and a donkey (*Equus asinus*) will interbreed. But, as you can see, they are not called the same species. Find out why this is so.
4. Borrow a copy of *Native Trees of Canada* by R. C. Hosie from your teacher or the library.
   a) Find out which trees in the key of Section 3.7 are native species and which are non-native.
   b) Describe the distribution of each of those native species in Canada. (You may wish to draw simple maps.)

Unit 2

Cells: The Building Blocks of Life

CHAPTER 4
The Microscope

CHAPTER 5
The Cell and Its Parts

CHAPTER 6
Cells at Work:
How Substances Enter and Leave Cells

CHAPTER 7
Cells at Work:
Photosynthesis and Respiration

CHAPTER 8
Cells at Work:
How Living Things Grow

All living things are made of tiny building blocks called **cells**. Some living things, like the amoeba, are made of just one cell. Others have many cells. For example, you have about sixty million million cells (60 000 000 000 000).

What are cells? What do they look like? Why are cells so small? Why are large organisms made of tiny cells instead of just one big cell? Why aren't there any giant 100 kg amoebas in your favourite pond? Why aren't there any one-celled cows on our farms? In this unit you will spend many interesting hours at the microscope seeking answers to questions such as these.

Fig. 4-0 Cells are the building blocks of living things. We study cells with a microscope.

4 The Microscope

4.1 Early Microscopes and the Cell Theory
4.2 Activity: Parts of the Microscope
4.3 Activity: Preparing the Microscope for Use
4.4 Activity: Preparing a Wet Mount
4.5 Activity: Focusing the Microscope

Scientists first used microscopes in the 1600s. With their simple microscopes they developed the idea that all living things are made of cells.

The microscope you will use this year is much better than the ones early scientists used. Therefore, you should be able to see cells more clearly than they did. However, you must use the microscope properly. If you don't, you could damage it. If you do, you will discover an exciting new world — the world of the living cell.

4.1 Early Microscopes and the Cell Theory

In the 1600s scientists learned how to grind glass to make lenses. What a discovery this was! Galileo put some lenses together and made a telescope. Now humans could study distant objects such as the moon, sun, and planets. Other scientists used lenses to make eyeglasses. Now some people who were almost blind could read again. Still other scientists used lenses to make microscopes. Let us see what these scientists discovered.

Discovery of the Cell

Robert Hooke (1635-1703) was an English scientist. He built one of the first microscopes in the world (Fig. 4-1). One day he had to give a demonstration at a meeting of the Royal Society of London. He wanted his demonstration to be new and exciting. Therefore he decided to use his microscope in a demonstration to show why cork floats. He cut a thin slice of cork and put it under his microscope. When the light shone through the cork, Hooke saw an amazing thing. The cork was not solid. Instead, it was made of what Hooke called "little boxes". These little boxes were full of air. That is why the cork floats.

These little boxes reminded Hooke of the cells in honeycombs. Therefore he called them **cells**. This was the first time that the word "cell" was used in biological studies. And a new scientific field, the field of cell biology, was born.

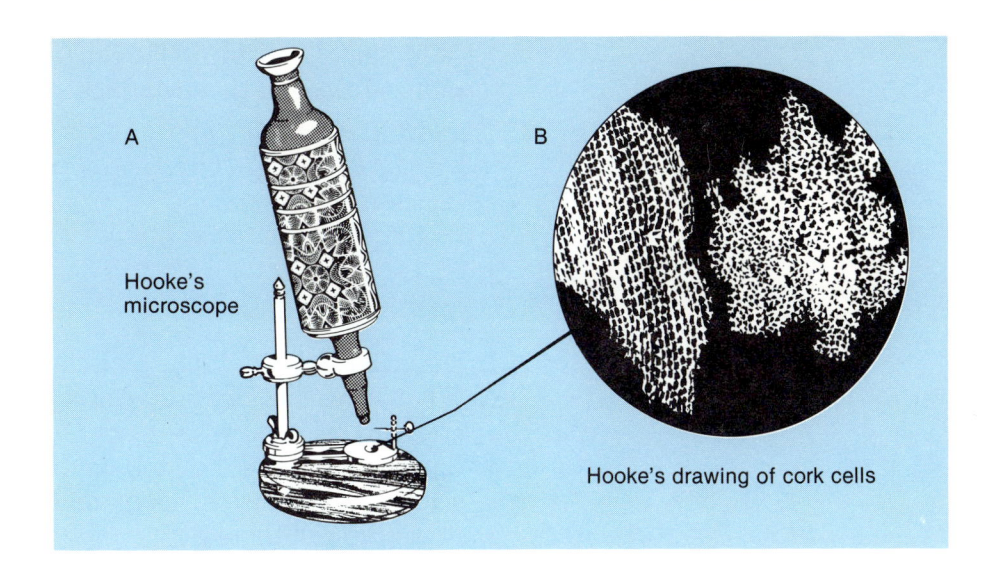

Fig. 4-1 Hooke's microscope and his drawing of cork cells

Van Leeuwenhoek's Work

Many decades passed before scientists realized how important Hooke's discovery was. During that time Antony van Leeuwenhoek helped keep interest in microscopy alive. He built about 250 different microscopes in his lifetime. One was especially built for looking at organisms in pond water. He saw and drew pictures of many microscopic organisms. He never used the word "cell". But he did draw the many cells that he saw in his studies.

The Cell Theory

In 1805 a German scientist, Lorenza Oken, stated that all organisms are made of cells and that all organisms come from cells. This statement later became known as the cell theory. At the time, however, no one paid much attention to Oken.

In 1824, a French scientist, Henri Dutrochet, studied many plant and animal cells. He even discovered that plants grow by producing more cells. From his studies he concluded that "the cell is truly the fundamental part of the living organism".

By now many scientists were studying cells. Many new discoveries took place. In 1831, a Scottish scientist, Robert Brown, saw a dark body in the centre of a cell. He called it the **nucleus**. In 1835, another French scientist, Félix Dujardin, saw a moving fluid in cells. He called it **protoplasm** [PROH-tow-plaz-em].

In 1838, Matthias Schleiden, a German botanist, concluded that all plants were made of cells. He concluded, also, that cells were alive. And he decided that cells were responsible for the operation of the organisms of which they were a part. In 1839, Theodor Schwann, a German zoologist, came to the same conclusions about animals. For the first time, people knew that animals and plants had much in common. In 1858, Rudolf Virchow, another German scientist, concluded that all cells came from other living cells.

Section 4.1

Oken said it all in 1805. But 53 years passed before scientists recognized the cell theory which says: *Modern Cell Theory*

1. *All living things are made of cells.*
2. *Cells carry out the functions of the organisms of which they are a part.*
3. *All cells come from living cells.*

Section Review

1. Describe Robert Hooke's discovery of the cell.
2. What contributions did Antony van Leeuwenhoek make to microscopy?
3. Seven scientists helped develop the cell theory. Name each scientist and state his contribution to the theory.
4. State the cell theory.

4.2 ACTIVITY Parts of the Microscope

There are many kinds of microscopes (Fig. 4-2). The kind you will use is properly called a **monocular compound microscope**. (We will just call it a microscope from now on.) One type is shown in Figure 4-2, A. Scientists usually use a **binocular compound microscope** (B). With it, you use both eyes. This helps prevent eye strain. A **dissecting microscope** (C) has a low magnification. It is used to study the features of whole objects.

Many of the activities in this text use the microscope. Therefore, you must be able to use it properly. But first, you need to know its parts.

Problem

What are the parts of a microscope called? What are their functions?

Materials

microscope

Procedure

Look at Figure 4-3 as you follow these steps.

a. Remove your microscope from its case or dust cover. Always grasp the **arm** firmly with one hand. Place the other hand under the **base**. This is the *only* way to lift and carry a microscope.
b. Place the microscope *gently* on the desk. The arm should point toward you. The **stage** should point away from you.
c. Copy Table 4-1 into your notebook.
d. Follow Figure 4-3 closely as your teacher describes the parts of the microscope and their functions.
e. Complete Table 4-1 in your notebook.

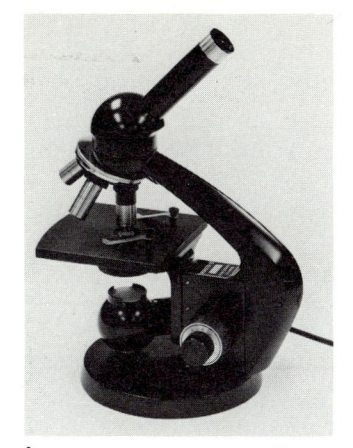

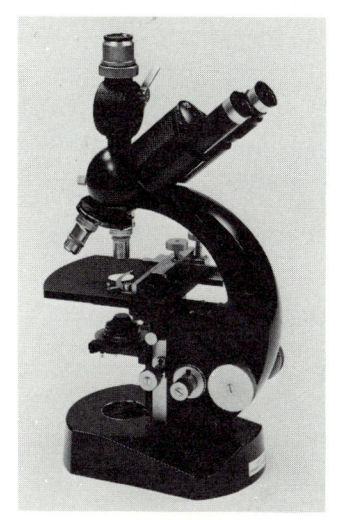

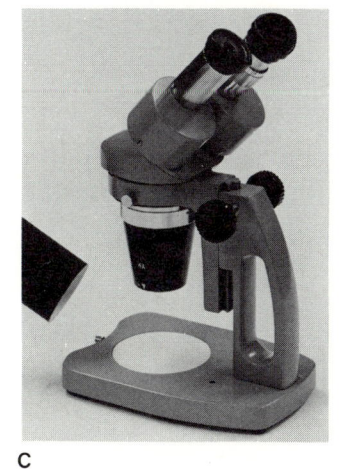

Fig. 4-2 Some kinds of microscopes

Fig. 4-3 Parts of a microscope

Table 4-1 Parts of a Microscope

Part	Function
Arm	
Base	
Stage	
Body tube	
Nosepiece	
Objectives	
Stage clips	
Diaphragm	
Lamp	
Ocular (eyepiece)	
Coarse adjustment	
Fine adjustment	

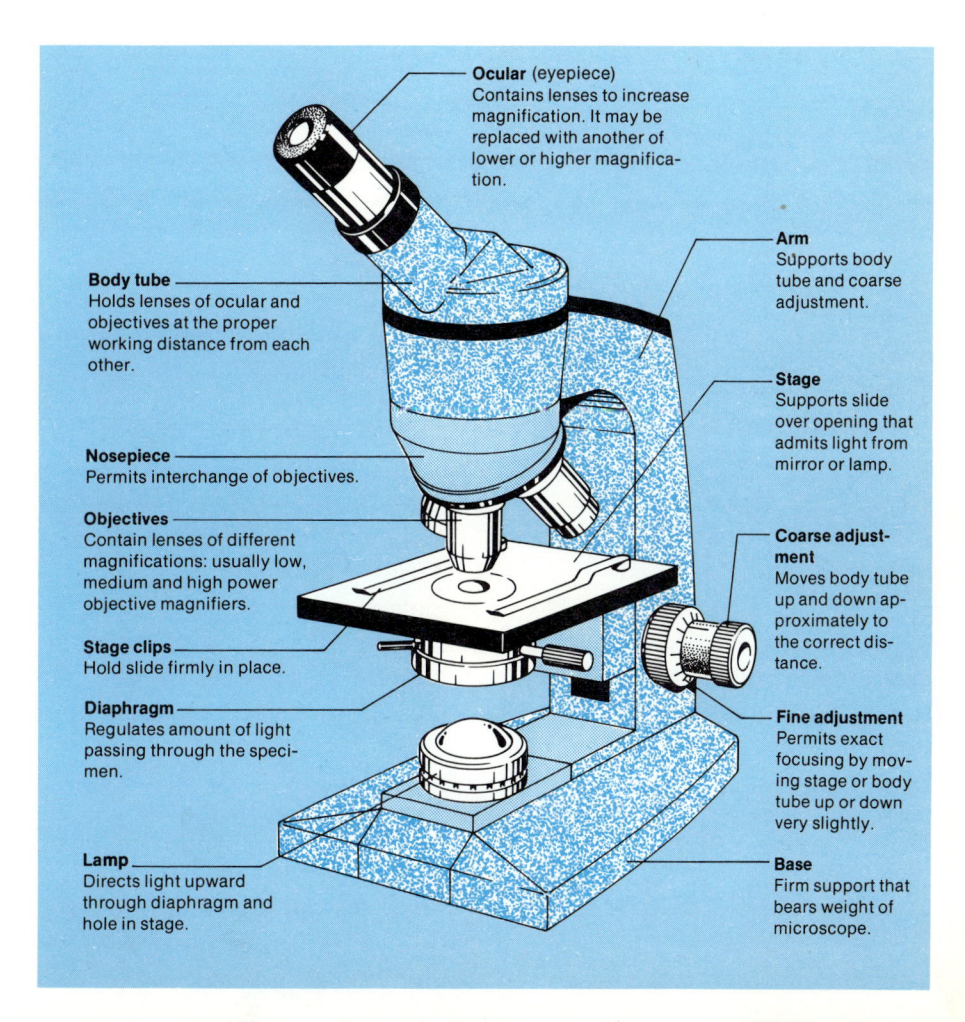

Ocular (eyepiece) Contains lenses to increase magnification. It may be replaced with another of lower or higher magnification.

Body tube Holds lenses of ocular and objectives at the proper working distance from each other.

Nosepiece Permits interchange of objectives.

Objectives Contain lenses of different magnifications: usually low, medium and high power objective magnifiers.

Stage clips Hold slide firmly in place.

Diaphragm Regulates amount of light passing through the specimen.

Lamp Directs light upward through diaphragm and hole in stage.

Arm Supports body tube and coarse adjustment.

Stage Supports slide over opening that admits light from mirror or lamp.

Coarse adjustment Moves body tube up and down approximately to the correct distance.

Fine adjustment Permits exact focusing by moving stage or body tube up or down very slightly.

Base Firm support that bears weight of microscope.

Section 4.2

Discussion

1. Why must a microscope be carried using only the arm and base?
2. Why must a microscope be handled gently?
3. Review the parts and their functions. When you know them, you are ready to do Activity 4.3.

4.3 ACTIVITY Preparing the Microscope for Use

This activity shows you how to get a microscope ready for use. Follow these steps carefully. If you do, your work will be very rewarding. Also, you will not damage this expensive instrument.

Problem

What is the proper way to prepare a microscope for use?

Materials

microscope   lens paper

Procedure

Every time you use a microscope, you should follow these steps:

a. Use the **coarse adjustment** to raise the **body tube** (or lower the stage) so the **objectives** will not hit the stage when the **nosepiece** is turned.
b. Turn the nosepiece so that the low power objective (the shortest one) is above the hole in the stage. You will hear a click when it is in the right place.
c. Close the **diaphragm** [DI-a-fram] to its smallest opening.
d. Adjust the **lamp** (or **mirror**) so that light goes up through the hole in the stage. *Do not use direct sunlight.* It could damage your eyes.
e. Look into the **ocular** [OK-you-lar]. If the **field of view** is not bright and evenly lit, adjust the lamp (or mirror) again.
f. Adjust the diaphragm until the field of view is bright but not glaring.
g. If the ocular and objectives are dirty, clean them. Use only lens paper. Use a gentle circular motion. Use a piece of lens paper only once. *Never* use any other material. You may scratch the lenses.

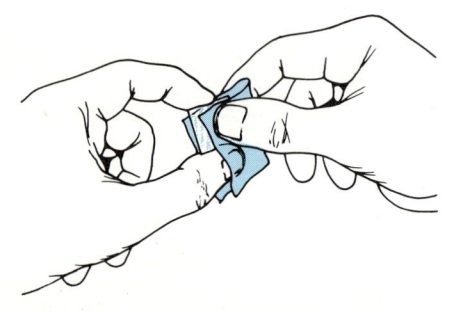

Fig. 4-4 A cover slip breaks easily. Wipe both sides at once so you won't break it.

Discussion

1. Describe the method for cleaning the lenses of the ocular and objectives.
2. What should the field of view look like when it is properly lit?
3. Review the steps for preparing the microscope for use. When you know them, you are ready to do Activity 4.4.

4.4 ACTIVITY Preparing a Wet Mount

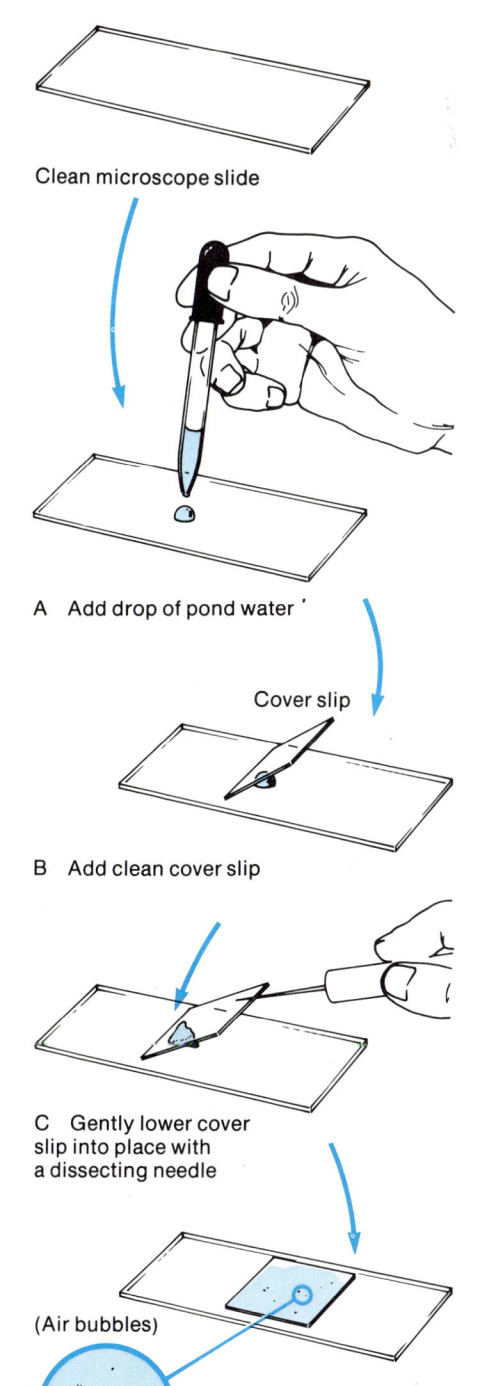

Sometimes during this course you will study prepared slides under your microscope. But, most often, you will have to prepare your own slides. Usually the specimen will be placed in a drop of water on a slide. A specimen prepared in this way is called a **wet mount**. In this activity you will learn how to prepare a wet mount.

Problem

How is a wet mount made?

Materials

microscope	paper towel	piece of newspaper
microscope slide	dissecting needle	dropper
cover slip	scissors	forceps (tweezers)

Procedure

Use this procedure every time you want to prepare a wet mount:

a. Hold a **microscope slide** by the edges. Wet it with water. Wipe both sides dry with a paper towel. Always hold the slide by the edges. If you don't you will get fingerprints on it.
b. Clean the **cover slip** the same way (or use a new plastic cover slip). However, wipe both sides at once as shown in Figure 4-4. Then you won't break the cover slip.
c. Use the **dropper** to place one drop of water near the centre of the microscope slide (Fig. 4-5,A). This water is called the **mounting medium**. It helps produce a clear image.
d. Cut out a small letter "a", "b", "d", "e", "g", or "h". (The smaller the letter the better.) Pick the letter up with the forceps. Then place it in the drop of water.
e. Pick up the cover slip by the edges. Lower one edge so that it touches one side of the drop of water at an angle of about 45° (Fig. 4-5,B). Now slowly lower the cover slip as shown in Figure 4-5,C. This will prevent the trapping of air bubbles under the cover slip. Large bubbles will interfere with your viewing. A few small ones, however, will not cause serious problems. Just be sure that you do not confuse them with your specimen. Air bubbles are circular. Also, they have thick, dark edges.
f. Place your wet mount on the stage. Make sure the letter is over the opening in the stage and the right way up for reading. The slide is now ready for viewing.

Discussion

1. Describe how to clean a microscope slide and a cover slip.
2. What is the purpose of the water?
3. Describe the proper way to make a wet mount.

A Add drop of pond water
B Add clean cover slip
C Gently lower cover slip into place with a dissecting needle
(Air bubbles) Microscope field
D Ready for observation

Fig. 4-5 Preparation of a wet mount

4.5 ACTIVITY Focusing the Microscope

By now you should know the parts of a microscope and their functions. You should also know how to prepare the microscope for use. Further, you should know how to prepare a wet mount.

If you know all these things, you are ready to use the microscope. Let's focus it and look at your wet mount and some other objects.

Problem

How do you focus and use a microscope?

Materials

microscope
wet mount from Activity 4.4
piece of coloured picture from a magazine

Procedure

Use this procedure every time you want to focus your microscope:

a. Using the **coarse adjustment**, turn the **low power objective** down (or raise the stage) as far as it will go. Watch what you are doing from the side, not through the ocular.
b. Look into the **ocular**. Slowly raise the objective (or lower the stage) by turning the coarse adjustment. Keep doing this until the letter comes into focus. You may have to move the slide around to find the letter. Try keeping both eyes open while looking into the microscope. This takes a little practice, but it prevents eyestrain and headaches.
c. Using the **fine adjustment**, focus as sharply as possible. If you cannot get a clear image, check your lighting.
d. Move the slide slightly in various directions. Note how the image moves in each case.
e. Now, view the letter under medium power as follows: Make sure the letter is, once again, in sharp focus at low power. Turn the nosepiece *carefully* so that the **medium power objective** is above the slide. Make sure that the objective *never* touches the slide or cover slip. It may be damaged if it does. Turn the fine adjustment slightly to focus the microscope. Then adjust the diaphragm for proper lighting.
Note: To be safe, *never use the coarse adjustment on medium power*.
f. Now, view the letter under high power as follows: Make sure the letter is still in sharp focus. Then, watch the slide as shown in Figure 4-6 and carefully turn the nosepiece so that the **high power objective** is above the slide. If necessary, focus using the fine adjustment. *Do not turn the objective down into the cover slip*. You could damage it. *Never use the coarse adjustment on high power.*

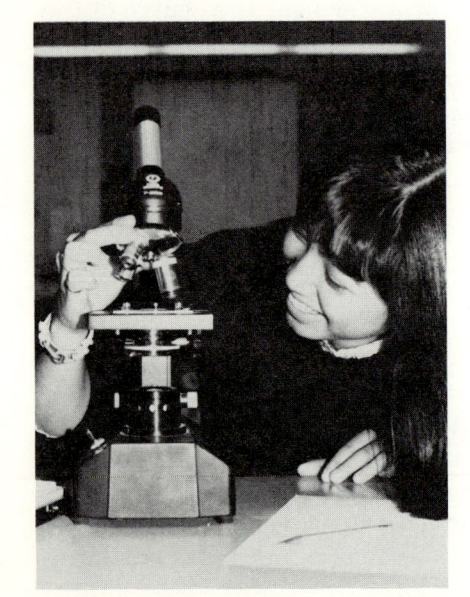

Fig. 4-6 When you are turning the nosepiece, watch what you are doing from the side. Then you will not damage the high power objective.

48 Chapter 4

g. Raise the body tube (or lower the stage) using the coarse adjustment. Remove the slide. Then turn the nosepiece until the low power objective is in place. The microscope is now ready for viewing another slide or to be put back in its case or dust cover. Make sure the low power objective is in place.

h. Prepare a wet mount using a small piece of the coloured picture. Repeat steps (a) to (f) to view it under low, medium, and high power. Record what you see for each magnification.

Discussion

1. Describe the steps you took to make sure you did not damage the medium and high power objectives.
2. The image was, of course, much larger than the actual letter. What else was different about the image?
3. When you moved the letter to the left, what did the image do?
4. When you moved the letter away from you, what did the image do?
5. A microscope has a greater **resolving power** than your eyes. This means that it can separate details better. How does step (h) prove this?

Main Ideas

1. All living things are made of cells.
2. Cells carry out the functions of the organisms of which they are a part.
3. All cells come from living cells.
4. Certain steps must be followed when preparing a microscope for use.
5. Certain steps must be followed to prepare a good wet mount.
6. Certain steps must be followed when focusing a microscope.

Glossary

cell		the basic unit of a living thing
diaphragm	DI-a-fram	the part of a microscope that controls the brightness of the field of view
mounting medium		a liquid, such as water, in which the specimen is placed
objective		the lower lens system of a microscope
ocular	OK-you-lar	the eyepiece of a microscope

Study Questions

A. True or False

Decide whether each of the following statements is true or false. If the sentence is false, rewrite it to make it true. (Do not write in this book.)
1. Lorenza Oken was the first person to use the word "cell" in biological studies.
2. Both eyes are used for viewing in a binocular compound microscope.
3. The diaphragm controls the amount of light entering the microscope.
4. A microscope slide is also called the mounting medium.
5. When viewing a specimen, you should start with the high power objective.

B. Completion

Complete each of the following sentences with a word or phrase that will make the sentence correct. (Do not write in this book.)
1. Robert Brown discovered the ▭ in the centre of a cell.
2. Cells have a moving fluid in them called ▭ .
3. All living things are made of ▭ .
4. When you lift and carry a microscope, graps the ▭ with one hand and the ▭ with the other hand.
5. The eyepiece of a microscope is called the ▭ .

C. Multiple Choice

Each of the following statements or questions is followed by four responses. Choose the correct response in each case. (Do not write in this book.)
1. The objectives of a microscope are mounted on the
   a) nosepiece **b)** diaphragm **c)** body tube **d)** stage
2. Compared to the object, the image in a microscope is
   a) the same way up, but reversed from side to side
   b) upside down and also reversed from side to side
   c) upside down only
   d) reversed from side to side only
3. When you are switching to high power, you should always
   a) open the diaphragm first
   b) close the diaphragm first
   c) watch what you are doing through the ocular
   d) watch what you are doing from the side

D. Using Your Knowledge

1. A hand lens (magnifying glass) is called a simple microscope. Why do you think your microscope is called a compound microscope?

2. The **magnification** of a microscope is expressed in **diameters**. Here is an example: Suppose an object is magnified ten diameters (written 10X). This means that the image is 10 times longer and 10 times wider than it would appear if the object were viewed from a distance of 25 cm with the unaided eye. The magnification is printed on each objective and ocular. Get these magnifications from your microscope. *The overall magnification of a microscope is the product of the magnifications of the objective and ocular being used.* Suppose, for example, that a 10X ocular is combined with a 40X objective. The overall magnification is 10 x 40, or 400 diameters (400X). Calculate all the magnifications your microscope can produce.

E. Investigations

1. Select one of the scientists mentioned in Section 4.1. Visit the library and collect more information on his life and work. Write a report of about 150 words on your findings.
2. Get small pieces of hair from several people. Prepare wet mounts of the samples. Examine the samples under low, medium, and high power. Look for split ends. See if you can discover what makes hair curl. What differences are there between hairs of different colours?
3. Prepare wet mounts of several fabrics. (Use just a few fibres from each fabric.) Use wool, cotton, nylon, polyester, dacron, and other common fabrics. Write a report on the differences you observe.

5 The Cell and Its Parts

5.1 Activity: Structure of a Plant Cell
5.2 Activity: Structure of an Animal Cell
5.3 The Cell as Seen with the Electron Microscope
5.4 Activity: Is Protoplasm Alive?
5.5 Protoplasm and its Activities
5.6 Building Organisms from Cells

There are many types of cells. For example, your body has blood cells, muscle cells, bone cells, skin cells, and many other types. You may think that these types of cells have little in common. It turns out, however, that the differences are not as great as you might suspect. In fact, they have several basic features in common.

In this chapter you will look at plant and animal cells. You will see how they differ. And, you will see what they have in common. Then you will look closely at the parts of the cell to see how cells do their jobs.

5.1 ACTIVITY Structure of a Plant Cell

All plant cells have three main parts: a **cell wall**, a **nucleus**, and **cytoplasm** [SI-toh-plaz-em]. Many plant cells also have **chloroplasts** [KLOR-o-plasts] and **vacuoles** [VAK-yoo-ohls]. All these parts are shown in Figure 5-1. In this activity you will look at onion cells to see how many of these parts you can find.

Problem
What does a plant cell look like?

Materials

onion bulb scale
microscope
microscope slides (2)
cover slips (2)
paper towel

lens paper
dropper
forceps (tweezers)
iodine solution
methylene blue solution

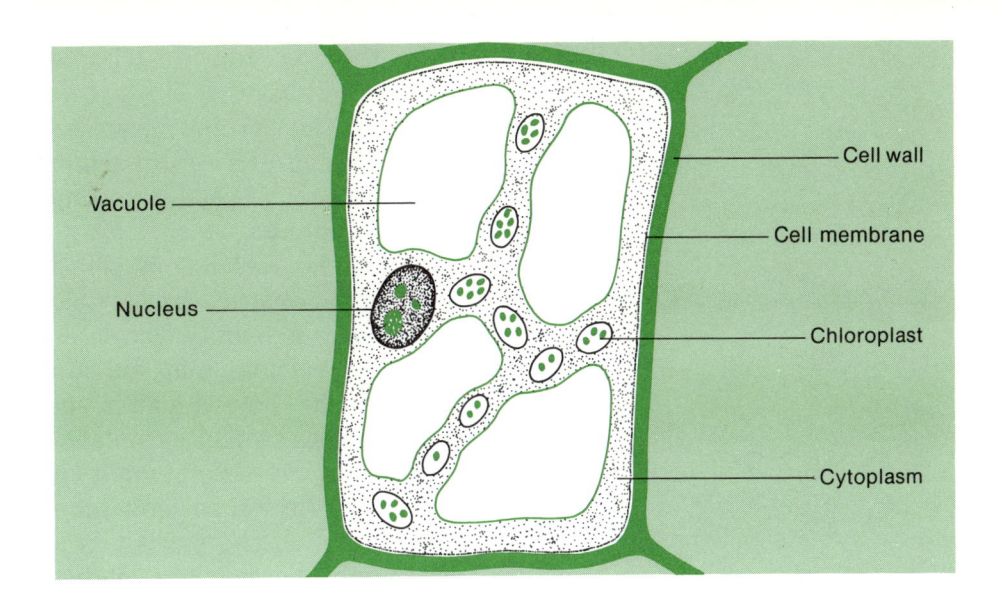

Fig. 5-1 Most plant cells have many of the parts shown here.

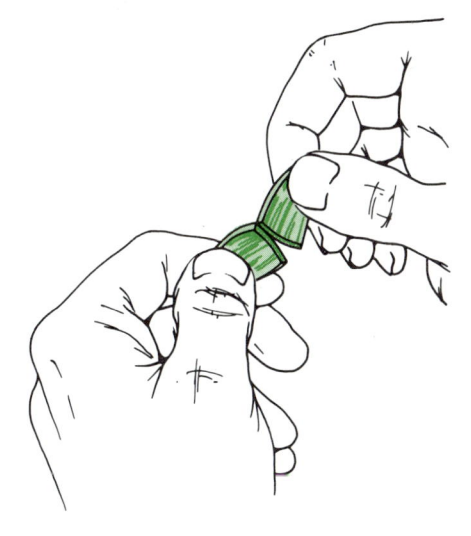

Fig. 5-2 Bend the onion bulb scale backwards to get a piece of "skin".

Procedure

a. Hold a piece of onion bulb scale so that the *inner* surface faces you. Then snap it backwards as shown in Figure 5-2. You should now be able to see a thin transparent "skin".
b. Use the forceps to pull off a small piece of this skin.
c. Prepare a wet mount of the piece. (If you have forgotten how, look back to Activity 4.4.) Try not to wrinkle the skin.
d. Look at the mount under low power. (Be sure to follow the steps you learned in Activity 4.5.) Answer as many of the Discussion items as you can.
e. Now pick one cell in which you can see the details clearly. Move it to the centre of the field of view. Switch to medium power. Look at the contents of the cell closely. Focus up and down slightly as you do so. (Use the fine adjustment.) Continue to work on the Discussion items.
f. Now switch to high power. BE CAREFUL! Refocus with the fine adjustment. Continue to work on the Discussion items.
g. Prepare a second wet mount of the onion skin. However, this time use iodine solution as the mounting medium instead of water. This solution is called a stain. It makes some parts stand out more clearly.
h. Repeat steps (d) to (f) using the stained skin. Continue to work on the Discussion items.
i. If you have time, repeat steps (g) and (h) using another stain, methylene blue.

Discussion

1. Describe the shape of a single cell of an onion skin.
2. Draw a group of 5 to 10 cells. Show how they are attached to one another. The "lines" that run between the cells are the **cell walls**.

Section 5.1

Label them on your drawing. Cell walls are non-living. They are made of cellulose.
3. Describe the **cytoplasm** of a cell. Include colour and clearness. Does it move? The outer edge of the cytoplasm is called the **cell membrane**. It is usually pushed out against the cell wall. Therefore it is hard to see.
4. Describe the **nucleus** of a cell. Where are they found in the cells?
5. The "empty spaces" you saw in the cytoplasm are called **vacuoles**. They contain mainly water and dissolved substances. In some cells one vacuole may fill almost the entire cell. Why is the nucleus near the cell wall in such cells?
6. The droplets in the cytoplasm are called **oil droplets**. Describe them. They give the onion its smell and make your eyes run.
7. Draw a single cell as seen under high power. Make it about 10 cm long. Label all the parts you saw: cell wall, nucleus, cytoplasm, cell membrane, vacuole, oil droplet.

5.2 ACTIVITY Structure of an Animal Cell

In this activity you will look at "skin" cells from the inside of your cheek. How are these animal cells like plant cells? How are they different?

Problem
What does an animal cell look like?

Materials
microscope
microscope slides (2)
cover slips (2)
paper towel
lens paper

toothpicks (4)
iodine solution
methylene blue solution
dropper

Procedure
a. Gently scrape the *inside* of your cheek with the blunt end of a toothpick. This should give you a sample of skin cells. They are always flaking off. You will likely get the best sample near your molars (Fig. 5-3).
b. Prepare a wet mount of the scrapings. Spread the scrapings through the drop of water before putting on the cover slip. Place the toothpick in the waste container.
c. Look at the scrapings under low power. Answer as many of the Discussion items as you can.

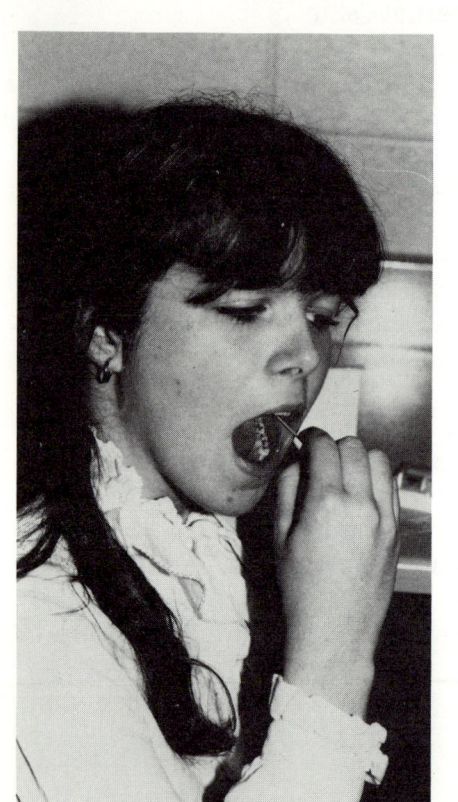

Fig. 5-3 Scrape the inside of your cheek with a toothpick. In this way you can collect some skin cells.

Note 1: You may have trouble telling the cells from food particles. The cells should look like irregular flagstones in a patio.

Note 2: These cells are very transparent. They cannot be seen if the light is too bright. Therefore, you may have to close the diaphragm more than usual.

d. Now switch to medium power. Study a single cell closely. Look for the parts that you found in plant cells. Continue to work on the Discussion items.

e. Now switch to high power. BE CAREFUL! Continue to work on the Discussion items.

f. Prepare a second wet mount of the skin cells. However, this time use methylene blue solution as the mounting medium. This stain should darken the cell contents. Study this slide as you did the first one. You should now find it easier to answer the Discussion questions.

g. If time permits, prepare a third wet mount. This time, use iodine solution as the mounting medium. Study this slide as you did the other two.

Discussion

1. Describe the shape of a single cheek skin cell.
2. Draw a group of 5 to 10 cells. Show how they are attached to one another.
3. Draw a single cell as seen under high power. Make it 5-10 cm across. Label all the parts you saw: nucleus, cytoplasm, cell membrane, vacuole.
4. List any differences between this cell and the plant cell.
5. List any similarities between this cell and the plant cell.

5.3 The Cell as Seen with an Electron Microscope

You have looked at plant and animal cells with your microscope. You saw that these cells have many parts. They have still more parts that you cannot see. However, these have been "seen" with the electron microscope. Let's summarize what you have discovered about cells. Then we will see what scientists have discovered using the electron microscope.

Comparison of Plant and Animal Cells

You learned that plant and animal cells differ in some ways. However, they have many basic features in common. Table 5-1 and Figure 5-4 summarize the differences and similarities.

PLANT CELL   **ANIMAL CELL**

- Cell wall
- Cell membrane
- Vacuole
- Nucleus
- Cytoplasm

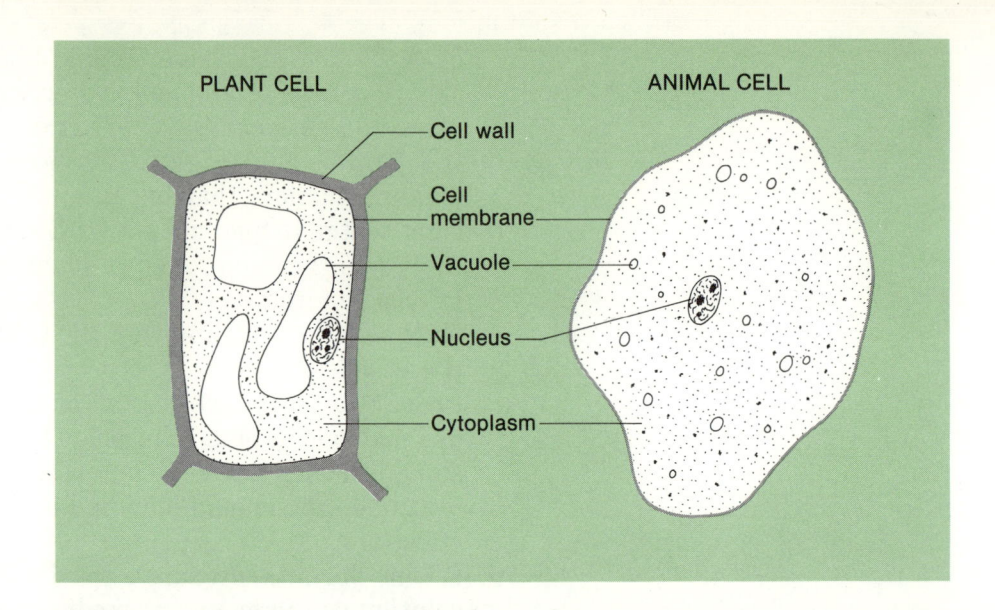

Fig. 5-4 Comparison of plant and animal cells

Table 5-1 Comparison of Plant and Animal Cells

Feature	Plant cell	Animal cell
Differences		
Outside support	Thick, hard cell wall	Thin, flexible cell membrane
Vacuoles	Often just a few large ones (or just one)	Usually many small ones
Similarities		
Nucleus present	Yes	Yes
Cytoplasm present	Yes	Yes
Cell membrane present	Yes	Yes
Vacuoles present	Yes	Yes

As you can see, there are more similarities than differences. Since this is so, scientists draw a "generalized" cell to represent all living cells. They do this with the help of the electron microscope (Fig. 5-5) which allows them to see many things in the cell that you haven't seen. Figure 5-6 is a drawing of a "generalized" cell.

Electron Microscope

This microscope does not use light as your microscope does. Instead, it uses a beam of electrons. Magnifications as high as 200 000 X can be reached. In fact, magnifications of 1 000 000 X can be reached by enlarging the resulting photograph.

The world's first electron microscope to magnify more than a

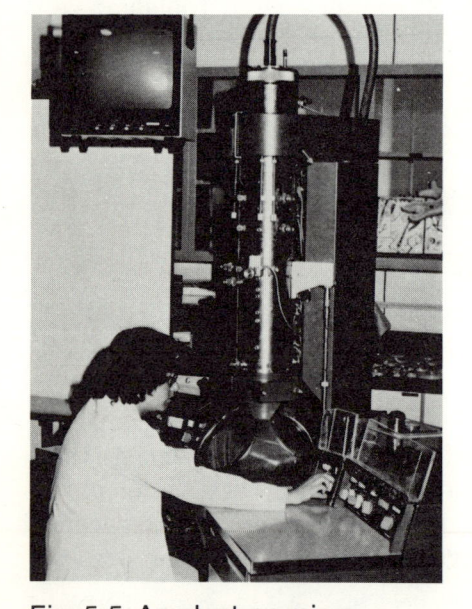

Fig. 5-5 An electron microscope. The image can be viewed on a television screen. However, it is usually photographed. The resulting picture is called an electron micrograph.

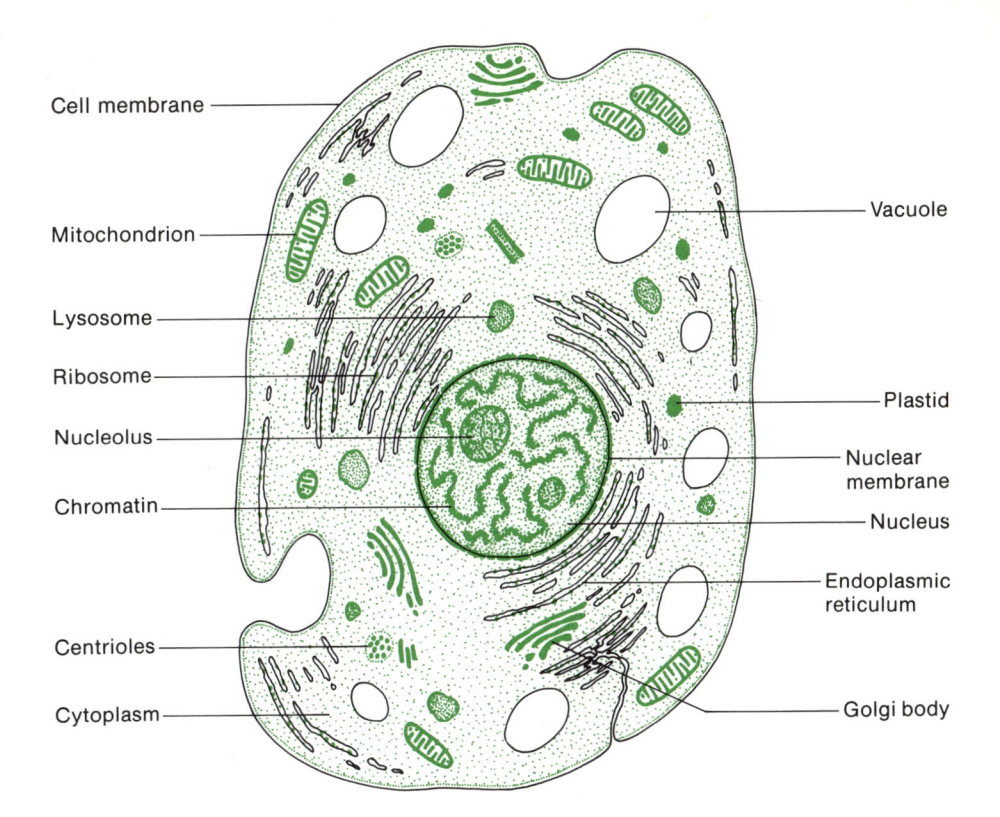

Fig. 5-6 Diagram of a generalized cell. This cell is neither plant nor animal. No cell looks exactly like this. This drawing was made from photographs taken with an electron microscope.

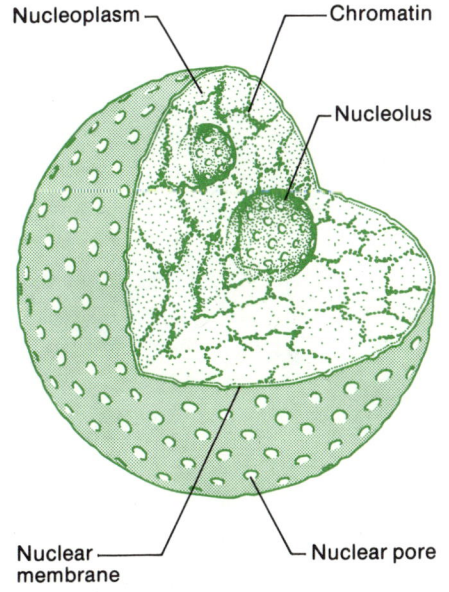

Fig. 5-7 This drawing of a nucleus was made from electron micrographs. A piece has been removed so you can see inside.

compound microscope was made at the University of Toronto in 1938. It was built by James Hillier and Albert Prebus. They were graduate students in the physics department.

The Parts of a Cell

Spend a few minutes looking closely at Figure 5-6. Then look ahead at the electron micrographs on the new few pages. These pictures look complicated, don't they? That's because an electron microscope "sees" much more detail than you can see with your microscope. However, things aren't as complicated as they appear. Let's study Figure 5-6 bit by bit.

The Nucleus When you looked at the nucleus of a cell, you saw little except a round body. However, the electron microscope shows some of the structure of the nucleus (see Figures 5-6, 5-7, and 5-8). The nucleus has a membrane around it. It is called the **nuclear membrane**. This membrane has pores in it. They allow contact between the nucleus and the cytoplasm. The nucleus also contains **nucleoli** (singular: **nucleolus**). Further, the nucleus contains strands of **chromatin** [KROME-a-tin]. Chromatin is the genetic material of the cell. This material forms **chromosomes**. You may have heard of them.

The Cytoplasm With your microscope, the cytoplasm looked clear but granular at times. The electron microscope shows that cytoplasm

Section 5.3

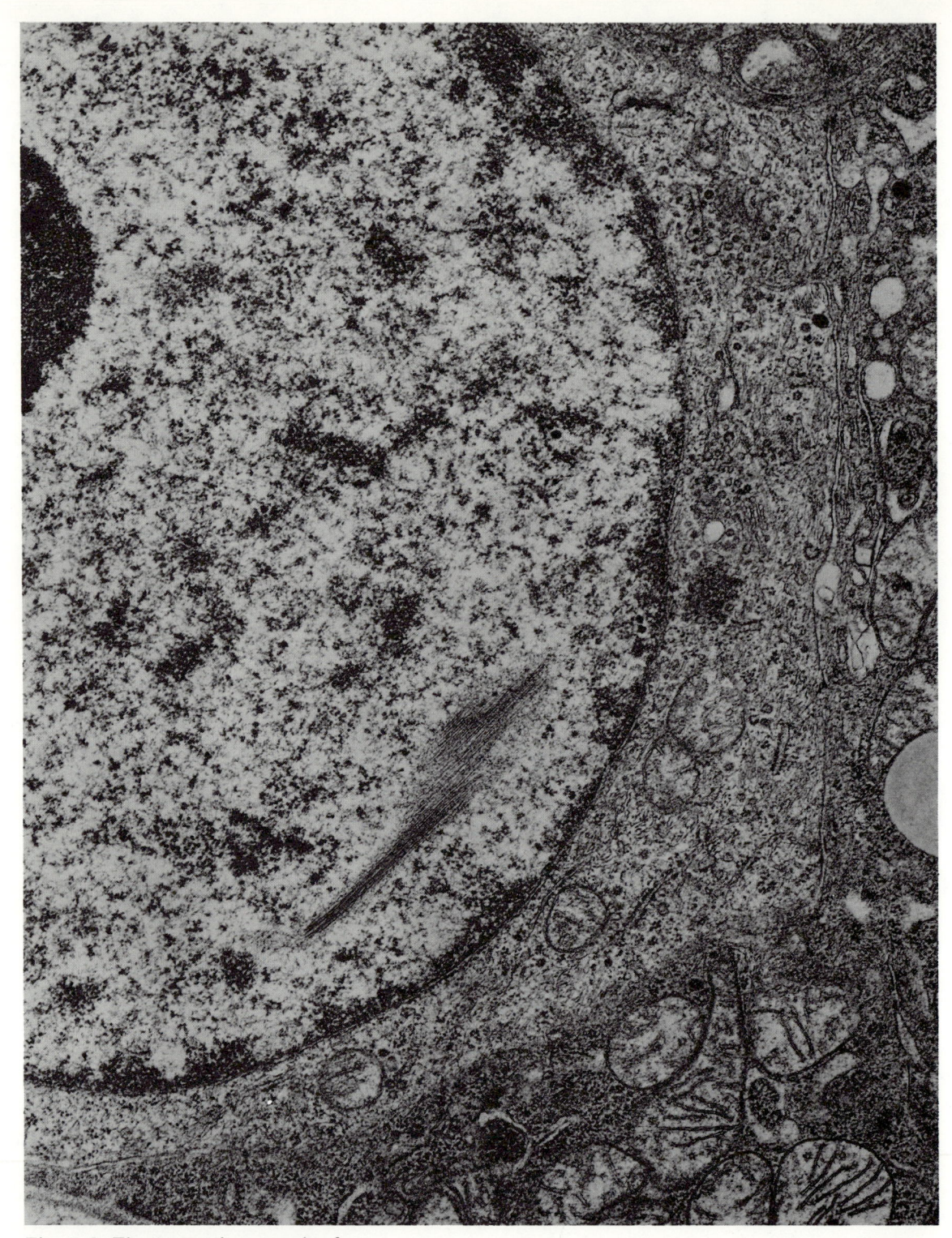

Fig. 5-8 Electron micrograph of part of a cell (6000X). The left half is the nucleus. Note the pores in the nuclear membrane. Note, also, the nucleolus, the dark area in the upper left.

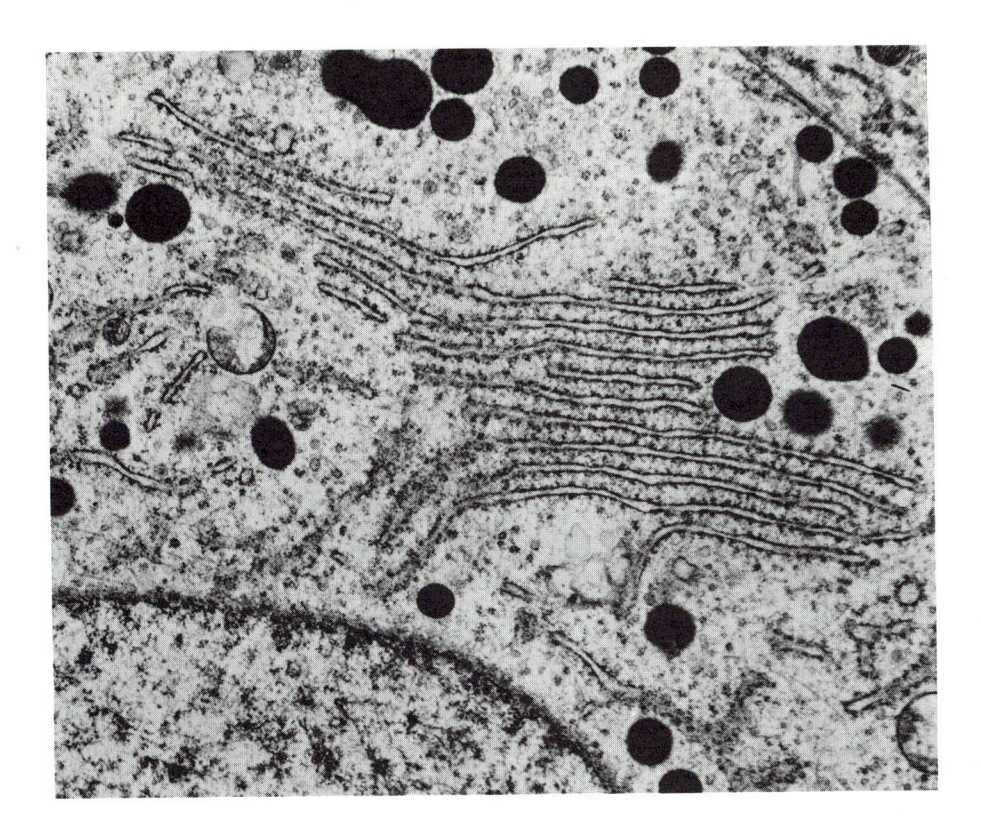

Fig. 5-9 The little tubes in the centre are part of the cytoplasm. They carry substances throughout the cell (13 000X).

has a **cell membrane** on its outer side. It also shows that cytoplasm has little tubes running through it (Fig. 5-9). Most of the other things you see in the cytoplasm are **organelles** ("little organs"). The organelles can be compared to the organs of your body. Your stomach, lungs, liver, and kidneys are organs. Each carries out certain functions. And they function together to make your body alive. In like manner, each organelle has certain functions. And all the organelles function together to make a cell alive.

The electron microscope shows many organelles. Among them are **mitochondria** (singular: **mitochondrion**), pronounced mi-toh-KON-dree-on, **plastids**, **lysosomes** [LI-so-somes], **ribosomes** [RI-bo-somes], **Golgi bodies**, and **centrioles** [SEN-tree-oles]. You will learn what some of these do later.

Vacuoles You saw **vacuoles** with your microscope. They looked like empty spaces. However, they contain water and dissolved substances. The electron microscope shows that each vacuole has a **membrane** around it.

Section Review

1. **a)** How are plant and animal cells alike?
   b) How are they different?
2. What is a generalized cell?
3. What parts of a nucleus does an electron microscope show us?

Section 5.3   59

4. What does cytoplasm look like under an electron microscope?
5. **a)** What are organelles?
   b) Name six organelles.

5.4   ACTIVITY   Is Protoplasm Alive?

Protoplasm is a granular, jelly-like substance. It makes up most of a cell. In fact, **the nucleus and ctyoplasm are protoplasm**. Protoplasm is about 70% water. Most of the remaining 30% is carbohydrates, lipids, and proteins.

Protoplasm is often called the **living material** of a cell. Yet some biologists do not agree that protoplasm is alive. Certainly no one part of it is alive. We would all agree that water, carbohydrates, lipids, and proteins are not alive. Yet, somehow, the protoplasm can put all these things together into a **"living condition"**. Then the cell can carry out life processes such as digestion and respiration.

One characteristic of life is movement. In this activity you will watch protoplasm to see if it moves. Then you will decide if you think protoplasm is alive.

Problem

Does protoplasm move? Is protoplasm alive?

Materials

Elodea leaf
microscope
microscope slides (2)
forceps

cover slips (2)
paper towel
dropper
lens paper

Fig. 5-10 *Elodea* is also called Canada waterweed. It grows in ponds and slow streams.

Procedure

a. Remove a young leaf from the tip of an *Elodea* plant (Fig. 5-10). Your teacher may have already removed the leaves. In that case, get one from your teacher.
b. Prepare a wet mount of the leaf.
c. Look at the leaf under low power. Move the slide around until you can clearly see cells.
d. Select one cell in which you can see the contents clearly. Move it to the centre of the field of view. Switch to medium power. Study the contents closely. Draw and label what you see.
e. Now switch to high power. BE CAREFUL! Refocus with the fine adjustment. Add any new details to your diagram.
f. The green objects in the cytoplasm are **chloroplasts**. They contain **chlorophyll**. Photosynthesis takes place in them. Watch them for several minutes to see if they move. If the cytoplasm (part of the protoplasm) moves, the chloroplasts will move.

60   Chapter 5

Discussion

1. Make a good drawing of an Elodea cell as seen under high power. Label these parts: cell wall, nucleus, cytoplasm, cell membrane, vacuole, chloroplasts.
2. Describe any evidence of movement that you saw.
3. Is protoplasm alive? Why do you say so?

5.5 Protoplasm and Its Activities

A cell contains both living and non-living materials (Fig. 5-11). The living material is called **protoplasm**. The introduction to Activity 5.4 described protoplasm. Read that description again.

Whether protoplasm is living or not, it does carry out life processes. Most of these processes are carried out by the organelles. However, other parts of the cell help with these processes. The following is a description of the functions of some main parts of the protoplasm.

The Nucleus

The nucleus is the control centre for all cell processes. The **nucleoli** in the nucleus store some proteins and make RNA (genetic material). The **chromatin** in the nucleus becomes **chromosomes** during cell division. Chromosomes are made of proteins and DNA (also genetic material). DNA is the hereditary material in cells. It carries the genetic code that controls all cell processes.

The Cytoplasm

The cytoplasm is all the protoplasm outside the nucleus. It has a **cell membrane** on its outside. This membrane controls what goes in and

Fig. 5-11 The living and non-living parts of a cell

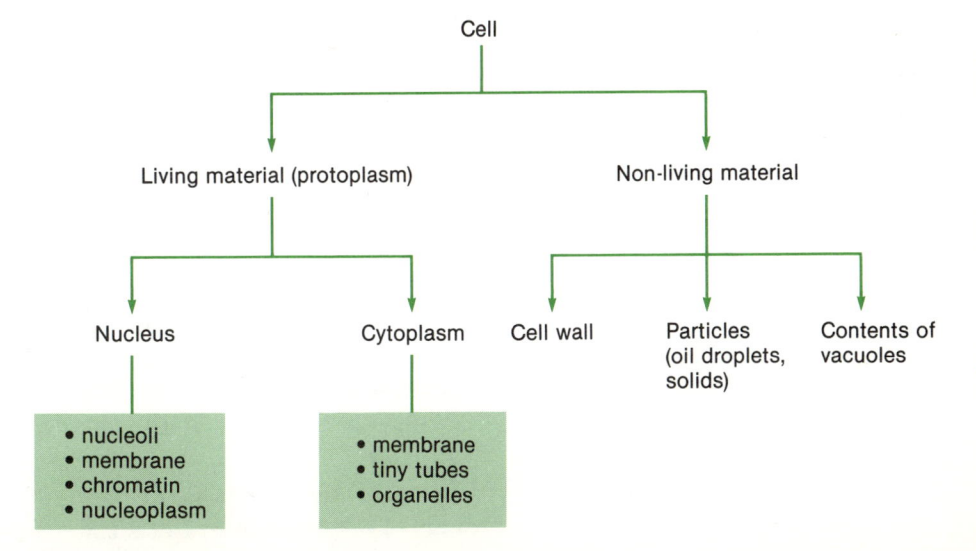

```
                              Cell
                               │
              ┌────────────────┴────────────────┐
   Living material (protoplasm)          Non-living material
              │                                  │
       ┌──────┴──────┐              ┌────────────┼────────────┐
    Nucleus      Cytoplasm       Cell wall   Particles     Contents of
                                             (oil droplets, vacuoles
                                              solids)
   • nucleoli      • membrane
   • membrane      • tiny tubes
   • chromatin     • organelles
   • nucleoplasm
```

Section 5.5

out of the cell. Tiny tubes (**endoplasmic reticulum**) make up much of the cytoplasm. They carry materials throughout the cell.

The cytoplasm contains **organelles**. Each type of organelle has a certain function. The functions of three organelles are discussed here.

Ribosomes These usually appear as small dots on the surface of the tiny tubes (endoplasmic reticulum). You can see them in Figure 5-9. **Ribosomes** are made mainly of RNA. They help make special proteins for the cell.

Mitochondria (singular: Mitochondrion) You can see **mitochondria** with a good compound microscope. They look like tiny specks, dots, or rods. However, they are very complex in structure (Fig. 5-12). Mitochondria are the centres of respiration for the cell. Here sugars and other substances are broken down. This releases energy to power all life processes.

Plastids These organelles are found mainly in plants. Some **plastids** make food. Others store food. You have seen the most common plastid, the **chloroplast**. It contains chlorophyll. Photosynthesis takes place in the chloroplasts. The chlorophyll assists with this process. The chloroplasts, then, make carbohydrates (food).

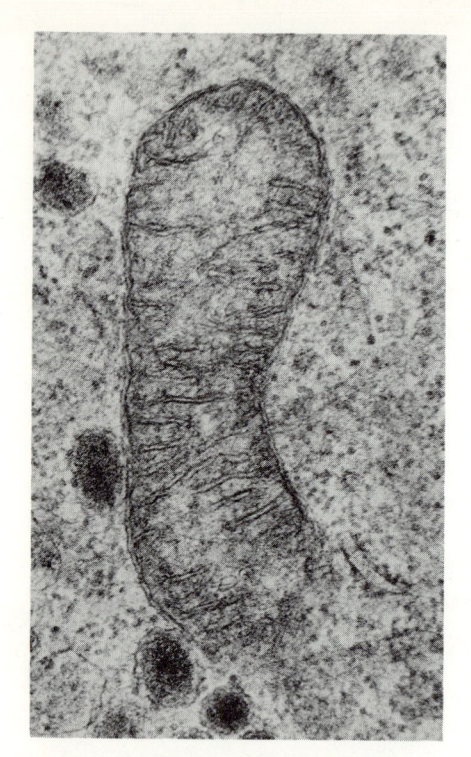

Fig. 5-12 Electron micrograph of a mitochondrion (42 000X)

Section Review

1. Name the living parts of protoplasm.
2. Name three non-living materials that may be part of a cell.
3. a) What is the function of the nucleus of a cell?
   b) What do the nucleoli do?
   c) What is the function of the chromatin?
4. State the functions of ribosomes, mitochondria, and chloroplasts.

5.6 Building Organisms from Cells

You are an organism. You are made of about sixty million million cells (60 000 000 000 000). You know that each of these cells can, by itself, carry out life processes. But how do all these cells "put their act together"?

The answer to this question has been known for some time. The cells are *organized* into groups. (That's why you are called an *organism*.) Each group performs certain tasks. Figure 5-13 shows how the **cells** are organized. Cells are organized into **tissues**. The tissues are organized into **organs**. The organs are organized into **organ systems**. And, finally the organ systems are organized into an **organism**.

Cells, tissues, organs, organ systems, and the organism are called **levels of organization**. Let's take a closer look at each level.

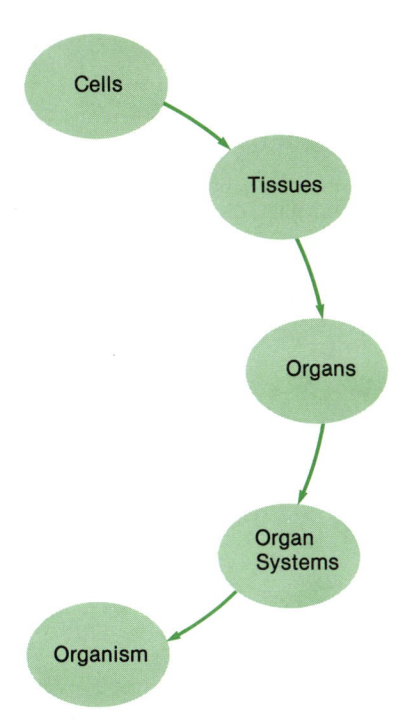

Fig. 5-13 Levels of organization

Muscle cell

Nerve cell

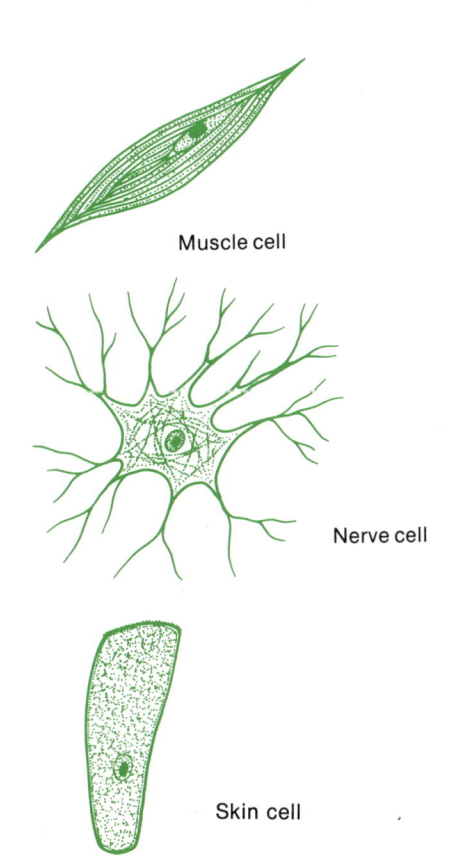

Skin cell

Fig. 5-14 Some types of cells. What functions do each of these carry out?

The Cell Level

*The **cell** is the basic unit of all living things.* Nothing smaller than a single cell can be called "living". One-celled organisms like the amoeba obviously don't have any levels of organization except the cell level. Their single cell must carry out all life processes.

Many-celled organisms (plants and animals) are much more complex. They have different types of cells to do different jobs. For example, your body has nerve cells, bone cells, muscle cells, skin cells, and many other types (Fig. 5-14).

The Tissue Level

In a big organism like you, one muscle cell could do very little. However, many muscle cells working together can move an arm. *A group of similar cells is called a **tissue**.* Thus a muscle is a tissue. Your body has many kinds of tissues (Fig. 5-15).

The Organ Level

Your heart is an organ. It needs several types of tissue before it can function. For example, it needs muscle tissue to make it pump; it needs nerve tissue to tell it when to pump; it needs connective tissue to hold various parts together.

Your stomach, brain, and kidneys are organs too. *An **organ**, then, is several tissues working together as a unit to do a certain job.*

The Organ System Level

Plants, as well as animals, have cell, tissue, and organ levels of organization. But only animals have organ systems. *An **organ system** is a group of organs that work together to do a certain job.* Your digestive system is an organ system. It is made of several organs that work together to digest food. Among these are salivary glands, stomach, intestine, and liver.

You have other organ systems. Some of these are the breathing system, circulatory system, nervous system, and skeletal system. In complex animals like us, there is an organ system for almost every life process. In other words, the organ systems go together to make an organism.

Section Review

1. Name, in order, the five levels of organization.
2. Name four types of cells found in your body.
3. Name four types of tissues found in your body.
4. a) Name four organs in your body.
   b) What tissues are in your arm?
5. a) Name five organ systems in your body.
   b) Name four organs that are in your digestive system.

Section 5.6

Muscle tissue

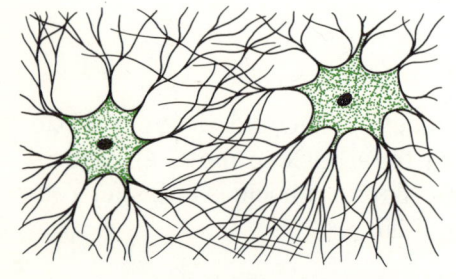

Nerve tissue

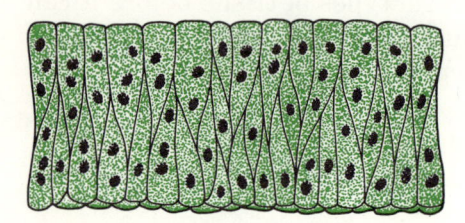

Skin tissue

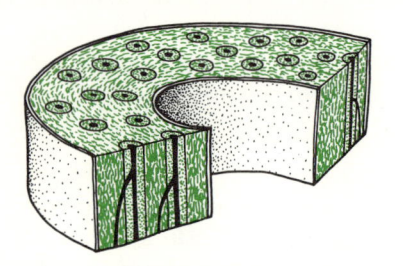

Bone tissue

Fig. 5-15 Some animal tissues. Note that each tissue is made of similar cells.

Main Ideas

1. All plant and animal cells have a nucleus, cytoplasm, cell membrane, and vacuoles.
2. The nucleus of a cell contains nucleoli, chromatin, and nucleoplasm.
3. Cytoplasm is surrounded by a cell membrane and contains tiny tubes (endoplasmic reticulum) and organelles.
4. Each organelle carries out certain life processes for the cell.
5. Protoplasm gives a cell its living condition.
6. The levels of organization are: cell, tissue, organ, organ system, organism.

Glossary

cell		the basic unit of all living things
chloroplast	KLOR-o-plast	a plastid containing chlorophyll; site of photosynthesis
chromatin	KROME-a-tin	the part of the nucleus that contains the genetic material
cytoplasm	SI-toh-plaz-em	all the protoplasm outside the nucleus
mitochondrion	mi-toh-KON-dree-on	an organelle that is the site of respiration
nucleolus	NOO-kli-ol-us	a part of the nucleus that makes proteins and RNA
nucleus		the control centre for cell processes
organ		several tissues working together as a unit
organ system		a group of organs working together
organelle		a small body in the cytoplasm that carries out some life processes
plastid		an organelle that makes or stores food
protoplasm		the living material of a cell
ribosome	RI-bo-some	an organelle that makes proteins

tissue		a group of similar cells
vacuole	VAK-yoo-ohl	a "sac" in a cell containing water and dissolved substances

Study Questions

A. True or False

Decide whether each of the following statements is true or false. If the sentence is false, rewrite it to make it true. (Do not write in this book.)
1. Both plant and animal cells have a cell wall.
2. Vacuoles have cytoplasm in them.
3. The nucleus of a cell contains chromatin.
4. Mitochondria are the sites of respiration.
5. A tissue is made of several organs that work together to carry out a certain function.

B. Completion

Complete each of the following sentences with a word or phrase which will make the sentence correct. (Do not write in this book.)
1. The living material of a cell is called _____ .
2. Photosynthesis takes place in plastids called _____ .
3. Protoplasm is made of two main parts, _____ and _____ .
4. A tissue is a group of similar _____ .
5. Your stomach, brain, kidneys, and _____ are organs.

C. Multiple Choice

Each of the following statements or questions is followed by four responses. Choose the correct response in each case. (Do not write in this book.)
1. Animal cells differ from plant cells in which way?
    a) They have cytoplasm.
    b) They have vacuoles.
    c) They do not have a cell wall.
    d) They do not have nucleoli.
2. Three organelles are:
    a) stomach, lungs, and liver
    b) nucleus, cytoplasm, and vacuole
    c) cell, tissue, and organ
    d) ribosome, mitochondrion, and plastid
3. A plant has a group of similar cells that work together to support the stem. These cells make up a
    a) tissue   b) organ   c) organ system   d) organism
4. Your red blood cells are an example of
    a) a tissue   b) an organ   c) an organ system   d) an organelle

D. Using Your Knowledge

1. Why is it better to say protoplasm creates a "living condition" than to say protoplasm is alive?
2. In some ways a one-celled organism has advantages over a many-celled organism. Explain why this is so.
3. The cells in the leg muscles of a runner have more mitochondria than other cells. Why is this so?
4. Suppose that a scientist mixed in a flask all the substances that make up protoplasm. Would the mixture now be a "living condition"? Explain your answer.

E. Investigations

1. Select one of the following organelles and research its structure and functions: lysosome, Golgi body, centriole.
2. Find out all you can about the structure of a chloroplast.

6 Cells at Work: How Substances Enter and Leave Cells

6.1 Demonstration: Diffusion in Air and Water
6.2 Explanation of Diffusion
6.3 Diffusion through Membranes
6.4 Activity: Osmosis
6.5 Explanation of Osmosis
6.6 Activity: Osmosis through an Animal Membrane
6.7 Activity: Osmosis through a Plant Membrane
6.8 Osmosis and Living Cells

The cells in your body need food, oxygen, water, and other substances. Also, they need to get rid of carbon dioxide and other by-products of metabolism. But every cell is surrounded by a cell membrane (Fig. 6-1). How, then, can substances enter and leave cells? How can the cell membrane let some substances in but not others? How can the

Fig. 6-1 This living cell needs water, oxygen and many other substances. It must get rid of the many by-products of its life processes. How do these materials get through the cell membrane?

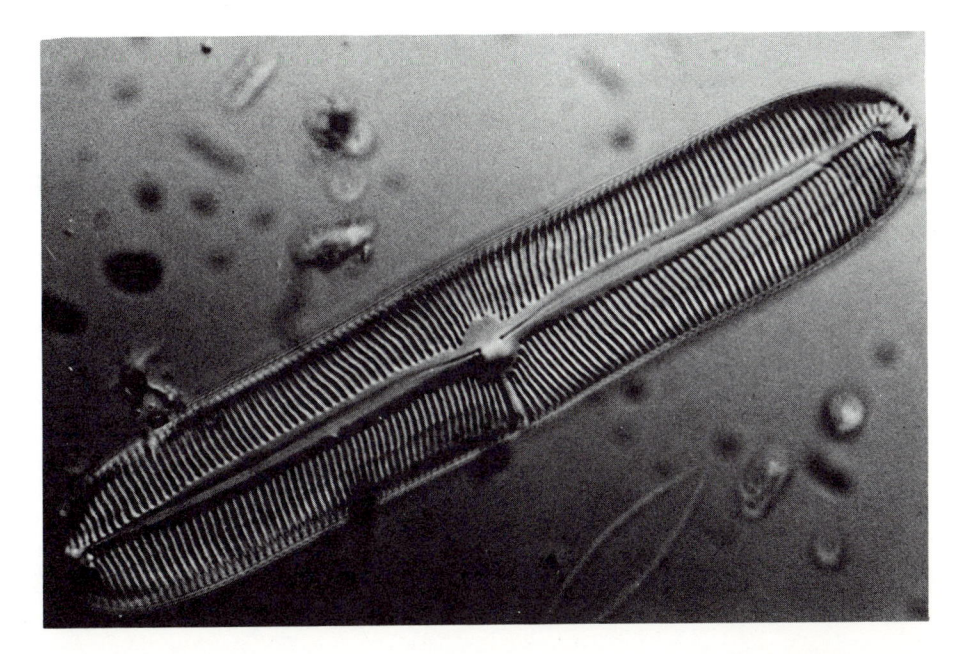

Introduction 67

cell membrane let some substances out without losing important cell contents?

To answer such questions, we must first understand a process called osmosis. What's osmosis? It's a special kind of diffusion. But what's diffusion? Well, that's where this chapter begins. We will study diffusion first. Then we will study osmosis. And, finally, we will figure out how substances enter and leave cells.

6.1 DEMONSTRATION Diffusion in Air and Water

Suppose you release some air freshener into one corner of a room. The odour does not stay there. It gradually spreads throughout the room. The odour seems to move by itself. You don't have to blow it around.

We call such movement **diffusion** [di-FEW-shun]. In this demonstration you will investigate diffusion. Let's see how much you can learn about it.

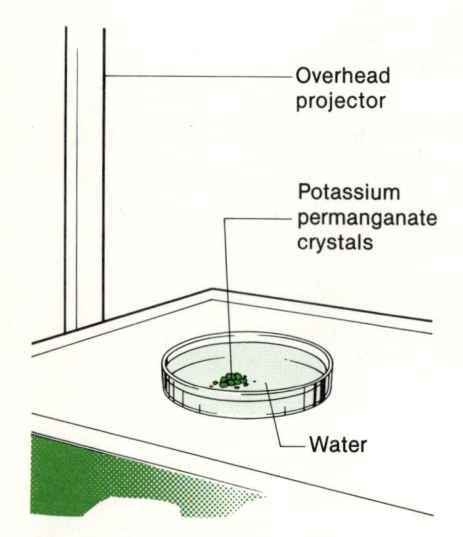

Fig. 6-2 Diffusion in water

Problem

What is diffusion?

Materials

bottle of cologne   overhead projector
petri dish   potassium permanganate crystals
water

CAUTION: Potassium permanganate can damage skin and clothes. Do not touch it.

Procedure A Diffusion in Air

a. Place some cologne in a dish in one corner of the classroom. (One class member will do this.)
b. Raise your hand when the odour reaches you at your desk.

Procedure B Diffusion in Water

a. Fill a petri dish with water. Place it on an overhead projector.
b. Place a few crystals of potassium permanganate in the centre of the dish (Fig. 6-2). Do not disturb the dish from this point on.
c. Observe the dish (or its image on the screen) for about 10 min. Make notes and sketches of the changes that take place.

Discussion A

1. In which part of the classroom did people smell the cologne first? last?
2. Where was the smell strongest? weakest?
3. Try to explain your observations.

68   Chapter 6

Discussion B

1. Where was the solution the darkest in colour? the lightest?
2. Try to explain your observations.
3. Is diffusion faster in a gas or liquid? How do you know?

6.2 Explanation of Diffusion

How Does Cologne Move Through Air?

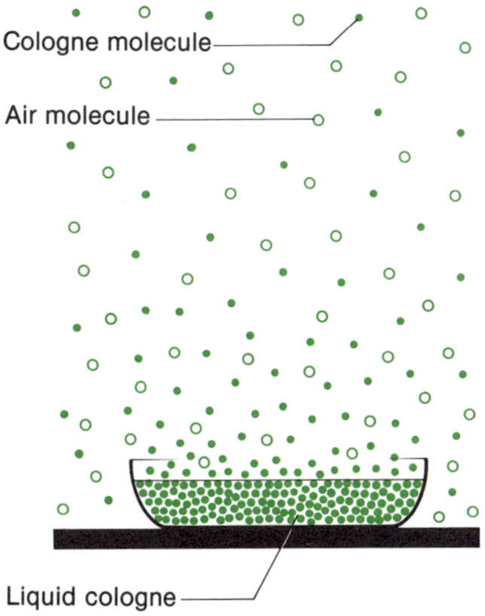

Your observations in Activity 6.1 can be explained using the **particle theory**. This theory says that all matter is made of particles. It also says that these particles are always moving.

The cologne is made of particles called **molecules**. The molecules are moving about in the dish. Some of them move out of the dish into the air (Fig. 6-3). They bump into air molecules (nitrogen, oxygen, etc.). The air molecules are also moving. Thus the cologne molecules are pushed around and spread throughout the air.

Definition of Diffusion

The cologne moved from an area where there was pure cologne to an area where there was a lower concentration. The potassium permanganate also moved from an area where there was pure potassium permanganate to an area where there was a lower concentration. This happens with all substances that diffuse. Thus **diffusion** *is the net movement of a substance from an area of high concentration to an area of low concentration. The movement is caused by the motion of particles.*

Fig. 6-3 Diffusion of cologne into air

Section Review

1. Use the particle theory to explain the diffusion of cologne into air.
2. Make up a similar explanation for the diffusion of potassium permanganate into water.
3. Define diffusion.

6.3 Diffusion Through Membranes

In this section we are going to do a "thought experiment". That's an experiment we do in our minds. Often such an experiment leads to a hypothesis. Then we can do a real experiment to test the hypothesis.

You know that particles diffuse from a region of high to low concentration. Our thought experiment deals with this problem: *What will happen if we put a membrane in the path of the diffusing particles?* (A membrane is a thin piece of substance.)

Look closely at Figure 6-4. All three cases have 10 particles of one substance on Side A and 10 particles of another substance on Side B. The three cases differ in only one way. Each has a different kind of

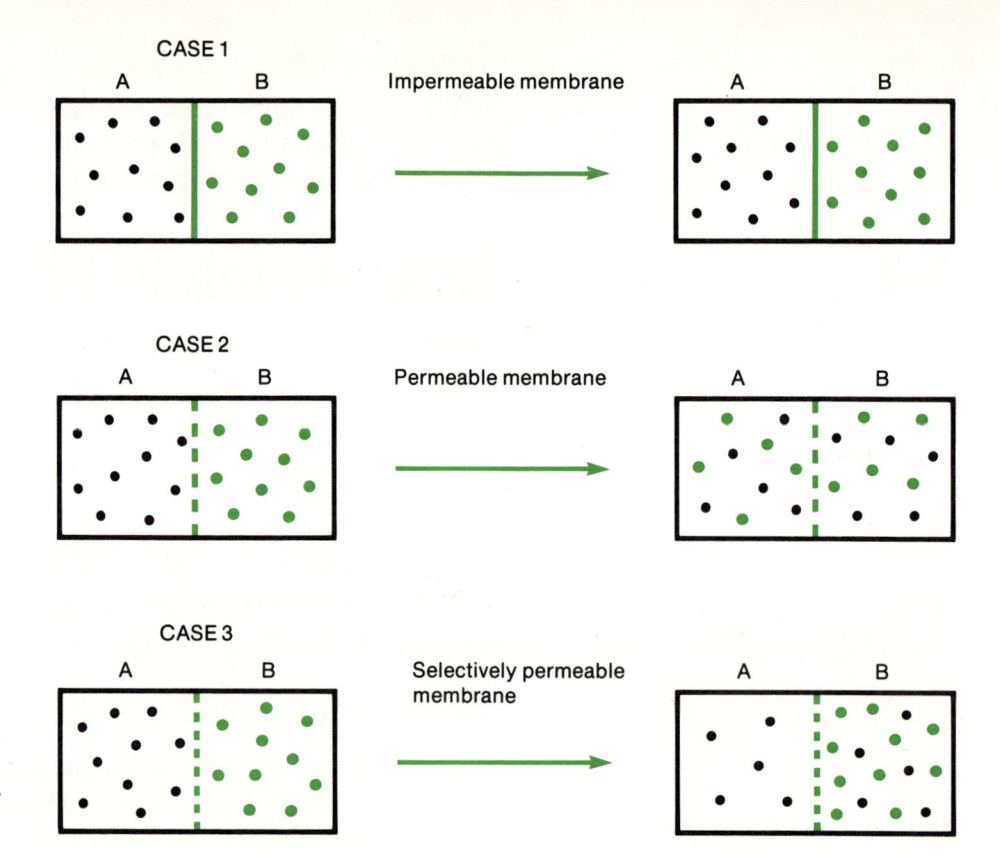

Fig. 6-4 Three types of membranes. How will they affect diffusion?

membrane between Sides A and B. Let's consider each of these cases.

Case 1: **Impermeable membrane.** An impermeable [im-PER-me-a-bel] membrane lets neither substance pass through. Rubber is impermeable to most substances. It does not let many types of particles through. Such a membrane stops the diffusion of one substance into the other.

Case 2: **Permeable membrane.** A permeable [PER-me-a-bel] membrane allows both substances to pass through. In other words, diffusion occurs almost as though the membrane was not there. Particles from Side A diffuse to Side B. And, particles from Side B diffuse to Side A. Filter paper is permeable to both water and substances dissolved in the water. For example, both the water and sugar molecules in a sugar solution go through filter paper.

Case 3: **Selectively permeable membrane.** Case 1 let no particles through. Case 2 let both types of particles through. Now, let's consider Case 3, which is in between.

Imagine a membrane that, somehow, can *select* the type of particle that can pass through. (That's why we call it *selectively* permeable.) Suppose, for example, that the pores in the membrane are the size of the A particles. This means that A particles can diffuse through. But the larger B particles can't. Eventually we end up, as shown, with A particles on both sides. Side A now has less material in it. And Side B has more material.

70   Chapter 6

Will this really happen? Is the hypothesis true? Try Activity 6.4 and see.

Section Review

1. What is an impermeable membrane?
2. What is a permeable membrane?
3. What is a selectively permeable membrane?

6.4   ACTIVITY   What Does a Selectively Permeable Membrane Do?

Look at Figure 6-5. The dialysis [die-AL-i-sis] tubing is a selectively permeable membrane. You will put sugar solution inside the membrane and pure water outside. Suppose the membrane lets water molecules through but not sugar molecules. What do you think will happen?

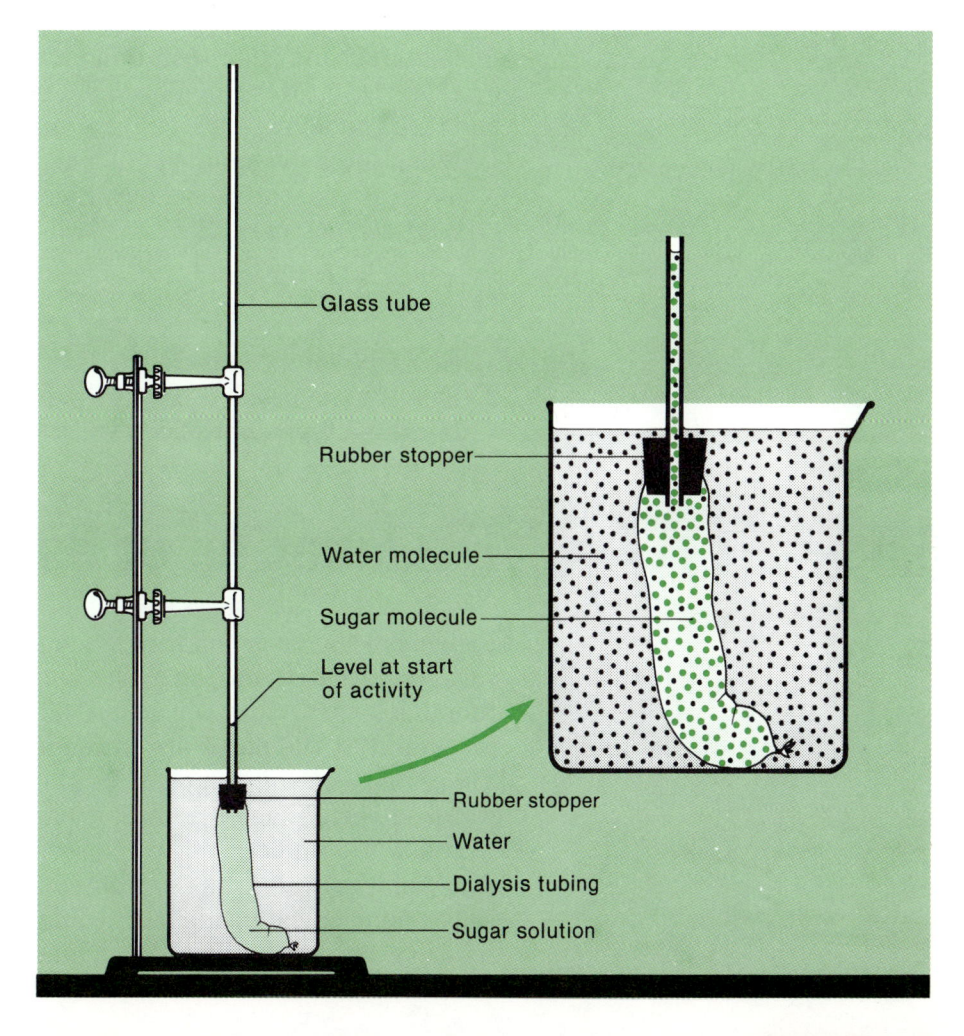

Fig. 6-5 Osmosis apparatus

Section 6.4   71

Problem

How does a selectively permeable membrane affect diffusion?

Materials

large beaker (600 mL or 1000 mL)
glass tubing (about 50 cm long)
dialysis tubing (20 cm long)
adjustable clamp
ring stand

sugar solution
piece of thread
rubber stopper
elastic band
marking pen

Procedure

a. Soak the dialysis tubing for 5 min. Then open it up by rolling an end between your thumb and first finger.
b. Tie a knot near one end of the tubing.
c. Fill the tube with sugar solution.
d. Put the rubber stopper with the glass tubing in it into the dialysis tubing as shown in Figure 6-5.
e. Make sure the tubing and rubber stopper fit tightly. Use the elastic band to hold them together.
f. Rinse off any sugar solution on the outside of the dialysis tubing.
g. Clamp the tubing in a beaker of water as shown.
h. Mark the level of the sugar solution in the glass tubing.
i. Mark the level again at the end of the period. If possible, mark it at the end of the school day. Leave the apparatus set up until your next science period.

Discussion

1. Describe any change in the level of the sugar solution.
2. Explain the results. (Read the introduction to this activity again. It will help you with the explanation.)

6.5 Explanation of Osmosis

Figure 6-6 shows an enlarged view of part of the dialysis tubing. The sugar solution is on the right and pure water is on the left. The tubing has pores in it. These are large enough to let water molecules through. But the pores are too small to let sugar molecules through. Therefore, the sugar molecules stay in the tubing. However, when a water molecule meets a pore, it moves through.

    The water molecules are closer together outside the tubing. We say that the water has a *high concentration* there. There are also water molecules inside the tubing. They are part of the sugar solution. However, they have a *low concentration* there. As a result, more water molecules meet pores from the water side than from the

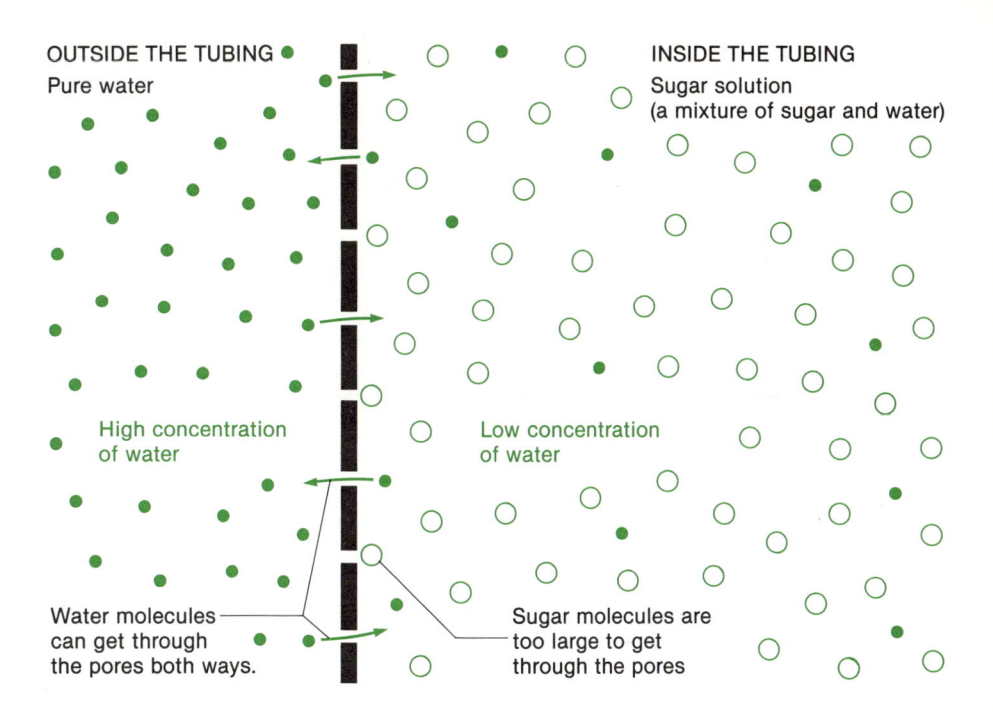

Fig. 6-6 An enlarged piece of the dialysis tubing. See if you can use this diagram to explain osmosis.

[Diagram labels: OUTSIDE THE TUBING — Pure water; INSIDE THE TUBING — Sugar solution (a mixture of sugar and water); High concentration of water; Low concentration of water; Water molecules can get through the pores both ways; Sugar molecules are too large to get through the pores.]

sugar solution side. Thus the *net* flow of water is inward. This makes the sugar solution rise in the glass tubing.

Definition of Osmosis

In our example, water moved from an area of high concentration of water to an area of low concentration of water. You may recall that the movement of molecules from an area of high to low concentration is called diffusion. The movement of water through a selectively permeable membrane is just a special type of diffusion. It is called osmosis [oz-MOH-sis]. **Osmosis** *is the diffusion of water through a selectively permeable membrane.*

Section Review

1. Define osmosis.
2. Explain osmosis using the particle theory.

6.6 ACTIVITY Osmosis through an Animal Membrane

The membrane just under the shell of an egg is selectively permeable. Oxygen must enter and carbon dioxide must leave. Otherwise, the egg cell inside will die. Is this membrane permeable to water? That is, will osmosis occur through this membrane? Try this activity and find out.

Note: Your teacher may suggest that you try this activity at home.

Section 6.6 73

Problem
Will osmosis occur through an animal membrane?

Materials
eggs (2)
dilute hydrochloric acid
  or vinegar
250 mL beakers or glasses (2)
distilled water
concentrated salt solution
  or molasses solution
spoon

CAUTION: Wear safety goggles if you use hydrochloric acid. Do not get the acid on your hands or clothing. If you do, report to your teacher at once and quickly wash with water.

Procedure
a. Remove the shells from the 2 eggs. Do this by covering them with dilute hydrochloric acid in a beaker. (If you are doing this at home, you can use vinegar.) The shells may dissolve in 30-40 min. Or they may take several hours to dissolve. Turn the eggs from time to time with a spoon. Tap them gently with the spoon to see if the shells are gone. Do not leave the eggs in the acid for long after the shells are gone. Some of the contents will be destroyed and the experiment will not work.
b. When the shells are gone, remove the eggs with a spoon. Gently rinse them with water. Handle them carefully. The membranes may break.
c. Place one egg in a beaker. Fill the beaker with distilled water (Fig. 6-7). This is Beaker A.
d. Place the other egg in a beaker. Fill the beaker with concentrated salt solution or molasses solution. This is Beaker B.
e. Observe the eggs at the end of the day and in the next science period. Record your observations.

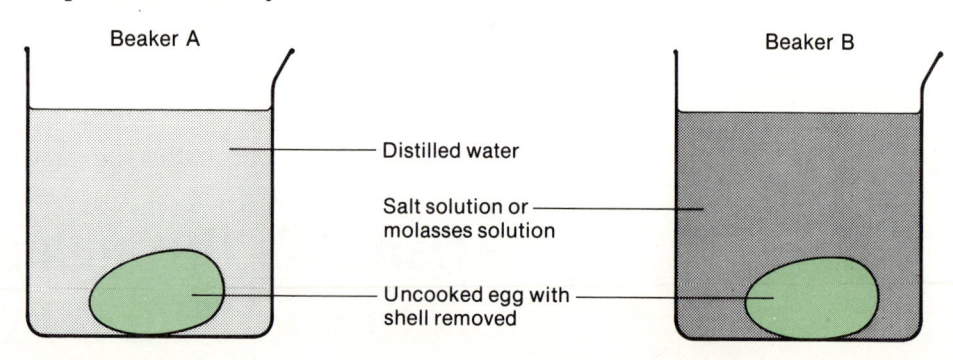

Fig. 6-7 Will osmosis take place through the membrane of an egg?

Beaker A — Distilled water
Beaker B — Salt solution or molasses solution
Uncooked egg with shell removed

Discussion
1. a) In what direction did the most water move in Beaker A? How do you know?
   b) Use your knowledge of osmosis to explain this result.
2. a) In what direction did the most water move in Beaker B?
   b) Use your knowledge of osmosis to explain this result.

6.7 ACTIVITY Osmosis Through a Plant Membrane

An uncooked potato is made of living cells. Do you think these cells have selectively permeable membranes around them? What is your proof? Will these membranes be permeable to water? Try this activity and find out.

Problem

Will osmosis occur through a plant membrane?

Materials

potato
250 mL beakers or glasses (2)
distilled water
concentrated salt solution
or molasses solution

Procedure

a. Cut 6 slices of potato that are all about 0.5 cm thick.
b. Place 2 slices in a beaker. Then fill the beaker with distilled water (Fig. 6-8).

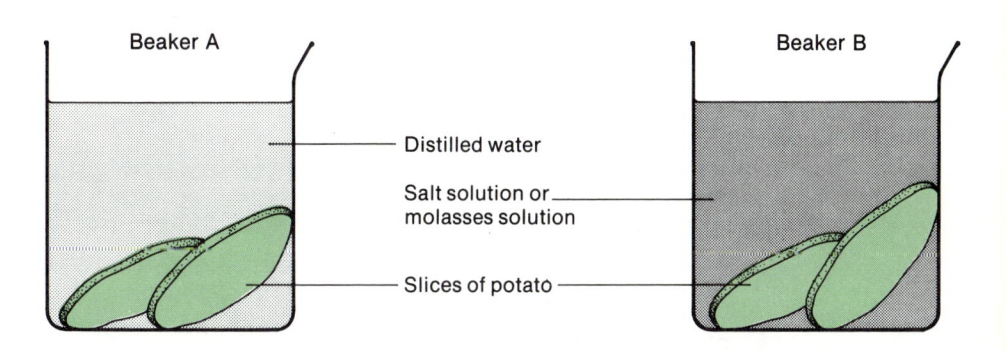

Fig. 6-8 Will osmosis take place through potato cells?

c. Place 2 slices in the other beaker. Then fill it with concentrated salt solution or molasses solution.
d. Study the remaining 2 slices closely. Make notes on what they feel like. Are they firm or limp? Are they hard or soft? This is your **control** for the experiment.
e. After 20-30 min, remove the slices from the distilled water. Repeat step (d) for these slices. Compare each of them to the control.
f. Now remove the slices from the salt (or molasses) solution. Repeat step (d) for these slices. Compare each of them to the control.
g. Return the slices to their proper beakers. Leave them overnight. Repeat steps (e) and (f) the following period.

Discussion

1. **a)** Consider the pieces of potato in the distilled water. Were they more or less firm than the control?
   b) Did the potato cells gain or lose water?
2. **a)** Consider the pieces of potato in the salt (or molasses) solution. Were they more or less firm than the control?
   b) Did the potato cells gain or lose water?
3. Use your knowledge of osmosis to explain these results.

6.8 Osmosis and Living Cells

You saw in both Activity 6.6 and Activity 6.7 that water moves through living cell membranes. In all cases the water moved from an area of high concentration of water to an area of low concentration of water. In other words, osmosis took place. Let's look at this process more closely.

Osmosis and the Cell Membrane

Most living things are over 70% water. Thus the movement of water into and out of cells is very common. Clearly the water must go through the cell membrane to get into or out of a cell.

The cell membrane must be a selectively permeable membrane. For example, it must let water molecules into the cell. But it must not let cell contents such as glucose leave the cell. Why?

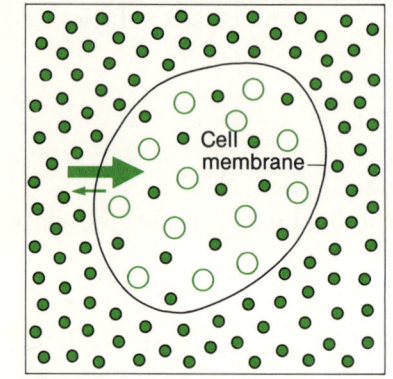

Osmosis with Cells in Pure Water

In your experiments you put living cells in pure water. In these cases the cells became firm. They took in water. Figure 6-9,A shows why. The concentration of water is higher outside than inside the cell. Thus osmosis occurs; water goes into the cell.

This osmosis builds up a pressure in the cell. This pressure can burst an animal cell. Plant cells, however, will not normally burst. They have rigid cell walls. Therefore the pressure just makes the cells more and more stiff. We say the plant becomes **turgid** [TUR-jid]. A healthy, properly watered plant will be turgid. Its leaves will be held stiffly out from the plant (Fig. 6-10,A).

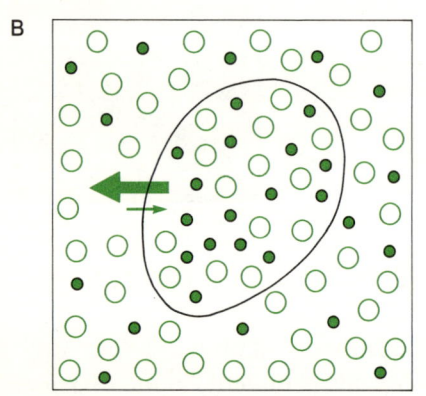

Osmosis with Cells in a Concentrated Solution

In your experiments you put living cells in concentrated salt or molasses solution. In these cases the cells became limp. They lost water. Figure 6-9,B shows why. The concentration of water is lower outside than inside the cell. Thus osmosis occurs; water leaves the cell.

A Cell in pure water

B Cell in salt (or molasses) solution

- ● Water molecule
- ○ Molecule dissolved in the water

Fig. 6-9 Osmosis takes place in both of these cases. Can you explain why? What will happen to the cell in each case?

Fig. 6-10 Can you explain the effect of the salt solution on the plant?

A. Properly watered plant

B. A similar plant watered with salt solution

This osmosis lowers the pressure in the cell. This is why the cells go limp. When the cells go limp, a whole plant can go limp (Fig. 6-10,B). We say the plant **wilts**.

Some people put salt on their sidewalks and driveways to melt ice. In the spring this salt dissolves. A concentrated salt solution is formed. This solution can cause enough osmosis to make the plant wilt permanently. You may have seen the dead grass along driveways and roads.

Section Review

1. a) What happens to a cell if it is placed in pure water?
   b) Explain the osmosis that occurs.
2. a) What happens to a cell that is placed in concentrated salt (or molasses) solution?
   b) Explain the osmosis that occurs.
3. Describe the difference between a turgid and a wilted plant.
4. Look at Figure 6-11,A. Will osmosis take place? Why?
5. Look at Figure 6-11,B. Will osmosis take place? Why?

A
Cell membrane
Cell in solution having the same concentration of water as the cell

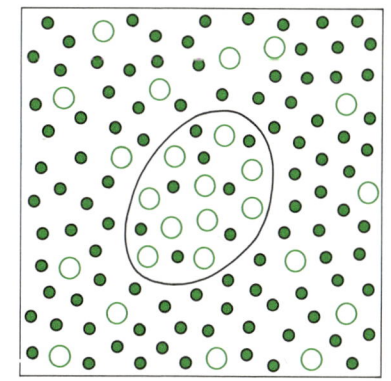

B
Cell in solution having a higher concentration of water than the cell

● Water molecule
○ Molecule dissolved in the water

Fig. 6-11 Will osmosis take place in these cases? Why?

Main Ideas

1. Diffusion is the net movement of a substance from an area of high concentration to an area of low concentration.
2. Diffusion is caused by the motion of particles.
3. Osmosis is the diffusion of water through a selectively permeable membrane.
4. The cell membrane of a living cell is selectively permeable.
5. Osmosis occurs through a cell membrane.

Main Ideas 77

Glossary

diffusion	di-FEW-shun	the movement of a substance from an area of high to low concentration
osmosis	oz-MOH-sis	diffusion of water through a selectively permeable membrane
selectively permeable membrane		lets some types of particles through but not others
turgid	TUR-jid	a stiff condition in an organism, caused by osmosis

Study Questions

A. True or False

Decide whether each of the following statements is true or false. If the sentence is false, rewrite it to make it true. (Do not write in this book.)
1. An odour moves across a room by a process called osmosis.
2. The cell membrane is selectively permeable.
3. Osmosis can occur through dialysis tubing.

B. Completion

Complete each of the following sentences with a word or phrase that will make the sentence correct. (Do not write in this book.)
1. Diffusion is caused by the motion of _____ .
2. Osmosis is the diffusion of _____ through a _____ permeable membrane.
3. Suppose that some plant cells are placed in pure water. Osmosis could cause these cells to become _____ .

C. Multiple Choice

Each of the following statements or questions is followed by four responses. Choose the correct response in each case. (Do not write in this book.)
1. During osmosis, water diffuses
    a) through a permeable membrane
    b) through an impermeable membrane
    c) from an area of high to low concentration of water
    d) from an area of low to high concentration of water
2. Dialysis tubing lets water molecules
    a) move in but not out
    b) move out but not in
    c) move neither in nor out
    d) move both in and out

3. If you put salt on a lawn the grass may die. This is because
   a) osmosis causes the grass to lose water
   b) osmosis causes the grass to gain water
   c) salt enters the cells of the grass and kills them
   d) the salt burns the grass

D. Using Your Knowledge

1. Explain why cooking odours from the kitchen can be smelled throughout the house.
2. A saltwater fish may die if placed in fresh water. Why?
3. A freshwater fish may die if placed in salt water. Why?
4. Explain why salt solution is a good antiseptic (kills bacteria).
5. A small amount of fertilizer will make the grass of a lawn grow. Too much fertilizer will kill the grass. Why?
6. What would happen if some of your red blood cells were placed in pure water? Why would this happen?
7. What would happen if some of your red blood cells were placed in concentrated salt solution? Why would this happen?

E. Investigations

1. An artificial kidney machine makes use of diffusion through a membrane to "clean" human blood. Find out how this machine works.
2. Do potato cells have to be living before osmosis will occur through them? Do an experiment to find out.
3. Do an experiment to show that too much fertilizer kills plants. Be sure to use proper controls in your experiment.
4. In most parts of Canada, salt is put on roads to melt snow. This salt often kills trees, shrubs, and other plants along highways. Find out any other harmful effects of road salt. Also, find out if road salt is really needed. Are there alternatives?

7 Cells at Work: Photosynthesis and Respiration

7.1 An Introduction to Photosynthesis and Respiration
7.2 Activity: Testing for Starch in Leaves
7.3 Activity: Is Chlorophyll Needed for Photosynthesis?
7.4 Activity: Is Light Needed for Photosynthesis?
7.5 Demonstration: Is Carbon Dioxide Needed for Photosynthesis?
7.6 Activity: Is Oxygen Produced during Photosynthesis?
7.7 A Summary of Photosynthesis
7.8 Activity: Does Respiration Produce Carbon Dioxide?
7.9 A Summary of Respiration

Photosynthesis [foh-tuh-SIN-thih-sis] and **respiration** [res-puh-RAY-shun] are two of the most important things that cells do. Only some kinds of cells carry out photosynthesis. But all kinds of cells respire. What are photosynthesis and respiration? And why are they so important?

7.1 An Introduction to Photosynthesis and Respiration

Photosynthesis Makes Food for Life Processes

The tree in Figure 7-1 has been growing in the same pot for many years. No new soil has been added to the pot. Yet the pot is full of soil. How did the tree get so big without using up much soil?

Plants carry out **photosynthesis**. This process forms foods such as glucose (a sugar). Some of this glucose is used, along with other things, for growth. Much of the rest of the glucose is used in respiration.

Respiration Gives Energy for Life Processes

The cow in Figure 7-2 eats grass for food. The cow gets energy from the grass. How did the energy get into the grass? How does the cow get energy from the grass?

Fig. 7-1 This plant has been growing in the same soil for many years. It started out just 10 cm tall. Now it is over 150 cm tall. Where did the plant get the atoms of which it is made? (*top left*)

Fig. 7-2 Grass gives this cow energy. How? (*top right*)

All living things need energy. This energy is needed for life processes such as movement. Most of the energy used by living things comes, in the first place, from the sun (see Figure 2-7, page 17). Photosynthesis stores some of the sun's energy in foods such as **glucose** [GLOO-kohs]. Plants break down some of the glucose during a process called **respiration**. This releases the energy the plants need to carry out life processes.

Some animals get their energy by eating plants. Others get their energy by eating other animals. In all cases, the energy is released from the foods by respiration.

In the rest of this chapter you will do activities to find out more about photosynthesis and respiration. What substances do they use up? What do they produce? Why are they so important?

Section Review

1. By what process do plants make food?
2. How do plants get the energy they need?
3. How do animals get the energy they need?

7.2 ACTIVITY Testing for Starch in Leaves

In the activities that follow, you will need a test to see if photosynthesis took place. This activity shows you that test.

Some of the glucose formed during photosynthesis is changed to starch. Starch is easy to detect. It turns iodine solution dark blue or purple. Therefore, to see if photosynthesis took place, we do a **starch test**.

Problem

How do we test for starch in leaves?

Materials

150 mL beaker
600 mL beaker
hot plate
crucible tongs
forceps

glass plate
ethyl or isopropyl alcohol
geranium plant that has been under bright light for 48 h
iodine solution

CAUTION: Wear safety goggles during this activity. There must be no flames in the room. The alcohol could catch fire.

Procedure

a. Put about 200 mL of water in the 600 mL beaker. Get this water boiling on the hot plate.
b. Remove a leaf from the geranium plant.
c. Put the leaf in the boiling water for a few seconds. Using the tongs, remove it as soon as it becomes limp.
d. Fill the 150 mL beaker half full of alcohol. Put the geranium leaf in the alcohol.
e. **Turn off the hot plate.** Now put the small beaker in the boiling water as shown in Figure 7-3. Use the tongs to do so.
f. When the leaf is white, take it out of the alcohol. (Use the forceps.)
g. Soften the leaf by dipping it in the hot water for 2-3 s.
h. Spread the leaf on a glass plate. Cover the leaf with iodine solution. Record your observations.

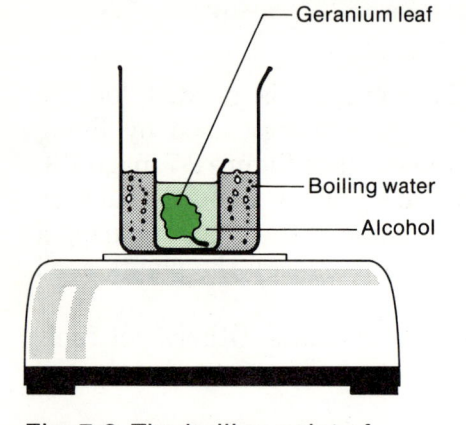

Fig. 7-3 The boiling point of alcohol is lower than that of water. Therefore water near its boiling point will make the alcohol boil.

Explanation

The starch is inside the cells of the leaf. The hot water breaks down the cell walls. Now the alcohol can get in to remove the green colour. The green colour is **chlorophyll** [KLOHR-o-fil]. If you don't remove the green colour, you may not see the final colour. The iodine solution reacts with the starch to form a dark blue or purple colour.

Discussion

1. Is chlorophyll soluble in water? How do you know?
2. Is chlorophyll soluble in alcohol? How do you know?
3. Make a summary of the method used to test for starch in a plant leaf.

7.3 ACTIVITY Is Chlorophyll Needed for Photosynthesis?

Many plants have leaves that contain chlorophyll in some areas but not others. The silver-leaf geranium and some *Coleus* plants are examples (Fig. 7-4). In this activity you will use one of these leaves to solve the problem.

Problem

Is chlorophyll needed for photosynthesis?

Materials

as in Activity 7.2   silver-leaf geranium or *Coleus* plant

CAUTION: Wear safety goggles during this activity. Turn off all flames.

Procedure

a. Leave the plant in bright light for 48 h.
b. Remove a leaf from the plant. Draw a "before" picture of it.
c. Remove the chlorophyll from the leaf as described in Activity 7.2.
d. Test the leaf for starch.

Discussion

1. Draw an "after" diagram of the leaf. In the "before" diagram, label where chlorophyll was present and where it was absent. In the "after" diagram label where the starch test worked.
2. Describe your results in words.
3. Draw a conclusion from this activity.
4. Good experiments always have a control. What was the control here?
5. Why is it important to begin this activity with a plant that has been in bright light for 48 h?

Fig. 7-4 A leaf from a silver-leaf geranium. The white part at the edge has no chlorophyll. Where will photosynthesis take place?

- White edge (contains no chlorophyll)
- Green centre (contains chlorophyll)

7.4 ACTIVITY Is Light Needed for Photosynthesis?

Is light needed for photosynthesis? If someone asked you that question, you would likely answer "yes". Plants don't live long without light, do they? But you don't really know that light is needed

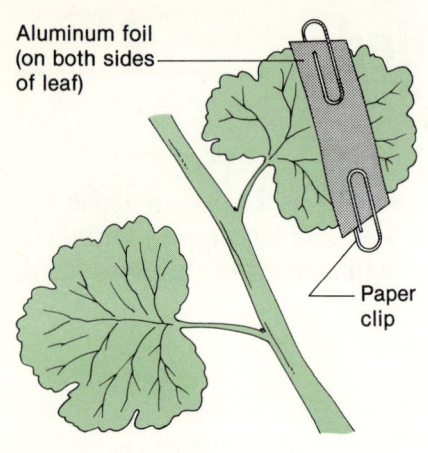

Fig. 7-5 Light is kept away from part of a leaf. Where will photosynthesis take place?

for photosynthesis. You just know that the plants die without light. So let's try this activity and solve the problem.

Problem

Is light needed for photosynthesis?

Materials

as in Activity 7.2
geranium plant
aluminum foil
paper clips (2)

CAUTION: Wear safety goggles during this activity. Turn off all flames.

Procedure

a. Leave the plant in the dark for 48 h. Any starch in the leaf will be used up in that time.
b. Cover part of one leaf with aluminum foil. Hold the foil in place with paper clips (Fig. 7-5).
c. Place the plant in bright light for 48 h.
d. Remove the leaf from the plant. Remove the foil.
e. Remove the chlorophyll from the leaf as described in Activity 7.2.
f. Test the leaf for starch.

Discussion

1. Describe the leaf after the foil was removed.
2. Describe the results of the starch test. Include a labelled diagram.
3. Draw a conclusion from this activity.
4. What was the control in this activity?

7.5 DEMONSTRATION Is Carbon Dioxide Needed for Photosynthesis?

Starch has carbon atoms in it. Where do the carbon atoms come from? We learned in Section 7.1 that a large tree can grow without using up much soil. Therefore the carbon probably does not come from the soil. Perhaps it comes from the air. Air contains carbon dioxide. Perhaps this carbon dioxide is made into starch during photosynthesis.

Problem

Is carbon dioxide needed for photosynthesis?

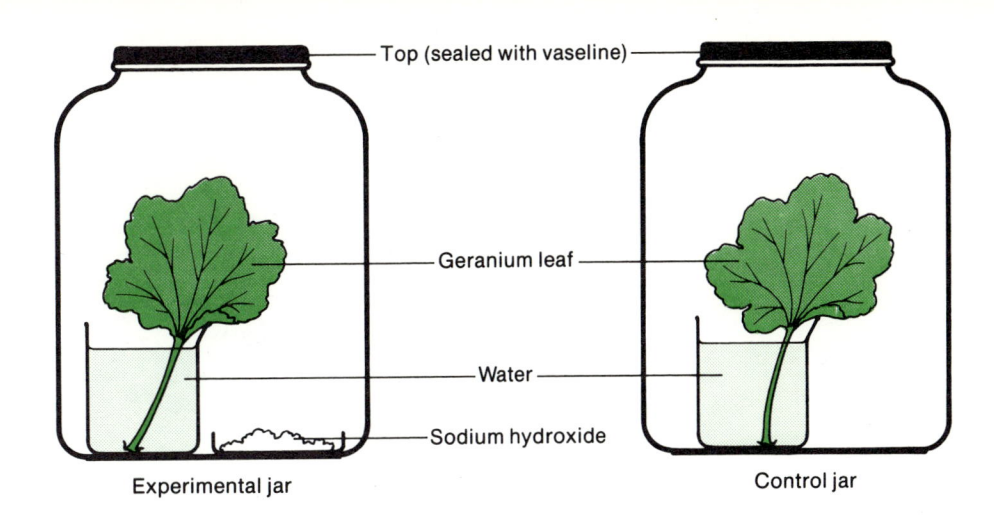

Fig. 7-6 Sodium hydroxide removes carbon dioxide from the air. Where will photosynthesis take place?

Materials

as in Activity 7.2
large jars (2)
small beakers (2)
geranium leaf (2)
3-4 cm³ of sodium hydroxide pellets
petri dish
water

CAUTION: Wear safety goggles during this activity. Turn off all flames. Sodium hydroxide will burn your skin. Do not touch it.

Procedure

a. Pick 2 large leaves from a geranium plant that has been in the dark for 48 h.
b. Set up the materials as shown in Figure 7-6. The sodium hydroxide in the experimental jar removes carbon dioxide from that jar.
c. Put the jars in bright light for 48 h.
d. Remove the chlorophyll from both leaves as described in Activity 7.2.
e. Test both leaves for starch.

Discussion

1. Describe the results of the starch tests.
2. Draw a conclusion from this activity.
3. Why was the control jar necessary?

7.6 ACTIVITY Is Oxygen Produced During Photosynthesis?

You have likely heard that plants produce oxygen. This activity will let you find out if what you heard is true.

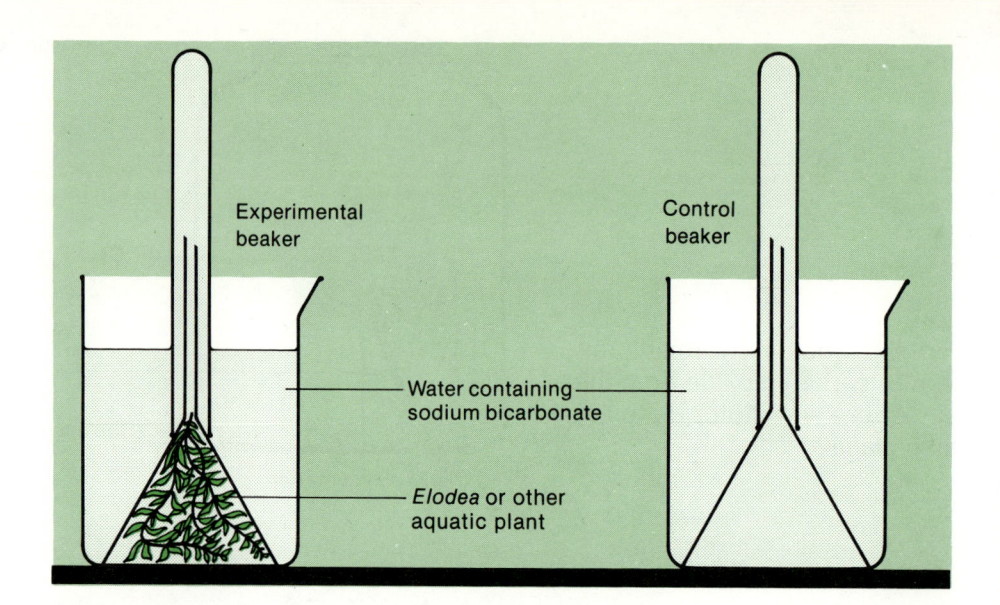

Fig. 7-7 If the plant produces a gas, the gas will collect in the test tube.

Problem

Is oxygen produced during photosynthesis?

Materials

test tubes (2)
100 mL beakers (2)
funnels
rubber stopper

sodium bicarbonate
wooden splint
aquatic plant such as *Elodea*

Procedure

a. Insert the cut ends of the *Elodea* shoots into the stem of one funnel (Fig. 7-7).
b. Fill both test tubes and funnels full of water. You can do this by putting the whole apparatus (beaker, funnel, and test tube) in a sink full of water.
c. Arrange the materials as shown in Figure 7-7.
d. Dissolve 2-3 cm^3 of sodium bicarbonate in the water of each beaker. This substance will supply the carbon dioxide needed for photosynthesis.
e. Shine a bright light on the setup for at least 48 h.
f. Test the gas that collects to see if it is oxygen. Use the glowing splint test. Oxygen is one of very few gases that cause a glowing splint to burst into flame. Do the test as follows:
  i. Fill the beaker full of water.
  ii. Lift the test tube off the funnel. But do not let any air in.
  iii. Put a stopper in the mouth of the test tube.
  iv. Lift the test tube out of the beaker.
  v. Turn the test tube upright. The gas should now be on top of the water as shown in Figure 7-8.

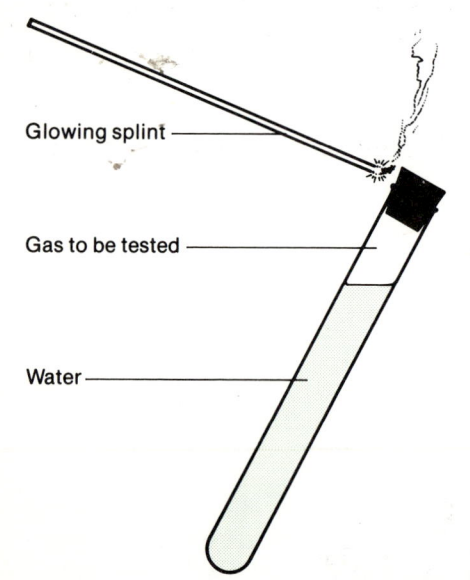

Fig. 7-8 The glowing splint test for oxygen. Have the glowing splint ready before you remove the stopper.

86 Chapter 7

vi. Remove the stopper and quickly put the glowing splint into the gas. (Have the glowing splint ready before you remove the stopper.)

Discussion

1. Describe the gas that collected in the experimental test tube.
2. What did this gas do to the glowing splint?
3. What gas do you think this is?
4. Why was a control needed?

7.7 A Summary of Photosynthesis

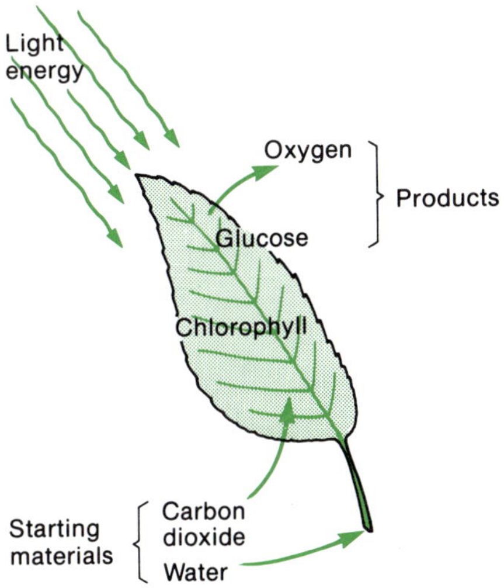

You have seen that photosynthesis requires chlorophyll, light, and carbon dioxide. It also requires water. But it is not easy for you to prove that.

You have also seen that photosynthesis produces oxygen and starch. (Actually, glucose is formed. Then some of it is changed to starch.)

The process of photosynthesis is summed up in this word equation:

$$\text{Carbon dioxide} + \text{Water} + \text{Light energy} \xrightarrow{\text{Chlorophyll}} \text{Glucose} + \text{Oxygen}$$

The carbon dioxide and water are combined to form glucose and oxygen. The light energy becomes chemical energy in the glucose molecules. The chlorophyll is not used up. It just helps make the reaction go (Fig. 7-9).

Fig. 7-9 This diagram sums up the process of photosynthesis.

Section Review

1. Write a word equation for photosynthesis.
2. Describe what happens during photosynthesis.

7.8 ACTIVITY Does Respiration Produce Carbon Dioxide?

Seeds are alive. Therefore they must respire. Life processes are most active in seeds while they are germinating. Therefore respiration must be at a peak during germination. This is because respiration provides the energy for life processes.

In this activity you will test germinating seeds to see if carbon dioxide is produced during respiration.

Problem

Do germinating seeds produce carbon dioxide?

Materials

250 mL flasks (2)
thistle tubes (2)
test tubes (2)
limewater (50 mL)
glass tubing (bent as shown in Figure 7-10)
absorbent cotton
germinating pea or bean seeds (soaked overnight in water)
2-hole rubber stoppers (2)
10% formalin solution
water

Procedure

a. Get about 100 mL of germinating seeds from your teacher. The seeds were soaked overnight in water. This starts them germinating.
b. Set up the materials (except the limewater) as shown in Figure 7-10.
   CAUTION: Be careful if you have to put the tubing through the stopper. Follow your teacher's instructions. Otherwise you could cut your hands.
c. Make sure the bottom of the thistle tube is covered with water. Do *NOT* cover the seeds.
d. Set up a similar apparatus as a control. However, this time use dead seeds. Your teacher will give you 100 mL of dead seeds. They were killed by soaking them for 1-2 h in 10% formalin.
e. Let the experiment stand for 24 h. If any carbon dioxide is formed, it will be in the flask.
f. Place a test tube containing limewater over the end of the glass tubing.
g. Pour 100 mL of water into the flask. This will force any gas in the flask to bubble through the limewater. If carbon dioxide is present, the limewater will turn milky. Record your results.
h. Repeat steps (f) and (g) for the control.

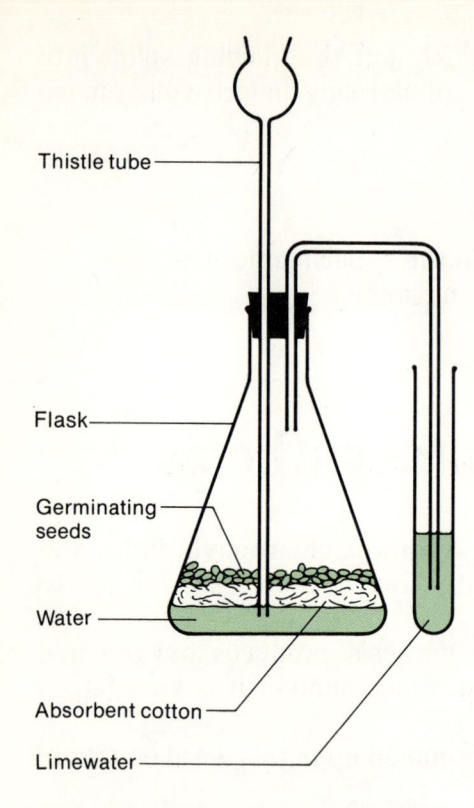

Fig. 7-10 Do germinating seeds respire? If so, what will they produce?

Discussion

1. Describe the limewater in the two test tubes at the end of the experiment.
2. What gas was produced?
3. Draw a conclusion from your results.
4. How could you prove that *you* produce carbon dioxide during respiration?

7.9 A Summary of Respiration

You have seen that germinating seeds give off carbon dioxide during respiration. Most living things do this. Even plants give off carbon dioxide during respiration. However, they also use it up during photosynthesis.

During respiration glucose is broken down. Oxygen is needed for this process. Energy is given off to support life processes. Water is produced, along with carbon dioxide.

The process of respiration is summed up in this word equation:

Glucose + Oxygen ⟶ Carbon dioxide + Water + Energy

Respiration and Breathing

Don't confuse respiration and breathing. **Respiration** is a chemical reaction. It occurs in cells. Its purpose is to release energy for life processes.

Breathing is simply the exchange of gases between an organism and its environment. It is not a chemical reaction.

Comparing Photosynthesis and Respiration

Study the summation equations for the two processes closely. How are they alike? How are they different?

Photosynthesis uses up carbon dioxide and water. It produces glucose and oxygen. Respiration does the reverse. It uses up glucose and oxygen. And it produces carbon dioxide and water. Photosynthesis uses light energy. Respiration gives off energy for life processes.

Photosynthesis changes light energy into chemical energy in glucose molecules. Respiration changes the chemical energy in glucose molecules into the energy needed for life processes.

Photosynthesis takes place *only* in cells that contain chlorophyll. And it takes place *only* in the light. Respiration takes place in *all* cells, *all* the time.

Section Review

1. Give the word equation for respiration.
2. What is the difference between breathing and respiration?
3. How are photosynthesis and respiration alike? How are they different?

Main Ideas

1. Photosynthesis changes light energy into chemical energy in glucose molecules.
2. Respiration changes chemical energy in glucose molecules into the energy needed for life processes.
3. The summation equation for photosynthesis is:

Carbon dioxide + Water + Light energy $\xrightarrow{\text{Chlorophyll}}$ Glucose + Oxygen

4. The summation equation for respiration is:

Glucose + Oxygen ⟶ Carbon dioxide + Water + Energy

5. Photosynthesis occurs only in cells containing chlorophyll. It requires light.
6. Respiration occurs in all cells all the time.

Glossary

chlorophyll	KLOHR-o-fil	a green substance needed for photosynthesis
glucose	GLOO-kohs	a sugar formed during photosynthesis
photosynthesis	foh-tuh-SIN-thih-sis	the changing of light energy into chemical energy
respiration	res-puh-RAY-shun	the changing of chemical energy into energy for life processes
starch		a substance formed from extra glucose in cells

Study Questions

A. True or False

Decide whether each of the following statements is true or false. If the sentence is false, rewrite it to make it true. (Do not write in this book.)

1. Chlorophyll is needed for the process of respiration.
2. Photosynthesis occurs in all cells all the time.
3. Respiration produces carbon dioxide.

B. Completion

Complete each of the following sentences with a word or phrase that will make the sentence correct. (Do not write in this book.)

1. Photosynthesis requires four things: ▇▇▇▇ ; ▇▇▇▇ ; ▇▇▇▇ and ▇▇▇▇ .
2. Respiration releases ▇▇▇▇ for life processes.
3. Photosynthesis changes light energy into ▇▇▇▇ .

C. Multiple Choice

Each of the following statements or questions is followed by four responses. Choose the correct response in each case. (Do not write in this book.)

1. Plants get their food
    a) by making it from carbon dioxide, water, and light energy
    b) from the soil
    c) from fertilizers in the soil
    d) from other organisms
2. The products of photosynthesis are
    a) energy and glucose
    b) carbon dioxide and water
    c) carbon dioxide and glucose
    d) oxygen and glucose
3. What would happen if all the plants on earth died?
    a) Only the animals that eat plants would die.
    b) All animals would eventually die.
    c) All animals that eat plants would begin eating other animals.
    d) Too much oxygen would be given off into the air.

D. Using Your Knowledge

1. Suppose that air pollution cut the sun's brightness in half. What do you think would happen to the living things on earth? Discuss both plants and animals.
2. Forests are being cut faster than they are being replanted. How might this affect the world's oxygen supply? Why?
3. How could astronauts get rid of carbon dioxide in a spaceship and, at the same time, produce oxygen and food?

E. Investigations

1. Do an experiment to prove that you give off carbon dioxide.
2. Design an experiment to show that a hamster gives off carbon dioxide. Do not do the experiment without your teacher's permission.
3. Can a plant make chlorophyll in the dark? Design and try an experiment to find an answer to this question.
4. Is a green leaf always green? Shine a red light on a green plant in a dark room. What colour are the leaves? Explain your observation.

8 Cells at Work: How Living Things Grow

8.1 The Need for Cell Division
8.2 Mitosis
8.3 Activity: Mitosis and Cell Division
8.4 Activity: Mitosis and Cell Division in Living Onion Root Tips

A tiny seed grows and develops into a huge maple tree (Fig. 8-1). One white egg develops into a marsh hawk. But another white egg develops into a chicken. A human egg only 0.2 mm across grew and developed into you.

Miracles? Not really. A miracle is a happening that cannot be explained by science. But the growth and development of living things can be explained by science. And that's what this chapter is all about.

8.1 The Need for Cell Division

Cells, you may remember, are the building blocks of living things. They are alive. They carry out life processes such as growth. But they also grow old, wear out, and die. For example, skin cells wear away from your hands and the rest of your body.

Fig. 8-1 How do living things grow?

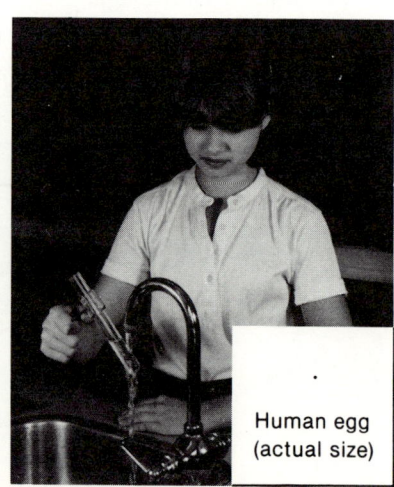

Human egg (actual size)

92 Chapter 8

Organisms must replace such worn-out cells. They must also repair damaged parts. And they must be able to grow. How do organisms do these things? The cells of the organisms make new cells. This process is called **cell division**. During cell division a cell divides to form two cells. Biologists call the original cell the **parent cell**. They call the cells that are formed the **daughter** cells. The daughter cells grow to a certain size. Then they may divide again. In this way, cell division makes new cells for growth. It also makes new cells to replace dead cells and repair damaged parts.

All cells are made from cells. More important, the daughter cells are just like the parent cell they came from. For example, muscle cells come from muscle cells. But skin cells come from skin cells. How do cells make exact copies of themselves? They do it through a process called *mitosis* [my-TOE-sis] and cell division. The next section explains how it works.

Section Review

1. What is cell division?
2. Why is cell division necessary?

8.2 Mitosis

The **nucleus** plays a key role in cell division. You learned in Chapter 5 that a nucleus contains **chromosomes**. This is the hereditary material of the cell. It determines the characteristics of the cell. The daughter cells are just like the parent cell. Therefore they must have chromosomes that are exactly like those of the parent cell. This is accomplished by mitosis. **Mitosis** [my-TOE-sis] *is the process by which a nucleus forms two exact copies of itself*. Once this has happened, the cell then divides into two (cell division). Let's see how mitosis works.

Stages in Mitosis

Mitosis is a continuous process. It does not occur in definite steps. However, five stages can be identified. Refer to Figure 8-2 as you read about these stages. The figure shows what happens in an *animal* cell.

Stage 1
- Most of a cell's time is spent in this stage.
- The cell is carrying out all cell activities except reproduction (growth, repair, metabolism, etc.).
- The **chromosomes** are spread through the nucleus in a thread-like form.
- The chromosomes are made of DNA and protein. **DNA** is the hereditary material of the cell.

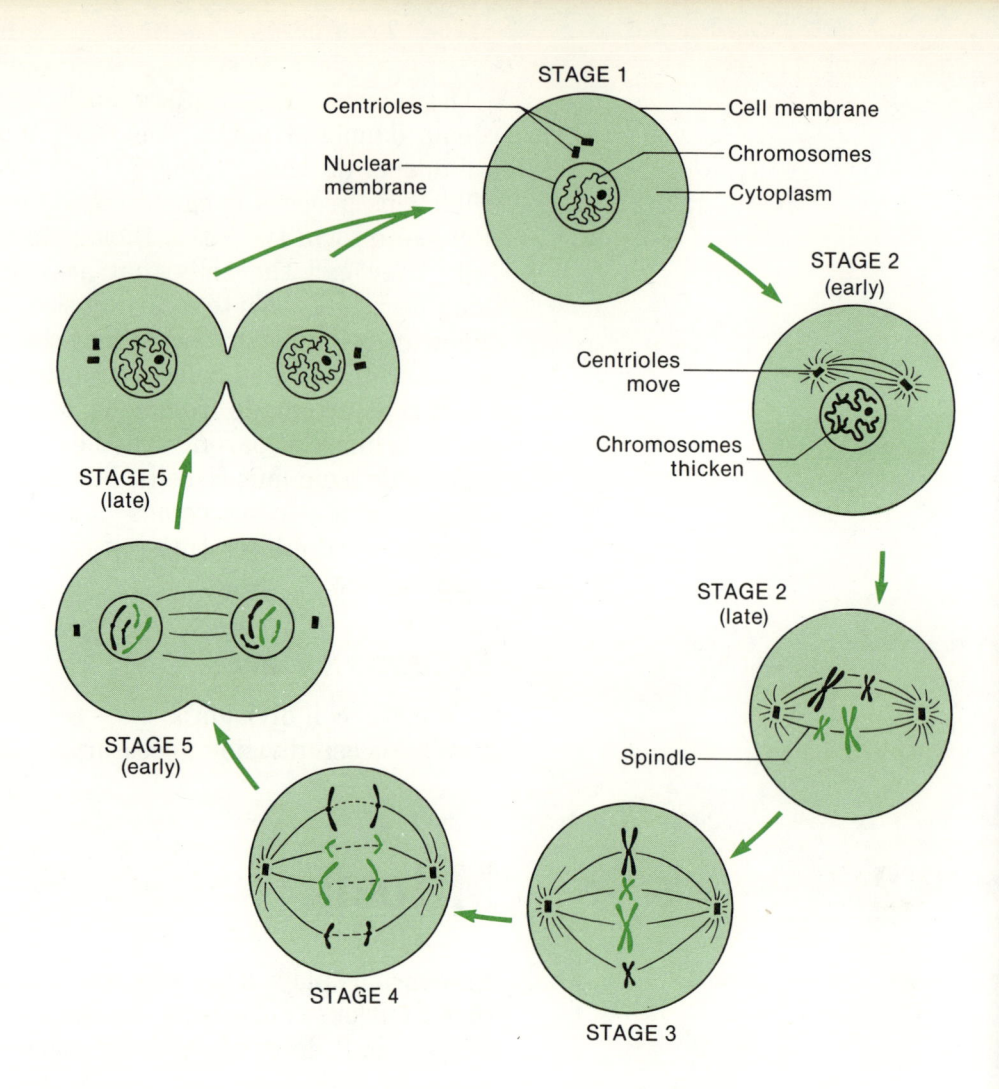

Fig. 8-2 Stages in mitosis of an animal cell

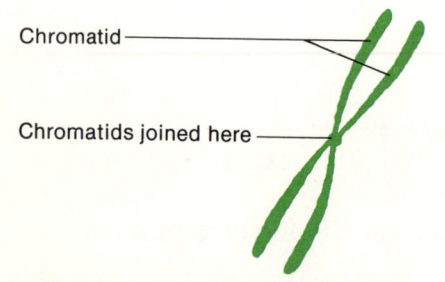

Fig. 8-3 A chromosome that has formed a pair of chromatids

- The daughter cells must get exact copies of the chromosomes of the parent cell.
- Mitosis begins near the end of this stage.
- The chromosomes make *two exact copies* of themselves. (They still appear as long thin threads.)

Stage 2
- The **centrioles** [SEN-tree-oles] move to opposite sides of the nucleus.
- Tiny fibres form around each centriole.
- The chromosomes appear to thicken and shorten.
- The nuclear membrane disappears.
- The two exact copies of each chromosome can be easily seen. They occur in pairs called **chromatids** [KROH-ma-tids]. The two chromatids are identical. They are joined at a point (see Fig. 8-3).

Stage 3
- The paired chromatids line up at the equator of the cell.
- Each chromatid pair separates into two identical chromatids, now known as chromosomes.

94 Chapter 8

- Each chromosome becomes attached to a fibre.

Stage 4
- The matching pairs of chromosomes separate.
- The two chromosomes of each pair move to opposite poles of the cell.

Stage 5
- The nuclear membrane reappears around each set of chromosomes.
- The chromosomes become thin threads again.
- The fibres disappear.
- Two daughter nuclei have been formed.
- Each centriole divides. Now each daughter cell has two centrioles.

Cell Division
- The cytoplasm also divides. The cell membrane pinches inward at the equator.
- Two **daughter cells** are formed. They then separate from one another.
- The daughter cells are now in stage 1. They are ready to go through mitosis again.

Notes

1. Mitosis ensures that each daughter cell has the *same number and kinds* of chromosomes as the parent cell. Therefore the daughter cells will be the same as the parent cell.
2. Mitosis and cell division in plant cells is much like that in animal cells. However, the cytoplasm does not pinch off. Instead, a new **cell wall** forms between the daughter nuclei. Also, there are no centrioles in plant cells.

Section Review

1. Why does the nucleus play a key role in cell division?
2. What is mitosis?
3. Make a summary of the five stages in mitosis. Include diagrams of each stage.
4. How does mitosis ensure that the daughter cells will be like the parent cell?

8.3   ACTIVITY   Mitosis and Cell Division

In Sections 8.1 and 8.2 you read about cell division and mitosis. In this activity you will study these processes using prepared slides and a microscope. The slides are of both animal and plant cells that are dividing.

Problem

How many stages in mitosis can you find and draw?

Materials

microscope
prepared slide of onion root tip
prepared slide of whitefish cells

Procedure A Mitosis and Cell Division in Plant Cells

a. Examine the slide of onion root tip cells under low power. Look for the section of the root tip that is most actively dividing. It will have the greatest number of nuclei.
b. Switch to medium power and focus. Now switch to high power and focus. Examine the section closely.
c. Find a cell that appears to have a nucleus in Stage 1. Sketch it. Label your sketch.
d. Seek out cells with nuclei in the other stages of mitosis. Sketch them. Label your sketches. Be patient. Sometimes these stages are difficult to find.

Procedure B Mitosis and Cell Division in Animal Cells

a. Examine the slide of whitefish cells under low power. Look for a section that shows mitosis.
b. Repeat steps (b) to (d) of Procedure A.

Discussion

1. Be sure your sketches are fully labelled. They are the main part of your report for this activity.
2. In what ways do dividing plant cells differ from dividing animal cells?

8.4   ACTIVITY   Mitosis and Cell Division in Living Onion Root Tips

The tip of a root is a place of active growth. Therefore there will be mitosis and cell division in the tip. In this activity an onion is placed in the dark with its base in water. After a few days new white roots appear. You will make a wet mount of these roots. Then you will look for mitosis and cell division.

Problem

Can you make a wet mount of onion root tips and find mitosis and cell division?

Materials

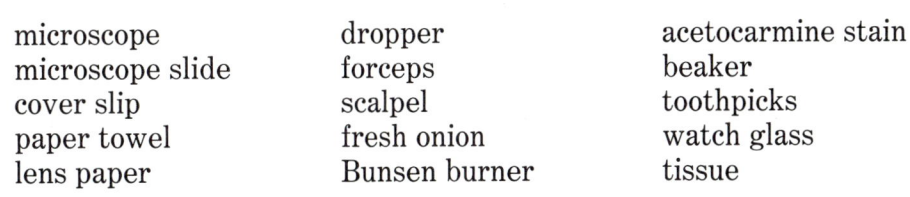

microscope	dropper	acetocarmine stain
microscope slide	forceps	beaker
cover slip	scalpel	toothpicks
paper towel	fresh onion	watch glass
lens paper	Bunsen burner	tissue

CAUTION: Wear safety goggles during this activity.

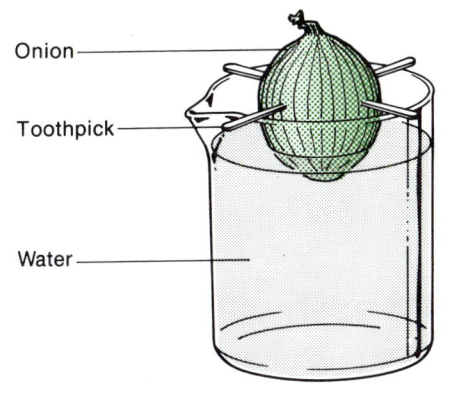

Procedure

a. Obtain an onion that is as firm and fresh as possible. Using a scalpel, cut off the dead roots close to the base of the onion.
b. Add water to a beaker until it is about 2/3 full.
c. Use 3 or 4 toothpicks to support the onion in the water as shown in Figure 8-4. You may have to add more water.
d. Place the beaker and onion in a cool dark place. Leave them there until new white roots form that are 1-2 cm long. Check the onion every day. Usually 2-4 d are required.
e. Half fill the watch glass with acetocarmine stain.
f. Remove the onion from the water. Cut each root about 0.5 cm up from the *tip*. Let the tips drop into the acetocarmine stain. Do *not* handle the tips with forceps.
g. Heat the stain plus tips *gently* for 3-5 min as directed by your teacher. *You must not boil the stain.*
h. Place a fresh drop of acetocarmine stain on a microscope slide.
i. Select 2 tips that seem well stained. Use the forceps to add them to the drop of stain on the microscope slide. Grasp the tips at the cut end, *not* the tip end.
j. Cut off the last 2 mm of each tip. Leave those pieces in the drop of stain. Remove the long pieces.
k. Cut the 2 tips into pieces that are smaller than a pin head.
l. Carefully place a cover slip over the pieces of root tip.
m. Tap the cover slip lightly with a pencil. This will separate and spread the cells of the root tips.
n. Make a "squash" preparation of the cells as directed in Figure 8-5. The intent is to squash the small pieces of root tip so that they are only one cell thick. *Do not slide or twist the cover slip.*
o. Examine your wet mount under low power, medium power, then high power. Make sketches of any stages in mitosis that you see. Label your sketches.

Fig. 8-4 A method for growing onion root tips that will show mitosis

Fig. 8-5 Making a squash preparation of onion root tip cells. *Do not slide or twist the cover slip.* Use these steps:
1. Fold a tissue several times.
2. Lay the tissue on the table.
3. Place your slide on the tissue.
4. Fold the tissue around the slide.
5. Press evenly downward with your thumb.
6. Remove the tissue from the slide.

Discussion

1. Why were just the tips used in this activity?
2. What did the acetocarmine stain do?
3. In what ways do you feel your wet mount was better than the prepared slides of Activity 8.3?
4. In what ways do you feel that the prepared slides were better than your wet mount?

Section 8.4

Main Ideas

1. Cell division makes new cells. These are needed for growth, repair of damaged parts, and replacement of dead cells.
2. Mitosis is the process by which a nucleus forms two exact copies of itself.
3. Chromosomes contain DNA, the hereditary material of cells.
4. During mitosis each daughter cell gets the same number and kinds of chromosomes as the parent cell.

Glossary

chromatid	KROH-ma-tid	an exact copy of a chromosome
chromosome	KROH-mo-sohm	a structure in the nucleus that is made partly of DNA
DNA		the hereditary material of a cell
mitosis	my-TOE-sis	the process by which a nucleus forms two exact copies of itself

Study Questions

A. True or False

Decide whether each of the following statements is true or false. If the sentence is false, rewrite it to make it true. (Do not write in this book.)

1. The only function of cell division is growth.
2. Cell division and mitosis are the same thing.
3. Chromosomes are made partly of DNA.
4. DNA is the hereditary material of a cell.

B. Completion

Complete each of the following sentences with a word or phrase that will make the sentence correct. (Do not write in this book.)

1. The two exact copies of a chromosome are called ▭ .
2. Mitosis ensures that each daughter cell has the same number and kind of ▭ as the parent cell.
3. After mitosis occurs in a plant cell, a ▭ forms between the daughter nuclei.

98   Chapter 8

C. Multiple Choice

Each of the following statements or questions is followed by four responses. Choose the correct response in each case. (Do not write in this book.)

1. Mitosis occurs
   a) in animal cells only
   b) in plant cells only
   c) in both plant and animal cells
   d) in stained plant and animal cells only
2. A daughter cell from mitosis will have
   a) the same number of chromosomes as the parent cell
   b) half the number of chromosomes as the parent cell
   c) twice as many chromosomes as the parent cell
   d) different kinds of chromosomes than the parent cell
3. The tip of a root is selected for studies of mitosis because
   a) the tip is easily stained
   b) the tip has stronger cells than the rest of the root
   c) mitosis occurs in the tip but cell division does not
   d) the tip is a place of active growth

D. Using Your Knowledge

1. Suppose your skin cells stopped dividing. What would happen? Refer to Figure 8-6 in your answer.
2. Some cancer is the result of uncontrolled mitosis in certain cells. What would such cancer be like?
3. Suppose that, during mitosis, the chromosomes did not form exact copies of themselves. What might the result be?

E. Investigations

1. Use coloured construction paper to make a working model of mitosis.
2. After cells divide, they sometimes have to differentiate. Find out what differentiation is.

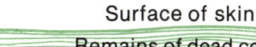

Fig. 8-6 Is cell division necessary in skin cells?

Unit 3
Monerans, Protists, and Fungi: Three Kingdoms of Living Things

CHAPTER 9
Monerans and Viruses

CHAPTER 10
Bacteria and Us

CHAPTER 11
Protists

CHAPTER 12
Fungi

In Chapter 3 you learned that all living things can be classified into 5 kingdoms. In this unit you will investigate three of those kingdoms: monerans, protists, and fungi.

These 3 kingdoms contain the simplest forms of life. Included among them are bacteria, amoebas, yeasts, and mushrooms. Though they are simple, they are still interesting and important.

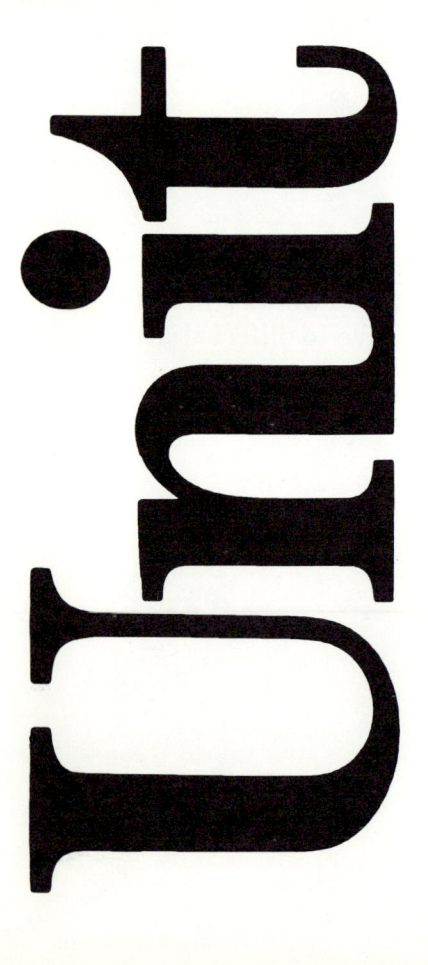

Fig. 9-0 These organisms are in the kingdom Fungi. Many years ago they were classified as plants. Now they are in a different kingdom from plants. How do fungi differ from plants?

9 Monerans and Viruses

9.1 What Are Monerans?
9.2 What Are Blue-Green Algae?
9.3 Activity: Study of Blue-Green Algae
9.4 What Are Bacteria?
9.5 Activity: Study of Bacteria
9.6 Life Processes of Bacteria
9.7 Viruses: Living and Non-living

The simplest living things...the earliest forms of life...the most numerous living things on earth...both friend and foe. That's a simple description of **monerans** [mahn-ER-ans].

The edge of life...both living and non-living...the key to the secret of life. That's how some biologists describe **viruses**.

Monerans and viruses are just as interesting as they sound. This chapter explains what they are and what they do.

9.1 What Are Monerans?

All monerans are either **bacteria** or **blue-green algae** [AL-jee]. About 3100 species of organisms are monerans **(kingdom Monera)**. Of these, about 1600 species are bacteria. The other 1500 species are blue-green algae (Fig. 9-1).

The number of species of monerans is small. The organisms themselves are very small (Fig. 9-2). But their importance is very great. Also, the number of individuals is enormous. Monerans are the most numerous living things on earth. They may also be the oldest. Fossils that look like monerans have been found that are over 3 000 000 000 a old!

Monerans live just about everywhere. They have been found at the tops of mountains and at the bottoms of oceans. They live in the arctic and in hot springs. They occur in fresh water and in salt water. They even live on rocks.

Just what are monerans? What do bacteria and blue-green algae have in common?

Monerans (3100 species)
Bacteria (1600 species)
Blue-green algae (1500 species)

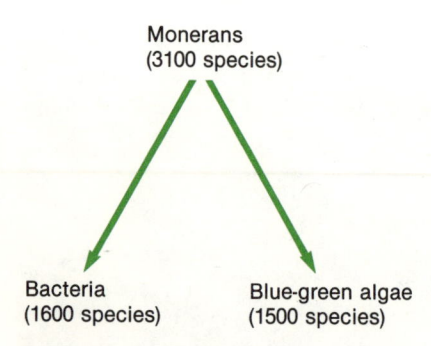

Fig. 9-1 Only 3100 species of organisms are monerans. (There are 2 000 000 known species of organisms.)

Characteristics of Monerans

1. *All monerans are one-celled.* As Figure 9-2 shows, the cells may occur singly or in groups or chains.

102 Chapter 9

A

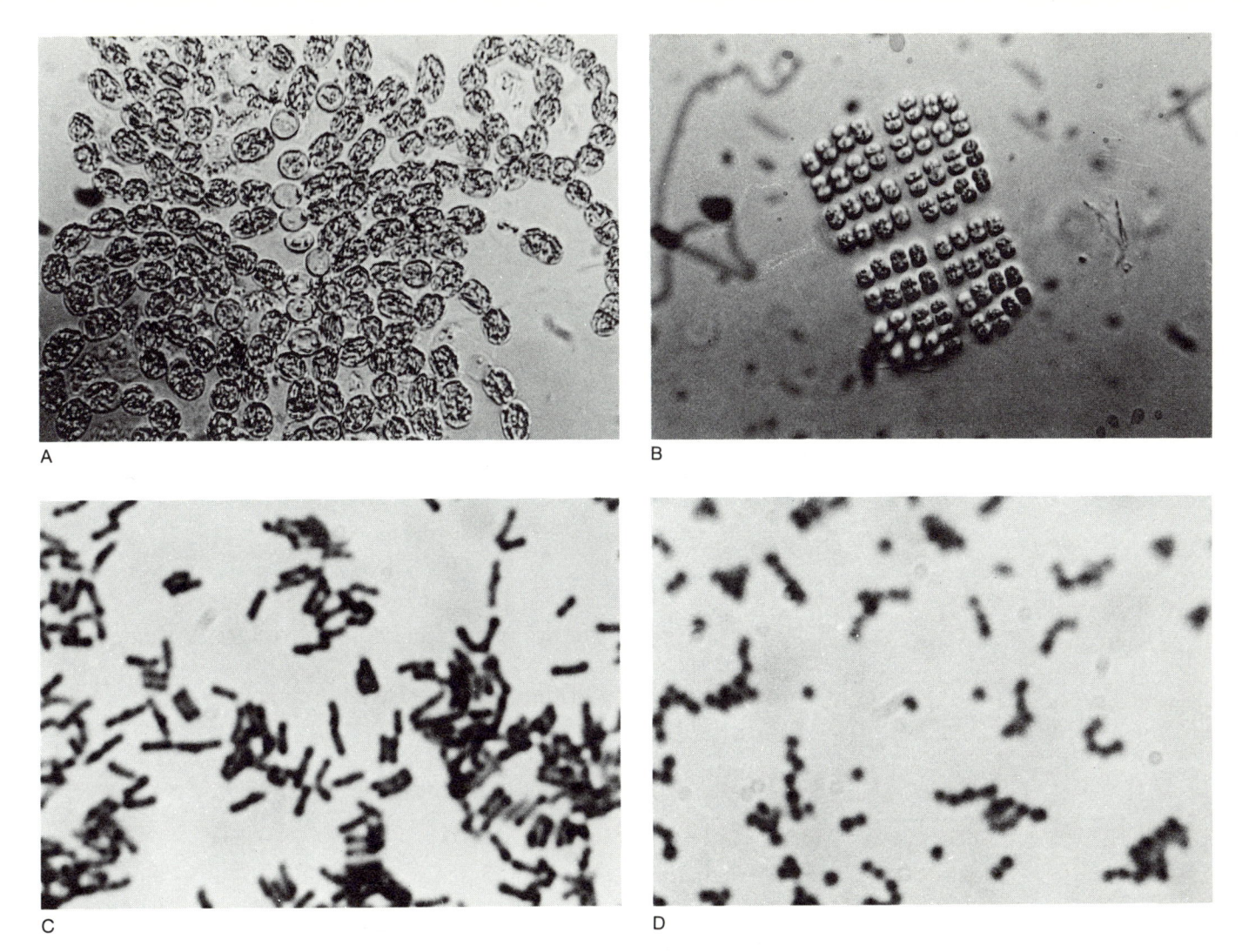

B

C

D

Fig. 9-2 Some monerans. The top two (A and B) are blue-green algae. The bottom two (C and D) are bacteria. Note that all of them are tiny cells.

2. *The cells of monerans lack a true nucleus.* They do not have a nuclear membrane. Therefore nuclear material such as DNA can spread throughout the cell (see Figures 9-5 and 9-9).
3. *The cells of monerans have no organelles.* The cells are very simple. They have no membranes to hold organelles together. Thus they do not have chloroplasts and other organelles (see Figures 9-4 and 9-5).
4. *The cells of monerans have stiff cell walls.* The material in these cell walls is special. It does not occur in any other organisms.
5. *Monerans often have a slime coating.* A jelly-like coating holds cells together. It also protects them.
6. *Many monerans can form spores.* They do this when the environment becomes too hot, too cold, too dry, or too wet. The spores have thick walls that protect the organisms from death.
7. *Most monerans reproduce by binary fission.* **Binary fission** [BY-na-ree FISH-uhn] is shown in Figure 9-3. Each cell divides into two identical cells.

Section 9.1

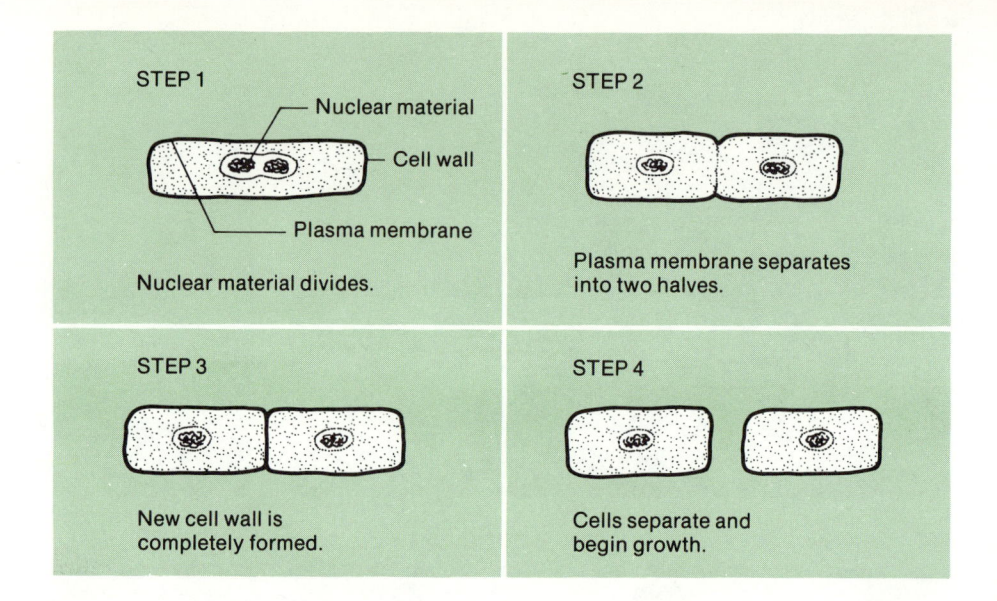

Fig. 9-3 Binary fission of a moneran

STEP 1 — Nuclear material, Cell wall, Plasma membrane. Nuclear material divides.

STEP 2 — Plasma membrane separates into two halves.

STEP 3 — New cell wall is completely formed.

STEP 4 — Cells separate and begin growth.

Section Review

1. What two groups of organisms make up the kingdom Monera?
2. Where are monerans found?
3. Make a summary of the characteristics of monerans.

9.2 What Are Blue-Green Algae?

Figure 9-4 shows some blue-green algae. Figure 9-5 shows the contents of one cell. Look at these figures as you read this section on blue-green algae.

Characteristics

1. *Blue-green algae are monerans.* Therefore they show the seven characteristics of monerans described in Section 9.1. Read them again as you look at Figures 9-4 and 9-5.
2. *Blue-green algae contain three pigments.* They contain the *green* pigment chlorophyll. (All plants and algae contain it.) They also contain a *blue* pigment. These two pigments often give the algae a blue-green colour. That's how these algae got their name. They also contain a *red* pigment. Therefore these algae are sometimes purple or even black.

Occurrence

Blue-green algae, like all monerans, are found just about everywhere. They have been found in polar icefields. In fact, they often make the ice and snow green or blue-green. They have also been found in deserts. Most species, however, live in or near water. For

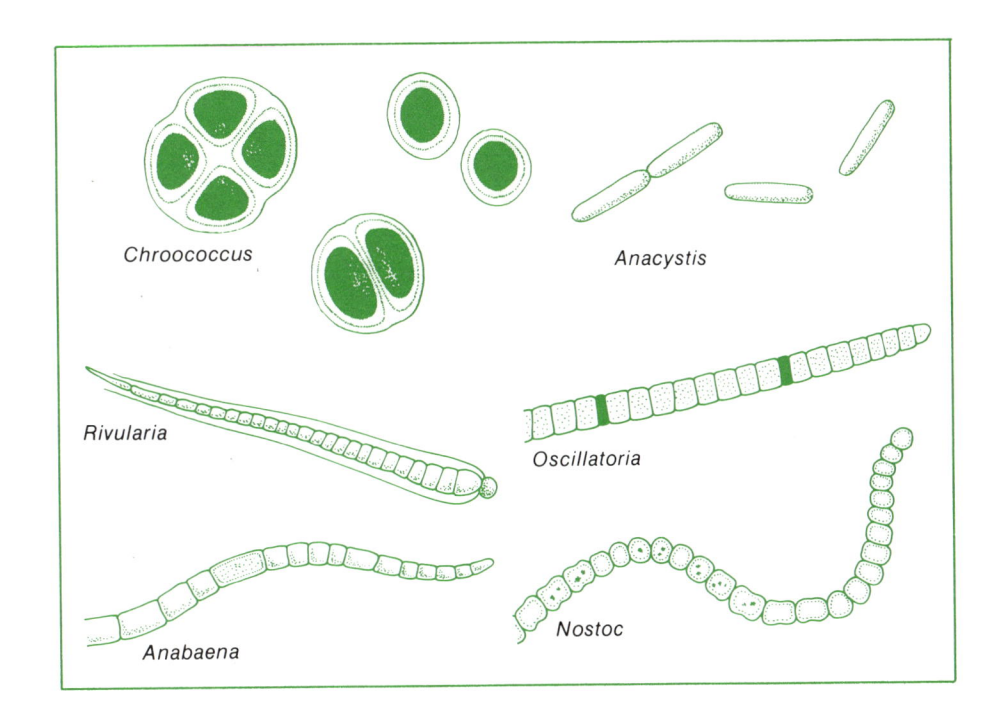

Fig. 9-4 Some blue-green algae. Note that some occur as single cells. Others occur as colonies or chains.

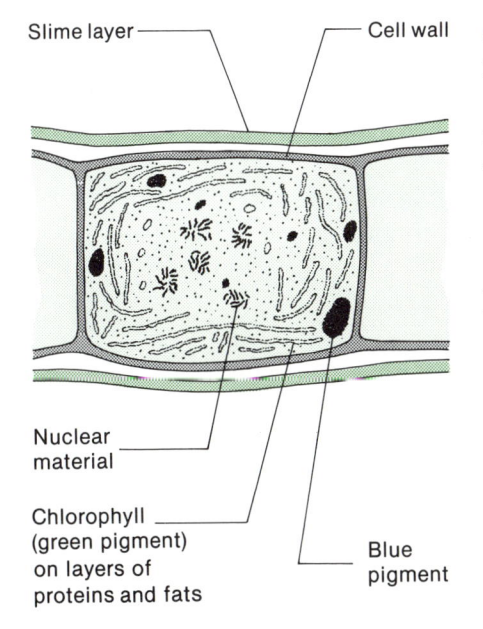

Fig. 9-5 A blue-green algal cell. Note that there is no nucleus. Also, organelles are absent.

example, *Rivularia* (see Figure 9-4) grows on rocks in streams. You may have seen the brown crusty mat of this alga on rocks.

Blue-green algae are also common on damp surfaces. These include moist rocks, moist soil, flower pots that are always wet, and tree trunks in a damp forest.

Importance

Blue-green algae are important for four reasons.

1. *They are **producers**.* All other algae and plants are also producers. This means they carry out photosynthesis. They *produce* food by photosynthesis using carbon dioxide, water, and light. As a result, they form the start of some food chains in ponds and lakes. Here is an example.

   Blue-green algae ⟶ **Tiny animals** ⟶ Minnows ⟶ Trout ⟶ Humans

   Humans are at the end of many food chains that start with blue-green algae.

2. *They make oxygen.* Since blue-green algae carry out photosynthesis, they add oxygen to water and air. This helps many other organisms to respire.

3. *They fix nitrogen.* All living things need nitrogen. They use it to make proteins. There's lots of nitrogen in the air. But few living things are able to use it. The nitrogen must be **fixed** first. This means the nitrogen must be changed into nitrates. This is a form of nitrogen that plants and other producers can use. Some blue-green algae can fix nitrogen.

Section 9.2   105

The water and soil of rice paddies contain blue-green algae. The algae fix nitrogen. This provides nitrate fertilizer to make the rice grow. Then the farmers do not have to buy fertilizer.

4. *They are often found in polluted water.* Many species grow well in polluted water. In fact, they sometimes grow too well (Fig. 9-6). The result is called an **algal bloom**. Algal blooms often ruin swimming beaches. They also clog filters in water purification plants. They can give water a bad taste and odour.

Some species give off **toxins** [TOX-ins]. These are poisons that can kill fish and even kill animals that drink the water.

When an algal bloom dies, the algae decay. This uses up oxygen in the water and, as a result, kills fish.

Fig. 9-6 This scum is part of an algal bloom.

Section Review

1. Make a summary of the characteristics of blue-green algae.
2. Describe the occurrence of blue-green algae.
3. Give four reasons why blue-green algae are important.
4. What is an algal bloom?

9.3 ACTIVITY Study of Blue-Green Algae

In this activity you will look at prepared slides of blue-green algae. Then you will make wet mounts of living algae.

Look for the characteristics listed in Section 9.2. You are not expected to learn the names of the algae you see.

Problem

How many characteristics of blue-green algae can you find?

Materials

microscope
prepared slides of blue-green algae
microscope slides (2)
cover slips (2)
paper towel

dropper
forceps
lens paper
living blue-green algae

Procedure

a. Your teacher will give you prepared slides of several species of blue-green algae. Look at each slide under low power. Then switch to medium, and finally, high power.
b. Make a sketch of each species.
c. Make a list of characteristics that all species have in common.

106 Chapter 9

d. Make a wet mount of a living blue-green alga. (See Activity 4.4, page 47 for the method.) The alga can be obtained from: the scum (brown or blue-green) on the glass in an aquarium; the green slime on rocks, tree trunks, and soil; the green film on clay flowerpots; blue-green or brown "blobs" floating in ponds.
e. Study your wet mount under low power. Then switch to medium and, finally, to high power.
f. Make a sketch of each species you see.

Discussion

1. Why were you not able to see all the characteristics of blue-green algae that are listed in Section 9.2?
2. What evidence did you see that blue-green algae are very simple organisms?

9.4 What Are Bacteria?

You have likely heard many bad things about bacteria. They cause tooth decay. They make us sick. They spoil our food. However, most species of bacteria are not harmful. In fact, life as we know it could not exist on earth without bacteria. Let's find out more about these organisms that are both friend and foe.

Characteristics

1. *Bacteria are monerans.* Therefore they show the seven characteristics of monerans described in Section 9.1. Read them again as you look at Figure 9-9.
2. *Bacteria are the smallest cells.* They are so small that they are measured in **micrometres**. (One micrometre = one millionth of a metre, or 1 μm = 0.000 001 m.) Bacteria range from 0.5 μm to 1 μm in diameter and from 1 μm to 10 μm in length. Many thousands of bacteria could fit on the smallest dot you can make with your pencil.
3. *Bacteria are of three basic shapes* (Fig. 9-7). Spherical bacteria are called **cocci** (singular: **coccus**). Rod-shaped bacteria are called **bacilli** (singular: **bacillus**). Spiral-shaped bacteria are called **spirilla** (singular: **spirillum**).
4. *Bacteria occur as single cells and in groups.* The groups may consist of just two cells, chains of cells, or clusters of cells. **Diplo-** refers to two cells. **Strepto-** refers to chains of cells. And **Staphylo-** refers to clusters of cells. Figure 9-8 shows how these three words are used to name groups of bacteria. Table 9-1 lists some diseases caused by bacteria and the shapes of their cells.

SPHERES
Singular:
Coccus (COCK-us)
Plural:
Cocci (COCK-ee)

RODS
Singular:
Bacillus (bah-SILL-us)
Plural:
Bacilli (bah-SILL-ee)

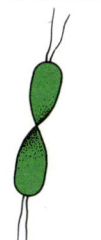

SPIRALS
Singular:
Spirillum (spir-IL-lum)
Plural:
Spirilla (spir-IL-lah)

Fig. 9-7 The three basic shapes of bacteria

COCCI

Coccus Diplococcus

Streptococcus Staphylococcus

BACILLI

Bacillus Diplobacillus

Streptobacillus

SPIRILLA

Spirillum

Fig. 9-8 Shapes and groups of bacteria

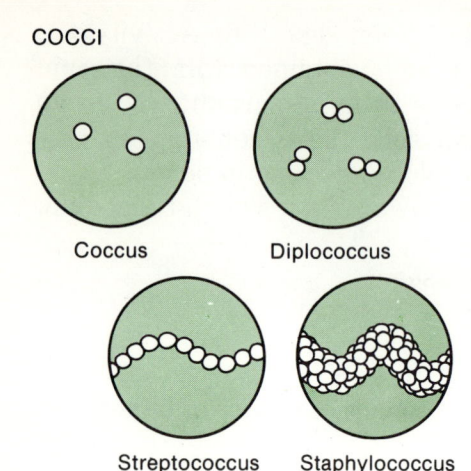

Table 9-1 Bacterial Diseases

Disease	Bacterium
Boils	*Staphylococcus*
Strept throat	*Streptococcus*
Gonorrhoea	*Diplococcus*
Syphilis	*Spirillum*
Whooping cough	*Bacillus*
Tetanus	*Diplobacillus*

Occurrence

Bacteria, like all monerans, live in almost all kinds of environments. They are present in large numbers in soil, plants, and animals. They abound in rotting wood and animal feces. Some species live in salt water; other species live in fresh water. Still others live in deserts. Some species need oxygen; others don't. Some live in icefields; others live in hot springs. Bacteria have been found on icy mountaintops and in the deepest parts of oceans. Bacteria live just about everywhere.

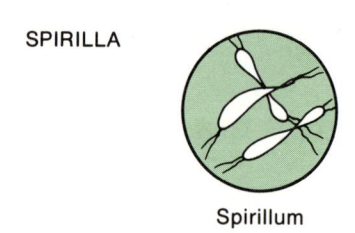

Structure of a Cell

A stiff **cell wall** gives a bacterium its shape (Fig. 9-9). The cell wall is usually surrounded by a **slime layer**. It may protect the cell against sudden changes in the environment. Bacteria with thick slime layers are usually hard to kill. Scientists have discovered that bacteria which cause serious diseases have thick slime layers.

Like all cells, a bacterial cell has **cytoplasm**. The cytoplasm is surrounded by a **cell membrane**. Some species have **flagella** (singular: **flagellum**) coming from the cell membrane. These move the bacteria by a whip-like action.

Like all monerans, bacteria have no nuclear membrane. However, the nuclear material is often near the centre of the cell.

Section Review

1. Describe the size of bacteria.
2. What are the three basic shapes of bacteria? Give the scientific names for these shapes.
3. Explain *diplo-*, *strepto-*, and *staphylo-*.
4. Describe the occurrence of bacteria.
5. Describe a single bacterial cell.

9.5 ACTIVITY Study of Bacteria

In this activity you will see the shapes of bacteria. You will also see groups of bacterial cells.

Problem
How many characteristics of bacteria can you find?

Materials
microscope
lens paper
prepared slices of cocci, bacilli, and spirilla

Procedure
a. Your teacher will give you prepared slices of cocci, bacilli, and spirilla. Single cells, pairs, chains, and clusters are included. Look at each slide under low power. Then switch to medium and, finally, to high power.
b. Make a sketch of each type. Look for special features such as flagella.

Discussion
1. Your laboratory sketches are your writeup for this activity. Be sure each type is labelled.

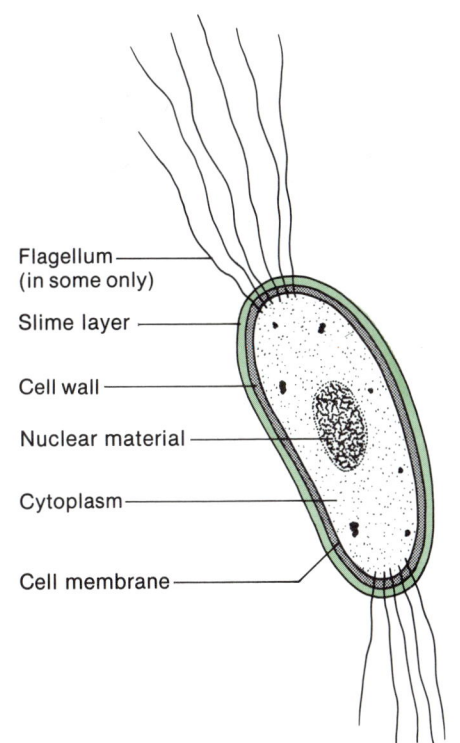

Fig. 9-9 Structure of a bacterium. Compare it to the blue-green alga in Figure 9-5. Both are monerans. How are they alike? How are they different?

Labels: Flagellum (in some only), Slime layer, Cell wall, Nuclear material, Cytoplasm, Cell membrane

9.6 Life Processes of Bacteria

Like all living things, bacteria carry out life processes. Some of these are movement, reproduction, and growth. To carry out life processes, bacteria need energy. This section looks at these life processes. It then explains how bacteria get their energy.

Movement
Some species of bacteria have **flagella** (singular: **flagellum**). A flagellum [fla-JEL-um] is a long hair-like structure. It whips back and forth to move the bacterium through a liquid. Figure 9-10 shows the flagella of *E. coli*. This species lives in your large intestine.

Species without flagella move very little. They may be moved about by water, wind, and animals. But, as you saw in Chapter 2, this is not true biological movement.

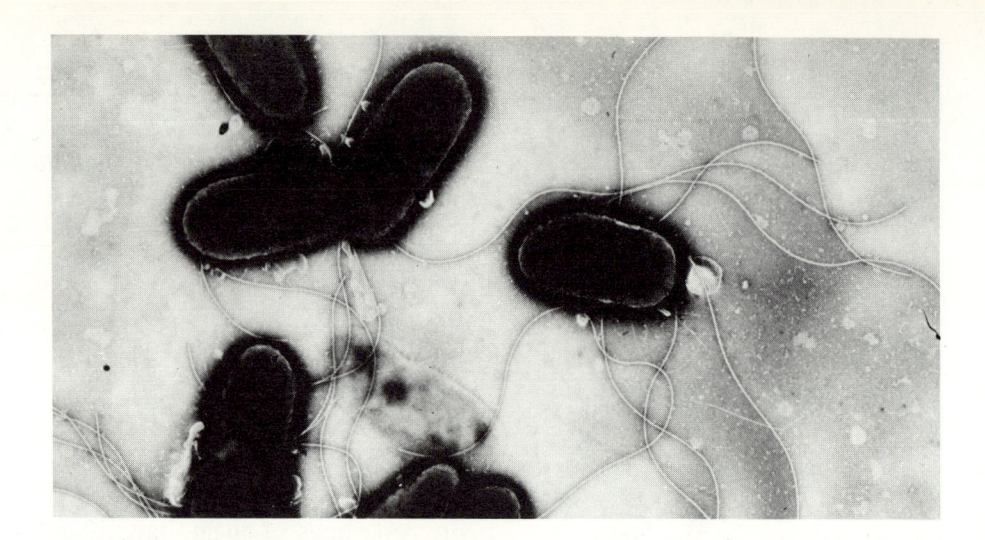

Fig. 9-10 Note the flagella on these bacteria. They help the bacteria to move. These bacteria are magnified 12 000 X.

Reproduction and Growth

Bacteria reproduce by **binary fission** (see page 104 and Figure 9-3). When conditions are suitable, the nuclear material divides into two identical parts. Then the cell membrane and wall grow across the centre of the cell. This forms two cells. The two cells eventually separate.

Under ideal conditions this process happens in about 20 min. Table 9-2 shows how the number of bacteria could grow under ideal conditions. If this continued for 48 h, about 20 000 000 000 000 000 000 000 000 000 000 000 000 bacteria would be formed! They would have a mass much greater than that of the earth!

Fortunately, ideal conditions don't last long. The bacteria use up their food. Also, they pollute their environment with their wastes. However, conditions are close to ideal for certain bacteria in warm salad dressing, warm milk, and even your body. Thus the bacteria can reproduce quickly in such places. This causes food spoilage and diseases. You will learn more about this in Chapter 10.

Table 9-2 Bacterial Reproduction and Growth

Time (min)	Number of bacteria
0	1
20	2
40	4
60	8
80	16

How Bacteria Get Food and Energy

Like all living things, bacteria need food. Some of this food is used to make new cells. The rest is used as a source of energy. It is broken down during respiration to release energy for life processes. Let us first see how bacteria get their food. Then we will see how they respire to get energy.

Some Bacteria Make Their Food A few species of bacteria make their food through **photosynthesis**. They combine carbon dioxide and water to make their food. Light energy is needed for this process.

A few other species of bacteria make their food without using light energy. They get energy by breaking down **chemical compounds**. Usually these are compounds of iron, sulfur, and nitrogen.

Most Bacteria Cannot Make Food Most species of bacteria cannot make their own food. Instead, they depend on other organisms for their food. There are two kinds of bacteria that do this: saprophytes and parasites.

The **saprophytes** [SAP-row-fites] get their food from *non-living* organic matter. That is, they feed on dead plants, dead animals, and animal wastes. Since they break down the organic matter, they are called **decomposers**. Without decomposers, the earth would soon be covered with dead plants and animals. However, not all decomposers carry out such useful jobs. Much of our food is dead plants and animals. Some decomposers break it down, causing it to spoil.

The **parasites** [PAR-a-sites] get their food by living inside or on *living* organisms. An organism that feeds parasites in this way is called the **host**. The parasites feed on the tissues of the host. If they cause a disease, these parasites are called **pathogens** [PATH-o-jens]. Tuberculosis is caused by a parasitic bacterium. It feeds on the lung tissue of the host.

Energy for Life Processes Like other living things bacteria need energy for life processes. They get it by breaking down foods through **respiration**.

Some species of bacteria *need oxygen* to break down their food (that is, to respire). When they respire, they form carbon dioxide, water, and energy. Other species do not need oxygen to break down their food. They live only where there is *no oxygen*. Most of these form carbon dioxide, alcohol, and energy when they respire. Others form just lactic acid and energy. Respiration without oxygen is called **fermentation**.

Section Review

1. How do bacteria move?
2. Describe how bacteria reproduce by binary fission.
3. Suppose one bacterium was placed in ideal conditions. How many bacteria would there be in 4 h if conditions stayed ideal?
4. Describe two ways by which some bacteria make their food.
5. a) How do bacteria that are saprophytes get their food?
   b) Why are these bacteria called decomposers?

6. a) How do bacteria that are parasites get their food?
   b) What is a host?
   c) What is a pathogen?
7. a) Why do bacteria need to respire?
   b) What is fermentation?

9.7 Viruses: Living and Non-Living

You have likely heard a sick person say, "I've got a virus." Just what does this person have? No one really knows. However, biologists have been trying to find out for many years. Therefore they do know some things about viruses.

Fig. 9-11 The tobacco mosaic virus. This was the first virus to be discovered. Wendell Stanley, an American, discovered it in 1935. This picture was taken with an electron microscope (77 000 X).

What Are Viruses?

Viruses are *not* monerans. In fact, if you look at the classification of living things in Chapter 3, you will not find viruses. Why, then, are they being discussed here?

A virus, alone, seems to be non-living. Figure 9-11 shows the tobacco mosaic virus. In a bottle, this virus seems to be lifeless crystals. These crystals can be kept on a shelf just like salt crystals in a bottle. But in a tobacco plant, the virus multiplies greatly and damages the plant. That is, it reproduces. This is a characteristic of living things. No ordinary crystals can reproduce.

Some of the time viruses seem to be non-living. And some of the time they seem to be alive. If viruses are alive, they are the simplest living things. That's why we are discussing them here.

Average animal cell (10 μm)

Scale: ⊢1 μm

Average bacterium (2.5 μm)

Smallpox virus (0.2 μm)

Polio virus (0.03 μm)

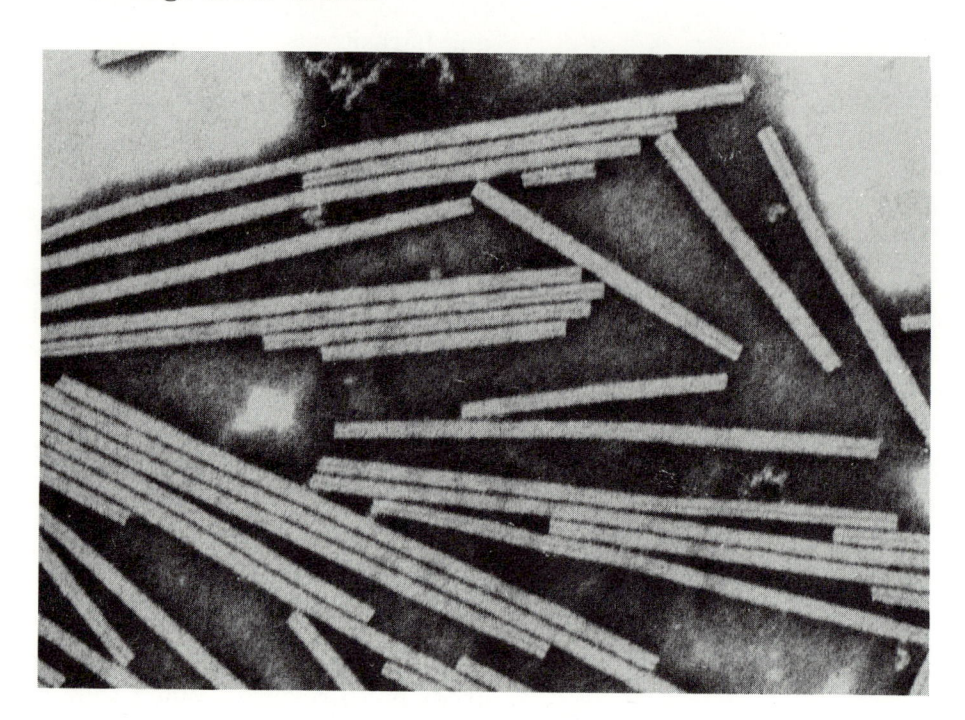

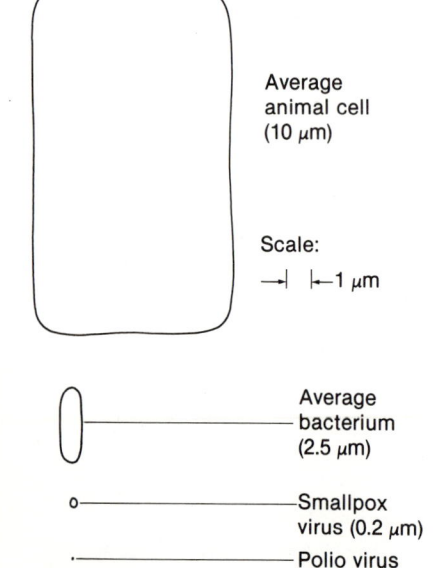

Fig. 9-12 Note how small viruses are, compared to bacteria and animal cells.

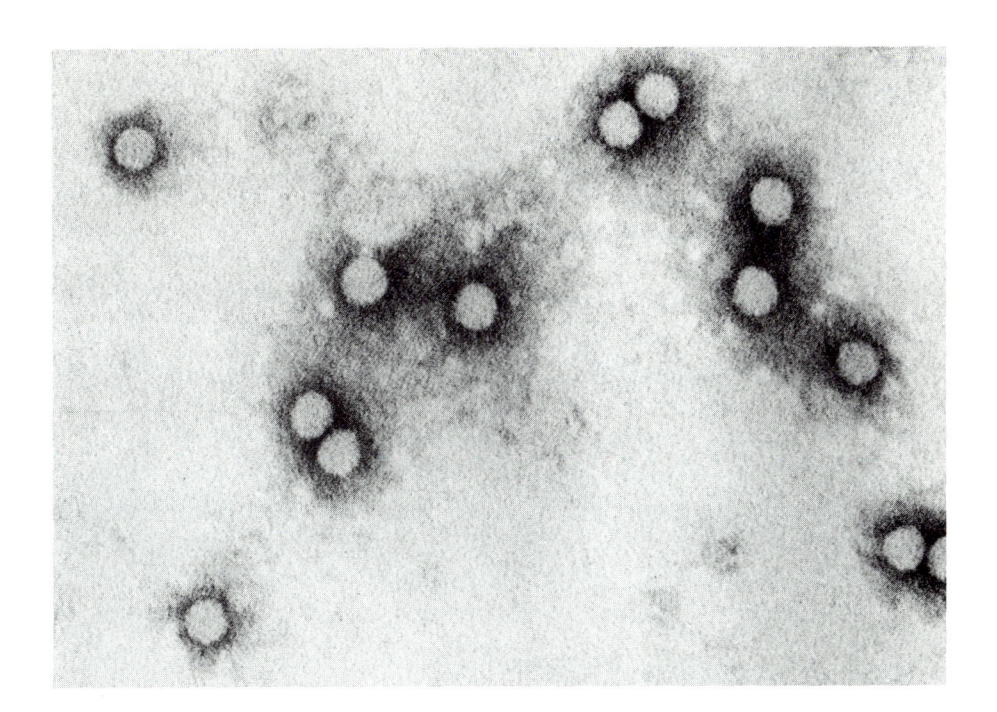

Fig. 9-13 The polio virus (200 000 X)

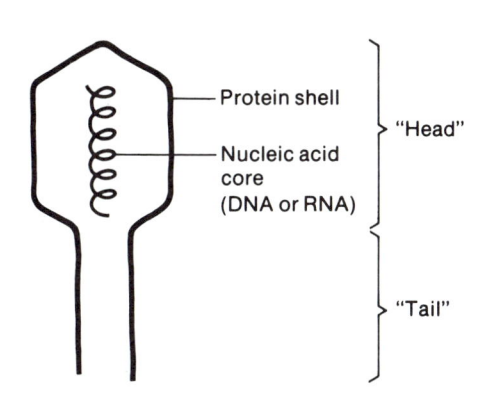

Fig. 9-14 The parts of a virus

Characteristics of Viruses

Size and Shape Viruses are much smaller than bacteria (Fig. 9-12). They range from 0.03 to 3 μm in diameter. They are so small that you cannot see them with an ordinary microscope. However, they can be seen with an electron microscope. Figure 9-11 shows tobacco mosaic viruses magnified 77 000 X. Figure 9-13 shows polio viruses magnified 200 000 X.

Viruses have many shapes. Some, like the tobacco mosaic virus, are rod-shaped. Others, like the polio virus, are spherical. Still others are cubic. And some are shaped like a tadpole, with a "head" and a "tail" region.

Structure A virus has two basic parts, a core and a shell (Fig. 9-14). The **core** is made of genetic material called **nucleic acid** (either **DNA** or **RNA**). The **shell** is made of protein. It protects the core.

Reproduction On its own, a virus shows no signs of life. Yet, within a host cell, the virus begins to reproduce. It doesn't actually reproduce by itself. Instead, it "directs" the host cell to make new viruses. The DNA of the virus takes over control of the host cell. It "tells" the cell to make viruses instead of cell materials. The cell becomes a factory for making viruses.

Some viruses attack bacteria. They are called **phages** [FAY-jes]. Figure 9-15 shows how a phage attacks a cell and directs it to make more phages.

Other viruses attack humans, animals, and plants. Some viral diseases of humans are the common cold, influenza, polio, rabies, smallpox, measles, warts, cold sores, chicken pox, and viral pneumonia.

Section 9.7

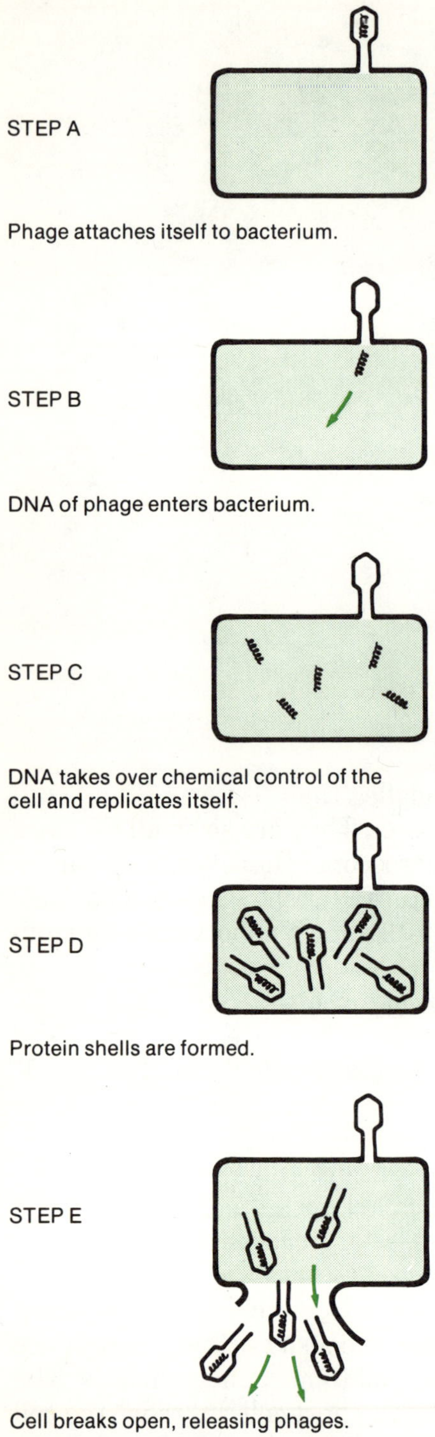

STEP A

Phage attaches itself to bacterium.

STEP B

DNA of phage enters bacterium.

STEP C

DNA takes over chemical control of the cell and replicates itself.

STEP D

Protein shells are formed.

STEP E

Cell breaks open, releasing phages.

Fig. 9-15 A phage directs the host cell to make more phages.

Section Review

1. What evidence do we have that viruses may be non-living?
2. What evidence do we have that viruses may be living?
3. Describe the size of viruses.
4. Describe the structure of viruses.
5. How do viruses reproduce?
6. What is a phage?
7. Name 10 viral diseases of humans.

Main Ideas

1. Monerans are the simplest living things.
2. There are two kinds of monerans: blue-green algae and bacteria.
3. Monerans have no nuclei or organelles.
4. Monerans reproduce by binary fission.
5. Blue-green algae are important producers in some lakes and ponds.
6. Some bacteria cause diseases. Others carry out useful functions such as decomposition.
7. Most bacteria are either saprophytes or parasites.
8. Viruses seem to be both living and non-living. They cause many human diseases.

Glossary

algal bloom		the rapid growth of algae
bacillus	bah-SILL-us	a rod-shaped bacterium
binary fission	BY-na-ree FISH-uhn	the dividing of a cell into two parts
coccus	COCK-us	a spherical bacterium
decomposer		an organism that breaks down organic matter
fermentation		respiration without oxygen
flagellum	fla-JEL-um	a hair-like structure that moves bacteria
moneran	mahn-ER-an	a one-celled organism with no nucleus; blue-green alga or bacterium
pathogen	PATH-o-jen	a disease-causing bacterium
parasite	PAR-a-site	an organism that feeds on a living host

114 Chapter 9

phage	fayj	a virus that attacks bacteria
saprophyte	SAP-row-fite	an organism that feeds on non-living organic matter
spirillum	spir-IL-lum	a spiral-shaped bacterium
toxin	TOX-in	a poison given off by some monerans
virus		the edge of life; shows living and non-living characteristics

Study Questions

A. True or False

Decide whether each of the following statements is true or false. If the sentence is false, rewrite it to make it true. (Do not write in this book.)

1. Monerans are the simplest form of life.
2. Blue-green algae have chloroplasts.
3. Staphylococcus bacteria occur in chains.
4. Viruses are smaller than any living thing.

B. Completion

Complete each of the following sentences with a word or phrase that will make the sentence correct. (Do not write in this book.)

1. There are two kinds of monerans, _____ and _____.
2. The cells of monerans have no nuclei or _____.
3. Most monerans reproduce by a process called _____.
4. Some species of bacteria use _____ to move about.
5. Organisms that feed on living hosts are called _____.

C. Multiple Choice

Each of the following statements or questions is followed by four responses. Choose the correct response in each case. (Do not write in this book.)

1. Most species of blue-green algae are found
   a) in polar icefields
   b) in deserts
   c) in soil
   d) in or near water
2. Rice paddies often require little fertilizer because
   a) blue-green algae in the water fix nitrogen
   b) rice grows well without fertilizer
   c) blue-green algae are producers
   d) water in rice paddies always has enough fertilizer
3. Diplococcus bacteria occur
   a) as single cells
   b) in pairs
   c) in chains
   d) in clusters

4. Bacteria that cause diseases are called pathogens. These bacteria could also be called
   a) saprophytes   **b)** hosts   **c)** decomposers   **d)** parasites

D. Using Your Knowledge

1. Suppose a bacterium divides every 20 min under ideal conditions. How long will it be before there are 1 000 000 bacteria? Assume that conditions stay ideal.
2. Draw sketches of the bacteria that cause these diseases: boils, strept throat, gonorrhoea, syphilis, whooping cough, and tetanus. (Hint: See Table 9-1, page 108).
3. The first monerans on earth were probably not saprophytes or parasites. Why?
4. **a)** If some apple juice is left exposed to air in a room, it will ferment. Why?
   b) If air is bubbled through the apple juice, it may not ferment. Why?

E. Investigations

1. Pick one bacterial disease. Research its cause, symptoms, treatment, and prevention.
2. Pick one viral disease. Research its cause, symptoms, treatment, and prevention.
3. Make an algal bloom in a container in your classroom.
4. Make a model that shows how a phage virus attacks a cell.

10 Bacteria and Us

10.1 Bacteria and Soil
10.2 Activity: Composting
10.3 Activity: The Decomposition of Litter
10.4 Food Spoilage
10.5 How Pathogenic Bacteria Harm Humans
10.6 The Spread of Diseases
10.7 Defences Against Diseases
10.8 Activity: Bacteria in Your Environment

Most things we hear about bacteria are bad: they make us sick; they spoil our food; they cause tooth decay. However, most species of bacteria are harmless. Indeed, many species are actually helpful. Even your body needs certain bacteria in order to function. In fact, the earth as we know it could not exist without bacteria (Fig. 10-1). Do you know why?

In this chapter you will study the importance of bacteria. Some bacteria are important because they are helpful. Others are important because they are harmful. Bacteria are both friend and foe.

10.1 Bacteria and Soil

Bacteria help keep soil fertile. They do this two ways: They decay organic matter and they fix nitrogen.

Decay of Organic Matter

Bacteria break down non-living organic matter such as dead plants, dead animals, and animal wastes. We call this process **decay** or **decomposition**. The bacteria that do this are called **decomposers** [dee-kum-POH-zurz].

The organic matter contains elements such as nitrogen, phosphorus and potassium. These elements are returned to the soil when the decomposers break down the organic matter. This makes the soil more **fertile**. The elements are now available to help plants grow. Animals get these elements by eating plants or other animals. When the plants and animals die, bacteria decompose them. Then the elements go back to the soil again. In this way, elements are recycled over and over again.

Bacteria do more than make the soil more fertile. They get rid of dead organisms and other wastes. If it were not for bacteria, dead animals, leaves, dead trees, and fecal matter would litter the earth. You will see in Chapter 12 that fungi help bacteria decompose organic matter.

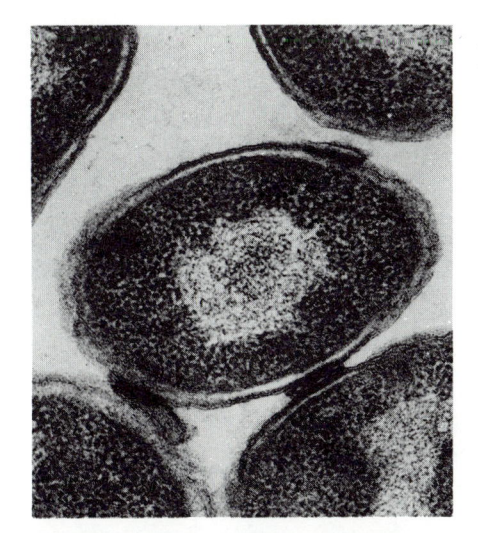

Fig. 10-1 Friend or foe? Most species of bacteria are friends. Only a few are foes.

Section 10.1 117

Fixing of Nitrogen

All plants need nitrogen. They use it to make proteins. However, plants cannot use the nitrogen that is in air. The nitrogen must be in the soil in the form of a nitrate. In this form, it can be absorbed by the plants. **Nitrogen-fixing bacteria** change nitrogen to nitrates. By doing this, they make the soil more fertile.

Some nitrogen-fixing bacteria live free in the soil. Others live on the roots of certain plants. Plants of the **legume family** (peas, beans, clover, alfalfa, soybeans) have lumps on their roots (Fig. 10-2). These lumps contain nitrogen-fixing bacteria. These bacteria enrich the soil with nitrates.

Crop Rotation Some farm crops such as corn and wheat need a great deal of nitrate. Therefore, many farmers grow legumes in a field the year before they grow corn or wheat in it. This practice is called **crop rotation**. For example, a farmer may grow clover one year and wheat the next. Or, he may grow beans one year and corn the next (Fig. 10-3).

Crop rotation saves money, because chemical fertilizers are expensive. Also, crop rotation is good for the soil. Growing the same crop for many years in the same field can damage the soil.

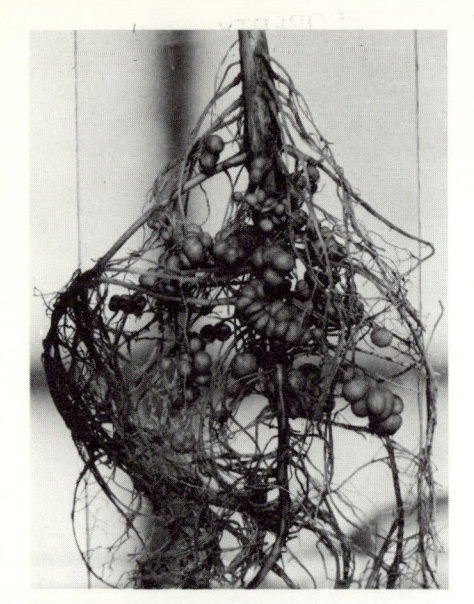

Fig. 10-2 The roots of a soybean plant. Note the lumps on the roots. They contain nitrogen-fixing bacteria.

Fig. 10-3 A crop of beans (left) adds nitrate to the soil. This nitrate makes corn (right) grow better.

Section Review

1. What are decomposers?
2. Explain how decomposers make soil more fertile.
3. Why do all plants need nitrogen?
4. Why are nitrogen-fixing bacteria important?
5. a) How are legumes different from most other plants?
   b) Name 5 kinds of legumes.
6. a) What is crop rotation?
   b) State two advantages of crop rotation.

10.2 ACTIVITY Composting

Leaves, grass clippings, and many kitchen wastes are organic matter. Therefore bacteria can decompose them. This decomposition forms fertile soil. Many of us waste this valuable organic matter. We rake leaves into the streets. We put grass clippings and kitchen wastes in the garbage. Then we go out and buy fertilizer for our lawns and gardens!

Fertilizer is becoming scarce and costly. But we can make our own fertilizer in our backyards by decomposing organic matter. This is called **composting**. The decaying organic matter is called a **compost heap**. Let's see how composting works.

Problem

Can you make a compost heap that works?

Materials

thermometer
large container (preferably glass)
a few handfuls of rich black soil
organic matter, such as leaves, grass, potato peelings, apple cores, banana peelings, and other plant wastes

Procedure

a. Collect a few handfuls of rich black soil. The soil should contain decaying organic matter. It should also contain animals such as sow bugs and earthworms.
b. Put the soil in a large container such as an old aquarium.
c. Add several handfuls of leaves, grass clippings, and other plant wastes (Fig. 10-4).
d. Moisten the plant material and soil.
e. Place a cover over the container to prevent the escape of moisture.
f. Check your compost heap every few days. Keep it moist but not soggy with water. Add new organic matter whenever there is room.
g. Make notes on the changes that occur over several months.
h. From time to time, measure and record the temperature of the decomposing organic matter.

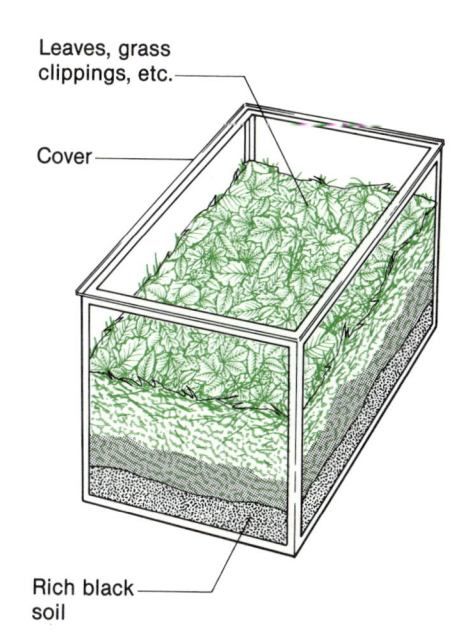

Fig. 10-4 A demonstration compost heap

Discussion

1. Why was rich black soil used?
2. Why should the soil contain some animals?
3. Why are animal wastes not used?
4. Which substances decay the fastest? the slowest?

5. Why must the compost heap be kept moist?
6. Make a summary of the changes that occur during composting.

10.3 ACTIVITY The Decomposition of Litter

What do an apple core, cigarette butt, wad of gum, and chocolate bar wrapper have in common? Just one important thing — they often become litter. Thoughtless people throw them on the streets, on lawns, and on the floors of public buildings.

Some types of litter last a long time. They simply don't decay. Other types decay quickly. We call them **biodegradable** [bi-oh-dee-GRAY-da-bel]. In this activity you will test several types of litter to see which are biodegradable.

Problem

What types of litter are biodegradable?

Materials

pieces of litter (e.g. cigarette butt, orange peeling, banana peeling, apple core, plastic cup, pull tab off a pop can, wad of gum, chocolate bar wrapper)

shallow tray
rich black soil
markers (e.g. popsicle sticks)

Procedure

a. Add at least 4-5 cm of the soil to the tray.
b. Bury the pieces of litter deep in the soil. Mark their positions with markers as shown in Figure 10-5. Write their names on the markers.
c. Moisten the soil until it sticks together when you pinch it. Keep it this moist throughout the experiment. Do not make it any wetter. Do not let it become any drier.
d. Every 2-3 d, dig up the pieces of litter. Make notes on their state of decomposition.
e. Continue the experiment until no further changes seem to occur.

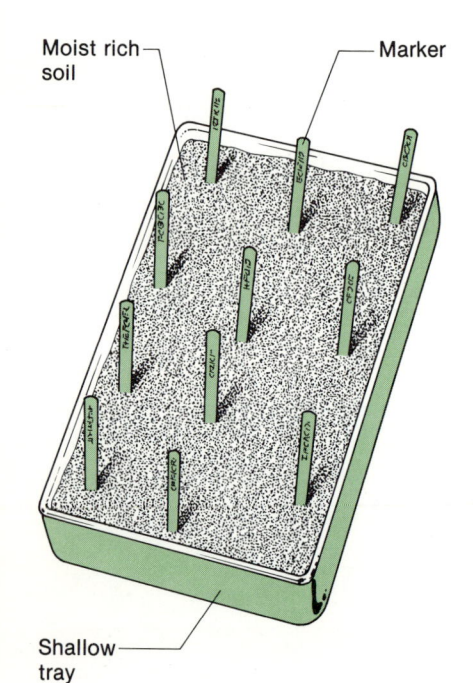

Fig. 10-5 What types of litter are biodegradable?

Discussion

1. Why is rich black soil used?
2. Why must the soil be kept moist but not too wet?
3. What types of litter are very biodegradable?
4. What types of litter are slightly biodegradable?
5. What types of litter are not biodegradable?
6. Why are some forms of litter biodegradable while others are not?

10.4 Food Spoilage

Most bacteria get their energy by breaking down non-living organic matter. Sometimes this action is good for us. For example, it forms cheese, butter, and silage. And it helps treat our sewage. But sometimes this action is harmful. The bacteria act on food and spoil it. They can even make it poisonous.

Sour milk, rotten fruits, and spoiled meat are caused by bacteria. Usually such spoilage is harmless to our health. You could normally drink a glass of sour milk or eat a rotten banana without getting sick. However, serious diseases can result from eating foods containing certain bacteria. These diseases are called **food poisoning**. Three types are discussed here.

Staphylococcus Food Poisoning

Cause This is the most common type of food poisoning. It is sometimes called ptomaine [toe-MAIN] poisoning. It is caused by the bacterium *Staphylococcus aureus*. This bacterium is often present on the skin and in the nose and throat of humans. It multiplies rapidly in many foods, particularly dairy products and cream-filled bakery goods. It gets into foods when they are handled in an unsanitary way. Then, if the food is not refrigerated, the bacteria multiply rapidly.

This type of food poisoning is caused by **toxins** (poisons). The bacteria secrete them while they are digesting the food.

Symptoms and Effects The symptoms are nausea, diarrhea, and abdominal cramps. Symptoms usually appear 1 to 5 h after infected food is eaten. Recovery usually occurs in a few days. Death seldom results.

Most people have likely had a mild case of this disease. Also, most people likely got the disease the same way. They ate chicken or egg salad sandwiches, potato salad, or any food with salad dressing in it. They may also have eaten cold meats or cream puffs. These foods were likely eaten at a picnic on a hot day. The foods had probably not been properly refrigerated. Chicken and dairy products, on a warm day, are almost ideal environments for *Staphylococcus aureus*.

Botulism

Cause This type of food poisoning is caused by the bacterium *Clostridium botulinum*. This bacterium is present in many soils. It seldom causes trouble when there is lots of air around. But, when there is little air, it produces dangerous toxins. They are more deadly than rattlesnake venom.

Most cases of botulism arise from eating home-canned vegetables that have not been properly cooked. String beans are the most ideal

Section 10.4 121

environment for this bacterium. Spores of the bacterium may get on the vegetables before they are canned. If the vegetables are poorly cooked, the spores may not be killed. When the jar is sealed, air is kept out. The spores become active bacteria and multiply rapidly. They secrete toxins when they feed on the vegetables. The toxins do not change the appearance or taste of the food.

Botulism can be prevented by heating canned foods before eating them. The toxins are destroyed by heat. *Never* buy a can of food with a bulge in it. This often indicates botulism.

Symptoms and Effects The symptoms are double vision, weakness, and paralysis. They begin 12-36 h after the infected food has been eaten. The disease is fatal in 7 out of 10 cases.

Salmonella Poisoning

Cause This disease is better called food infection than food poisoning. No poisons (toxins) are actually released into the food. Instead,

Fig. 10-6 *Salmonella* spreads from infected foods to humans.

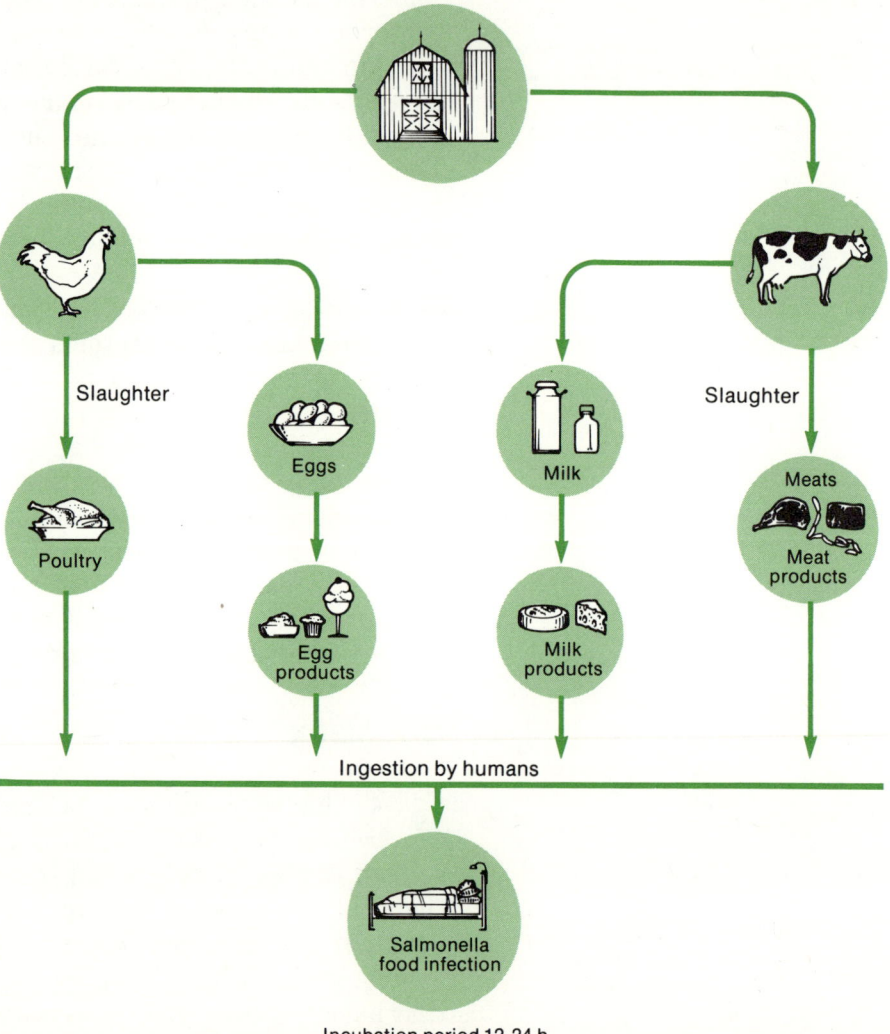

122   Chapter 10

food simply serves as a medium for transferring *Salmonella* bacteria from one person to another.

Almost all species of *Salmonella* cause diseases in humans. The best known is probably *Salmonella typhi*. It causes deadly typhoid fever. Many other species cause infections of the digestive system.

The bacteria get into food when the food is handled by an infected person. An infected cow or chicken may also be a source of the disease (Fig. 10-6). The foods involved are usually meat pies, cured meats, sausage, poultry, eggs, milk, and milk products.

Proper cooking kills the bacteria. However, many of the foods are often eaten uncooked or partly cooked. Eggnog, meringues, cream cakes, and custards are examples. Such foods should *never* be left in a warm place.

Symptoms and Effects Symptoms usually occur 12 to 24 h after infected food is eaten. They begin with a headache or chills. Then nausea, diarrhea, and abdominal pain follow. The disease is seldom fatal. Complete recovery usually occurs in 2 to 3 d, even without treatment.

Section Review

1. Make a summary of the cause, symptoms, and effects of *Staphylococcus* food poisoning.
2. Make a summary of the cause, symptoms and effects of botulism.
3. Make a summary of the cause, symptoms and effects of *Salmonella* poisoning.

10.5 How Pathogenic Bacteria Harm Humans

A **pathogen** is an organism that can cause a disease. Some species of bacteria are pathogens. Some protists, fungi, and viruses are also pathogens.

Most pathogenic bacteria harm humans by producing **toxins** (poisons). The toxins can travel through the body in the circulatory system. As a result, they often damage tissues far from the original infected area. For example, rheumatic fever is caused by a species of streptococcus bacterium. It lives in the human throat. But its toxin travels throughout the body. It often damages the heart valves and joints.

Tetanus bacteria enter the body through a cut made by a dirty nail or piece of glass. The toxin they produce travels throughout the body. It affects all muscles. This disease often causes paralysis of the jaw muscles. Therefore it is called lockjaw.

The toxins of typhoid fever bacteria destroy cells in the intestine. The toxins of tuberculosis destroy cells in the lungs. The toxins of cholera damage cells in the intestine. Then they spread and damage muscles and kidneys. Most bacterial diseases, then, are caused by toxins.

Section Review

1. What is a pathogen?
2. How do pathogenic bacteria harm humans?
3. Why can diseases often damage tissues far from the original infected area?

10.6 The Spread of Diseases

Diseases Spread by Air

Many diseases are spread through the air. Among these are colds, influenza, whooping cough, measles, and mumps. They are commonly spread by coughing and sneezing. These diseases enter the body by means of the breathing system. Therefore the infection usually begins in the nose, throat, and lungs.

Diseases Spread by Food and Water

Many diseases are spread through food and water. Most of these are diseases of the digestive system. An infected person handles food. Then the person eating the food gets the disease.

Many diseases can be spread through drinking water. In our society, we put sewage into lakes. Then we get our drinking water from the same lakes. That water may contain pathogens that were in the sewage. As a result, cities and towns chlorinate their drinking water. This kills pathogens. Never drink water from any source unless it is treated to kill pathogens. On a hike or canoe trip you can boil the water (Fig. 10-7). Or you can add chemicals that kill the pathogens.

Contact Diseases

Many diseases are spread by direct contact with an infected person. They may also be spread by contact with things handled by the infected person. For example, colds are commonly spread through the air by coughing and sneezing. But they can also be spread simply by shaking hands with infected people.

Among the serious contact diseases are **venereal diseases** [veh-NEER-ee-al]. The common ones are **syphilis** [SIF-ill-is], **gonorrhea** [gone-or-EE-ah], and **genital herpes**. The first two are bacterial diseases. The last one is a viral disease. All three are most commonly spread by sexual contact with infected persons.

Syphilis usually first appears as a hard sore on the gential organs. Then it may lie dormant for years. When it reappears, it can cause heart problems, insanity, and blindness.

Fig. 10-7 Water that looks clean can contain pathogens. Before you drink it, boil it or add special chemicals you can buy.

Anopheles mosquito (carrier of malaria, a protozoan)

Aedes mosquito (carrier of yellow fever, a virus)

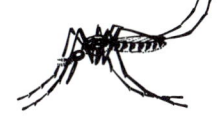

Culex mosquito (carrier of encephalitis, a virus)

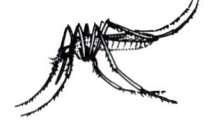

Human body louse (carrier of typhus fever, a bacterium)

Housefly (carrier of many diseases)

Tsetse fly (carrier of African sleeping sickness, a protozoan)

Fig. 10-8 These arthropods are carriers of human diseases.

Gonorrhea usually first appears as an inflammation of the genital organs. If left untreated, it can cause sterility.

Genital herpes is a close relative of the viruses that cause cold sores and shingles. The sores look much like cold sores. They first appear on the genital organs.

Medical attention should always be sought when symptoms of any venereal disease first appear.

Wound Diseases

Many diseases enter the body through breaks in the skin. Thus all wounds should be immediately cleaned and treated to prevent infection.

One of the most serious wound diseases is tetanus (lockjaw). Tetanus bacteria live in the soil. They commonly get into your body through a cut made by a dirty nail or piece of glass. Tetanus bacteria thrive in an oxygen-free area. Therefore a deep puncture is the most serious. Regular tetanus shots are the best protection against lockjaw.

Carriers

Humans, as you have seen, carry and spread many diseases. Some humans are **immune carriers**. These people carry a disease but show no symptoms. Typhoid fever, polio, and diphtheria are often spread by immune carriers.

Animal carriers are another serious problem. Fortunately, most diseases of mammals (including pet dogs and cats) do not affect humans. Arthropods, however, spread many diseases. Figure 10-8 shows six arthropods and the serious diseases they carry.

Section Review

1. Describe how diseases are spread through air.
2. Describe how diseases are spread through food and water.
3. What is a contact disease?
4. Name three venereal diseases. How is each one caused? What are the first symptoms of each disease?
5. What is a wound disease? Describe one example.
6. What is an immune carrier?
7. What is an animal carrier?

10.7 Defences Against Diseases

The human body has 3 ways of fighting diseases: it can stop bacteria from entering the body; it has special cells that attack bacteria; it makes substances called antibodies that act against diseases. In recent years scientists have added a fourth method: the use of chemicals such as drugs. And there is still another method: your lifestyle. Let's see how these 5 methods work.

Keeping Bacteria out of the Body

The **skin** is one of the body's main defences against bacteria. Most bacteria cannot get through skin. Some try to get in through the pores. But the salt and acid in sweat usually kill them.

Mucus in your breathing system traps bacteria that enter by your nose and mouth. Enzymes in the mucus kill most bacteria. Also, your breathing system is lined with hair-like objects called **cilia**. They sweep mucus-covered bacteria and dirt up the throat. Then you can cough them out or swallow them. If you swallow them, your stomach acid will kill the bacteria.

Tears wash bacteria out of your eyes. Also, enzymes in the tears kill many types of bacteria.

Special Cells That Attack Bacteria

Your body has white blood cells called **leucocytes** [LOO-ko-sites]. These cells "eat" and digest bacteria. Often, a mass of dead bacteria, dead leucocytes, and blood serum collects at the site of an infection. This mixture is called **pus**. You have likely seen pus form at cuts and scrapes in your skin.

Have you ever had "swollen glands"? The lymph glands under your arms and under your skin often swell. They swell when leucocytes are fighting bacteria in them.

When you have an infection, you often get a **fever**. Often the fever is caused by your body to help it fight infections. Therefore, a low fever can be helpful. However, a high fever may destroy body cells, especially brain cells, as well as bacteria.

Production of Antibodies

Foreign proteins that enter your body are called **antigens** [AN-tih-jens]. Bacteria and their toxins are both antigens. To fight an antigen, your body makes a special protein called an **antibody**. The antibody destroys the antigen. If the antigen is a toxin, the antibody is called an **antitoxin**.

Have you been **vaccinated** for polio? When this is done, a dead or weakened antigen is put into your body. Your body makes an antibody to fight the antigen. Then you have protection against the disease for many years.

Use of Chemicals

Disinfectants are one kind of chemical used to kill pathogens. They are very strong and cannot be used on humans. Instead they are used to scrub sources of bacteria such as hospital floors, restaurant counters, and barns.

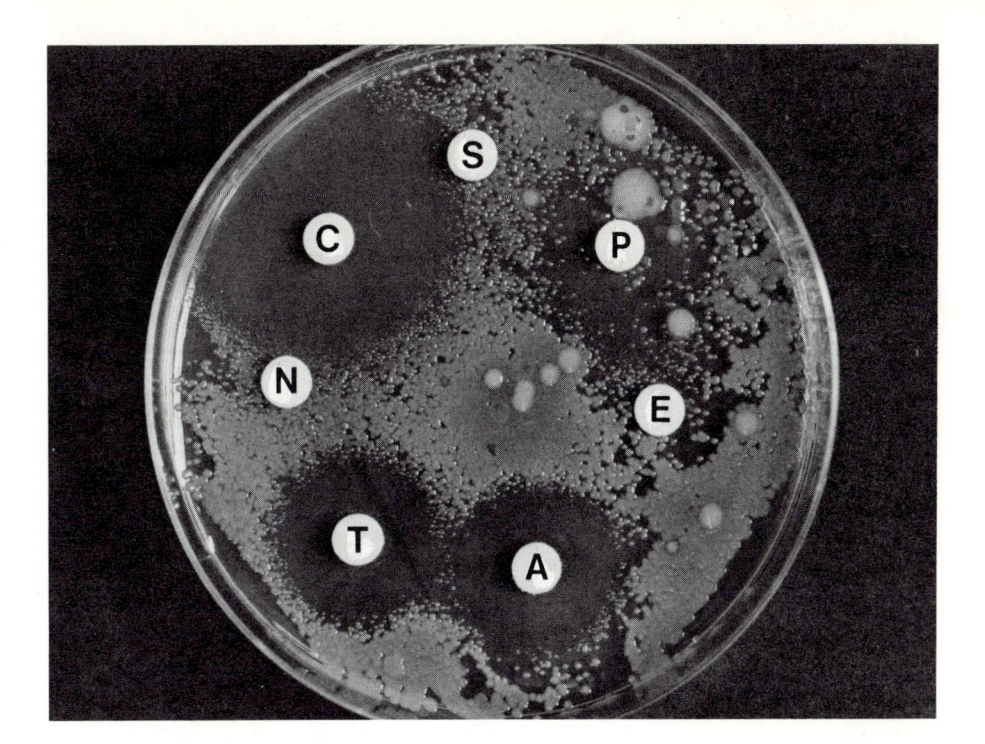

Fig. 10-9 Seven antibiotics were used: neomycin (N), terromycin (T), aureomycin (A), erythromycin (E), penicillin (P), streptomycin (S), and chloromycetin (C). In this case, chloromycetin was the most effective.

Antiseptics are weaker chemicals. They slow down the growth of pathogens. However, they may not kill them. Antiseptics can be used on human flesh. Ethyl alcohol and boric acid are good antiseptics.

Antibiotics are chemicals that kill or slow the growth of bacteria. Penicillin, streptomycin, terramycin, and ampicillin are common antibiotics. One antibiotic may be more effective than another against a certain bacterium. As a result, doctors often perform a test to see which antibiotic to use. Some of the infection is spread on a culture plate. Then antibiotic discs are placed on the plate. Each disc contains a different antibiotic (Fig. 10-9). The most effective antibiotic will kill more bacteria near it.

Sulfa drugs have been around since the 1930s. They kill many types of bacteria. However, antibiotics are usually more effective. Sulfa drugs are still best for most diseases of the excretory system.

Your Lifestyle

One of the best defences against diseases is a **healthy lifestyle**. Much will be said about this in Unit 6. However, we would like you to think about it now. Perhaps your class would like to discuss these questions:

- What diseases can be prevented by washing your hands before meals?
- How can you prevent food poisoning?
- How can you avoid getting venereal diseases?
- What "shots" do doctors say you should have?
- How should you clean a wound?
- What diseases can be prevented by proper nutrition?

Section 10.7

- Will regular exercise help prevent disease?
- What diseases are caused by smoking? (Diseases need not be caused by bacteria or viruses.)
- What are the effects of diet, exercise, and smoking on diseases of the circulatory system (heart attacks and strokes)?

Section Review

1. Describe how the skin, mucus, and tears keep bacteria out of your body.
2. Explain these terms: leucocytes, pus, swollen glands, fever.
3. Explain these terms: antigen, antibody, antitoxin.
4. Explain the difference between a disinfectant and an antiseptic.
5. a) What is an antibiotic?
   b) Name 4 antibiotics.
6. Write answers to the questions asked under the heading "Your Lifestyle."

10.8   ACTIVITY   Bacteria in Your Environment

In this activity your class will test different parts of your school for bacteria.

Problem

What parts of your environment have high bacteria counts? Will disinfectants and antiseptics kill the bacteria?

Materials

nutrient agar plate
sterile cotton swabs (2)
marking pen

transparent tape
antiseptics and disinfectants

CAUTION: Treat all plates as though they contained pathogens (disease-causing organisms). NEVER remove the tops after bacteria have been placed on the agar. Follow your teacher's directions closely.

Fig. 10-10 The bacteria on the swab are transferred to the agar. The nutrient in the agar feeds them. Each bacterium grows into a colony you can see.

Procedure

a. Your teacher will prepare a control plate. How do you think it should be prepared?
b. Pick an area of your environment that you think may contain bacteria. Some possible areas are your locker, the drinking fountain, lunch counter, door knobs, a coin, a window ledge.

   c. Run a sterile cotton swab over this area.
   d. Lightly brush the swab over the entire surface of the agar (Fig. 10-10).
   e. Dip the second swab in one of the antiseptics or disinfectants provided by your teacher. Pick one that you think would be a good one for controlling bacteria in the area you are testing.
   f. Rub this swab over one half of the agar. Put a mark on this side of the dish. Then you can tell later which half had the chemical added.
   g. Put the cover back on the plate. Write your name on the cover.
   h. Seal the cover on with transparent tape.

   CAUTION: **Never** remove the top. The plate may contain disease-causing organisms. Your teacher will dispose of the plates at the end of the activity.

   i. Incubate the plate as follows. Place the plate in an incubator at 25°C to 35°C. Place it upside down. This keeps moisture in the agar.
   j. Check the plate every day for the next 3-4 d. Make notes and diagrams of the shape, size, structure, texture, and colour of any colonies that appear.

Discussion

1. How large was each colony when it started?
2. By what process did the colonies grow?
3. Mould colonies are usually fuzzy and larger than bacterial colonies. How many kinds of moulds are in your plate?
4. Compare your results to those of your classmates.
5. What areas of the school had the largest number of bacteria?
6. What areas of the school had the greatest variety of bacteria?
7. How effective were the antiseptics and disinfectants used by your class?

Main Ideas

1. Bacteria decompose soil and make it fertile.
2. Some bacteria can fix nitrogen.
3. Composting gets rid of wastes and forms soil.
4. Some litter is biodegradable; some is not.
5. Bacteria can cause food spoilage and food poisoning.
6. Most pathogenic bacteria harm humans by making toxins (poisons).
7. Diseases can be spread by air, water, food, contact, wounds, and carriers.
8. The human body has 3 ways of fighting diseases; chemicals and lifestyle are 2 more ways.

Glossary

antibody		a protein made by the body to fight an antigen
antigen		a foreign protein (e.g. bacterium or its toxin) that enters the body
antitoxin		an antibody produced when the antigen is a toxin
biodegradable	bi-oh-dee-GRAY-da-bel	a substance that decays quickly
carrier		an organism that carries a disease but shows no symptoms
decomposers	dee-kum-POH-zurz	bacteria that break down non-living organic matter
pathogen	PATH-oh-jen	a disease-causing organism
toxin		a poison produced by a bacterium

Study Questions

A. True or False

Decide whether each of the following statements is true or false. If the sentence is false, rewrite it to make it true. (Do not write in this book.)

1. Bacteria that break down non-living organic matter are called nitrogen-fixing bacteria.
2. Leaves are biodegradable.
3. *Salmonella* poisoning is usually fatal.
4. A disinfectant is stronger than an antiseptic.

B. Completion

Complete each of the following sentences with a word or phrase that will make the sentence correct. (Do not write in this book.)

1. A can with a bulge in it may be a sign of ▨▨▨▨.
2. An organism that causes a disease is called a ▨▨▨▨.
3. Most pathogenic bacteria harm humans by making ▨▨▨▨.
4. Three venereal diseases are ▨▨▨▨, ▨▨▨▨, and ▨▨▨▨.

C. Multiple Choice

Each of the following statements or questions is followed by four

responses. Choose the correct response in each case. (Do not write in this book.)
1. Which one of the following is most likely to cause botulism?
   a) eating dairy products that have been left in warm air
   b) eating home-canned vegetables that were not properly cooked
   c) eating cream-filled bakery goods
   d) eating uncooked foods such as eggnog
2. One disease that can be spread by an immune carrier is
   a) polio
   b) *Salmonella* food poisoning
   c) influenza
   d) the common cold
3. The vaccine for polio contains a dead or weakened
   a) toxin   b) antitoxin   c) antibody   d) antigen
4. Bacteria in a kitchen sink are best killed with a
   a) antiseptic   b) antibiotic   c) disinfectant   d) antitoxin

D. Using Your Knowledge

1. Your body could contain atoms that were once part of a dinosaur. How could this happen?
2. Very few homes have compost heaps. Do you think they should have them? Why do you suppose they don't?
3. Why are regular tetanus shots the best protection against lockjaw?
4. Why do smokers get colds more often than do non-smokers?
5. What precautions would you take to prevent food poisoning at a pot-luck dinner?

E. Investigations

1. Select one of the arthropods in Figure 10-8, page 125. Find out where it lives, how it spreads diseases, how serious the disease is, and methods for controlling the disease.
2. The first antibiotic to be discovered was penicillin. Find out how it was discovered.
3. Make a compost pile at home. Keep it going for several months. Make notes on how you maintained it. Write a report on your findings that is about 300 words long.
4. Bacteria have many important uses. Among these are cheese-making, butter-making, silage production, vinegar production, and sewage treatment. Find out all you can about one of these uses. Then write a report of about 300 words. Include diagrams, if they are helpful.
5. Make some sauerkraut or yogurt.
6. Carefully dig up a legume plant. Look for the lumps on the roots where the nitrogen-fixing bacteria are. Press the plant and bring it to class.
7. Make some butter.
8. Food spoilage can be controlled by pasteurization, refrigeration, freezing, dehydration, canning, smoking, or by the use of chemical preservatives. Write a report of about 50 words on how one of these methods controls food spoilage.

11 Protists

11.1 The Protozoans
11.2 Activity: Investigating Some Protozoans
11.3 Two Common Protozoans
11.4 Activity: Investigating the Amoeba
11.5 Activity: Investigating the Paramecium
11.6 The Algal Protists
11.7 Activity: Investigating Some Algal Protists

The protists are simple organisms (Fig. 11-1). They are, however, more complex than the monerans (Chapter 10). Like the monerans, they are one-celled. And they live in water or in wet land environments. But unlike the monerans, most of their cells have nuclei and organelles. Of the 5 kingdoms of living things, the **kingdom Protista** (the protists) is the second simplest.

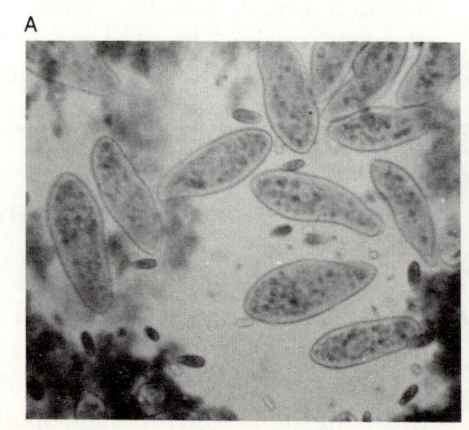

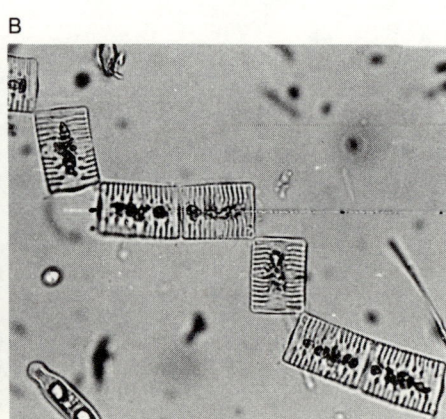

There are two kinds of protists: **protozoans** [pro-toe-zo-ans] and **algae** [AL-jee]. There are several thousand species of each (Fig. 11-2). We usually call the algae in this kingdom **algal protists**. That's because there are also algae in the moneran kingdom (Chapter 10) and in the plant kingdom (Chapter 13).

11.1 The Protozoans

Protozoans are important to us in many ways. Some are food for small animals. In this way, they support food chains of which we may be a part. Other protozoans break down sludge in sewage treatment plants. Others break down and recycle dead organic matter in ponds and lakes. Unfortunately, some species cause deadly diseases. Malaria is an example.

Fig. 11-1 Protists are one-celled. Some, like the paramecium (A), can move. Others, like the diatom (B), cannot.

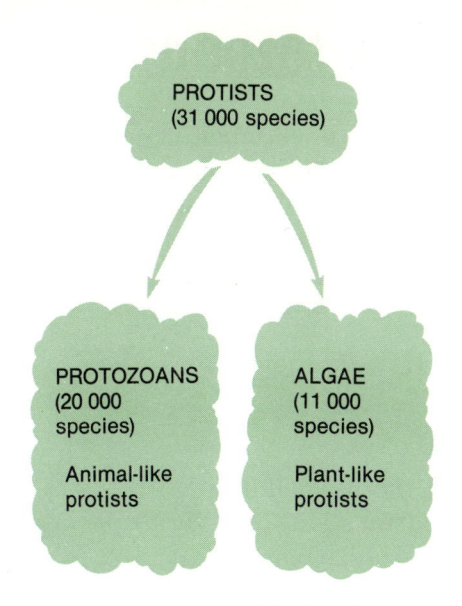

Fig. 11-2 The two kinds of protists

Fig. 11-3 Some common protozoans

Protozoans used to be considered the simplest animals. In fact, the word *protozoan* means "first animal". However, they are no longer classified as animals. You will see why after you read the following characteristics and look at them in Activity 11.2.

Characteristics

1. Protozoans are one-celled. Some of these cells, however, live together in colonies.
2. A few species can change shape. However, most have a definite shape (Fig. 11-3).
3. Most cells have a nucleus and some organelles.
4. Most protozoans can move. They use **pseudopods** (false feet), **cilia** (tiny hairs), or **flagella** (whip-like tails).
5. Protozoans reproduce by binary fission, or splitting into two parts. Some species can reproduce sexually. That is, genetic material is exchanged between two cells.
6. Protozoans are consumers. They feed on bacteria, yeasts, algae, other protozoans, and dead organic matter.
7. Most protozoans are microscopic. Some are only 2 μm (micrometres) long. Most are less than 250 μm long. A few are over 3 mm long. These can be seen without a microscope.

Section Review

1. a) What is a protist?
   b) Name the 2 types of protists.
2. List 4 ways in which protozoans are important to us.
3. Make a summary of the characteristics of protozoans.

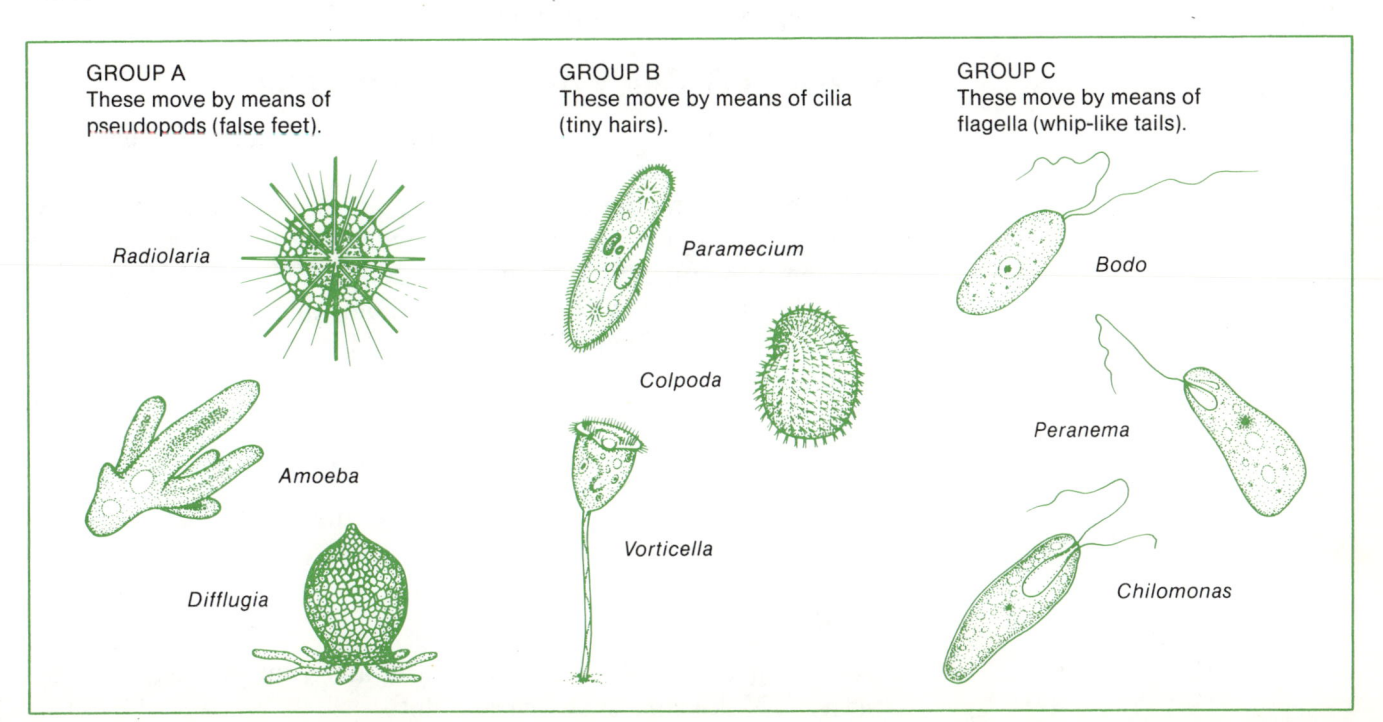

GROUP A
These move by means of pseudopods (false feet).

Radiolaria
Amoeba
Difflugia

GROUP B
These move by means of cilia (tiny hairs).

Paramecium
Colpoda
Vorticella

GROUP C
These move by means of flagella (whip-like tails).

Bodo
Peranema
Chilomonas

Section 11.1

11.2 ACTIVITY Investigating Some Protozoans

You learned in Section 11.1 that protozoans have much in common. Yet they also differ in many ways. In this activity you will look at several species of protozoans. Note how they are alike and how they are different.

Problem

What do protozoans have in common? How do they differ?

Materials

microscope
microscope slides (2)
cover slips (2)
paper towel
lens paper
dropper
identification books
1.5% methyl cellulose
prepared slides of several species of protozoans
one or more of:
  mixed protozoan culture
  hay infusion culture
  pond water culture

Procedure A Study of Prepared Slides

a. Read again the characteristics of protozoans in Section 11.1.
b. Look at as many prepared slides of protozoans as possible. Observe them first under low power. Then switch to medium and, finally, high power.
c. Sketch the shape of each cell. Draw in any parts you can find. Label such parts as the cell membrane, cytoplasm, nucleus, pseudopods, cilia, and flagella.

Procedure B Study of Living Protozoans

a. Use the dropper to get a drop of the culture. Follow your teacher's directions.
b. Prepare a wet mount of the culture. But do *not* add a cover slip.
c. Look at the culture under low power. The protozoans may be moving too quickly. If they are, add a drop of 1.5% methyl cellulose. It will slow them down without killing them.
d. Add a cover slip. Switch to medium and then to high power. Study each species closely.
e. For each species, sketch the shape of the cell. Then draw in any other features you see.
f. Make notes on how each species moves.
g. Try to identify each species to the genus level. Use the books provided by your teacher.

Discussion

1. Your diagrams and laboratory notes are your writeup for this activity. Make sure they are complete.

11.3 Two Common Protozoans

This section describes the features of two common protozoans: the **amoeba** and the **paramecium**. Then in the following activity, you will use your microscope to study those features.

Fig. 11-4 The amoeba

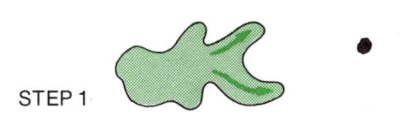

STEP 1
Amoeba senses food.

STEP 2
Pseudopods extend toward food.

STEP 3
Pseudopods surround food.

STEP 4
Food is engulfed by pseudopods.

STEP 5
Food vacuole formed; it circulates through cytoplasm as digestion occurs.

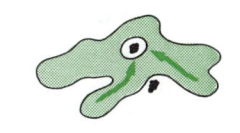

STEP 6
Indigestible material is left behind.

Fig. 11-5 Feeding by an amoeba. Less than 10 min is needed for all the steps shown here.

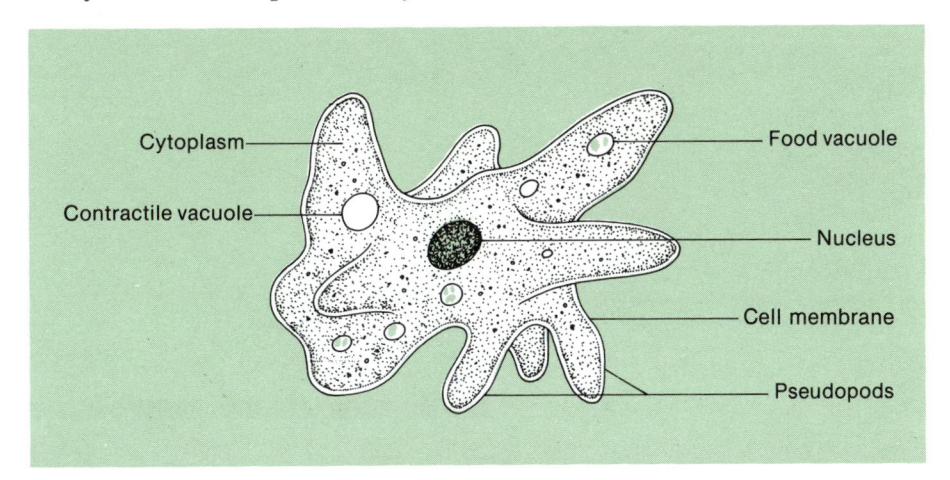

The amoeba was selected because it is one of the simplest protozoans. The paramecium was selected because it is one of the most complex.

Both organisms can move, reproduce, feed, digest food, excrete wastes, and respire. They can do almost everything you can do! Yet they don't have organs like your stomach, heart, and kidneys. Let's see how they can do this.

The Amoeba

An amoeba is a mass of jellylike protoplasm (Fig. 11-4). Most species can just be seen with the naked eye. The amoeba does not have a fixed shape. It changes shape by sending out **pseudopods** (false feet). Pseudopods [soo-doe-pods] are part of the **cytoplasm**. An amoeba moves by sending pseudopods out in the direction it wants to move.

Within the cytoplasm are the **nucleus** and two kinds of vacuoles [VAK-yoo-ohls]. The **food vacuoles** contain food that is being digested. An amoeba takes in food by engulfing it (flowing all around it). Figure 11-5 shows the steps in feeding.

The **contractile vacuole** contains mainly water. Its function is to remove excess water from the amoeba. It begins as a tiny sphere. It grows as it fills with water. Then it moves to the cell membrane. It now bursts, getting rid of the excess water.

An amoeba reproduces by **binary fission** (Fig. 11-6). The nucleus divides by mitosis. Then the cytoplasm splits into two parts.

An amoeba gets the oxygen it needs by diffusion from the water through the cell membrane. Carbon dioxide, urea, and other metabolic wastes diffuse out through the cell membrane.

An amoeba responds to many stimuli in its environment. Figure 11-7 shows how it responds to four stimuli.

Section 11.3   135

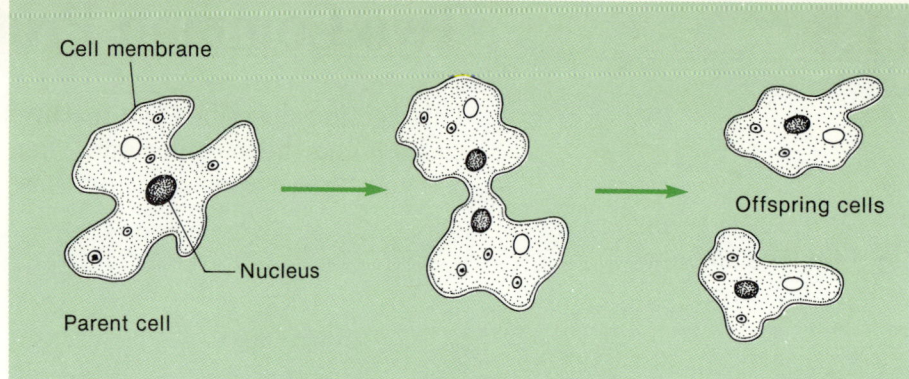

Fig. 11-6 Binary fission of an amoeba

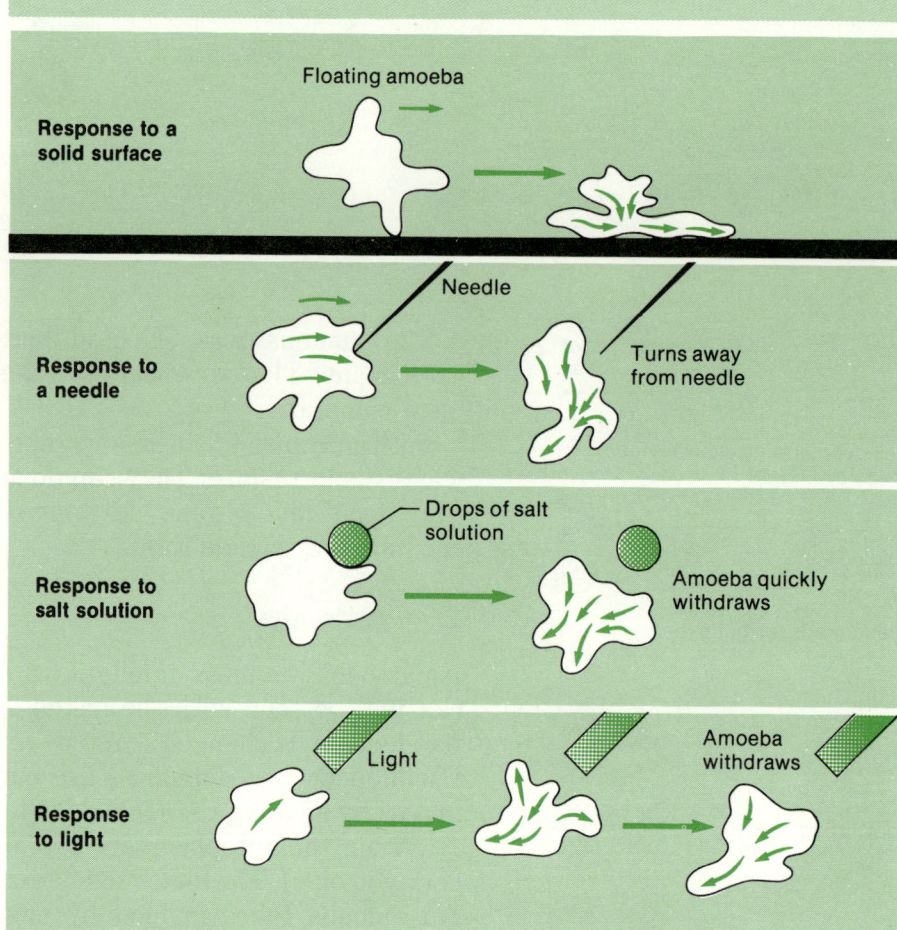

Fig. 11-7 Like all living things, an amoeba responds to stimuli.

The Paramecium

The paramecium has a definite shape. It is rounded at the front end and pointed at the rear end (Fig. 11-8). The outer surface is covered with an elastic membrane called the **pellicle**. The pellicle is covered with tiny hairs called **cilia**. They beat backward to push the paramecium forward. The extra cilia around the **oral groove** cause the paramecium to move as shown in Figure 11-9.

A paramecium feeds on bacteria, algae, yeasts, and small protozo-

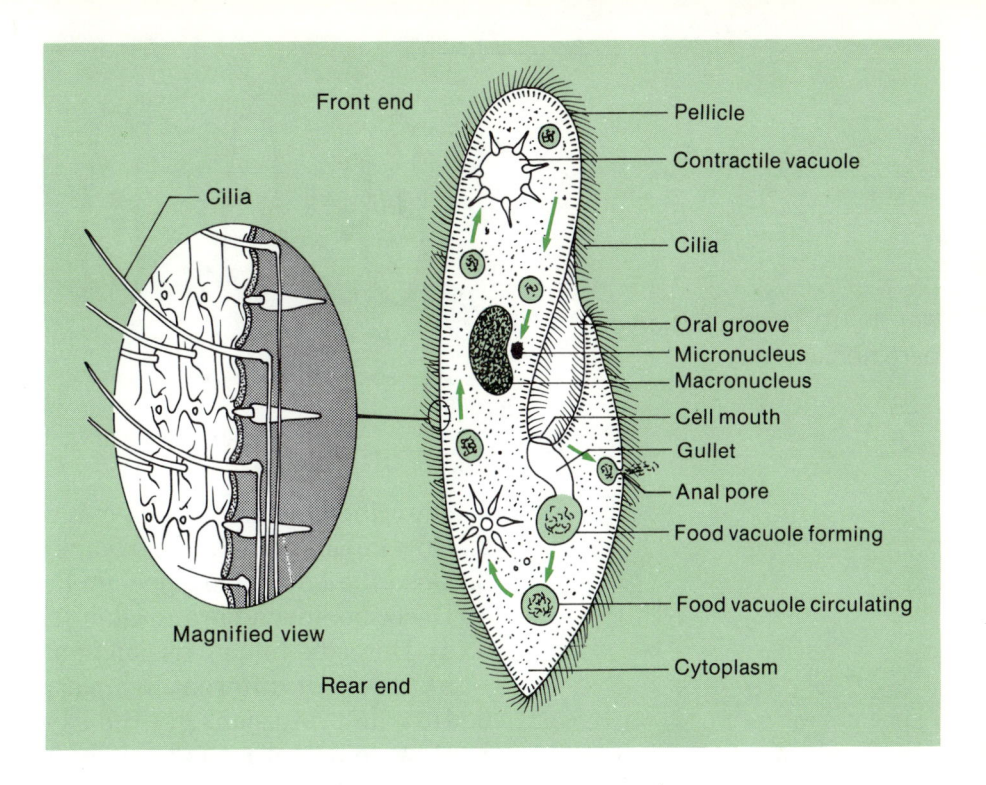

Fig. 11-8 The paramecium

ans. The cilia in the oral groove sweep the food into the **cell mouth**. From there the food moves to the **gullet**. At the end of the gullet, a **food vacuole** forms. The vacuole circulates around the cell as the food is being digested. Undigested waste is expelled at the **anal pore**.

A paramecium has two **contractile vacuoles**. These have canals that help with the gathering of excess water.

Like the amoeba, the paramecium reproduces by **binary fission**. A paramecium has two nuclei. The larger one, the **macronucleus**, directs most cell functions. The smaller one, the **micronucleus**, directs reproduction. As Figure 11-10 shows, binary fission is more complex in a paramecium than in an amoeba. From time to time, paramecium reproduce sexually. Two organisms join by their oral grooves. Then they exchange material from their micronuclei. After the exchange, the organisms divide by binary fission.

A paramecium gets the oxygen it needs by diffusion from the water through the pellicle. Carbon dioxide, urea, and other metabolic wastes diffuse out through the pellicle.

A paramecium responds to many stimuli in its environment. Figure 11-11 shows how it responds to five stimuli.

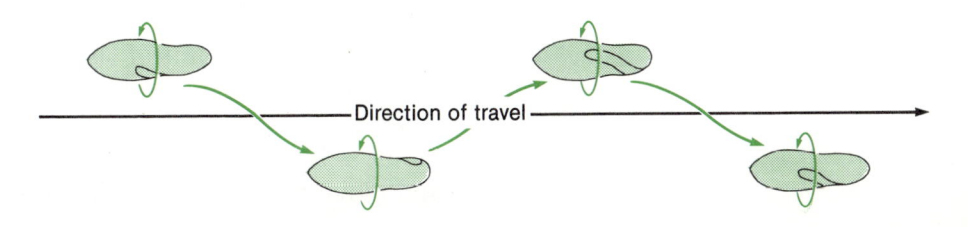

Fig. 11-9 Locomotion of a paramecium

Section 11.3   137

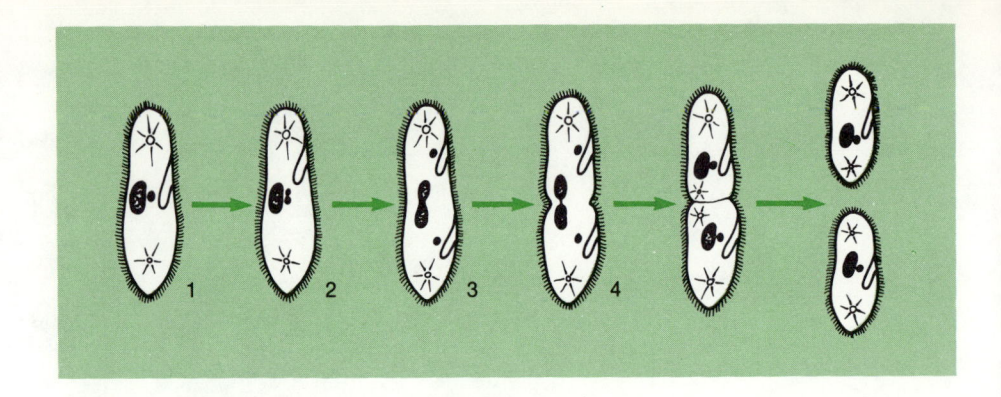

Fig. 11-10 Binary fission of a paramecium

Section Review

1. Describe how an amoeba moves.
2. How does a contractile vacuole work?
3. Describe how a paramecium moves.
4. Describe how a paramecium feeds.
5. **a)** Describe binary fission in an amoeba.
   b) How is it different in a paramecium?
6. How do protozoans get the oxygen they need?
7. How do protozoans get rid of metabolic wastes?
8. What is meant by "protozoans respond to stimuli"?

11.4   ACTIVITY   Investigating the Amoeba

It's easy to read about an organism as you did in Section 11.3. But, can you find an amoeba and see the things you read about?

Problem

Can you observe the structure and behaviour of an amoeba?

Materials

microscope   paper towel   dropper
microscope slide   lens paper   amoeba culture
cover slip

Procedure

a. Suck up a drop or two of the amoeba culture with the dropper. The cloudy sludge at the bottom of the culture jar is usually best.

b. Prepare a wet mount of the sample. Lower the cover slip slowly. Then you won't squash the amoebas.

c. Scan the sample slowly under low power. It may take a few minutes to find an amoeba. Start at one corner of the cover slip and move back and forth in parallel lines.

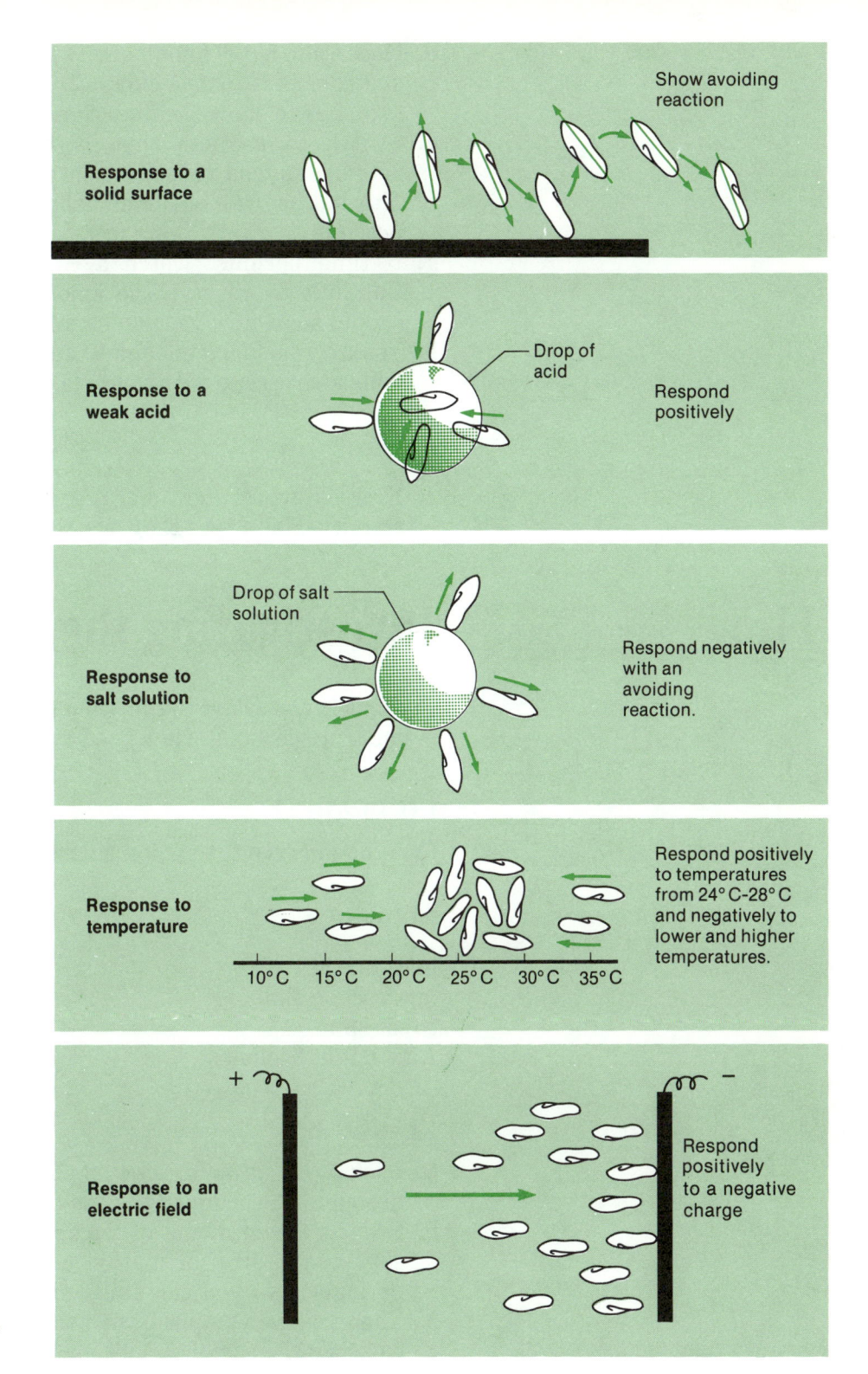

Fig. 11-11 Like all living things, paramecia respond to stimuli.

Section 11.4 139

- d. Once you have found one, study its locomotion closely. Make sketches of its shape every 30 s for 2 or 3 min. Put arrows on the sketches to show the direction in which the cytoplasm is moving.
- e. Switch to medium then high power. Look for the nucleus, cytoplasm, food vacuoles, contractile vacuole, and pseudopods.
- f. Make a diagram of the amoeba you saw. Label as many parts as possible.
- g. If time permits, watch a contractile vacuole until it bursts. Continue to observe the amoeba for a few more minutes. You should see a new vacuole form.
- h. You may be lucky enough to see an amoeba feed. If you do, share your microscope with your classmates.

Discussion

1. Your sketches and laboratory notes are your writeup for this activity. Make sure they are complete.

11.5 ACTIVITY Investigating the Paramecium

Paramecia are easier to find than amoebas. However, they don't stay in one place for long. How good are you at "paramecium hunting"?

Problem

Can you observe the structure and behaviour of a paramecium?

Materials

microscope
microscope slide
cover slip
1.5% methyl cellulose
paper towel
lens paper
dropper
paramecium culture

Procedure

- a. Suck up a drop or two of the paramecium culture with the dropper. The scum on the top of the culture is best.
- b. Prepare a wet mount of the sample. Before you put on the cover slip, add a drop of methyl cellulose. This will slow down the paramecia so you can study them. Lower the cover slip slowly. Then you won't squash the paramecia.
- c. Observe the sample under low power. Make notes on the locomotion of paramecia.
- d. Zero in on one organism. Switch to medium and then to high power.
- e. Look for the macronucleus, cilia, oral groove, pellicle, cytoplasm, and contractile vacuoles.

f. Make a diagram of this paramecium. Label as many parts as possible.

Discussion

1. Your sketches and laboratory notes are your writeup for this activity. Make sure they are complete.

11.6 The Algal Protists

You may recall that algae are spread over three kingdoms. Some (the blue-green algae) are in the kingdom Monera (see Chapter 9). Some are in the kingdom Plantae (see Chapter 13). The rest are in the kingdom Protista. This section is about the algae in this last group, the **algal protists**.

The algal protists vary greatly in form and function. Yet they have much in common.
1. All are one-celled. (In some cases these cells are in colonies.)
2. Practically all contain chlorophyll.
3. All live in water or in wet land environments.
4. All have nuclei and organelles.

There are about 11 000 species of algal protists. Of these, about 500 species are **coloured flagellates**. About 1000 species are **dinoflagellates**. And most of the rest are **diatoms**. Let's take a brief look at these three groups of algal protists.

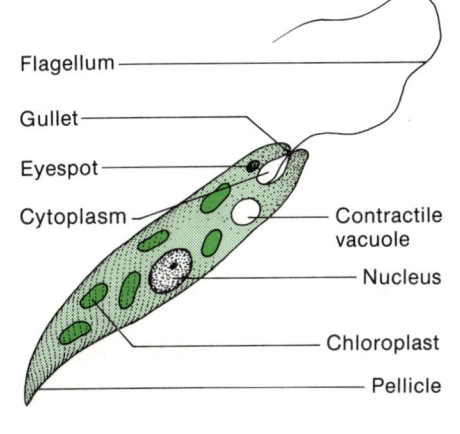

Coloured Flagellates

Characteristics These one-celled organisms are active swimmers (Fig. 11-12). They move themselves with one or two **flagella**. That's why they are called flagellates. They are called *coloured* flagellates because they are green, due to the **chlorophyll** they contain. If you look back to Figure 11-3, you will see that some protozoans are also flagellates. However, they are not coloured.

Each cell is covered by an elastic **pellicle** like that of the paramecium. Food is usually taken in at the **gullet**.

Most species have an **eyespot**. It is sensitive to light. It is usually red.

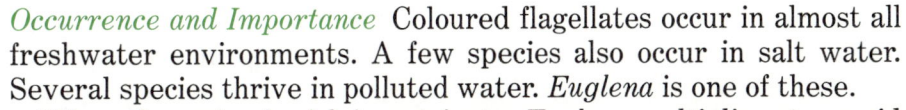

Occurrence and Importance Coloured flagellates occur in almost all freshwater environments. A few species also occur in salt water. Several species thrive in polluted water. *Euglena* is one of these.

When the water is rich in nutrients, *Euglena* multiplies at a rapid rate. It does so by binary fission. Sometimes the water turns bright green because of all the *Euglena*. Such a state is called an **algal bloom**. Sometimes the water turns red or brown. This is caused by *Euglena*'s red eyespots.

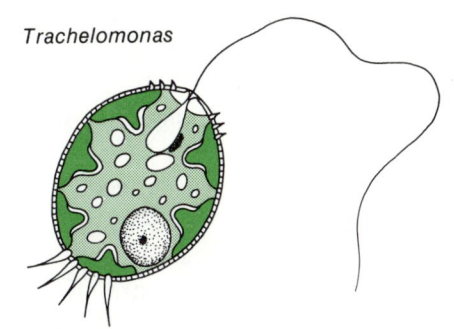

Fig. 11-12 Two typical coloured flagellates. What do they have in common?

Euglena (labels: Flagellum, Gullet, Eyespot, Cytoplasm, Contractile vacuole, Nucleus, Chloroplast, Pellicle)

Trachelomonas

Peridinium

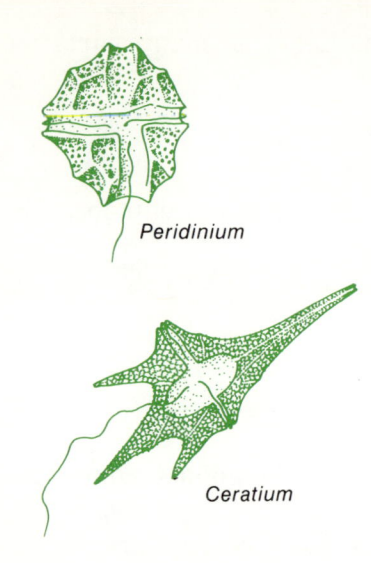

Ceratium

Fig. 11-13 Two typical dinoflagellates

Coloured flagellates are often the first step in food chains in polluted water. They store energy from the sun through photosynthesis. Then they are eaten by small animals. These animals, in turn, are eaten by fish. Therefore, without the flagellates, the water would contain fewer fish. Also, the flagellates add oxygen to the water.

Dinoflagellates

Characteristics The dinoflagellates [die-no-FLA-jell-ates] are one-celled (Fig. 11-13). They use *two* **flagella** to move about.

These organisms have a true cell wall. Like plant cells, it is made of **cellulose**. The cell wall is made of overlapping plates. This often makes the cell wall look like a suit of armour. It also places dinoflagellates among the most beautiful of all organisms to observe under the microscope.

Dinoflagellates contain chlorophyll and some orange colour as well. Therefore they usually look brown.

Occurrence and Importance Dinoflagellates are common in both fresh water and the oceans. The ocean species are most common in warm regions.

Dinoflagellates are the second most important algae in the world. They are the first step in many ocean and lake food chains. As a result, they help to produce many of the fish we eat by providing food for them. They also add oxygen to the atmosphere.

Many species of dinoflagellates form algal blooms easily. In some blooms there are 20 000 000 dinoflagellates in just one litre of water! The water often turns reddish during such blooms. As a result, these blooms are called "red tides". They are common along all the shores of North America.

Some dinoflagellate blooms produce deadly toxins. Thousands of millions of fish are killed each year by some of these toxins. Other toxins build up in shellfish such as clams and oysters. Humans can die if they eat these shellfish.

Diatoms

Characteristics Like dinoflagellates, diatoms are beautiful cells to study. They come in all kinds of interesting shapes (Fig. 11-14). Diatoms are one-celled. Most species cannot move. Those that can move do so by sliding along on a solid surface.

Diatoms have cell walls. The walls are very hard since they contain **silica** (the glassy substance in sand).

Occurrence and Importance Diatoms are the most important of all algae. In fact, they are among the most important of all living things. They are the first step in many food chains in the oceans. As a result, they help to produce much of our food. They also make much of the earth's oxygen.

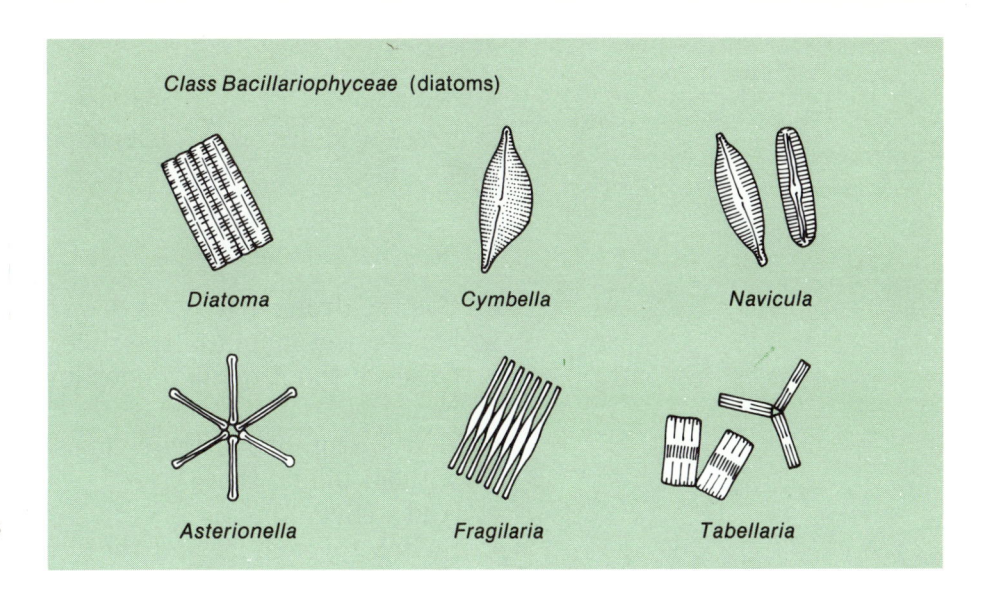

Fig. 11-14 Some typical diatoms

Diatoms occur in fresh water, salt water, and wet land environments. Next to bacteria, there are more diatoms on earth than any other type of organism.

When diatoms die, everything decays except the silica cell walls. This material settles to the bottom of the lakes or ocean. Over the years it forms a deep layer of fine solid. This solid is called **diatomaceous** earth [die-ah-tom-AY-shus]. We now mine diatomaceous earth from areas that used to be lakes and seas. It is used for filtering sugar, gasoline, and water. It is also used as an abrasive in polishing preparations. It is even used in some types of insulation.

Section Review

1. What do all algal protists have in common?
2. Name 3 main groups of algal protists.
3. Make a summary of the characteristics, occurrence, and importance of coloured flagellates.
4. Make a summary of the characteristics, occurrence, and importance of dinoflagellates.
5. Make a summary of the characteristics, occurrence, and importance of diatoms.

11.7 ACTIVITY Investigating Some Algal Protists

In this activity you will study the structure of *Euglena*, a coloured flagellate. Then, if time permits, you will look at some diatoms.

Problem

Can you find and draw some algal protists?

Materials

microscope	paper towel	1.5% methyl cellulose
microscope slides (2)	lens paper	*Euglena* culture
cover slips (2)	dropper	diatom culture

Procedure A *Euglena*: A Coloured Flagellate

a. Use the dropper to get a drop of the *Euglena* culture.
b. Make a wet mount of the slide. But do not add the cover slip.
c. Observe the *Euglena* under low power. Make notes on how they move.
d. Add a drop of methyl cellulose. It will slow down the *Euglena* without killing them.
e. Add a cover slip.
f. Switch to medium and then to high power. Study one *Euglena* closely.
g. Make sketches of *Euglena* about every 30 s. Do this until you have 5 or 6 sketches. What happened to the shape of the cell? How does *Euglena* move?
h. Make a large drawing of *Euglena*. Draw in and label these parts (if you can find them): pellicle, chloroplast, nucleus, contractile vacuole, gullet, eyespot, flagellum, front end, rear end.

Procedure B Diatoms

a. Prepare a wet mount of the culture.
b. Observe the culture under low power. Scan the culture and note the number of different species.
c. Study a cell of each species under medium and then high power. Sketch the shape of the cell. Draw in any features you can see.

Discussion

1. Your laboratory notes and drawings are your writeup for this activity. Make sure that they are complete.

Main Ideas

1. Protists are the second simplest kingdom of living things.
2. There are two kinds of protists: protozoans and algae.
3. Protozoans are one-celled organisms that move with pseudopods, cilia, or flagella.
4. The amoeba and paramecium are common protozoa.
5. Algal protists are one-celled organisms containing chlorophyll.
6. Coloured flagellates, dinoflagellates, and diatoms are three main groups of algal protists.

Glossary

cilia	SILL-ee-ah	short hairs on the surface of some protozoans and various other cells
contractile vacuole		a structure that pumps water out of some protists
diatom		an algal protist with a silica cell wall
dinoflagellate	die-no-FLA-jell-ate	an algal protist with a cellulose cell wall
flagellum	fla-JEL-um	a hair-like structure that moves protists
pellicle		the elastic membrane around some protists
protist		a one-celled organism; usually has a nucleus and organelles
protozoan	pro-toe-ZO-an	a one-celled animal-like organism
pseudopod	soo-doe-pod	false foot of an amoeba

Study Questions

A. True or False

Decide whether each of the following statements is true or false. If the sentence is false, rewrite it to make it true. (Do not write in this book.)
1. All protists contain chlorophyll.
2. Protozoans reproduce by binary fission.
3. An amoeba moves by using pseudopods.
4. A paramecium feeds by using its contractile vacuoles.
5. Diatoms are the most important algae on earth.

B. Completion

Complete each of the following sentences with a word or phrase that will make the sentence correct. (Do not write in this book.)
1. There are two kinds of protists, ▓▓▓▓ and ▓▓▓▓ .
2. Reproduction of a paramecium is directed by the ▓▓▓▓ .
3. Coloured flagellates are green because they contain ▓▓▓▓ .

4. Dinoflagellates have a cell wall made of ▮▮▮▮ .
5. Diatoms have a cell wall made of ▮▮▮▮ .

C. Multiple Choice

Each of the following statements or questions is followed by four responses. Choose the correct response in each case. (Do not write in this book.)

1. A paramecium moves by using
   a) cilia   **b)** flagella   **c)** pseudopods   **d)** the pellicle
2. An algal bloom is
   a) a flower that forms on algae   **c)** a sign of clean water
   b) a rapid growth of algae   **d)** a swelling of an algal cell
3. Fish are often killed by
   a) diatom blooms   **c)** dinoflagellate blooms
   b) protozoan blooms   **d)** coloured flagellate blooms
4. Without diatoms, the oceans would
   a) be much healthier   **c)** lose their bluish colour
   b) have more oxygen   **d)** support fewer fish

D. Using Your Knowledge

1. Amoebas that live in salt water have no contractile vacuole. Why? (Hint: Think back to osmosis in Chapter 6.)
2. **a)** In what ways is a paramecium like an amoeba?
   b) In what ways does a paramecium differ from an amoeba? Do any of these differences give the paramecium an advantage over the amoeba? Explain your answer.
3. When it comes to feeding, coloured flagellates like *Euglena* have an advantage over protozoans and the other algal protists. What is this advantage?
4. *Euglena* is studied by biologists who are seeking a link between the plant and animal kingdom. Why?

E. Investigations

1. Several serious diseases are caused by protozoans. Among them are amoeboid dysentery, malaria, and African sleeping sickness. Select one of these diseases. Find out its causes and effects, as well as methods of treatment.
2. Find out how to make a hay infusion culture of protozoans. Make one. Then do a study of the protozoans in it.
3. Collect some pond water. Include some plants and bottom material such as decaying leaves. Do a study of the protists in the pond water.

12 Fungi

12.1 What Are Fungi?
12.2 The Club Fungi
12.3 Activity: Structure of a Mushroom
12.4 The Black Mould Fungi
12.5 Activity: Investigating Bread Mould
12.6 The Sac Fungi
12.7 Activity: Investigating Yeast

How many of these fungi have you seen?

- mushrooms
- puffballs
- shelf fungi
- wheat rust
- corn smut
- mildews
- slime moulds
- yeast
- bread mould
- white pine blister rust
- apple scab
- *Penicillium*
- blue-green moulds
- water moulds

Just 3 or 4, you say? Well, that's not bad. However, there are 80 000 species of **fungi** (singular: **fungus**). And they are among the most important living things on earth. Perhaps we should get to know them better.

12.1 What Are Fungi?

Characteristics

Take a moment and look at the photographs in this chapter. These are all photographs of fungi. They don't look much alike, do they? Yet fungi have two important common characteristics.

1. *All fungi lack chlorophyll.* They cannot make their own food. Therefore they must get it from an outside source. Fungi get their food in two ways (Fig. 12-1). Some fungi are **parasites**. They get their food from *living* organic matter. The remaining fungi are **saprophytes** (decomposers). They get their food from *non-living* organic matter such as dead plants, dead animals, and animal wastes.

    Fungi feed by secreting enzymes into the food. The enzymes break the food down into small particles. These are then absorbed by the fungus and used for nutrition.

2. *Most fungi are made of hyphae* (singular: hypha). Each **hypha** [HIGH-fah] is a hair-like structure. The **hyphae** [HIGH-fee] often appear as a fuzzy mass of separate tiny hairs. Moulds are an example. Sometimes the hyphae are woven together to form a fleshy mass. Mushrooms are an example.

```
                    ALL FUNGI
                   /        \
              PARASITES    SAPROPHYTES
           Feed on living   Feed on non-living
           organic matter   organic matter
```

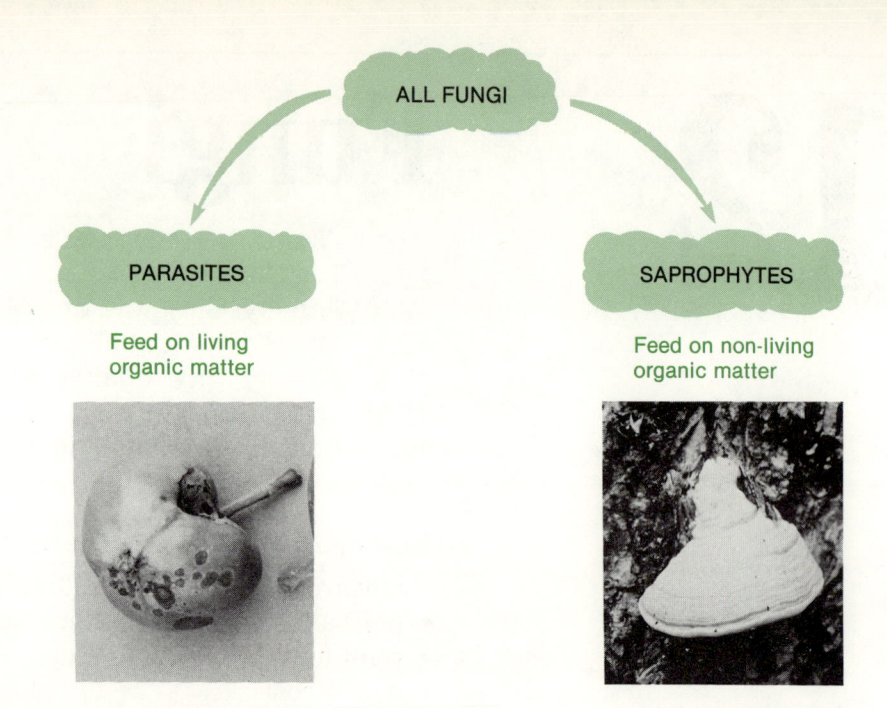

Fig. 12-1 Fungi cannot make their own food.

Occurrence

Fungi live in almost all environments. However, they are found most often in moist, organic-rich areas. Some species live in water. Some live on dead trees. Some live on soil. And some live on living plants and animals, even humans. Fungi must have organic matter to survive. That organic matter may be a dead tree, a slice of bread, an orange, or a human foot.

Importance

Some fungi are important because they are helpful. Others are important because they are harmful. Let's take a brief look at the importance of fungi. More will be said about this throughout the chapter.

Decomposers Many fungi are saprophytes. They feed by decomposing (breaking down) non-living organic matter. By doing this, they return valuable substances to the soil. These substances are now available to help plants grow. Fungi share with bacteria this important role of decomposers.

Unfortunately not all decomposer action is helpful. Fungi also decompose foods, leather, cloth, wood, and paper.

Parasites Many serious diseases of plants and animals are caused by parasitic fungi. For example, most grain crops (corn, wheat, oats, barley) can be attacked by fungi. Also cattle, horses, and even humans can be attacked by the ringworm fungus.

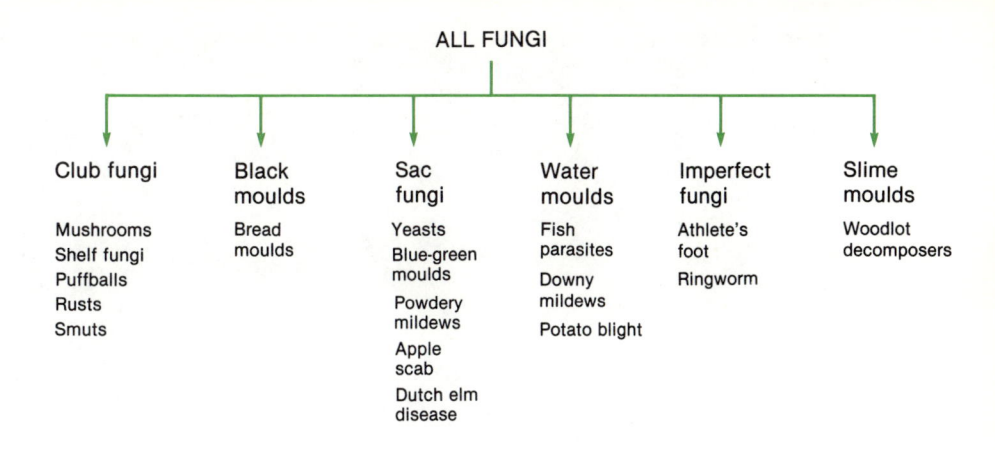

```
                              ALL FUNGI
        ┌──────────┬──────────┬──────────┬──────────┬──────────┐
        ↓          ↓          ↓          ↓          ↓          ↓
    Club fungi  Black      Sac         Water      Imperfect   Slime
                moulds     fungi       moulds     fungi       moulds
```

Club fungi	Black moulds	Sac fungi	Water moulds	Imperfect fungi	Slime moulds
Mushrooms	Bread moulds	Yeasts	Fish parasites	Athlete's foot	Woodlot decomposers
Shelf fungi		Blue-green moulds	Downy mildews	Ringworm	
Puffballs		Powdery mildews	Potato blight		
Rusts		Apple scab			
Smuts		Dutch elm disease			

Fig. 12-2 The 6 main groups of fungi

Food for Us Most people have eaten mushrooms. Many other fungi are also edible. Among them are puffballs, morels, truffles, and yeasts. Also, many cheeses owe their special flavour and colour to moulds.

Industrial Uses The yeasts are widely used in the baking industry. They are also used in the making of alcoholic beverages. Still other fungi make antibiotics, vitamins, and other important chemicals.

Classification of Fungi

Fungi are divided into six main groups. Figure 12-2 gives an overview of these groups. In the rest of this chapter you will study fungi from some of these groups.

Section Review

1. What are the two main characteristics of fungi?
2. Some fungi are parasites; others are saprophytes. What does this mean?
3. In what kinds of environments are fungi usually found?
4. Make a summary of the importance of fungi.

12.2 The Club Fungi

About 14 000 species of fungi are **club fungi** (Fig. 12-3). They are called club fungi because they make their spores at the end of club-shaped stalks (see Fig. 12-4). You have probably seen some club fungi — mushrooms, puffballs, and shelf fungi. They are common decomposers in woodlots. But, unless you live on a farm, you may not have seen rusts and smuts. These parasites can cause serious crop damage. Rusts attack wheat, coffee, and white pine trees. Smuts attack corn, oats, barley, rye, and some pasture grasses.

In this section you will learn about the mushroom, one of the most important club fungi.

White pine blister rust | Shelf fungus | Corn smut
Wheat rust | Mushroom | Puffball

Fig. 12-3 Some club fungi. They don't look much alike. But all form spores on club-shaped stalks.

Structure of a Mushroom

You probably have not seen the largest part of a mushroom. It is in the organic matter in which the mushroom is growing (Fig. 12-5). It is a tangled mass of **hyphae**. Even for just one mushroom, this mass can spread several metres through the organic matter. These hyphae secrete enzymes that break down the organic matter into small particles. Then the hyphae absorb these particles. This is how the mushroom gets its food.

The part of the mushroom above ground consists of a **stalk** and a **cap**. Under the cap are the **gills**. The gills are very delicate. Therefore they are protected by a **veil** as the mushroom is pushing up through the ground (see Fig. 12-5). When the cap grows, the veil breaks. This leaves a **ring** on the stalk.

The gills produce **spores**. The spores are on the ends of **club-shaped stalks**. When the spores are mature, they are shot out of the gills. They drift in the wind. If they land where conditions are suitable, they germinate. And a new mushroom bed begins.

A single mushroom can make over two thousand million (2 000 000 000) spores! These are shot into the air from time to time at the rate of a few million per minute for several days.

150 Chapter 12

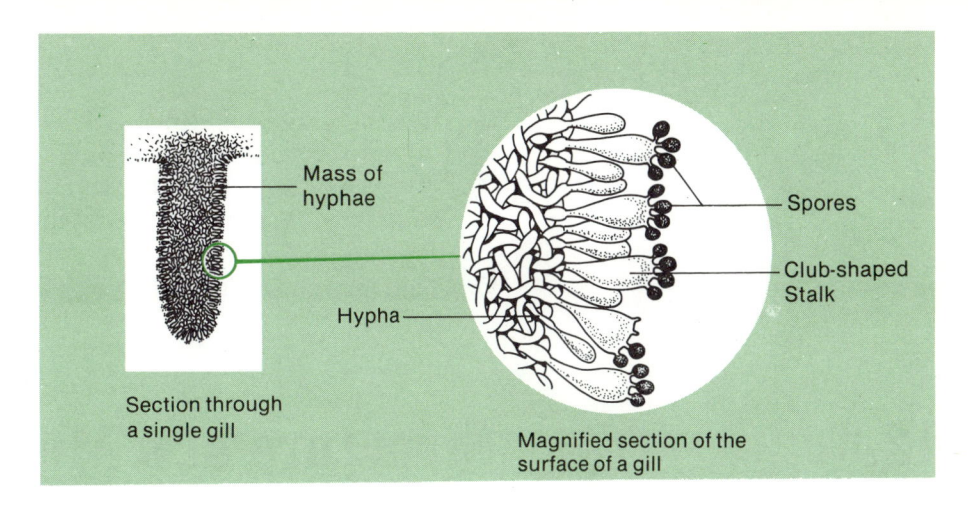

Fig. 12-4 Structure of the gills of a mushroom

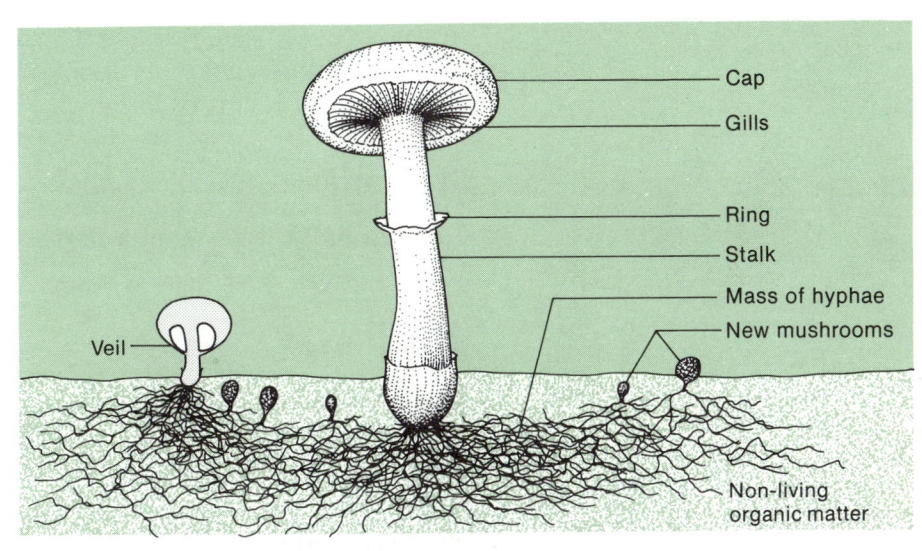

Fig. 12-5 Structure of a mushroom

Fig. 12-6 Some poisonous mushrooms look very much like some edible ones.

Edible and Poisonous Mushrooms

CAUTION: Never eat any mushroom or other fungus unless you are absolutely certain it is edible.

About half of all mushrooms are edible. Of course, this means the rest are poisonous or distasteful. The most poisonous mushrooms belong to the genus *Amanita*. Small amounts of these mushrooms cause hallucinations and even death. In fact, the most poisonous of all fungi is *Amanita verna*, the "destroying angel". One mushroom can kill several people. The victims suffer cramps, vomiting, and diarrhea for 3 to 4 d before dying.

There are no simple tests or rules that you can use to tell poisonous from edible mushrooms (Fig. 12-6). You must be able to identify the species. Therefore, unless you are a mushroom expert, you should eat only one kind of mushroom — the kind you buy in stores.

A large industry exists to grow edible mushrooms. They are eaten "as is", put in soup, and added to pizzas. They are a good source of many minerals and vitamins.

Section 12.2

Section Review

1. How did club fungi get their name?
2. Name three types of club fungi that are decomposers in woodlots.
3. Name two types of club fungi that are plant parasites.
4. Make a summary of the structure of a mushroom. Include a diagram.
5. What precaution should you take if you plan to eat some mushrooms?

12.3 ACTIVITY Structure of a Mushroom

In this activity you will use a hand lens and microscope to study a common edible mushroom. Look back to Section 12.2 when you need help with any terms used here.

Problem

Can you find the parts of a mushroom as shown in Figures 12-4 and 12-5?

Materials

fresh mushroom (edible)	paper towel
scalpel	lens paper
hand lens	dropper
microscope	forceps
microscope slides (2)	*optional:* prepared slide
cover slips (2)	of mushroom gills

Procedure

a. Copy Table 12-1 into your notebook.
b. Hold the mushroom in your hand. Look at it with the hand lens. Record in your table a description of the stalk, cap, ring, and gills.
c. Break the stalk. Look at the broken area with the hand lens.
d. Using the forceps, remove a small piece of the stalk from the broken area. Make a wet mount of this piece. Look at it under low, medium, and high power with your microscope. Write a description of what you saw.
e. Make a cut from top to bottom through the cap. Look at the cut area with the hand lens.
f. Using the forceps, remove a small piece of the cap from the surface. Study it as described in step (d).
g. Using the forceps, remove a piece of gill. Study it as described in step (d). Try to find the club-shaped stalks and the spores.
   Note: Your teacher may give you a prepared slide of gills to study.
h. Hold the cap of a mature mushroom over a microscope slide. Tap

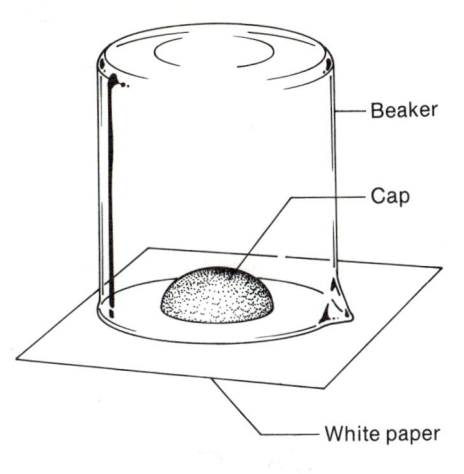

Fig. 12-7 Making a spore print of a mushroom

it gently. This should knock out some spores. Look at them under low, medium, and high power. Draw one spore.

i. (Optional) If enough mushrooms are available, make a **spore print** of the mushroom as follows: Cut off the cap of a fresh mature mushroom. Place it, gills down, on a piece of white paper. Cover it with a large beaker or jar (Fig. 12-7). Let it sit untouched for a day or two. Then remove the beaker. Now carefully lift the cap from the paper. Look at the spore print with the hand lens. Record your observations.

Table 12-1 Parts of a Mushroom

Part	Description
Stalk	
Cap	
Ring	
Gills	

Discussion

1. What is each of the following parts made of? the stalk; the cap; the gills.
2. Why are mushrooms called club fungi?

12.4 The Black Mould Fungi

About 1000 species of fungi are **black moulds**. They occur in most land environments. They are excellent decomposers.

You have likely seen several black moulds. Bread mould is probably the one you know best (Fig. 12-8). This fungus attacks some fruits, as well as bread. This section describes this important black mould.

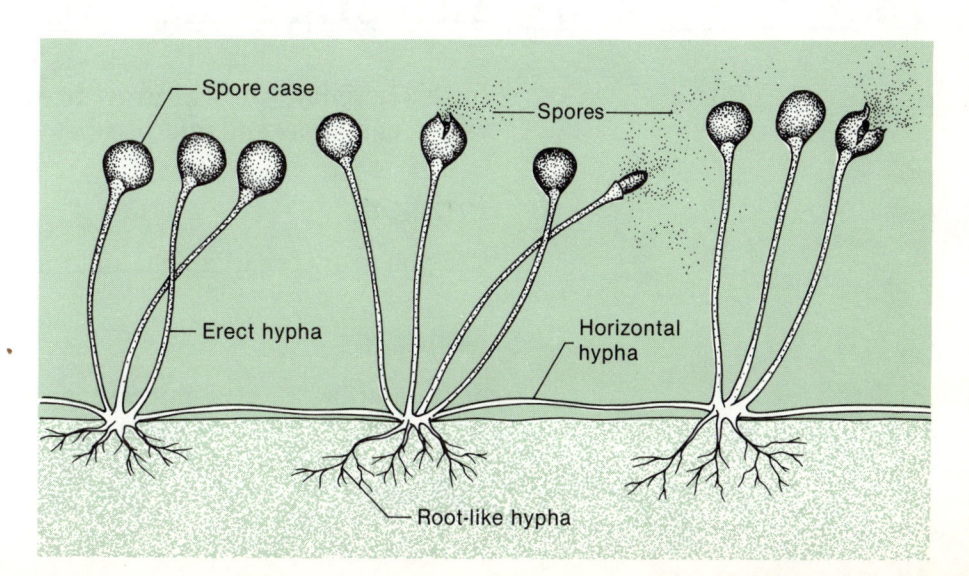

Fig. 12-8 The structure of bread mould. Note the three types of hyphae: root-like, erect, and horizontal.

Section 12.4   153

Structure of Bread Mould

The common bread mould can be found in the upper centimetre of almost any kind of soil anywhere on earth. Its spores are in the air at most times. It is a **saprophyte** and grows rapidly on moist non-living matter. It is also a **parasite** at times. It attacks fruits such as grapes, strawberries, and cantaloupes.

The mould first appears as a fluffy mass of fine white threads. When the spores mature, the mass turns black. As a result, such moulds are called **black moulds**. Within a few days, a piece of bread or fruit can be completely covered with this fungus.

Bread mould consists of a branching mass of **hyphae** (Fig. 12-8). There are three types of hyphae:

1. **Horizontal hyphae** run across the surface of the substance on which the mould is growing. They help spread the mould.
2. **Root-like hyphae** develop along the horizontal hyphae. These hyphae grow down into the substance. They secrete enzymes that break it down. Then they absorb the nutrients the mould needs. This, of course, destroys the substance. We say the bread or fruit "spoiled". It was actually digested by the mould.
3. **Erect hyphae** grow up from the horizontal hyphae. Each one of these has a **spore case**. Each spore case contains up to 70 000 **spores**. The spore cases are white at first. They turn black when they are mature and ready to release spores.

Section Review

1. Why are these moulds called "black moulds"?
2. There are three types of hyphae in bread mould. Describe each type and state its function.
3. Explain how bread mould spoils bread and fruits.

12.5 ACTIVITY Investigating Bread Mould

In this activity you will grow bread mould. Then you will study it with a hand lens and microscope.

Problem

Can you grow bread mould and find the parts shown in Figure 12-8?

Materials

piece of bread	microscope slide	dropper
petri dish	cover slip	forceps
hand lens	lens paper	dissecting needle
microscope	paper towel	

Procedure A Growing Bread Mould

a. Line the bottom of a petri dish with paper towelling.
b. Wet the paper. Then pour off any excess water.
c. Wipe a piece of bread over a table top or floor to collect bread mould spores.
d. Place the bread on the paper towelling in the petri dish.
e. Add several drops of water to the bread. *Do not soak it.* Place the cover on the petri dish.
f. Store the petri dish in a dark cool place for several days. Look at it every day with a hand lens. Record your observations. *Do not remove the cover at any time.* If you do, you will destroy the hyphae.
g. Continue your daily study of the mould until it turns black.

Procedure B Structure of Bread Mould

a. Carefully remove the cover from the petri dish.
b. Using the forceps, remove a sample of the mould. Try to get all the parts shown in Figure 12-8.
c. Prepare a wet mount of this sample. Spread the mass of hyphae with a dissecting needle. Then you will be able to see them better.
d. Copy Table 12-2 into your notebook.
e. Look at the mount under low, medium, and high power. Find and describe (in your table) each of the following: spores, spore case, erect hyphae, horizontal hyphae, root-like hyphae. (In order to see root-like hyphae, you may have to include a piece of bread in your sample.)

Table 12-2   Structure of Bread Mould

Part	Description
Spore	
Spore case	
Erect hyphae	
Horizontal hyphae	
Root-like hyphae	

Discussion

1. If bread is left exposed to the air for an hour, it will often go mouldy. Why?
2. Describe the development of the mould from the day it began until it turned black.
3. Describe the odour given off when you took off the lid. What do you think caused it?
4. What evidence did you see that bread mould is a saprophyte?

12.6 The Sac Fungi

About 15 000 species of fungi are **sac fungi**. They are called sac fungi because their spores are in **sacs**. You have probably seen many sac fungi. Some common ones are yeasts, blue-green moulds, morels, and powdery mildews. Apple scab and many other plant diseases are also sac fungi. This section looks at yeasts in detail and then briefly describes other sac fungi.

The Yeasts

Yeasts are very common in substances that contain a great deal of sugar. Thus decaying fruits and grains contain yeasts. Fruit orchards and vineyards are excellent habitats for these fungi. The soil and air near your home probably contain them.

Structure Yeast cells are very small. Just 1 g of the yeast you can buy in a grocery store contains over 6 000 000 000 cells! Still, these cells are larger than most bacteria. However, you can see little detail with an ordinary microscope.

Yeasts are one-celled, but **chains** of cells often form when the yeast is multiplying rapidly (Fig. 12-9). The cells are usually egg-shaped. Each cell has a **cell wall** made of cellulose. Within the cell wall are the **nucleus** and **cytoplasm**. The cytoplasm usually has one large **vacuole** and some smaller ones.

How Yeasts Get Food and Energy Most yeasts are saprophytes. They get their food from non-living organic matter. They use enzymes to break down sugars to get energy. They do this without oxygen. This type of respiration is called **fermentation**. Fermentation of sugars produces ethyl alcohol and carbon dioxide. Energy is released for life processes.

$$\text{Sugars} \xrightarrow{\text{Enzymes}} \text{Ethyl alcohol + Carbon dioxide + Energy}$$

Reproduction Yeasts reproduce by **budding** (Fig. 12-10). When a yeast cell is mature, a bulge called a **bud** forms on it. Mitosis occurs and one nucleus moves into the bud. The other nucleus stays in the parent cell. Budding often occurs so quickly that buds form on buds before they can break away from the parent cell. A **chain** of cells results (see Fig. 12-9).

Yeasts can also reproduce sexually. Two cells unite and exchange genetic information. Then **spores** form. The spores are held in a **sac**.

Importance Yeasts are used in baking, brewing, and wine-making. They are used because they act on sugars to form ethyl alcohol and carbon dioxide.

Baker's yeast gives off carbon dioxide in dough. Bubbles of this gas make the dough rise. The ethyl alcohol produced is driven off by the heat.

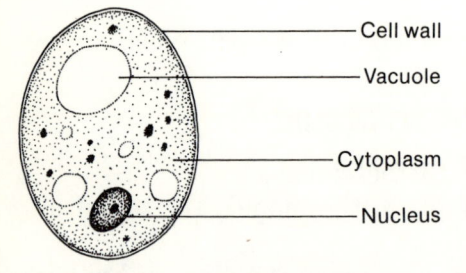

MATURE YEAST CELL

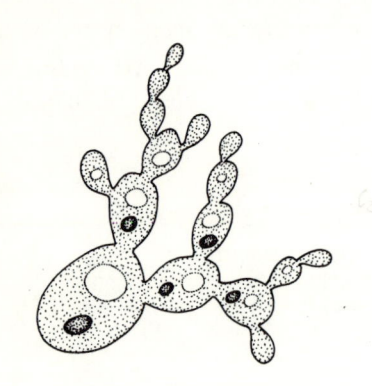

CHAIN OF YEAST CELLS

Fig. 12-9 Yeasts are one-celled. However, long chains of cells sometimes form.

Fig. 12-10 Budding of yeast

Fig. 12-11 Apple scab. Serious infections destroy the shape and stunt the growth of the apple.

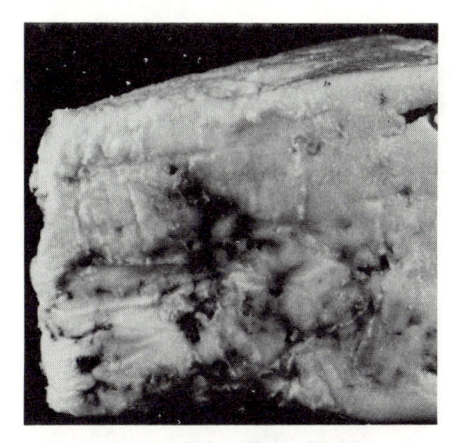

Fig. 12-12 Roquefort cheese. It is also called blue cheese because of the fungus that gives it its special taste.

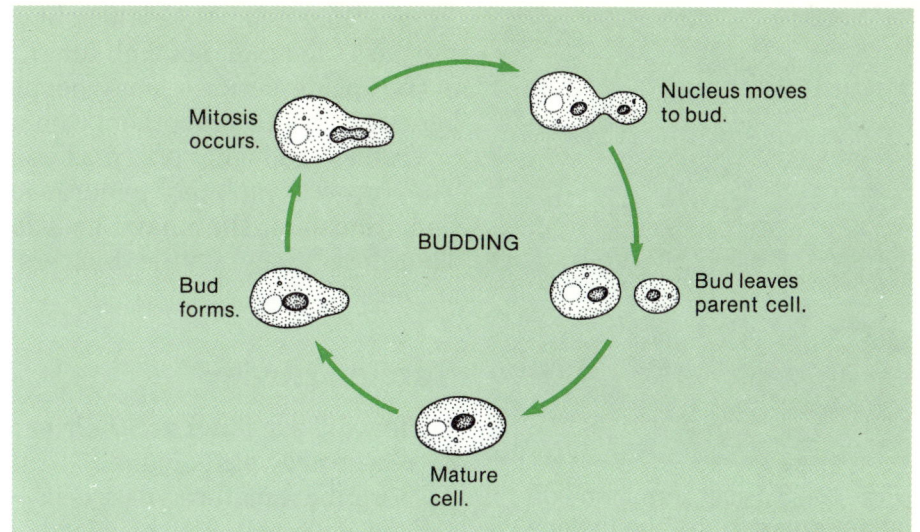

Brewer's yeasts act on sugars that are formed by the digestion of starches in grains such as barley. This forms ethyl alcohol. Wine-making yeasts act on sugars in grapes and other fruits to form ethyl alcohol.

Other Sac Fungi

Apple Scab Apple scab is a parasitic sac fungus. It destroys many apple crops (Fig. 12-11). It appears as scabs on the fruit and leaves. The scabs often crack the fruit. This lets bacteria in, and the fruit rots.

In most orchards, apple scab is controlled by spraying the trees with a **fungicide** (fungus killer). Just 4 or 5 sprayings would prevent most scabs. But consumers won't buy apples with any scabs at all. Therefore, many growers use up to 15 sprayings during the season.

Here's something for you to think about. A few scabby apples never hurt anyone. But fungicides are harmful to humans. Why do we insist on scab-free apples?

Blue-Green Moulds *Penicillium* is a well-known genus of blue-green moulds. One species makes the antibiotic penicillin. Another species, *Penicillium roqueforte* makes the special taste and colour of roquefort cheese (Fig. 12-12). The same species, however, destroys the taste of other cheeses.

Other blue-green moulds destroy leather, cloth, bread, meat, oranges, lemons, and other fruits (Fig. 12-13).

Powdery Mildews These fungi attack food crops such as wheat and barley. They also attack fruits such as grapes and apples. They even attack ornamental plants such as roses, and lilac trees (Fig. 12-14).

Powdery mildews grow rapidly if the leaves of the plant are damp. As a result, crop losses are often high in rainy years. You should never water roses, grapes, or lilacs in the evening, since the leaves could stay wet all night. This is an ideal environment for powdery mildews.

Section 12.6 157

Fig. 12-13 Blue-green moulds attack some fruits. The hyphae penetrate into the fruit. Then enzymes digest the fruit.

Edible Sac Fungi **Morels** look like mushrooms (Fig. 12-15). However, they are sac fungi, not club fungi. They are collected from the woods in the spring because many people like their special flavour. No one has been able to grow them commercially.

Truffles are highly prized as gourmet foods, particularly in Europe. They grow completely underground. As a result, they are hard to find. However, they have an odour that dogs and pigs can smell. Therefore many truffle hunters train dogs and pigs as "truffle sniffers".

Section Review

1. How did sac fungi get their name?
2. Where are yeasts found?
3. Describe a mature yeast cell.
4. How do yeasts get their food and energy?
5. What is budding?
6. Describe the importance of each of the following: yeasts, apple scab, blue-green moulds, powdery mildews.

12.7 ACTIVITY Investigating Yeast

In this activity you will grow yeast. You will also study the cells, budding, and fermentation.

Problem

How do yeast cells grow? What is fermentation?

Materials

test tube rack
test tubes (4)
10% sugar solution
granular yeast
iodine stain
methylene blue stain
limewater
gas collecting assembly

oil (e.g. cooking oil)
microscope
microscope slide
cover slip
dropper
lens paper
paper towel

Fig. 12-14 The gray patches on these leaves are a powdery mildew.

Procedure

a. Fill two test tubes about two-thirds full with 10% sugar solution. Label one test tube "Control" and the other "Experimental".
b. Add a pinch of granular yeast to the "Experimental" test tube. Shake the test tube to mix the yeast well with the sugar solution.

Fig. 12-15 The morel is highly prized for its flavour.

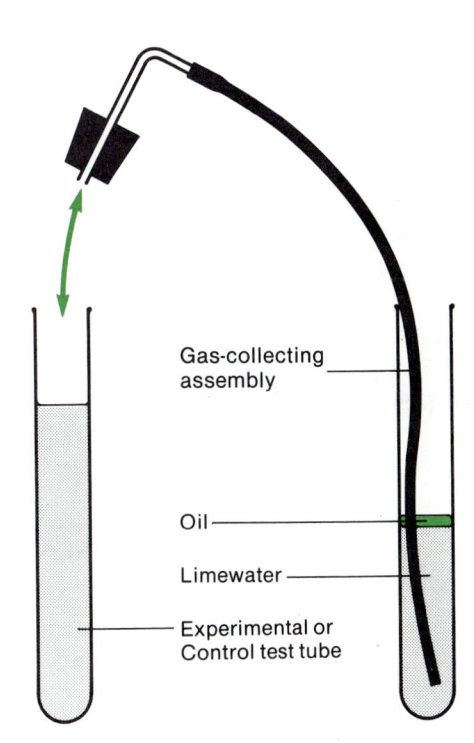

Fig. 12-16 Studying the properties of yeast

c. Look at a drop of the mixture from each test tube under low, medium, and high power. Note the structure of the cells.
d. Stain a fresh sample of the mixture from the "Experimental" test tube with iodine solution. Note any additional structures that you can see.
e. Repeat step (d) using methylene blue stain.
f. Connect each test tube to the assembly shown in Figure 12-16. The end of the rubber tubing should be covered by about 5 cm of limewater. Cover the limewater with a thin layer of oil. This will keep air away from the limewater.
g. Stand the test tubes in a test tube rack. Store them in a warm place, preferably at 25°C to 30°C. Note any changes that occur during the next two days. Record any changes in appearance and odour.
h. After two days, look at a drop of the mixture from the "Experimental" test tube under low, medium, and high power. Look at an unstained mount, a mount stained with iodine solution, and a mount stained with methylene blue. Draw any changes that took place in the appearance of the yeast cells.

Discussion

1. What is the function of the "Control" test tube in this activity?
2. a) Describe yeast cells as they appear in the unstained mount.
   b) What structures were made visible by methylene blue stain?
   c) What further structures were made visible by methylene blue stain?
3. a) Limewater turns cloudy in the presence of carbon dioxide gas. (Excess carbon dioxide may turn it clear again.) What do you conclude from the change that occurred in the limewater in this activity?
   b) What other evidence did you see to support your conclusion?
4. Describe the change in odour that occurred in this activity. What caused it?
5. What changes took place in the appearance of the yeast cells after two days?

Main Ideas

1. All fungi lack chlorophyll.
2. Most fungi are made of hyphae.
3. Some fungi are parasites; others are saprophytes.
4. Fungi are important decomposers of organic matter.
5. Some parasitic fungi cause serious plant and animal diseases.
6. Fungi are divided into six main groups: club fungi, black moulds, sac fungi, water moulds, slime moulds, imperfect fungi.

Section 12.7

Glossary

budding		a method of reproduction of yeast
fermentation		respiration without oxygen; usually forms ethyl alcohol and carbon dioxide
fungicide	FUN-ji-side	a fungus killer
fungus		an organism without chlorophyll and made of hyphae
hypha	HIGH-fah	a hair-like part of a fungus

Study Questions

A. True or False

Decide whether each of the following statements is true or false. If the sentence is false, rewrite it to make it true. (Do not write in this book.)

1. Fungi are in the plant kingdom.
2. All fungi are saprophytes.
3. Fungi are decomposers of organic matter.
4. All mushrooms are edible.

B. Completion

Complete each of the following sentences with a word or phrase that will make the sentence correct. (Do not write in this book.)

1. Fungi are made of hair-like structures called ▒▒▒▒ .
2. Mushrooms make their spores in ▒▒▒▒ .
3. When bread mould attacks living fruits it is said to be a ▒▒▒▒ .
4. Yeasts reproduce by two methods, ▒▒▒▒ and ▒▒▒▒ .
5. Fermentation releases energy and forms ▒▒▒▒ and ▒▒▒▒ .

C. Multiple Choice

Each of the following statements or questions is followed by four responses. Choose the correct response in each case. (Do not write in this book.)

1. A fungus is found growing on a slice of bread. This fungus is best called
    a) a sac fungus   **b)** a parasite   **c)** a saprophyte   **d)** a club fungus
2. The blister rust that attacks white pine trees is a
    a) club fungus   **b)** sac fungus   **c)** black mould   **d)** water mould

D. Using Your Knowledge

1. Describe the role of fungi as decomposers. Include both good and bad aspects.
2. Are most mushrooms parasites or saprophytes? How do you know?
3. Explain how bread mould destroys fruits and bread.

E. Investigations

1. Find out how mushroooms are grown commercially. Get some commercial mushroom culture and try to grow mushrooms.

   CAUTION: Do not try to grow wild mushrooms. You may grow some poisonous ones.

2. In Activity 12.5, you assumed that bread mould grows best in a moist, cool, dark place. Are these really the best conditions? Design and try an experiment to find out.
3. Get an apple that is infected with apple scab. (Wild trees are almost always infected.) Study the infection with a hand lens and microscope. Then read about apple scab in some books. Write a report of your findings. Include a description of apple scab, its effects, and methods of control.
4. Research the causes and effects of athlete's foot or ringworm. Also, find out how the disease can be cured and prevented.
5. Collect 3 or 4 different lichens. Write a report on the structure of a lichen.
6. Conduct a debate on the use of fungicides for controlling apple scab.
7. Write a paper of about 300 words on the causes, effects, life cycle, and control of one of these fungus diseases of plants: corn smut, wheat rust, white pine blister rust, apple scab, potato blight, Dutch elm disease.

Unit 4: The Plant Kingdom

CHAPTER 13
A Survey of the Plant Kingdom

CHAPTER 14
Reproduction in Flowering Plants

CHAPTER 15
Roots, Stems, and Leaves

CHAPTER 16
Growing Plants

The plant kingdom includes over 300 000 species of living things. What are the members of this kingdom like? In this unit you will discover that plants vary in appearance a great deal. But all plants have some features in common. For instance, most of them are able to use light energy to make their own food.

In this unit you will study the largest and most important groups of plants. You will learn about their structure and how they reproduce. You will also find out how easy it is for gardeners, farmers, and other "plant lovers" to grow and reproduce the plants they desire.

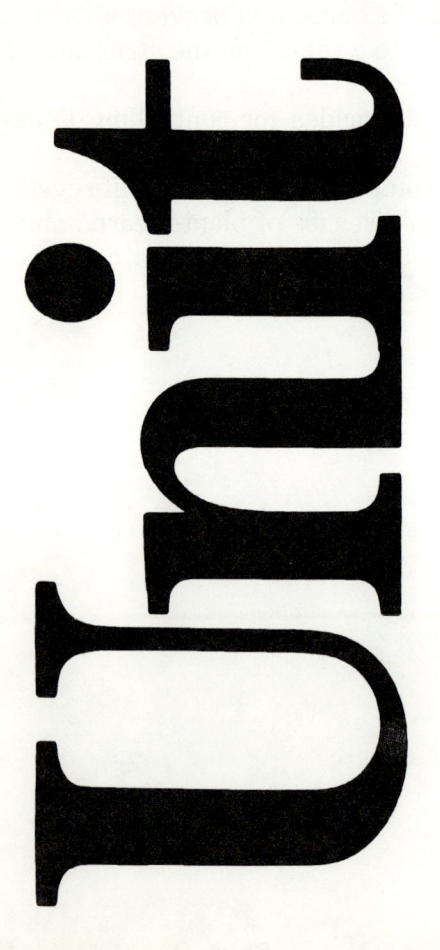

Fig. 13-0 You would find a great variety of plants in a scene like this one. Different types of plants grow in the forest, the swamp, and the lake.

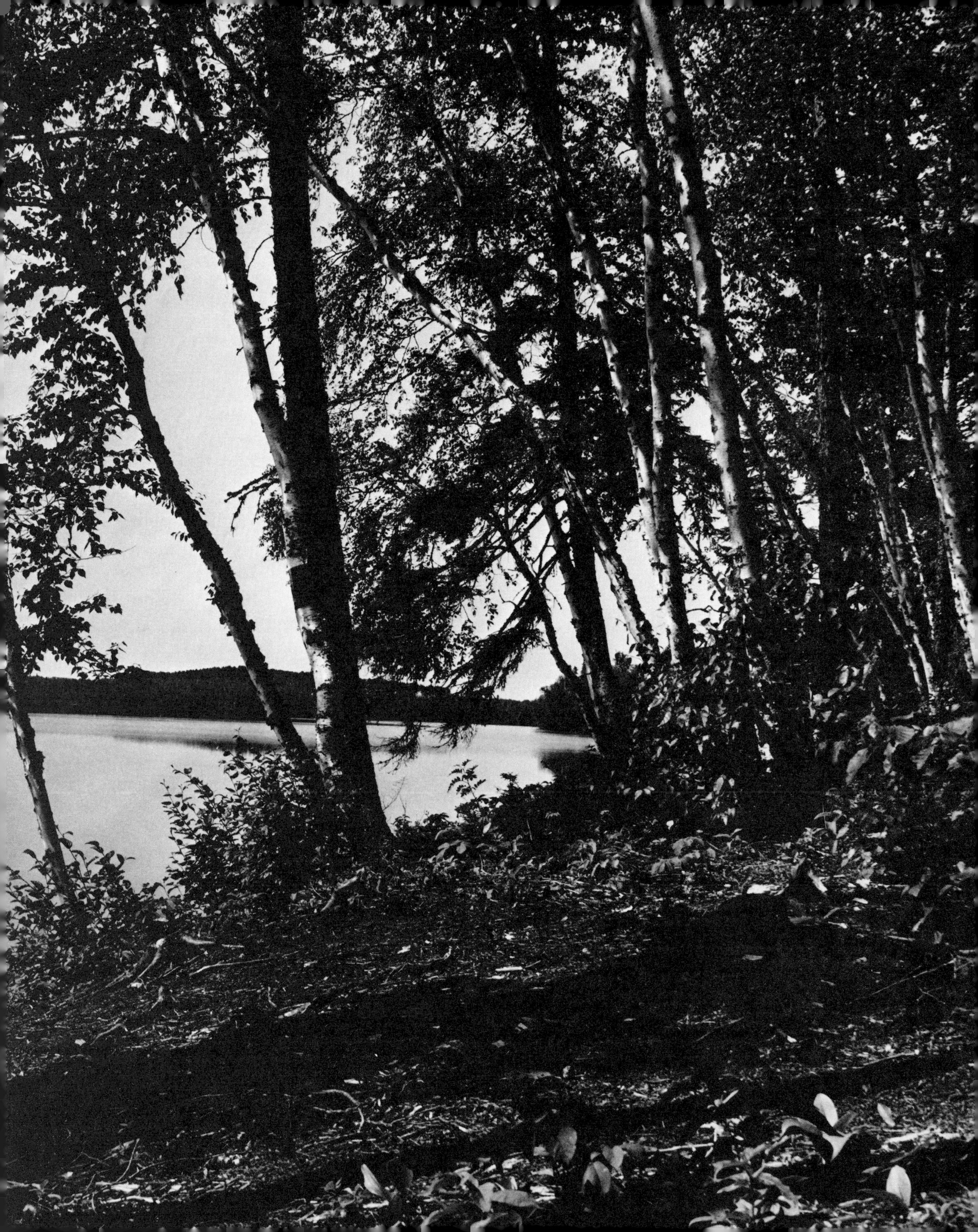

13 A Survey of the Plant Kingdom

13.1 What Is a Plant?
13.2 What Type of Plant Is It?
13.3 Activity: Green Algae
13.4 Activity: The Ferns
13.5 The Winners on Land: The Seed Plants

How do you know a plant is a plant? How can you tell one plant from another? The plant kingdom includes many peculiar types. A few, such as many algae, are small and simple. Others, such as some trees, are much larger than any other living thing on earth. Yet all plants have some features in common. In this chapter, you will be introduced to the many life forms that biologists call "plants".

13.1 What Is a Plant?

Even biologists argue about what is a plant and what isn't. Fungi, bacteria, and some protists used to be called "plants". But they are now classified in separate kingdoms. Some of the algae are included in the plant kingdom, but others are not. At least no one has ever tried to say that a maple tree isn't a plant! Ferns and mosses and roses are all members of the plant kingdom, too. The variety of types is truly amazing.

The Characteristics of Plants

What are the characteristics that separate plants from other types of organisms? Plants have all or most of the following chracteristics:
1. *Plants are made of cells.* Some simple plants are one-celled. Most are made of many cells. These cells may be all alike. Or, they may be specialized for different functions (Fig. 13-1).
2. *The cells contain specialized parts such as a nucleus.*
3. *Plant cells have walls that are made of* **cellulose**. This is a tough material. It keeps the cells rigid and protects their contents.
4. *Most plant cells contain a green pigment called* **chlorophyll**. The chlorophyll is found in small organelles called **chloroplasts**. Chlo-

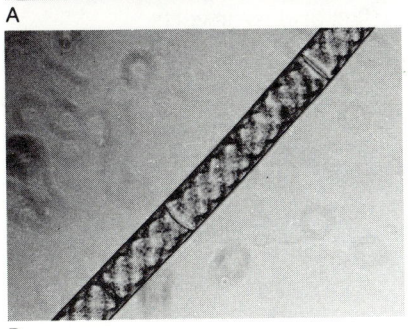

Fig. 13-1 Simple plants like the desmid (A) consist of only one cell. The green alga (B) contains several similar cells. Most flowering plants (C) contain many different cell types.

rophyll allows plants to use sunlight to produce food energy. This process is called **photosynthesis**.
5. *Most plants are anchored in place by roots or other structures.*
6. *Like other living things, plants are able to grow and reproduce.* Many reproduce **asexually**. This means that part of the plant, such as a leaf, breaks off, and develops into a new plant on its own. Most plants also reproduce **sexually**. This means that they produce sex cells called **gametes** [GAM-eets]. The gametes join together into one cell called a **zygote** [ZY-goat]. This cell then develops into a new plant by cell division.

Section Review

1. Describe three features shared by most plant cells.
2. Why is it important that most plants contain chlorophyll?
3. What is asexual reproduction?
4. What is a zygote?

13.2 What Type of Plant Is It?

Plants are grouped into ten phyla [FI-la] according to their structure and colour. Table 13-1 lists these ten phyla. It also lists the approximate number of species known for each phylum. Since many species probably haven't been discovered yet, these numbers cannot be definite.

Table 13-1 The Main Phyla of the Plant Kingdom

Common name of the phylum	Number of species
Green algae	6 000
Red algae	2 500
Brown algae	1 500
Stoneworts	100
Mosses and liverworts	24 000
Club mosses	1 100
Horsetails	30
Ferns	10 000
Conifers	550
Flowering plants	250 000

1. Green Algae

Some green algae exist as a single cell; others are many-celled. Some are made of long chains of cells. Most green algae look green because of the chlorophyll in their cells. Some types grow in freshwater streams and lakes. Others prefer the salt water of the oceans.

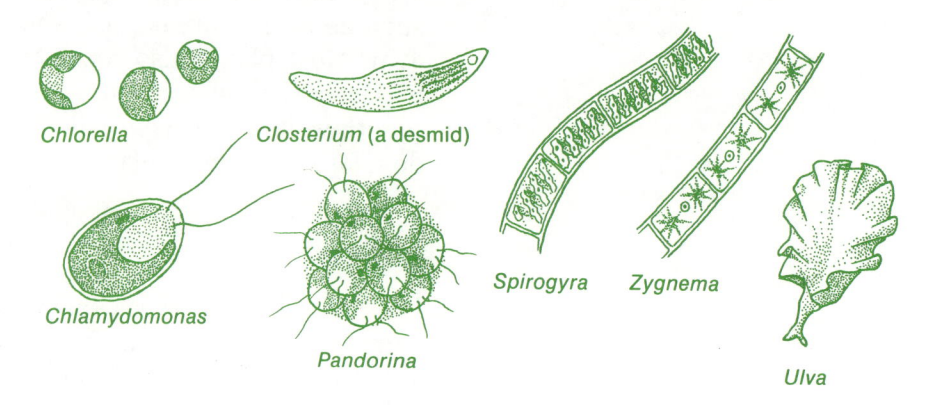

Chlorella *Closterium* (a desmid)

Chlamydomonas *Pandorina* *Spirogyra* *Zygnema* *Ulva*

2. Red Algae

Red pigments in the cells of these plants make them appear red or brownish in colour. Chlorophyll is also present, but its green colour cannot be seen. Red algae may be one-celled or many-celled. Some are very branched and feathery. Most grow in warm ocean waters. They may be attached to the rocks near the shore, or floating. They are often called **seaweeds**. Some species are harvested by people in North America, Europe, and Japan for food and other products. One substance that comes from red algae is used as a thickener in ice cream, toothpaste, and chocolate milk.

Chondrus *Porphyra* *Polysiphonia* *Corallina*

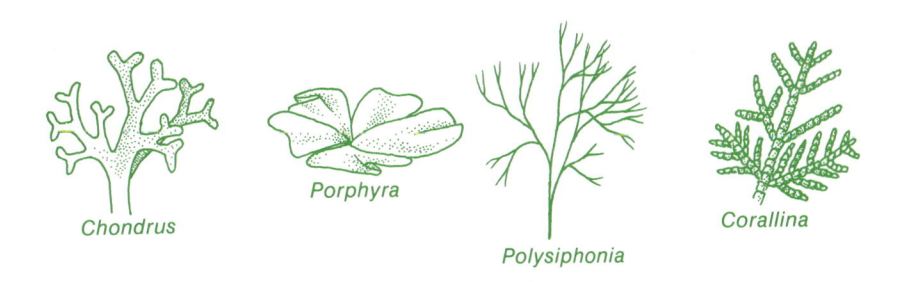

3. Brown Algae

Brown algae are called seaweeds too. The large ones, which may reach 30 m long, are known as **kelps**. They usually grow attached to rocks in cold oceans. Their cells contain brown pigments as well as chlorophyll.

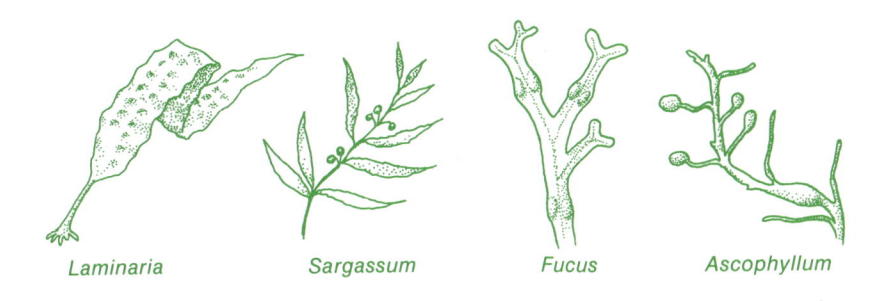

Laminaria *Sargassum* *Fucus* *Ascophyllum*

4. Stoneworts

Stoneworts are strange plants that live in freshwater streams and ponds. The main shoot has many circles of short branches on it. The plants are anchored to the mud or sand. Stoneworts often feel quite gritty to touch. That's because they are usually covered with pieces of lime.

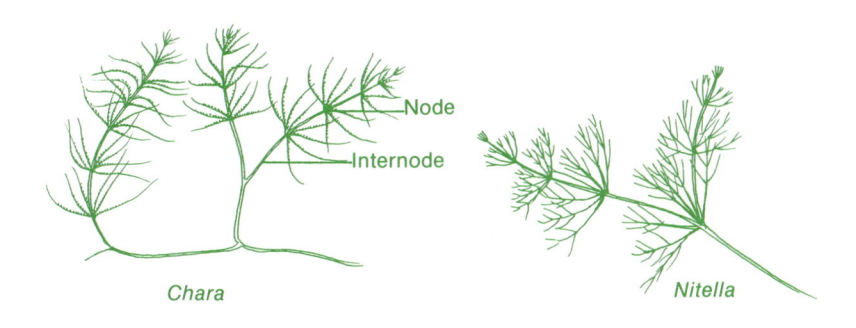

Chara (Node, Internode) *Nitella*

5. Mosses and Liverworts

Mosses and liverworts usually grow in damp shady places on rocks or soil. They are small greenish plants that are rarely more than 20

Section 13.2

cm in height. This is because they have no conducting tissue to carry water very far up from the ground. The plant body may be leafy in appearance, or flat and leathery. Biologists believe mosses were the first plants to grow on land.

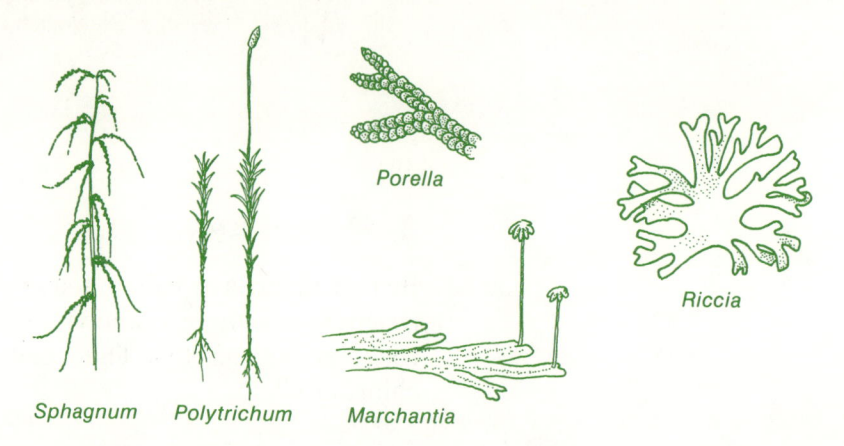

Sphagnum *Polytrichum* *Porella* *Marchantia* *Riccia*

6. Club Mosses

These are small leafy plants. They spread by horizontal stems that run along the surface of the ground, or just underneath it. The upright shoots are less than 40 cm tall. They are **evergreen**. Some of them have club-shaped cones at the tips of the shoots. These contain **spores**. The spores are released into the wind. When they land in a suitable place, they grow into new plants.

Club mosses grow in shady moist places. Some are bushy and look like tiny pine or cedar trees. For this reason, they are called ground pines and ground cedars.

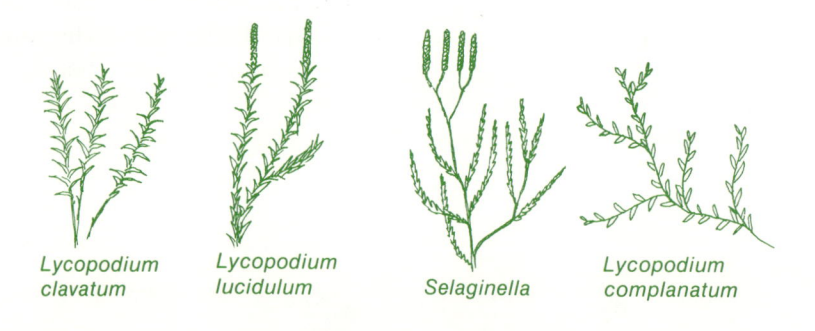

Lycopodium clavatum   *Lycopodium lucidulum*   *Selaginella*   *Lycopodium complanatum*

7. Horsetails

Horsetails grow in clumps near lakes and streams, and in damp fields and ditches. Most are about 20 cm to 50 cm tall. Horizontal stems that grow underground send up hollow, jointed shoots. The shoots often have circles of soft, feathery branches (resembling a horse's tail, perhaps!). In the spring, the tips of some shoots develop cone-like structures. These release spores for reproduction.

Horsetails feel gritty to touch. That's because their cells contain a hard substance called **silica**. (Sand has silica in it.) The pioneers used these gritty branches to scour (clean) out pots and pans. Thus, horsetails are also known as "scouring rushes".

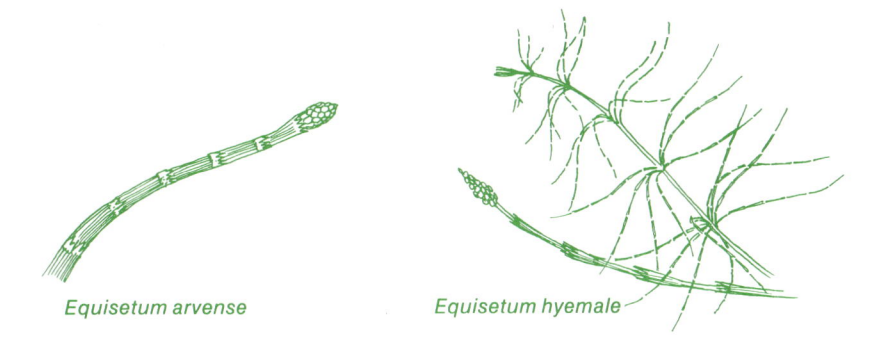

Equisetum arvense *Equisetum hyemale*

8. Ferns

Most ferns grow in shady, moist places such as forests and the cracks between rocks. They are leafy plants that are usually less than 1 m tall. A few tropical ones, known as "tree ferns", are much larger.

Ferns often have underground stems that send up new leaves every spring. The leaves are usually divided into many leaflets. On the backs of some of the leaves are tiny spore cases. They release spores into the air. The spores grow into new plants when conditions are right.

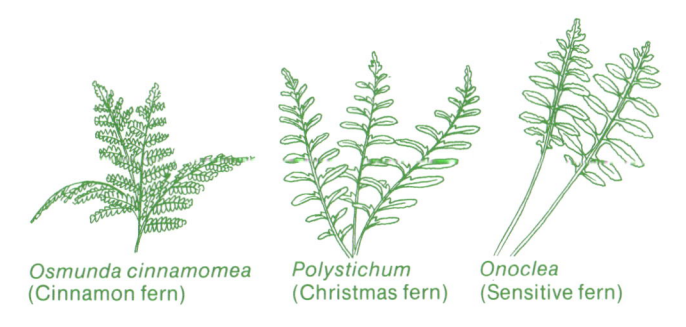

Osmunda cinnamomea (Cinnamon fern) *Polystichum* (Christmas fern) *Onoclea* (Sensitive fern)

9. Conifers

"**Conifer**" means cone-bearer. All conifers are woody trees and shrubs. They reproduce by seeds that develop inside cones on the branches. The leaves of conifers are usually very small and evergreen. They may be needle-like or scale-like.

Pines, spruces, cedars, and firs are conifers. Forests of them cover large areas of Canada. They are very valuable for their lumber, pulp and paper, and resins.

Juniperus virginiana
(Red cedar)

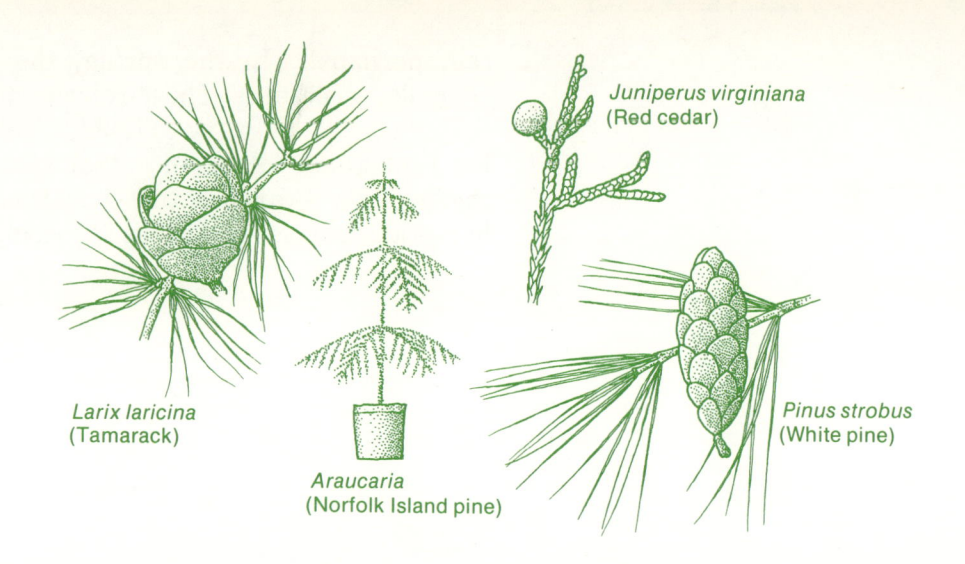

Larix laricina
(Tamarack)

Araucaria
(Norfolk Island pine)

Pinus strobus
(White pine)

10. Flowering Plants

This large phylum includes a wide variety of trees, shrubs, and soft-stemmed plants. All of them produce some kind of flower. The flowers may not be very noticeable. They develop into **fruits** that contain one or more **seeds**.

Any tree that is not a conifer is a flowering plant. For example, maples, oaks, beeches, and willows are flowering plants. So are grasses, vegetables, water lilies, milkweeds, and poison ivy. Most human food plants are flowering plants too.

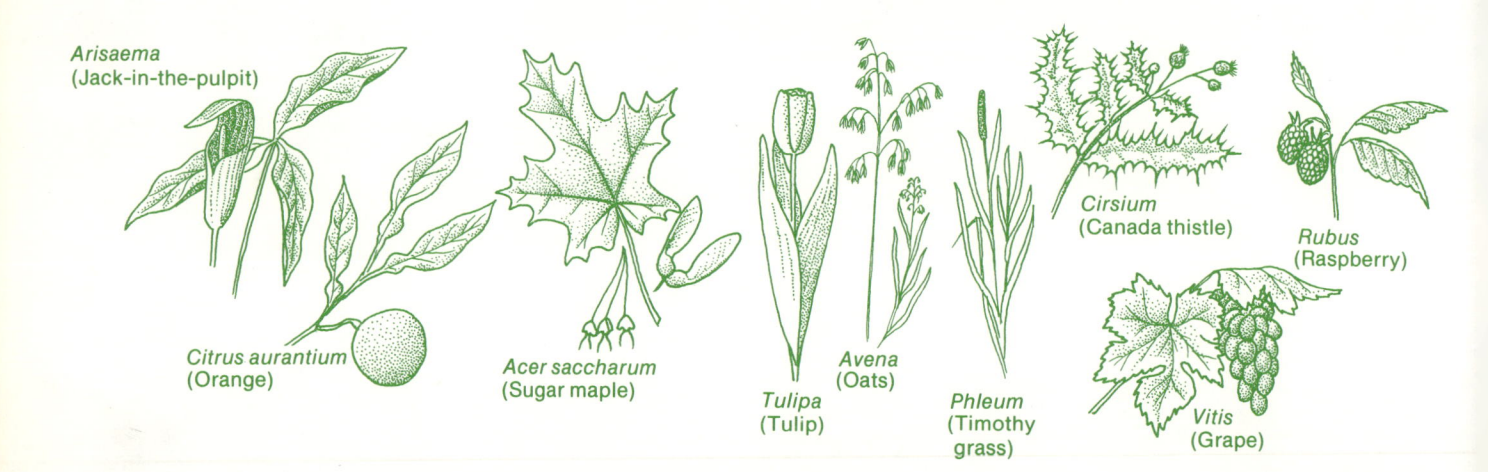

Arisaema
(Jack-in-the-pulpit)

Citrus aurantium
(Orange)

Acer saccharum
(Sugar maple)

Tulipa
(Tulip)

Avena
(Oats)

Phleum
(Timothy grass)

Cirsium
(Canada thistle)

Rubus
(Raspberry)

Vitis
(Grape)

Section Review

1. Which phylum of plants includes the most species?
2. Why do green algae appear green?
3. Brown and red algae contain chlorophyll. Why do they not appear green?

4. "Seaweeds" include red and brown algae. List two differences between them.
5. Stoneworts and horsetails both feel gritty to touch, but for different reasons. What is the reason in each case?
6. Organize the ten phyla into two lists, as follows:
   a) plants that grow in water;
   b) plants that grow on land
7. Which phylum of plants is thought to be the first phylum to survive on land?
8. a) Name three groups of plants that reproduce by spores.
   b) Name two groups of plants that reproduce by seeds.

13.3   ACTIVITY   Green Algae

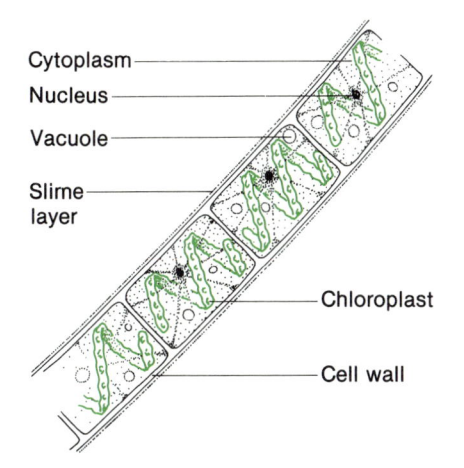

Fig. 13-2 *Spirogyra*. The chloroplasts in *Spirogyra* cells are spiral in shape.

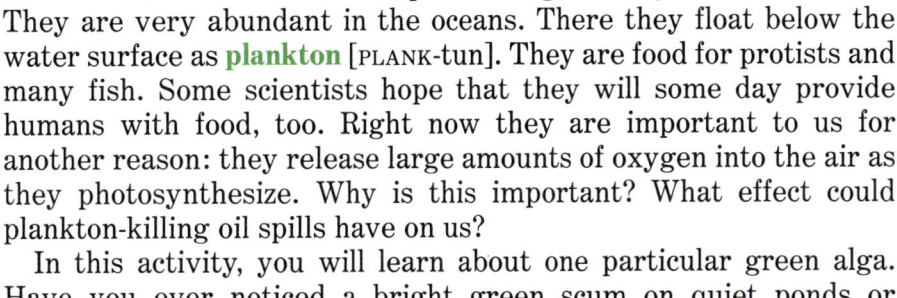

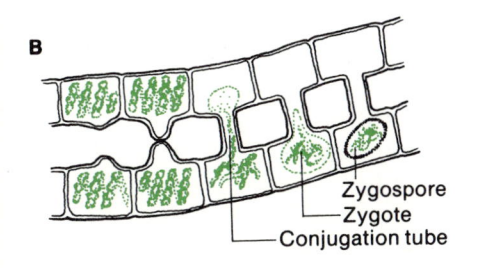

Fig. 13-3 Conjugation in *Spirogyra*

There are several thousand species of green algae in this world. They are very abundant in the oceans. There they float below the water surface as **plankton** [PLANK-tun]. They are food for protists and many fish. Some scientists hope that they will some day provide humans with food, too. Right now they are important to us for another reason: they release large amounts of oxygen into the air as they photosynthesize. Why is this important? What effect could plankton-killing oil spills have on us?

In this activity, you will learn about one particular green alga. Have you ever noticed a bright green scum on quiet ponds or streams? This "pond scum" is known as *Spirogyra* [SPY-row-JI-ruh]. It is one of the easiest green algae to study because it is easy to find and its cells are quite large.

The cells of *Spirogyra* are arranged in strings called **filaments**. A filament may be 30 cm long. The filaments are covered with a jelly-like substance. It protects them from injury.

Each cell contains a **nucleus** and one or two **chloroplasts**. These chloroplasts are spiral and ribbon-like in shape (Fig. 13-2). They spiral around a large **central vacuole** that contains a watery fluid.

Algal filaments usually reproduce asexually. The filaments simply break apart. This could be due to water currents, or to the nibbling of fish. The broken pieces grow by cell division into complete new filaments.

Spirogyra also reproduces sexually. Two filaments line up side by side. Each cell in the two filaments grows a small projection (Fig. 13-3). These projections join to form a tube between each pair of cells. Then the contents of the cells in one filament move through the tubes and unite with the contents of the cells in the other filament. This process is called **conjugation** [kon-juh-GAY-shun]. Each union produces a cell called a **zygospore**. Zygospores form thick coatings that protect them. They can live through the winter and grow into new filaments in the spring.

Section 13.3   171

Problem

What is the nature of *Spirogyra*?

Materials

microscope
fresh or prepared slide of *Spirogyra*

Procedure

a. Mount the slide on the microscope stage on low power. Focus on several filaments.
b. Observe a filament. Find out where one cell ends and the next begins. Describe the shape of each cell.
c. Make a large labelled diagram of a filament as it appears on low power.
d. Switch to medium or high power. Focus on one cell. Use the fine adjustment to observe the spiral arrangement of one chloroplast.
e. Draw one cell in detail. Label the cell wall, vacuole, chloroplast, and nucleus (if visible).
f. Your teacher may have slides of conjugating filaments. If so, observe them on low and medium power. Draw several conjugating cells. Label the zygospores, tubes, and filaments.

Discussion

1. What is plankton?
2. State two reasons why green algae are important.
3. List two characteristics of *Spirogyra* that would help you to identify it in a stream.
4. How does *Spirogyra* get its name?
5. Describe asexual reproduction in *Spirogyra*.
6. Describe the process of conjugation.
7. How does *Spirogyra* survive the winter?

13.4 ACTIVITY Ferns

In northern climates like Canada, most ferns are less than 1 m tall. They grow in damp forests and along shaded streams. They help hold the soil in place and help prevent flooding.

Fern plants are made of true roots, stems, and leaves. Tree ferns have strong, erect stems. But most ferns have underground stems called **rhizomes** [RI-zomes]. These grow horizontally just beneath the soil surface. They usually live from year to year. The rhizomes develop roots that grow downward to absorb water from the soil. Each spring the rhizomes send up clusters of large leaves known as **fronds** (Fig. 13-4). The fronds are usually divided into many small

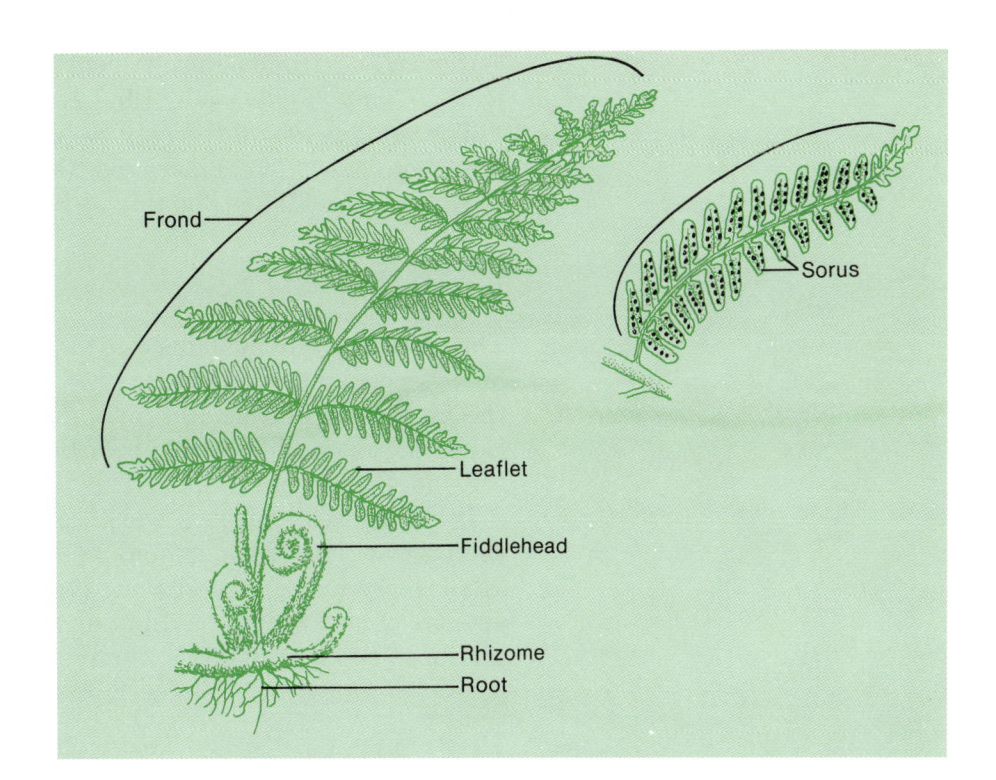

Fig. 13-4 The parts of a spore-producing fern plant

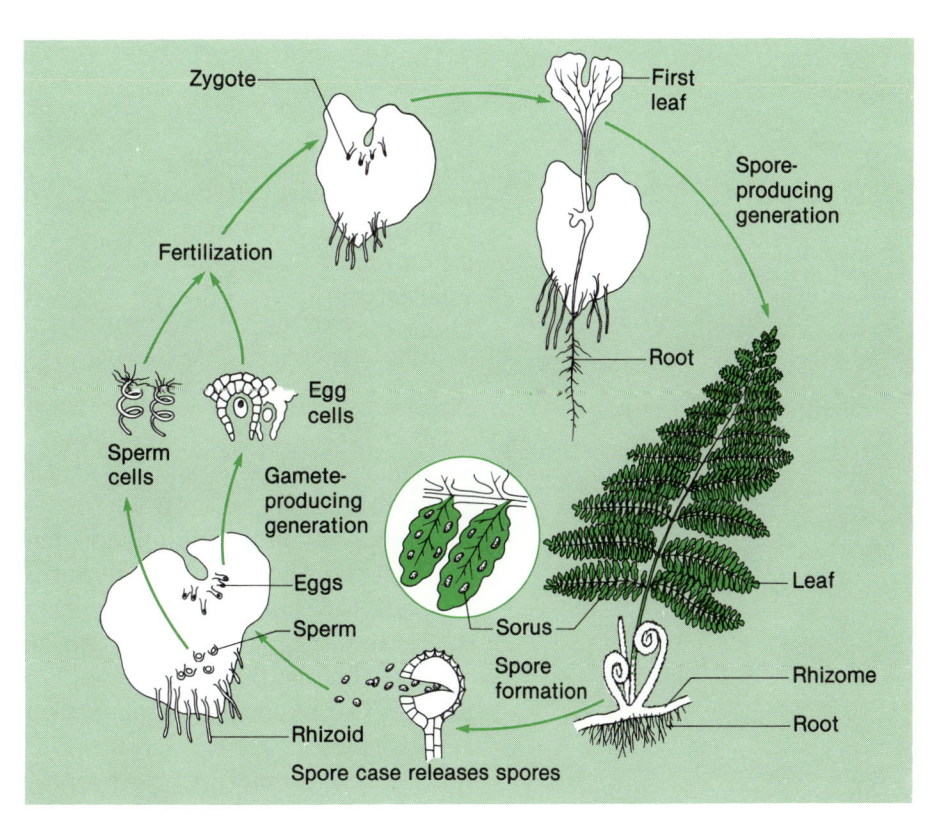

Fig. 13-5 Life cycle of a fern

Section 13.4

leaflets. The fronds unroll from tightly coiled buds called **fiddleheads**. Can you guess why they are called that?

The fiddleheads of the ostrich fern are edible. When cooked, they taste somewhat like asparagus. Other fern fiddleheads are thought to cause cancer. They should not be eaten.

Life Cycle of a Fern

The fern life cycle includes two different generations (Fig. 13-5). The leafy fern plant is the **spore-producing generation**. On the backs of the mature leaves, small brown dots appear (Fig. 13-6). These dots, called **sori** (singular: **sorus**) contain spore cases. When they dry out, the spore cases burst open. This releases the spores into the air.

If the spores land in a warm, moist place, they germinate. They then grow into tiny, green, heart-shaped plants. These are the **gamete-producing generation**. They produce eggs and sperm. The sperm swim through a film of water to fertilize the eggs. The fertilized eggs, or zygotes, then grow into new fern plants.

Problem

Can you find the parts of a fern?

Materials

a potted fern plant, or a freshly collected specimen with fertile fronds	cover slip
	water
	dissecting needle
a hand lens	dropper
microscope	glycerin
slide	

Procedure

a. Wash away the soil from the plant. Observe the roots and rhizome.
b. Look for young fiddleheads coming up from the rhizome.
c. Observe the fronds and note the arrangement of the leaflets. Look for the sori (brown dots) on the backs of them.
d. Make a large diagram of the fern plant. Label the roots, rhizome, fiddleheads, fronds, leaflets, and sori.
e. Use the hand lens to observe the sori. Describe their shape and position.
f. Using a dissecting needle, gently put one sorus into a drop of water on a slide. Separate it with the needle. Observe it under the microscope on low power. Draw what you see. Label the sorus and spores.
g. Add one drop of glycerin to the sorus. Glycerin will cause the spore cases to dry out. Observe what happens.

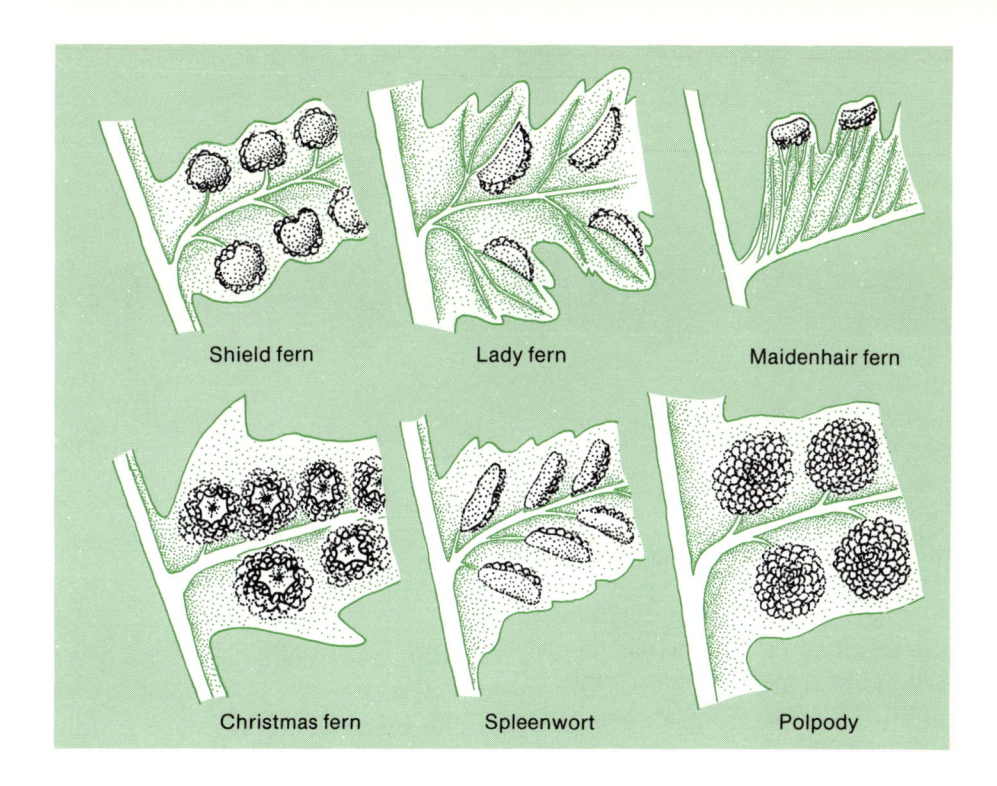

Fig. 13-6 Sorus types

Discussion

1. What is a rhizome?
2. Describe the shape of a young fern frond.
3. Describe the structure of a typical fern frond.
4. Describe the contents of a sorus.
5. What happens when glycerin is added to a sorus?

13.5 The Winners on Land: The Seed Plants

Why are Seed Plants So Successful?

The **conifers** and **flowering plants** are the best adapted to live on land. They are very abundant all over the world. There are several reasons for this. First, they have well-developed tissues to carry water up from the soil. They can survive in dry places by sending roots deep down for water. Second, some of the tissues are also very strong. They support the leaves and branches far above the ground.

The third reason that conifers and flowering plants are successful has to do with their life cycle. Unlike ferns, they do not require water for fertilization. The male sperm cells can be carried to the female eggs by wind, insects, or other animals. Also, the life cycle includes a seed. A **seed** contains a tiny plant (embryo) and stored food inside a protective coating. It can remain in a resting state for a long period of time. Seeds are easily spread around by wind, water,

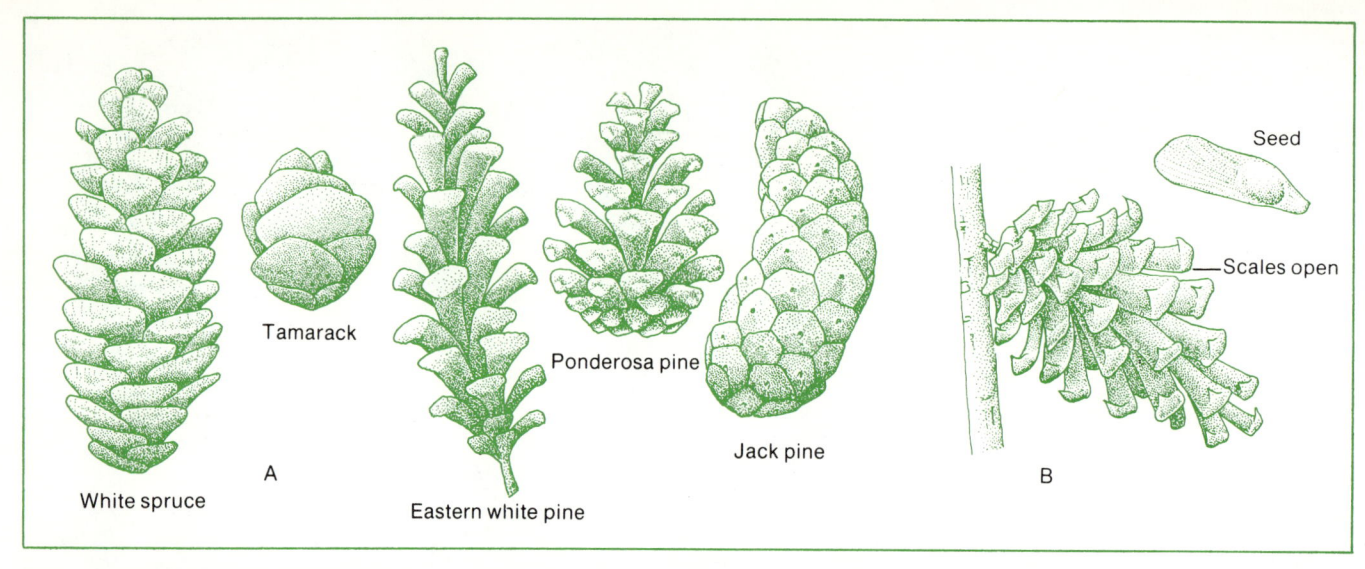

Fig. 13-7 The seeds of conifers are produced in cones (A). When the scales of a pine cone open, the seeds are released to the wind (B).

and animals. When they land in a suitable place, they germinate into new plants.

The seeds of conifers are produced in **cones** (Fig. 13-7). Flowering plants produce seeds inside **fruits**. These develop from **flowers** (Fig. 13-8). Because fruits protect the seeds and help them to get spread around, flowering plants are found everywhere.

Monocots and Dicots

Flowering plants are separated into two groups. This is done by their seed structure. One group is called the **monocot** group. The

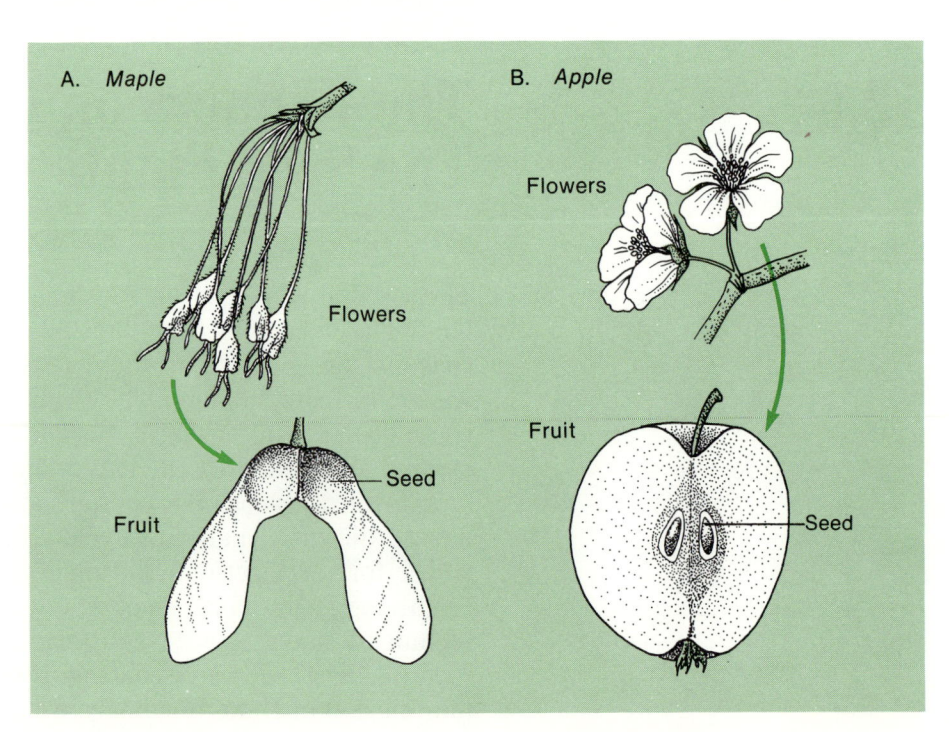

Fig. 13-8 Seeds develop inside flowers. The flowers produce fruits to protect the seeds and to disperse them.

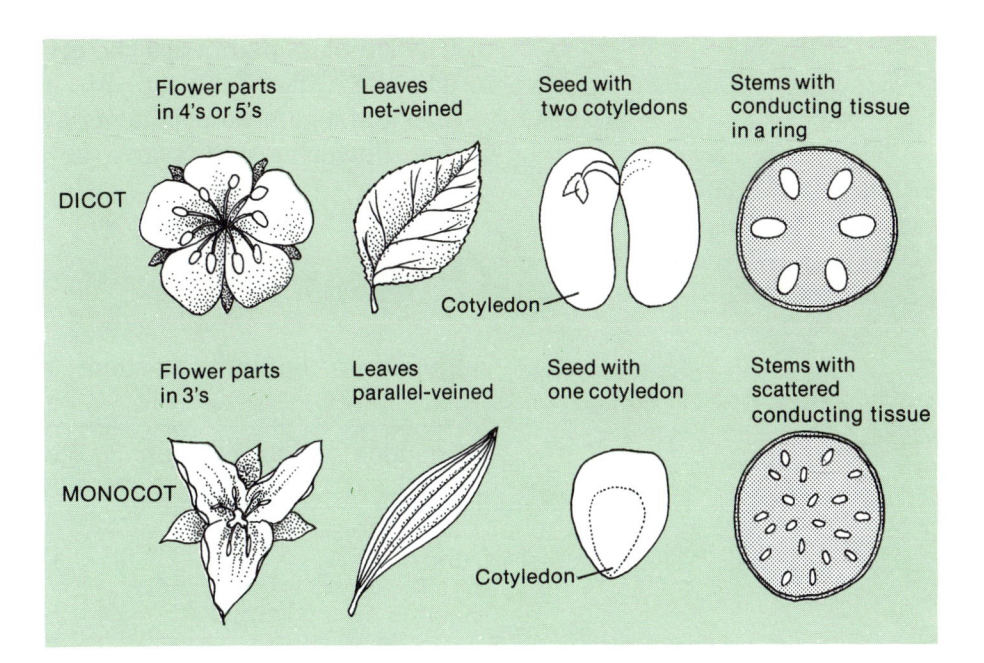

Fig. 13-9 Monocots and dicots differ in four basic ways.

other is the **dicot** group. These words refer to the number of **seed leaves**, or **cotyledons** [kot-i-LEE-donz] inside the seeds. Monocots have one cotyledon; dicots have two.

There are other differences between monocots and dicots, too (Fig. 13-9). Monocots have flowering parts in groups of three. Dicot flower parts are in groups of four or five. The leaves of monocots have many parallel veins. The leaves of dicots are net-veined. The stems are different, too. The conducting tissue in a dicot stem is arranged in a ring inside the stem. In monocots it is scattered throughout the stem. Dicot stems often become woody, but monocot stems never do. Some examples of monocots include grasses, lilies, and tulips. Dicots include sunflowers, beans, roses, and all flowering trees. The dicot group is the largest, most successful group of plants existing today.

Section Review

1. Give two reasons why plants with conducting tissues are very successful.
2. List two differences between the life cycle of a fern and the life cycle of a flowering plant.
3. What is a seed?
4. Why are flowering plants more successful than conifers?
5. List four differences between monocots and dicots.

Main Ideas

1. Plants are made of cells that contain chlorophyll.
2. Plants are grouped into ten main phyla.

3. The simplest plants are the green, red, and brown algae.
4. The life cycle of most plants includes a spore-producing generation and a gamete-producing generation.
5. The flowering plants are the largest, most successful group of plants.

Glossary

conjugation	kon-juh-GAY-shun	sexual reproduction between two algal filaments
cotyledons	kot-i-LEE-donz	seed leaves used for energy
filament	FILL-a-ment	a chain of algal cells
frond		the leaf of a fern
rhizome	RI-zome	horizontal underground stem
sorus	SORE-us	a spore case on the back of a fern leaf

Study Questions

A. True or False

Decide whether each of the following statements is true or false. If the sentence is false, rewrite it to make it true. (Do not write in this book.)
1. Kelps are large brown algae that grow in the ocean.
2. Stoneworts grow in shady woodlands.
3. Conjugation is a type of sexual reproduction in green algae.
4. Some ferns in Canada grow as large as trees.
5. The purpose of a flower is to develop into a fruit containing seeds.

B. Completion

Complete each of the following sentences with a word or phrase that will make the sentence correct. (Do not write in this book.)
1. To photosynthesize, plants require the green pigment _____ .
2. The fusion of two gametes produces a cell called a _____ .
3. The green alga that forms pond scum is called _____ .
4. Ferns grow from underground stems called _____ . These send up large leaves known as _____ .

C. Multiple Choice

Each of the following statements or questions is followed by four responses. Choose the correct response in each case. (Do not write in this book.)

1. The phylum of plants that includes more species than all the others put together is
   a) the mosses and liverworts
   b) the ferns
   c) the algae
   d) the flowering plants
2. Which of the following plants reproduces by seeds?
   a) fern   b) conifer   c) moss   d) horsetail
3. The most likely location of ferns is in
   a) the desert
   b) freshwater ponds
   c) damp, shady forests
   d) fields and roadsides
4. Algae often grow as chains of cells called
   a) filaments   b) fiddleheads   c) rhizoids   d) rhizomes

D. Using Your Knowledge

1. What part of the life cycle of a fern depends on water?
2. List three reasons why seed plants have adapted to land better than ferns.
3. Why are algae important?
4. How does a fern ensure that its spores are spread far away from the parent plant? Why is this important?

E. Investigations

1. Shake some spores from a mature fern leaf onto some moist soil in a container. Cover them and keep them in a warm, lighted place for several days. Look for the growth of tiny, heart-shaped plants. Transfer one to a microscope slide. Observe it under low power on a microscope.
2. The dinosaur age is sometimes called the **Age of Ferns**. Look up this period of time in an encyclopedia. What was the climate like in the dinosaur age? What groups of plants were most abundant? Much of the coal we use today for fuel was formed then. Find out how coal is formed.
3. A terrarium is a clear glass or plastic container that can be used to grow mosses and ferns. Set up a terrarium as follows: Cover the bottom with a layer of gravel or sand. Add several centimetres of sterilized soil. Transplant some clumps of moss and some small ferns into it. Water the plants and cover the top. The plants should grow if they receive moisture, warmth, and light. Experiment with several terraria to see what conditions are best.

14 Reproduction in Flowering Plants

14.1 Flowers
14.2 Activity: Flower Structure
14.3 Pollination and Fertilization
14.4 Fruits
14.5 Activity: Fruit Structure
14.6 Activity: Seeds
14.7 Reproduction without Seeds

Flowering plants reproduce **sexually** by seeds. The sexual organs are the flowers. Flowers produce fruits, in which the seeds for the next generation develop.

Most flowering plants also reproduce **asexually** from the vegetative parts. These are the roots, stems, and leaves.

14.1 Flowers

Flowers differ a great deal in appearance. However, they all have some or all of the parts shown in Figure 14-1.

Parts of a Flower

Corolla The colourful part of most flowers is the **corolla**. It is made of several **petals**. Their colour and fragrance attract insects for pollination. Some corollas produce a sweet liquid called **nectar**. Nectar attracts insects too.

Calyx Below the corolla are green, leafy structures called **sepals**. Together the sepals form the **calyx**. The calyx protects the inner flower parts while the flower is in bud (not fully opened).

Stamens Inside the corolla of most flowers there are several stamens. Stamens produce pollen grains, or male spores (Fig. 14-2). A stamen is composed of two parts: an anther and a filament. The anther is a sac-like structure that produces pollen. The filament is the thin stalk that supports the anther.

Pistil The **pistil** is the female part of the flower that produces seeds. It is usually vase-shaped. Its three parts are the ovary, style, and stigma. The **ovary** is the large part at the base of the pistil. It

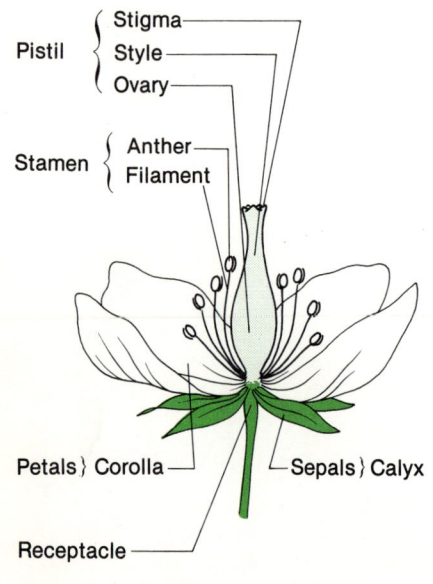

Fig. 14-1 Parts of a flower

contains one or more **ovules**, which contain the eggs. These may develop into seeds if they are fertilized. Above the ovary is the **style**. It is a stalk that joins the ovary to the **stigma** at the top. The surface of the stigma is often sticky or feathery to catch pollen.

Receptacle The base of the flower is called the **receptacle**. It supports the flower parts and the fruit that develops later. It may even become part of the fruit.

Section Review

1. **a)** Which part of the flower attracts insects?
   b) List three ways it does this.
2. Which two parts of a flower are needed for sexual reproduction?
3. What is the purpose of the anther?
4. Which part of the flower catches pollen?

14.2　ACTIVITY　Flower Structure

In this activity you will examine the structure of one or more flowers.

Problem

How are flowers alike? How are they different?

Materials

| fresh flower of petunia, toadflax, buttercup, morning glory, lily, snapdragon, touch-me-not, or other simple flower | hand lens
forceps
microscope slide
cover slip | water
dropper
microscope
scalpel |

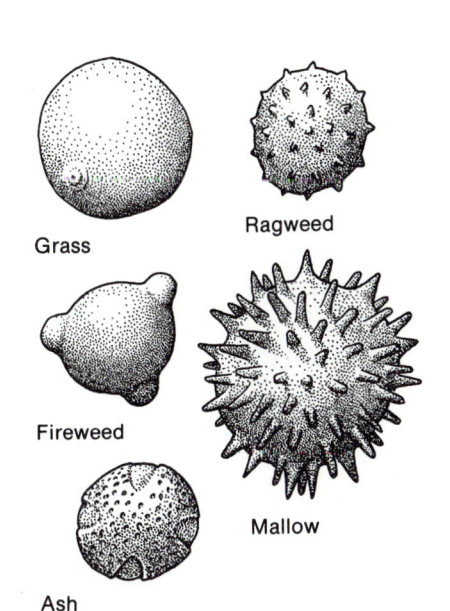

Fig. 14-2 Pollen grains. Smooth pollen grains are often carried by wind. The others are usually carried by insects.

Procedure

a. Without damaging any of the flower parts, observe an entire flower. Look for the calyx, corolla, stamens, pistil, and receptacle.
b. Make a large diagram of the whole flower. Show its shape and the relative sizes of its parts. Label all the parts you can see.
c. Observe the calyx. Are the sepals separate or joined together? How many sepals are there? Describe the shape and colour of the calyx.
d. Observe the corolla. Describe its shape. Are the petals separate or joined? How many petals are there? Record their colour. Note whether or not they are scented.
e. To observe the stamens, you may have to use the forceps to remove part of the corolla. Count the stamens. Are they attached to the petals? Observe the anthers with the hand lens. Is any dust-like pollen visible?

f. Remove one anther with the forceps. Put it into a drop of water on a microscope slide. Crush it gently to remove the pollen from the sacs. Cover with a cover slip. Observe the pollen under the microscope. Make a large diagram of a few pollen grains.
g. Locate the pistil in the centre of the flower. Describe its shape. Observe the ovary, the style, and the stigma. Is the stigma sticky? Observe it closely with the hand lens. Is there any pollen on it?
h. Inside the ovary are the ovules. Cut the ovary carefully downward through the centre with the scalpel. Look for ovules inside the ovary with the hand lens. Describe their appearance. Estimate how many are present.
i. Make a final diagram of all the parts of a flower as they appear with some of the petals removed and the ovary cut open. Title this diagram "A Vertical Section of a Flower". Label all the parts.
j. If a second type of flower is available, repeat steps (a) to (i). Compare the parts to the first flower as you go along.

Discussion

1. Complete a table like the following in your notebook. (Do not write in this book.)

Table 14-1   Structure of a Flower

Flower name: _____

	Number of parts	Are the parts joined?	Where are the parts attached?	Description
Calyx				
Corolla				
Stamens				
Pistil				

2. a) How do the number of sepals and petals compare?
   b) How do the number of petals and stamens compare?
   c) Is the flower a monocot or a dicot? Explain.
3. If all the ovules were fertilized, how many seeds would this flower produce?

14.3 Pollination and Fertilization

How do flowers produce seeds? Since a seed contains a zygote, one step in the process must be fertilization. Before that can occur, the male sperm have to get to the female eggs. The sperm are produced inside pollen grains on the anthers. The eggs are produced in ovules inside the pistils. The transfer of pollen grains from an anther to a pistil is called **pollination**.

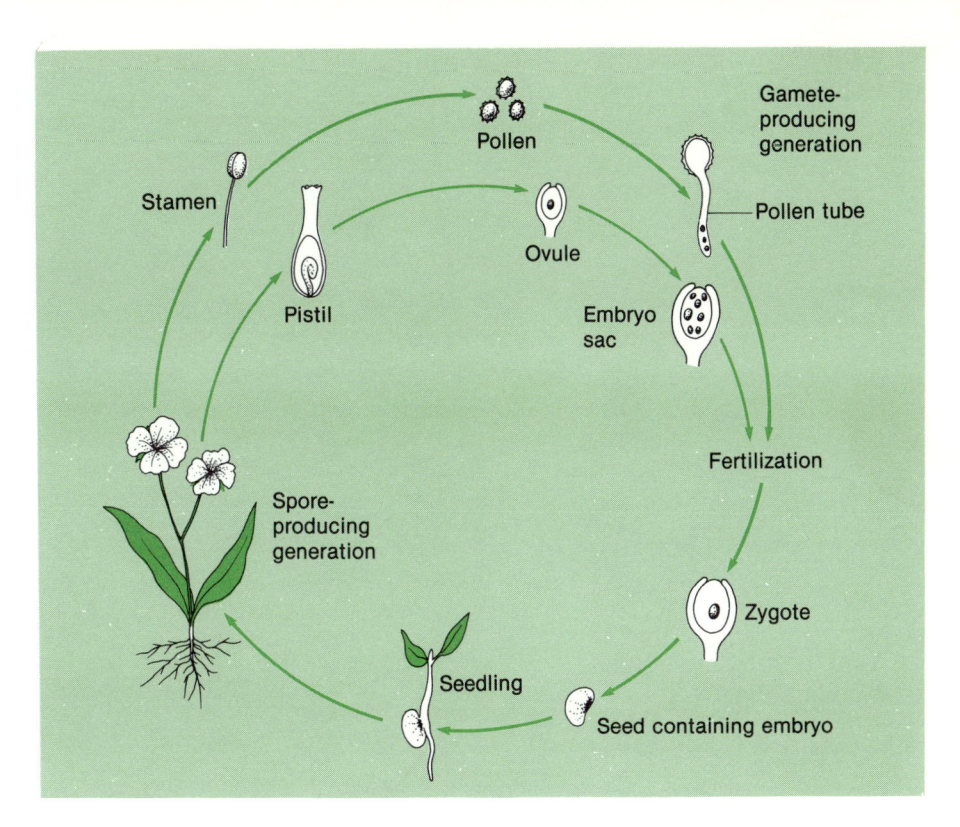

Fig. 14-3 The life cycle of a flowering plant

The life cycle of a flowering plant is shown in Figure 14-3. As in ferns, a spore-producing generation alternates with a gamete-producing one. Unlike ferns, these two generations are not separate. The gamete-producing generation is produced right in the spore-producing generation. However, the gamete-producing generation can hardly be seen.

The first step in seed production is the development of male and female spores in the flower. Most flowers produce both types of spores. Female spores develop inside the ovules in the pistil (Fig. 14-4). The pistil may contain one ovule or many. Inside each ovule, one spore develops into an **embryo sac**. This is a tiny gamete-producing plant. One nucleus inside it becomes an **egg**.

Meanwhile, the male spores, or **pollen grains**, develop in the

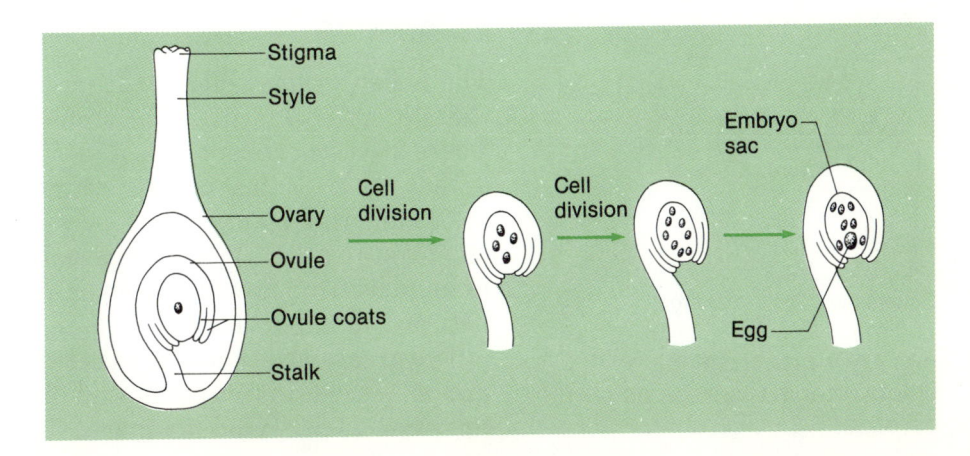

Fig. 14-4 Development of an embryo sac inside the ovule of a flower

Section 14.3   183

Fig. 14-5 Development of a pollen tube

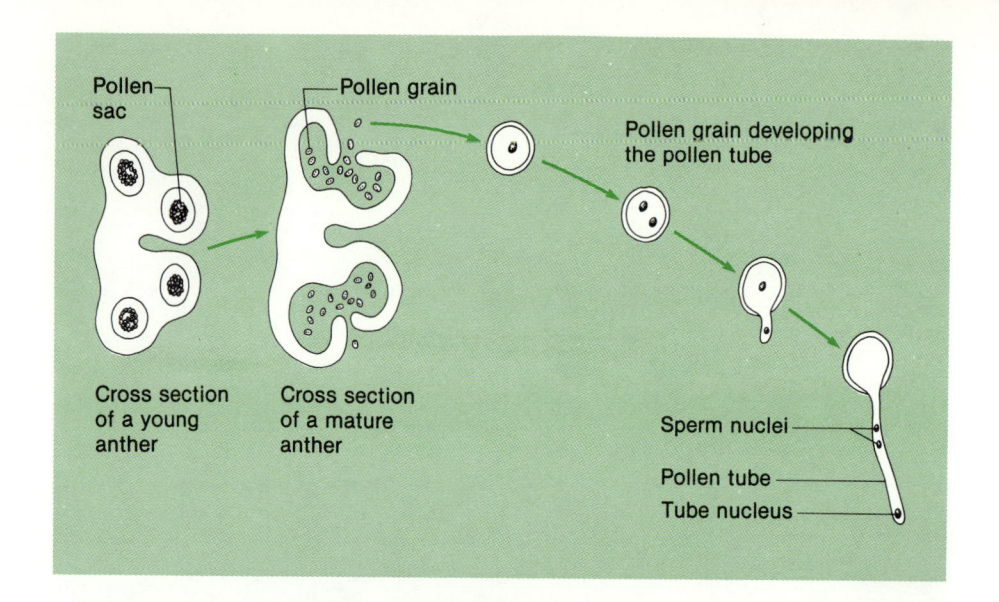

Fig. 14-6 Self-pollination and cross-pollination

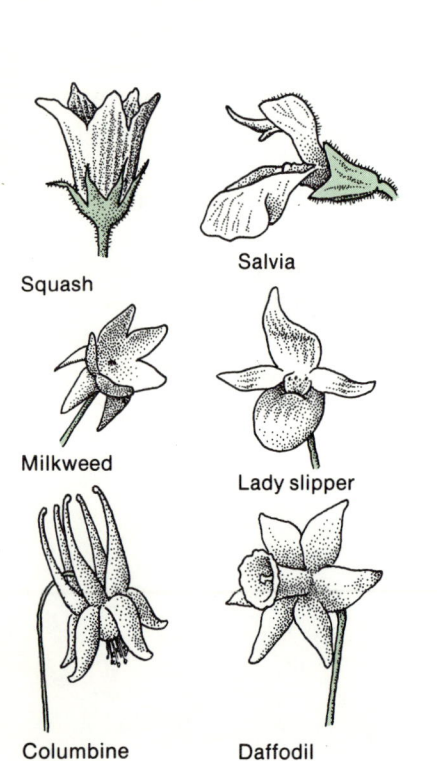

Fig. 14-7 Insect-pollinated plants often have large colourful corollas.

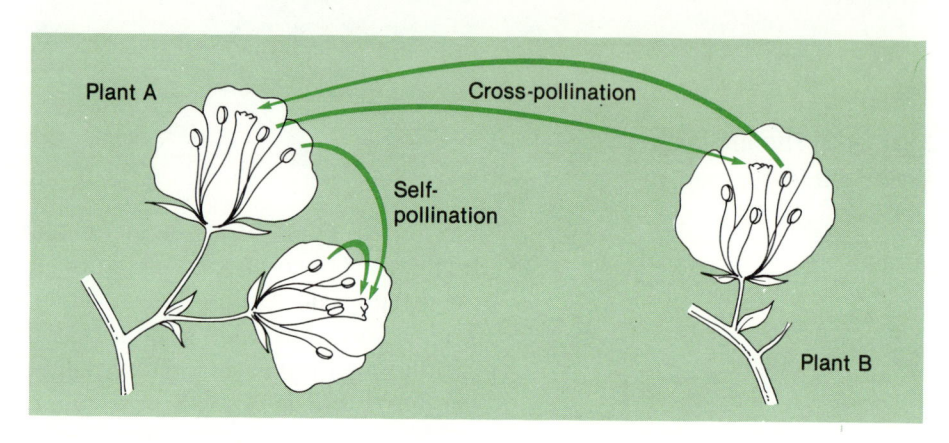

anthers of the stamens (Fig. 14-5). Originally, a pollen grain contains only one cell with one nucleus. Divisions of the nucleus result in a group of several cells. This group of cells is the gamete-producing generation. It produces two **sperm nuclei**, and a **tube nucleus**. When the pollen grains are mature, they are released from the anther.

Pollination

The pollen grains released from the anther must be transferred to the stigma of a pistil. The transfer of pollen is called **pollination**. There are two types. **Self-pollination** is the transfer of pollen from an anther to a stigma on the same plant (Fig. 14-6). Tomatoes are self-pollinating plants. **Cross-pollination** is the transfer of pollen from one plant to another. In this case, the pollen must be carried by wind or animals from flower to flower. Bees, butterflies, moths, birds, and bats are all important pollinators.

The great variety in the structure of flowers is related to how they are pollinated. Insect-pollinated flowers are usually large and colourful (Fig. 14-7). Wind-pollinated flowers are usually smaller. They

184 Chapter 14

Fig. 14-8 Wind-pollinated flowers are small. They are often missing parts that would block the wind.

often lack parts such as sepals and petals that would block the wind (Fig. 14-8).

Cross-pollination produces more varied offspring than self-pollination. This is because the genetic information from two parents is mixed together. The offspring will be slightly different from each other and from the parent plants. This means some of them may be able to survive conditions that would kill the others. Variability increases the chances of survival for the species.

For this reason, cross-pollination is preferred over self-pollination. Some plants cannot self-pollinate at all. Their anthers may release pollen long before the pistil is ready to receive it. In others, the pistil produces a chemical that kills any pollen from the same plant. Some plants even produce male and female flowers on completely separate plants. Willows are an example of this.

The Pollen Tube

After a pollen grain lands on a stigma, it absorbs water and begins to grow into a **pollen tube** (Fig. 14-9). This tube digests its way to the ovules in the ovary. It is guided by chemicals produced by the ovules. It carries the two sperm nuclei to an ovule for fertilization. The tube enters the ovule and releases the two sperm nuclei into the embryo sac.

Fertilization

At fertilization, one of the sperm nuclei joins with the **egg nucleus** to form a **zygote**. The other sperm nucleus joins with two other nuclei

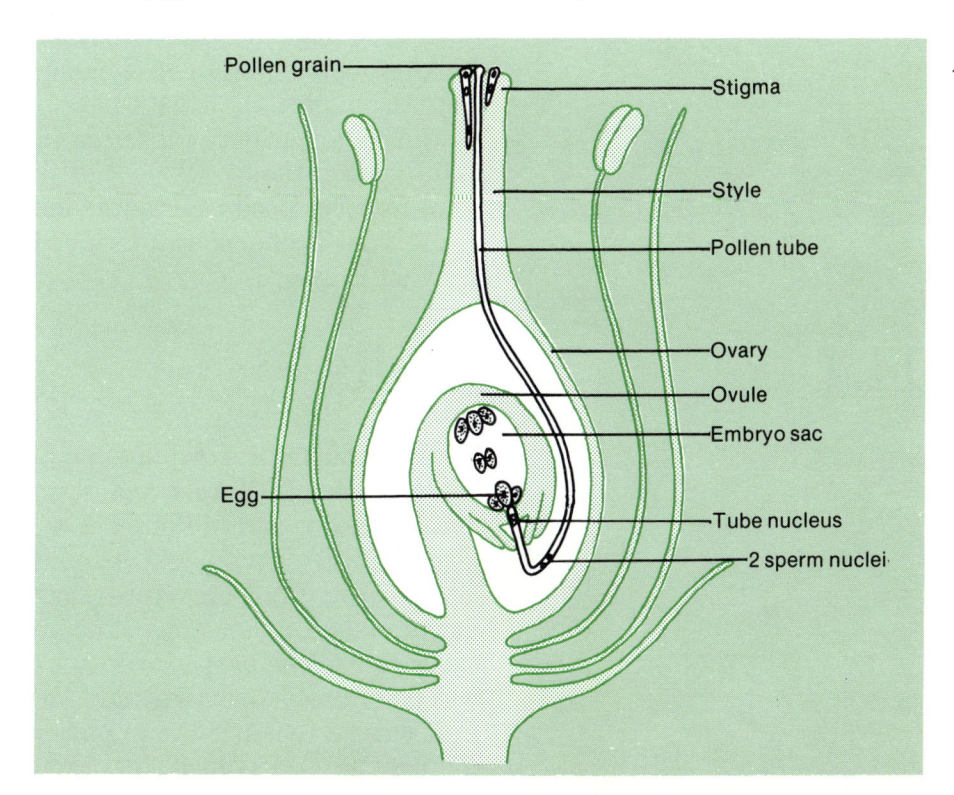

Fig. 14-9 Vertical section of a flower ready for fertilization

Section 14.3

in the embryo sac to form an **endosperm nucleus**. This endosperm nucleus divides to form a tissue called **endosperm**. Endosperm becomes a food supply for the developing embryo.

Because there are two sperm nuclei involved in the fertilization of an ovule, the whole process is called **double fertilization**. Double fertilization occurs only in flowering plants. Without it, no seeds are produced.

Development of the Seed

After fertilization, the zygote begins to develop into a young plant, or **embryo**. The embryo uses the endosperm tissue as a source of energy to grow. Around the embryo and the endosperm, the ovule coats begin to harden into **seed coats**. The seed that finally forms is composed of three parts: the embryo, the endosperm, and the seed coats.

Section Review

1. **a)** Where in a flower are pollen grains produced?
   b) Where in a flower are eggs produced?
2. Define pollination.
3. **a)** What is an embryo sac?
   b) How many sperm nuclei does a mature pollen grain contain?
4. Explain the difference between self-pollination and cross-pollination.
5. How do insect-pollinated and wind-pollinated flowers differ in appearance?
6. **a)** Explain why cross-pollination is preferable to self-pollination.
   b) List three ways self-pollination can be prevented by plants.
7. After a pollen grain has landed on a stigma, how do the sperm get to the ovule?
8. **a)** Explain why fertilization in flowering plants is called double fertilization.
   b) What is produced by the union of an egg and sperm?
   c) What is endosperm tissue? How is it formed?
9. What is a seed?

14.4 Fruits

While seeds are developing in a flower, other changes are occurring too. The sepals, petals and stamens are no longer needed. They usually wither and fall off. The stigma and style usually dry up too. However, the ovary begins to enlarge as the seeds develop. It grows into a protective structure that biologists call a **fruit**. In some flowers, such as an apple, the receptacle enlarges along with the ovary to become part of the fruit as well.

A fruit always has two scars on the outside. One scar is where the fruit was attached to the parent plant. A stalk may still be joined to the fruit at this point. The other scar is that left by the withered

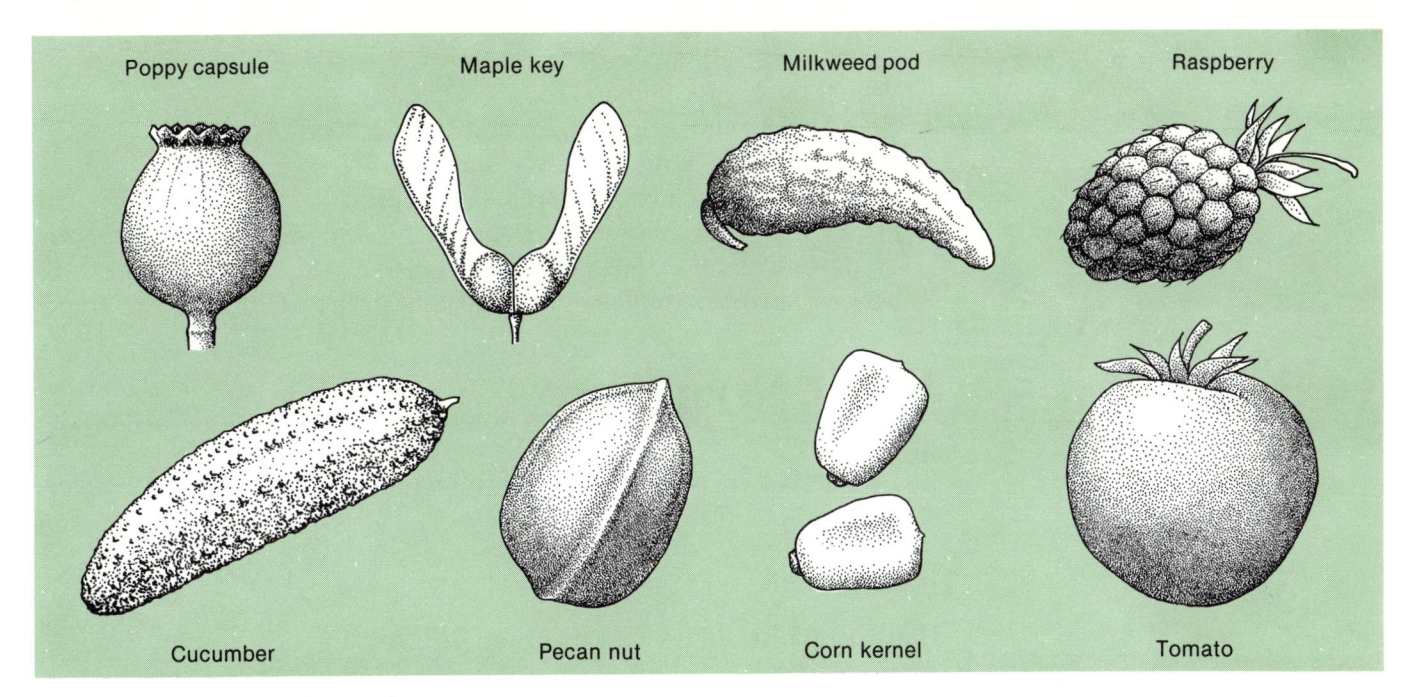

Poppy capsule   Maple key   Milkweed pod   Raspberry

Cucumber   Pecan nut   Corn kernel   Tomato

Fig. 14-10 Fruits are ripened ovaries that contain seeds.

style and stigma. Unlike fruits, seeds have only one scar. It occurs where the seed was attached to the inside of the ovary wall.

A fruit is not necessarily something that is fleshy and edible! Certainly, oranges, raspberries, cherries, and peaches are fruits. But so are poppy capsules, maple keys, milkweed pods, cucumbers, pecan nuts, and corn kernels (Fig. 14-10)! Most of us make the mistake of calling some fruits, such as tomatoes, vegetables! *All ripened ovaries containing seeds are fruits.*

Functions of Fruits

Fruits have two main functions. First, they help to protect the seeds from damage as they develop. Second, they also help to scatter the seeds away from the parent plant. Imagine what would happen if all the seeds dropped to the ground in one spot and began to grow. The young plants would crowd each other and probably die from lack of sun, water, or nutrients.

Fruits have many adaptations to help scatter the seeds around. Some, like burrs, have hooks or spines that catch on the fur of animals or on the shoes of humans. Coconuts float away on water. Maple keys have wings that carry them in the wind. Dandelion fruits have a parachute of hairs that float them through the air. Fleshy fruits like cherries get eaten by birds and animals. The cherry seeds pass through the digestive systems and are dropped far from the tree that produced them.

Some fruits are dry and others are fleshy. Some are made from one ovary. Others are made of several ovaries fused together. Some contain one seed, while others contain hundreds. Since all 250 000 or so flowering plants in the world produce fruits, the variety is truly amazing.

Section 14.4

Section Review

1. Which parts of a flower produce a fruit?
2. How could you distinguish a fruit from a seed?
3. Name two dry fruits and two fleshy fruits.
4. What are the two main functions of fruits?
5. **a)** Why is it important for seeds to be scattered away from the parent plant?
   b) List three examples of adaptations that scatter fruits around.

14.5 ACTIVITY Fruit Structure

In this activity you will be able to observe the structure of three different types of fruit.

Problem

How are fruits alike? How are they different?

Materials

scalpel or knife
apple
orange or tomato
bean or pea pod

Procedure

a. Observe the end of an apple opposite the stalk. Is there any evidence of flower parts there? What are they? Draw the apple as it appears whole.

b. Cut the apple in half vertically from top to bottom. Examine the flesh, core, and seeds. The core is the ovary wall. The fleshy part that we would eat is the receptacle of the flower. Make a diagram of the sectioned apple. Label the receptacle, ovary wall, and seeds.

c. Remove the seeds and record how many there are. Suggest how apple seeds may be spread in nature.

d. Obtain an orange or a tomato. Can you tell where it was attached to the plant? Cut it open crosswise. This type of fruit has no core. The skin and flesh are both part of the ovary wall. In oranges, the skin is extra thick and is called a rind. Draw a diagram of the inside of the fruit and label the parts.

e. Examine the outside of a bean or pea pod. Look for a dried stigma and style at the end opposite to the stalk. Consider what part of the flower produces the pod. Then draw the pod and label the parts.

f. Split the pod open from one end to the other along one edge. Spread the two halves out flat. Draw the inside of the pod and label the parts.

g. How many seeds does the pod contain? Note where each seed is attached.

Discussion

1. a) What flower parts remain at the end of an apple opposite the stalk?
   b) Why do the flower parts dry up as the apple develops?
2. a) Which part of the flower produces the fleshy part of an apple?
   b) Which part produces the core?
   c) Which part produces the seeds?
3. a) What is the function of a seed?
   b) What is the function of the flesh and core of an apple fruit?
4. a) What differences did you note between the inside of a tomato (or orange) and the inside of an apple?
   b) How many seeds are contained in this fruit?
   c) Which part of the flower develops into the flesh of this fruit?
5. a) How do you know that a bean or pea pod is a fruit?
   b) Which part of the flower produces the pod?
   c) How are the seeds released from a pod?

14.6 ACTIVITY Seeds

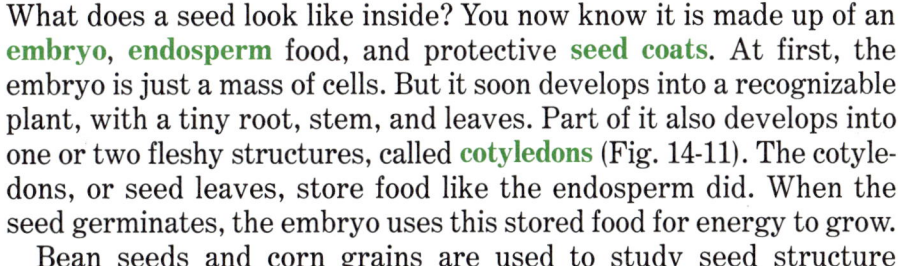

What does a seed look like inside? You now know it is made up of an **embryo**, **endosperm** food, and protective **seed coats**. At first, the embryo is just a mass of cells. But it soon develops into a recognizable plant, with a tiny root, stem, and leaves. Part of it also develops into one or two fleshy structures, called **cotyledons** (Fig. 14-11). The cotyledons, or seed leaves, store food like the endosperm did. When the seed germinates, the embryo uses this stored food for energy to grow.

Bean seeds and corn grains are used to study seed structure because they are large and easy to obtain. They both need to be soaked in water overnight. Soaking softens the seed coats. Then the seeds will open easily.

The stored food in a seed is usually in the form of starch. You can tell if starch is present by using an iodine test. A drop of iodine will make starch turn bluish-black. If no starch is present, the black colour does not appear. You will be able to test the seeds with iodine to see if any parts contain starch.

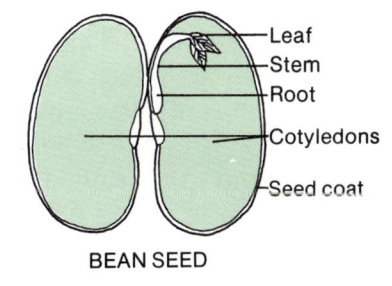

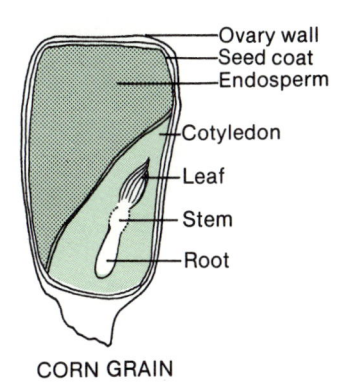

Fig. 14-11 The embryo inside a seed consists of a tiny plant and one or two cotyledons.

Problem

What is the structure of a seed?

Materials

bean seed, soaked overnight in water
bean seed, dry
corn grain, soaked overnight
hand lens
dropper
iodine solution
scalpel

Section 14.6

Procedure A Study of a Bean Seed

a. Observe the bean seed that has been soaked in water. Describe its shape, colour, and size.
b. Describe the seed coats that cover the seed. Along the concave edge of the seed, locate the small oval scar. Why do all seeds have this scar?
c. Just above the scar is a tiny pore. This is where the pollen tube entered the ovule. Find the pore. If you gently squeeze the seed, a drop of liquid should come out of this pore. What is this liquid?
d. Draw a labelled diagram of the concave side of the unopened seed.
e. Break the seed coats open along the convex side of the seed. Remove the seed coats. Inside you will find two fleshy cotyledons. Gently separate them. Look for the rest of the embryo near one end. Use the hand lens to observe the embryo closely. Look for a tiny root, stem, and two tiny leaves.
f. Draw and label the interior of a bean seed, as it appears with the cotyledons laid open.
g. Put a drop of the iodine solution on the inner surface of one cotyledon. Observe any changes. What does this tell you?
h. Observe a dry (unsoaked) seed. What differences do you notice in the seed coat? What is the function of the seed coat?

Procedure B Study of a Corn Grain

a. A corn grain is a fruit. The seed is inside it. There should be two scars on the grain. Look for them. Why are there two scars?
b. Look for a lighter-coloured area on one side of the grain. This is where the embryo is. The rest of the space is filled with endosperm. Draw the grain as it appears from the outside. Label all the parts.
c. Cut the grain lengthwise with a scalpel. Make the cut parallel to the broad side of the grain, so that the embryo is cut in half. Observe the ovary wall and seed coat around the edges of the cut surface. Note how they are firmly attached to each other.
d. Test the cut surfaces with a drop of iodine solution. Which parts turn dark?
e. Observe the embryo with the hand lens. Most of it is the cotyledon. Try to locate the tiny young plant.
f. Make a diagram of the grain as it appears when cut. Label all parts.

Discussion A

1. a) Explain how to test for the presence of starch.
   b) Which part of a bean seed contains starch?
   c) What is the starch used for?
2. What differences did you observe between a soaked and an unsoaked bean seed?

Fig. 14-12 Asexual reproduction from leaves. This *Bryophyllum* leaf produces plants along its edge.

190 Chapter 14

Discussion B

1. Why are there two scars on a corn grain?
2. Which part of a corn grain contains starch?
3. List three differences in the structure of a bean seed and a corn grain.

14.7 Reproduction without Seeds

Many flowering plants reproduce without forming seeds. They simply form new plants from pieces of root, stem, or leaf. This type of reproduction is called **vegetative propagation**. Often gardeners grow copies of their favourite plants this way. The new plants are always genetically identical to the original one. Why would this be so?

Reproduction from Leaves

In a few species, the leaves can produce young plants. One example is *Bryophyllum* (Fig. 14-12). This plant grows tiny plants between the teeth along its leaf edges. When these plants fall onto soil, they soon develop into new *Bryophyllum* plants.

Reproduction from Roots

Poplar trees often send up shoots around the trunk of the tree (Fig. 14-13). These shoots are called **suckers**. They develop from the roots. They are more likely to grow if the main tree has been damaged in some way.

Fig. 14-13 Poplar trees propagate by suckers.

Reproduction from Stems

There are five different methods of producing another plant from the stem of a parent plant:

1. The strawberry is a good example of a plant that produces **runners**. These are thin, horizontal stems that grow near the surface of the soil. Some distance from the parent plant, the tip of a runner produces a new plant (Fig. 14-14). Some runners spread just below the soil. Mint is an example of a plant that spreads by underground runners.
2. Some shrubs develop new plants when their lower branches just bend over and touch the soil. This is called **layering**. Currants and raspberries often reproduce this way (Fig. 14-15).
3. Many plants spread by underground stems called **rhizomes** (Fig. 14-16). These fleshy stems produce new shoots all along their length. Irises, ferns, and crab grass are examples.
4. If bits of stem are accidentally torn or chewed off a plant, they may be able to root. These broken pieces of stem are called

Fig. 14-14 Strawberry plants spread by runners.

cuttings (Fig. 14-17). Willow trees will often grow from cuttings. So will many house plants. Begonias, coleus, geraniums, and ivies are usually propagated by cuttings.

5. Some types of plants produce thick underground stems that are used for storage of food. These modified stems are called tubers, corms, and bulbs. Each is able to sprout into one or several new plants.
   a) A **tuber** is a thick fleshy stem that has "eyes", or buds, which can develop new shoots (Fig. 14-18). Potatoes and dahlias are examples.
   b) A **corm** is a short, thick stem covered with a few scales (Fig. 14-19). Gladioli and crocus plants grow from corms.
   c) A **bulb** is actually a tiny stem surrounded by many layers of thick, fleshy leaves (Fig. 14-20). The leaves contain stored food. Examples of bulbs are onions, tulips, and daffodils.

Section Review

1. What is vegetative propagation?
2. Why are plants that are produced by vegetative propagation genetically identical to the original plant?
3. Describe one plant that reproduces asexually by leaves.
4. Explain how poplar trees reproduce asexually.
5. Describe the difference between runners and rhizomes, and state one example of each.
6. What is a cutting?
7. State the differences among tubers, corms, and bulbs, and name one example of each.

Fig. 14-15 Layering. When a branch bends over and touches the soil, a new plant develops.

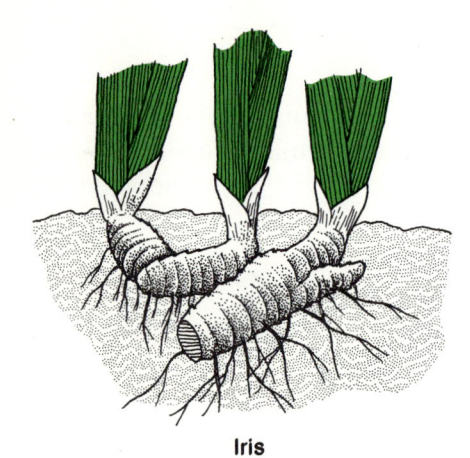

Fig. 14-16 Thick underground stems that grow horizontally are called rhizomes.

Fig. 14-17 Many house plants reproduce easily by cuttings.

COLEUS   BEGONIA   GERANIUM

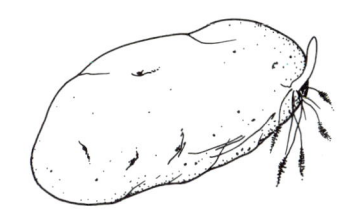

Fig. 14-18 Tubers of potato

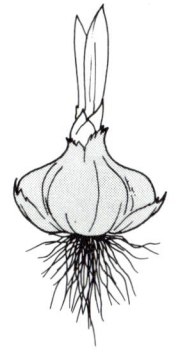

Fig. 14-19 Corm of crocus

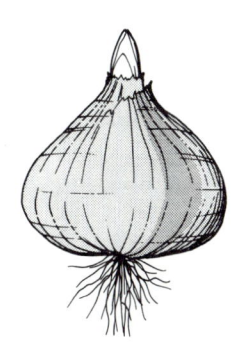

Fig. 14-20 Bulb of an onion

Main Ideas

1. The purpose of flowers is to produce fruits that contain seeds.
2. The transfer of male gametes to female gametes in flowering plants is called pollination.
3. The colour, shape, and size of flowers is related to how they are pollinated.
4. Pollen grains and seeds are spread from place to place by wind, water, insects, birds, and other animals.
5. A fruit is a ripened ovary in which one or more seeds develop.
6. A seed is composed of an embryo, a food supply, and a protective coating.
7. Many flowering plants reproduce asexually from roots, stems, and leaves.

Glossary

calyx	KAY-licks	the green sepals around a flower
corolla	kuh-RAW-luh	petals of a flower
cotyledon	kot-i-LEE-don	seed leaf
endosperm	EN-doh-sperm	tissue that supplies food to an embryo
fruit		a ripened ovary containing seeds
ovule	O-vyool	female part of a flower that produces a seed
pistil	PIS-til	female organ of a flower
pollination	pol-i-NAY-shun	transfer of pollen from anther to stigma
seed		fertilized ovule in a protective coat
stamen	STAY-min	male organ of a flower

Study Questions

A. True or False

Decide whether each of the following statements is true or false. If the sentence is false, rewrite it to make it true. (Do not write in this book.)

1. Insects are attracted to flowers by the colour and fragrance of the stamens.
2. In flowering plants, the sperm are carried inside pollen grains to the female pistil.

3. Self-pollinated plants produce more varied offspring than cross-pollinated plants.
4. At fertilization, one sperm nucleus joins with the egg nucleus to form endosperm.

B. Completion

Complete each of the following sentences with a word or phrase that will make the sentence correct. (Do not write in this book.)
1. The young plant that develops from a zygote inside a seed is called the ▨ .
2. The transfer of pollen from an anther of one plant to a stigma of another plant is called ▨ .
3. Flowers that have small petals and sepals, and large, feathery stigmas, are probably pollinated by ▨ .
4. The ripened ovary of a bean or pea flower is called a ▨ .
5. Some plants can be propagated easily from broken pieces of stems called ▨ .

C. Multiple Choice

Each of the following statements or questions is followed by four responses. Choose the correct response in each case. (Do not write in this book.)
1. Stamens are
    a) male flower parts composed of anthers and filaments
    b) female flower parts composed of anthers and filaments
    c) male flower parts composed of stigmas, styles, and ovaries
    d) female flower parts composed of stigmas, styles, and ovaries
2. Fertilization in a flowering plant produces
    a) a pollen tube and an embryo sac
    b) a gamete and an ovule
    c) a zygote and an endosperm nucleus
    d) pollen grains and two cotyledons
3. The five petals of a petunia flower together form the
    a) calyx   b) pistil   c) sepals   d) corolla
4. From which part of a flower does the fleshy part of an apple develop?
    a) receptacle   b) ovary   c) seed   d) ovule
5. Rhizomes, corms, and tubers are
    a) roots   b) stems   c) leaves   d) fruits

D. Using Your Knowledge

1. The life cycle of a flowering plant includes a gamete-producing generation and a spore-producing generation. Describe how these two generations alternate in the life cycle.
2. a) Why is cross-pollination preferred over self-pollination?
   b) List three ways how plants prevent self-pollination.

3. From which two sources does a young plant inside a seed get the energy to grow?
4. The seed coats of most seeds are very hard and dry. Suggest why this might be an advantage.
5. Discuss the importance of vegetative propagation to gardeners and farmers.

E. Investigations

1. Research the relationship between the appearance of flowers and their methods of pollination. The colour, shape, odour, and size of the flowers are important factors. Determine which types of flowers attract
   a) bees **b)** butterflies **c)** moths **d)** bats **e)** hummingbirds
2. Flowers have many adaptations to prevent all the seeds from falling to the ground right underneath the parent plant. Germinate a large number of seeds in a pot of soil at home. Observe the growth of the crowded seedlings over several weeks. Are there any advantages to spreading seeds over a large distance?
3. In the autumn, make a collection of fruits from both wild and garden plants. Include fruits such as milkweeds, poppies, sunflowers, acorns, grapes, peanuts, irises, cucumbers, cherries, plums, corn grains, peaches, pears, beggar's-ticks, and burdocks. Then construct a classification scheme for fruit types. Use characteristics such as number of seeds, and whether the fruits are dry or fleshy, edible or nonedible, and split open or closed.
4. Find out what plant cloning is.

15 Roots, Stems, and Leaves

15.1 Vegetative Tissues
15.2 Roots
15.3 Activity: Root Hairs
15.4 Stems
15.5 Activity: Stem Tissues
15.6 Leaves
15.7 The Stomata of a Leaf

Flowering plants produce flowers and fruits for sexual reproduction. They also produce vegetative, or non-flowering, parts. Most plants have three vegetative organs: **roots**, **stems**, and **leaves** (Fig. 15-1).

15.1 Vegetative Tissues

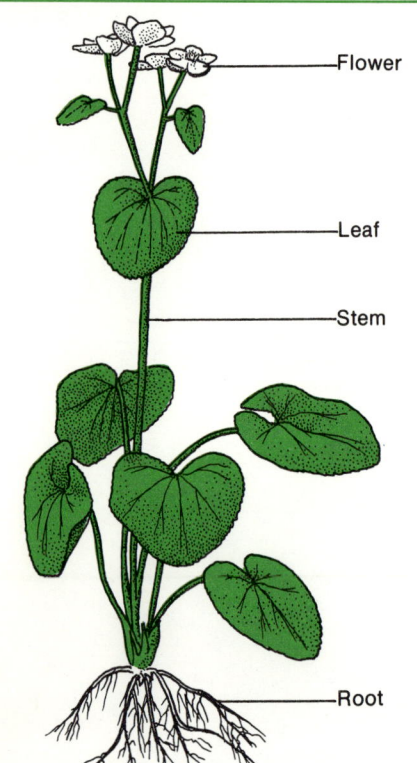

Fig. 15-1 The reproductive and vegetative parts of a flowering plant, the marsh marigold

Flowering plants are highly organized. The vegetative organs are specialized for different functions. The **roots** absorb water and minerals from the soil. They also anchor the plant in place. Some roots store food for the plant until it is needed. **Stems** carry water and food up and down between the roots and the leaves. They also support the leaves, flowers, and fruits. **Leaves** produce food energy for the plant by photosynthesis. They also exchange oxygen and carbon dioxide with the air.

All these vegetative parts grow from small groups of cells located at the tips of the roots and shoots. These cells are called **meristems**. The meristems make new cells by cell division. As the new cells are formed, they start to become specialized.

Some cells become an outer protective layer, called an **epidermis** [ep-uh-DUR-mis]. Others become **vascular tissue** [VAS-cue-ler] to carry water and food. Vascular tissue includes two types of cells. The first type are water-carrying cells called **xylem** [ZY-lem]. They are tubular cells with thick walls (Fig. 15-2). They provide a great deal of strength and support to the plant. When mature, the cytoplasm in xylem cells dies, and, in some cells, the end walls disappear. This leaves some hollow tubes in which water can flow.

The second type of vascular cells are food-carrying cells called **phloem** [FLO-em]. They are tubular like xylem. However, their walls are not as thick (Fig. 15-3). Their ends have holes to allow food to pass through.

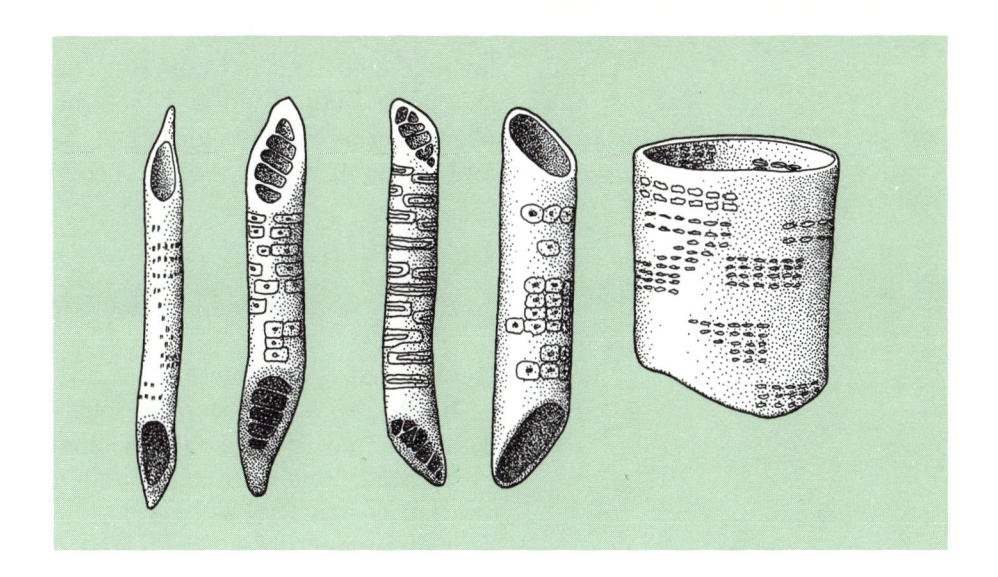

Fig. 15-2 Xylem cells. The cell walls are thickened for strength. The end walls have one or several holes through which water passes.

The bulk of most roots and stems is made up of cells that store food. These large cells form the **cortex**. In stems, they often contain chlorophyll for photosynthesis.

Leaves also contain cells that photosynthesize. The chloroplasts are located in a tissue called the **mesophyll** [ME-so-fill]. "Meso" means middle. Mesophyll cells form layers in the middle of a leaf.

Table 15-1 lists the major types of plant tissues and their functions. Notice that the last tissue in the table is **cambium**. This is a special

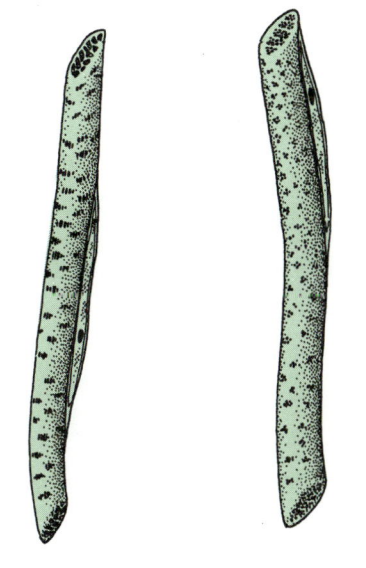

Fig. 15-3 Phloem cells. The holes at the ends of these long tubular cells allow food to move from one cell to the next.

Table 15-1 The Tissues of a Flowering Plant

Tissue type	Description	Function
Epidermis	A single layer of cells forming an outer covering	Protective layer
Vascular: a) xylem	Thick-walled, tubular cells, dead at maturity, with holes in the end walls	To carry water and minerals and to provide strength and support
b) phloem	Tubular, thin-walled cells with holes in the end walls	To carry food
Cortex	Large, thin-walled cells in roots and stems; some have chlorophyll	To store starch; photosynthesis
Mesophyll	Large, thin-walled cells in leaves; contain chlorophyll	Photosynthesis
Meristem	Tiny unspecialized cells at the root and shoot tips	To keep dividing to produce the other tissues during growth of the plant
Cambium	A single layer of tiny cells between the xylem and phloem in roots and stems that become woody	To keep dividing to produce more xylem and phloem

Section 15.1

layer of cells that is found only in dicot stems and roots. The cells of this layer keep dividing throughout the life of the plant. They keep producing new xylem and new phloem cells. Because of this, the roots and stems of woody plants become thicker in diameter every year.

Section Review

1. State two functions for each of the three vegetative organs of a plant.
2. a) What is a meristem?
   b) Where are meristems located on a plant?
3. a) Name the two types of cells that transport materials in a plant.
   b) State one similarity and one difference in the structure of these two types of cells.
4. a) In which plant tissue is food stored?
   b) Where in a plant is this tissue located?
   c) Most food is produced by photosynthesis in the leaf. Name the leaf tissue that produces food.
   d) How does this food get from the leaf to the site of storage?
5. a) Name the tissue that allows woody plants to increase in diameter.
   b) Explain how this occurs.

Fig. 15-4 Tap roots and fibrous roots

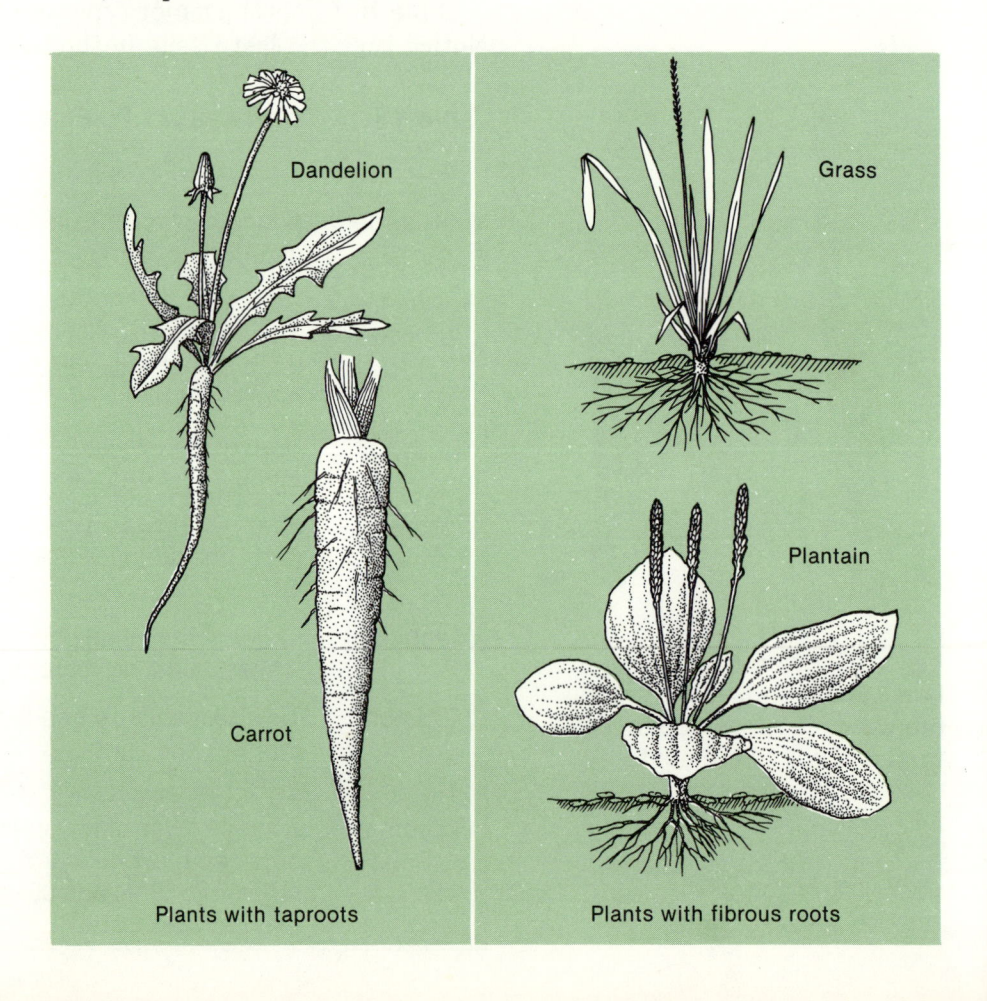

15.2 Roots

Root Types

As a seed germinates, the first part to emerge from the seed coats is the root. In some plants, such as carrots and dandelions, this **primary root** keeps growing to form a large **tap root** (Fig. 15-4). Tap roots are able to reach deep into the soil for water. They are also effective for the storage of food. In other plants, the first root branches into many little roots. These form a mass called a **fibrous root** system (Fig. 15-4). Most grasses have fibrous roots. Their roots spread outward just beneath the surface. They help hold the soil together. They also absorb surface water quickly after rains.

Root Tissues

Roots are composed of many types of cells. Right at the tip of the root is a **root cap**. It protects the root as it pushes through the soil (Fig. 15-5). Behind the cap are the **meristem cells** that keep dividing to produce new cells. This is how the root grows. A few millimetres from the tip, the cells mature into **xylem** and **phloem**, **cortex**, and

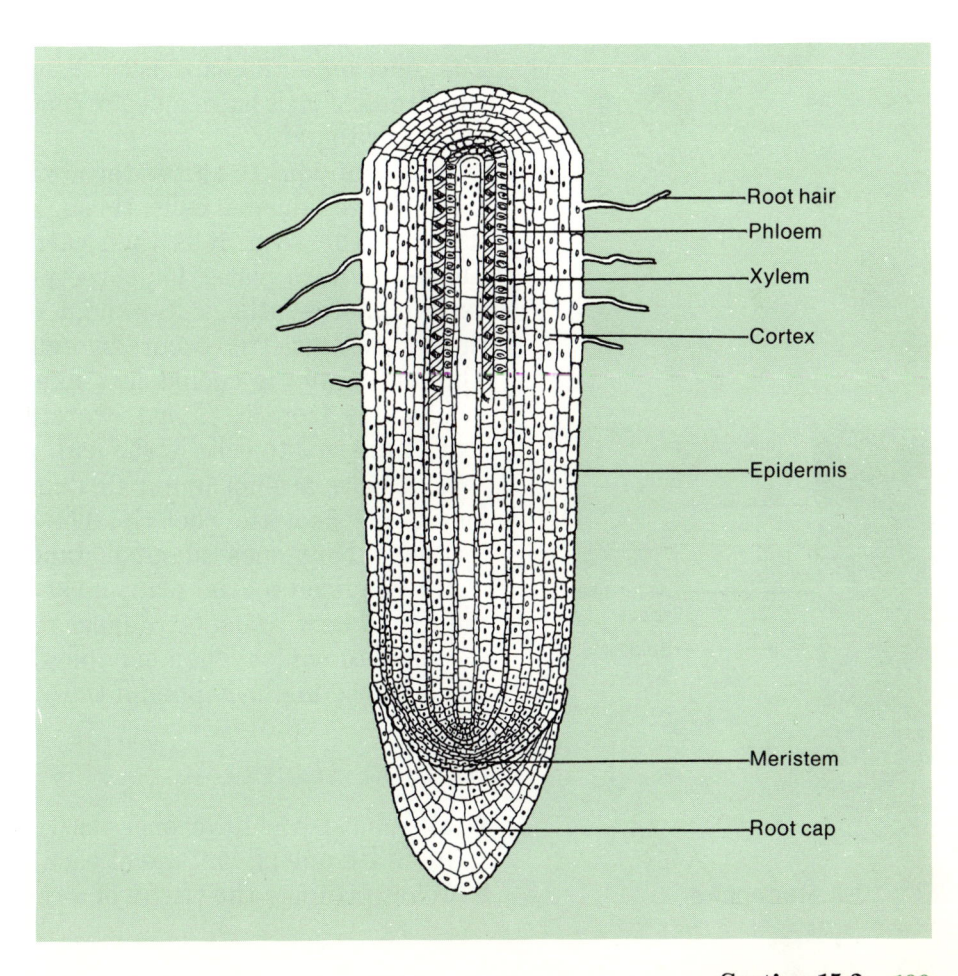

Fig. 15-5 The tissues of a root are protected by the root cap. The root tip is shown here in longitudinal section.

Section 15.2 199

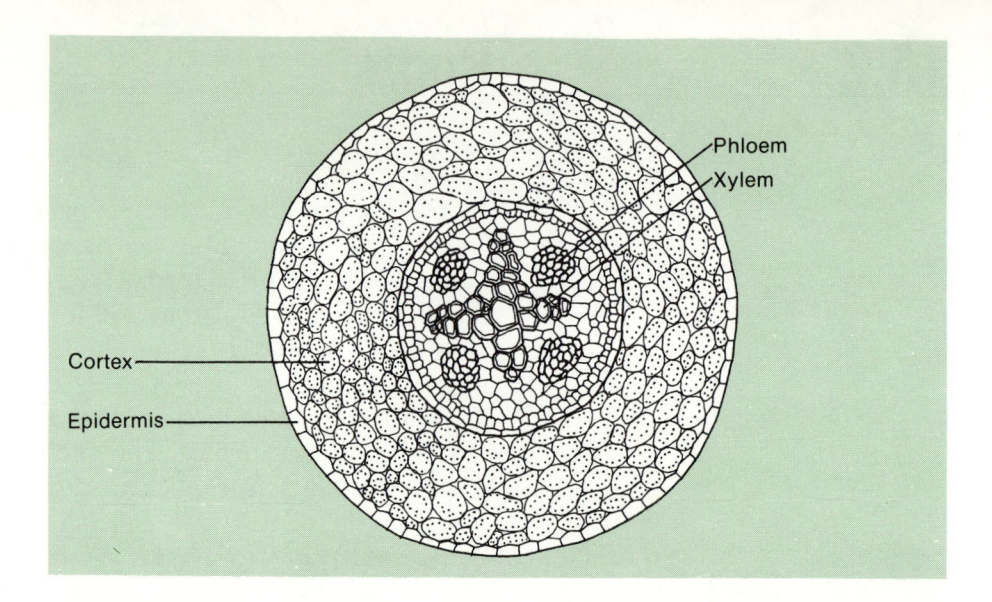

Fig. 15-6 Cross section of a dicot root

epidermis. Figure 15-6 shows these tissues in a cross-sectional view. Do you remember what the functions of these four tissues are?

Root Hairs

In young roots, the epidermal cells produce **root hairs** (Fig. 15-7). The function of the hairs is to absorb water from the soil by osmosis. Hundreds of root hairs on each root provide a very large surface area for water to enter.

What happens to the water absorbed by the root hairs? It moves across the epidermal cells, through the cortex, and into the vascular tissue of the root. It is then carried upward to the leaves. Why? Leaves require water for three purposes. First, a small amount is used in photosynthesis. Second, a much larger amount is used to cool the leaf tissues. This occurs by evaporation of the water into the air. The leaf tissue is cooled just like your skin is cooled when water evaporates from it. Third, water must move through the leaf to carry minerals to cells in the leaf.

Root hairs are not found all along the length of the root. They are found only near the root tip. They are very fragile, and last only a few days. New ones must constantly be produced. Almost all of the water required by the plant must be absorbed through the hairs at the tips of new roots. Gardeners must be very careful not to disturb the roots too much when moving plants from place to place. What do you suppose might happen if the young roots were torn off?

Section Review

1. a) What is the difference between a tap root and a fibrous root?
   b) State one advantage of each type.
2. What protects the tissue of a root as it pushes through the soil?

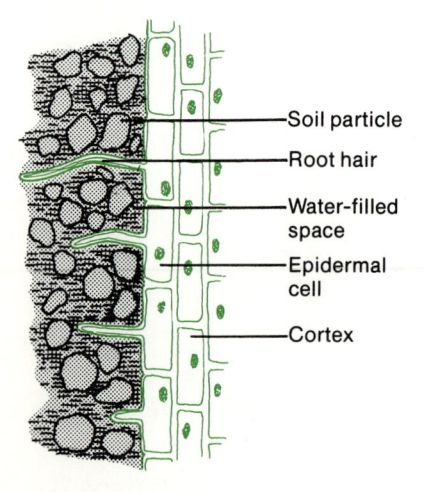

Fig. 15-7 Root hairs

200 Chapter 15

3. The cells in a root tip mature into four types. List the four types and state their functions.
4. a) What is a root hair?
   b) On which part of a root are root hairs found?
   c) What is the purpose of root hairs?
5. a) How does the water absorbed by root hairs reach the leaves?
   b) What is the water used for in the leaves?
6. Why is it important to keep a ball of soil around the roots of a plant when it is moved?

15.3 ACTIVITY Root Hairs

Root hairs are easy to observe on fresh young seedlings of radishes or mung beans.

Problem

Can you find root hairs?

Materials

radish or mung bean seeds (5)   hand lens   masking tape
paper towelling   petri dish

Procedure

a. Cut three pieces of a paper towel into a circular shape that will fit into a petri dish. Put the papers into the dish. Then soak them with water. Pour off the excess water, leaving the papers wet but not floating.
b. Put five seeds onto the paper. Cover them with the lid. Seal the dish with masking tape to keep the moisture in. Keep the seeds warm for 3 to 5 d.
c. When the roots are about 2 cm long, examine them carefully with a hand lens. Describe their appearance.
d. Look for root hairs a short distance from the tip. Measure this distance in millimetres. Describe the colour, location, and length of the root hairs.
e. Make a large drawing of a seedling, showing where the root hairs are. Label the root cap, root hairs, seed, and shoot.

Discussion

1. How far from the tip of the root are the root hairs produced?
2. What is the function of root hairs?
3. Radish and bean seedlings have a very large number of root hairs. Why would this be an advantage?

15.4 Stems

Stems and roots are similar in many ways. They are made of the same types of tissues. They share two important functions. They transport water, food, and nutrients. They also store food materials in their cells.

Stem Types

The growth of stems varies from species to species. Trees such as pines and spruces have one main trunk with short side branches. Others, such as maples, have several large branches that spread outward (Fig. 15-8). Some stems are not upright at all. Cucumber and strawberry stems trail along the surface of the ground. Potato tubers and iris rhizomes are stems that grow underneath the soil instead of above it.

Growth of Stems

The points along a stem at which the leaves attach are called **nodes** (Fig. 15-9). The piece of stem between each node is called an **internode**. The internodes are very long in some plants and short in others.

The angle formed between a node and a leaf stalk is called a **leaf**

Fig. 15-8 Stem types

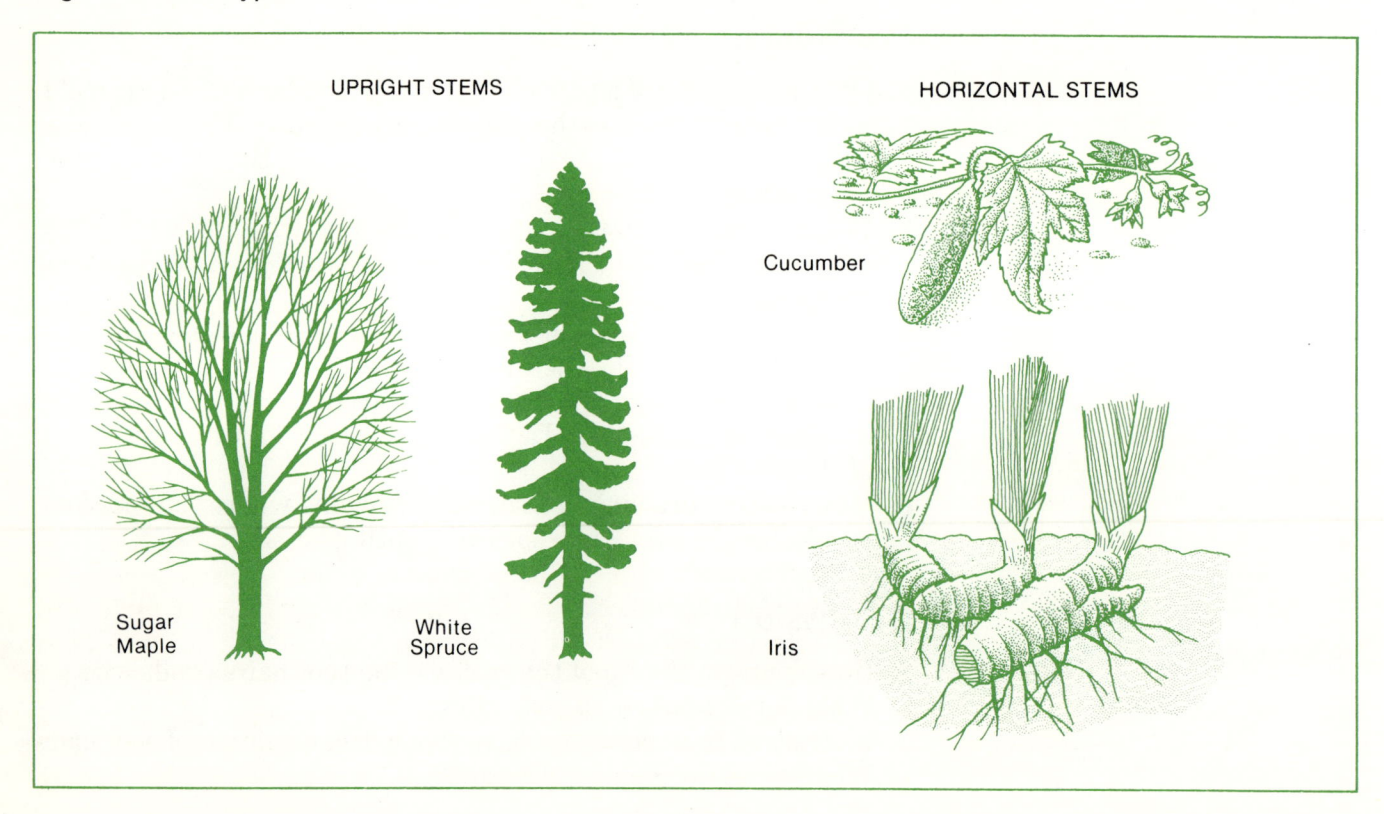

UPRIGHT STEMS — Sugar Maple, White Spruce

HORIZONTAL STEMS — Cucumber, Iris

202   Chapter 15

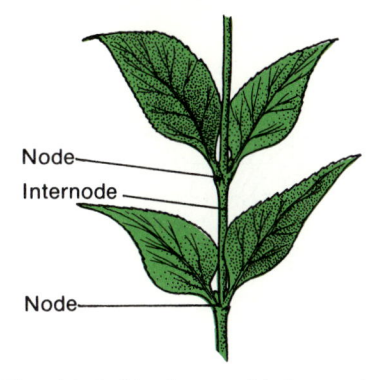

Fig. 15-9 Nodes and internodes of a stem

Fig. 15-10 A vascular bundle (monocot)

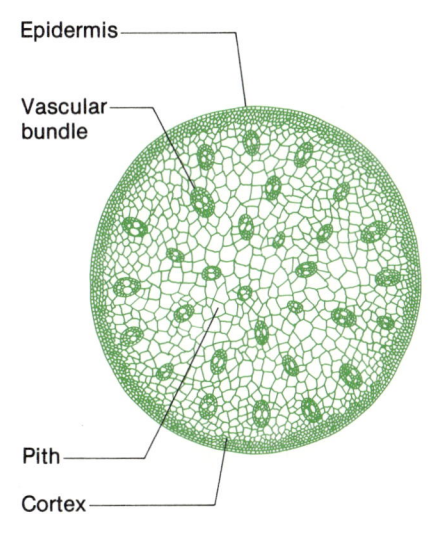

Fig. 15-11 A cross section of a monocot stem

axil. In a leaf axil is a **bud**. A bud can grow into either a branch or a flower.

At the tip of most stems is a bud that is called a **terminal** (end) **bud**. A stem will keep growing upward as long as the terminal bud is not damaged. In many plants, the terminal bud releases a chemical hormone that prevents the buds farther down from growing into branches. If the terminal bud is cut off, the flow of hormones is reduced. Then the other buds start to grow. Gardeners often make plants become branched and bushy by purposely cutting off the tips of the stems. This process is called **pruning**.

Stem Tissues

Stem tissues are very similar to those of roots. The tip of the stem contains a **meristem**. There the cells keep dividing into new cells. These new cells increase in size, causing the stem to grow longer. A stem will keep growing near the tip as long as it is alive. The tip is not protected by any covering similar to a root cap. Can you explain why?

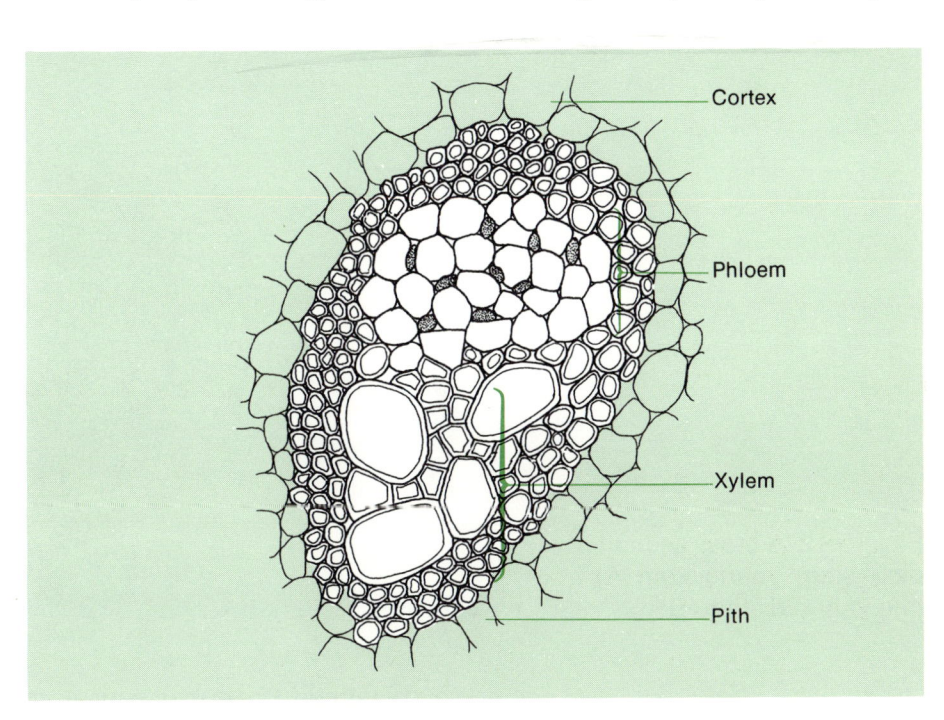

As in the root, a stem has several layers of tissue. On the outside is an **epidermis**. It protects the stem. Under the epidermis are large **cortex** cells that are used for storage of food. In green stems, the cortex cells contain chloroplasts. This allows them to photosynthesize. The **xylem** and **phloem** cells are arranged in small groups called **vascular bundles** (Fig. 15-10). The bundles surround a central region of large storage cells called a **pith**.

In monocot stems such as corn, vascular bundles are scattered through the cortex and pith (Fig. 15-11). In dicot stems, the bundles

Section 15.4 203

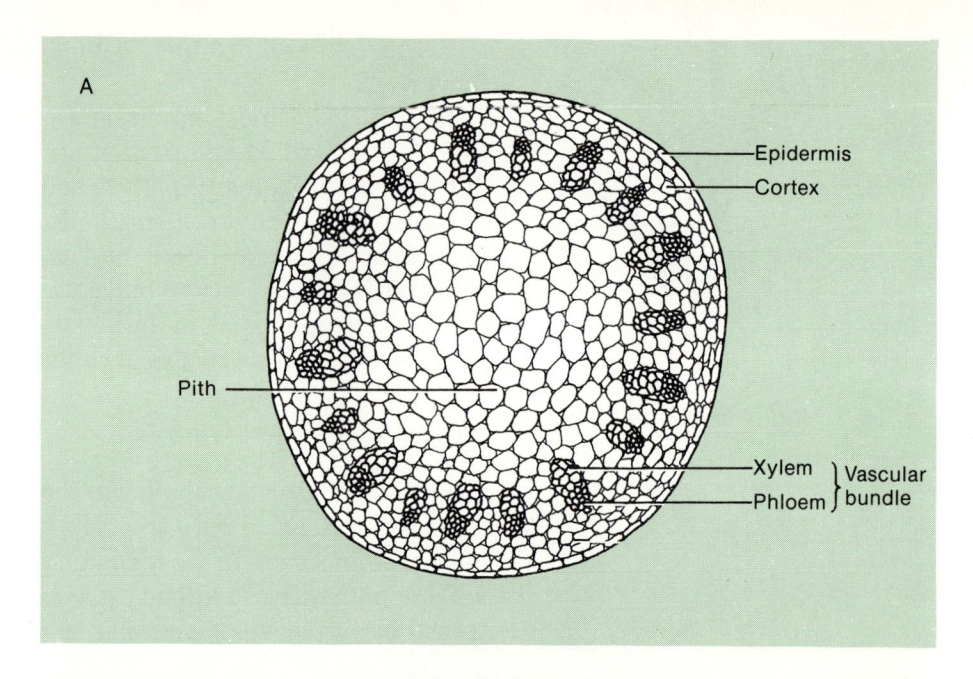

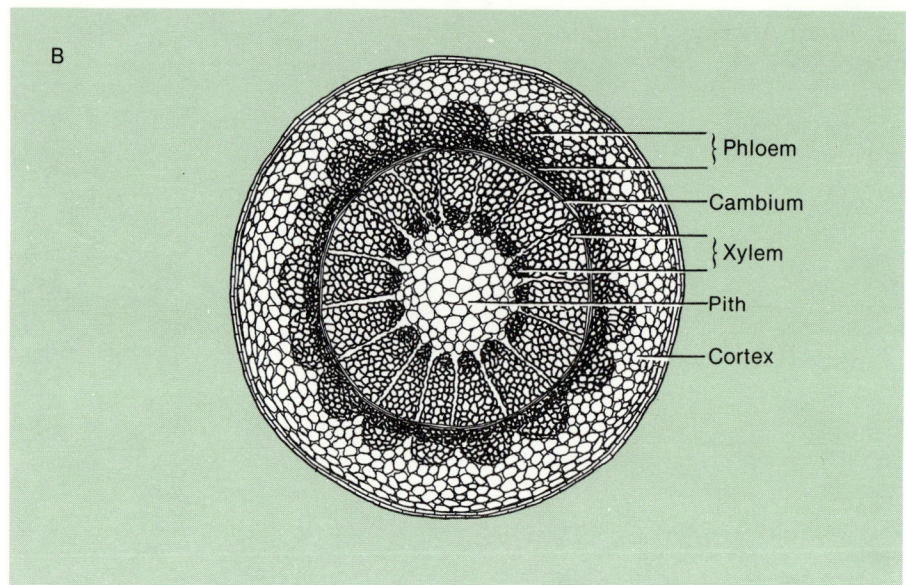

Fig. 15-12 A cross section of a dicot stem: young stem (A); a one year old stem (B)

are arranged in a broken ring (Fig. 15-12). When dicot stems get old, the vascular bundles join together in a complete circle. This is because the tissue called a **cambium** forms between the xylem and phloem. The cambium cells divide to produce new circles of xylem and phloem every year. The circles of xylem are known as **wood**.

It is easy to tell the age of a tree by its wood. If the tree trunk is cut crosswise, it shows **annual rings** (Fig. 15-13). Each ring represents one year's growth. The rings form because the xylem cells of the wood grow differently in the spring than they do in the summer and fall. In spring, the cells that develop from the cambium are large. That's because there is lots of water available for growth. In

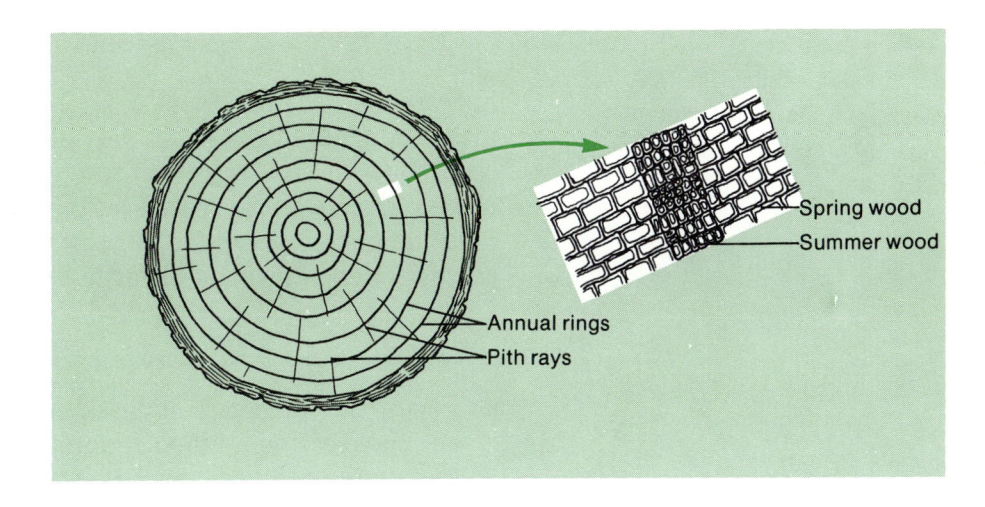

Fig. 15-13 Annual rings in the cross section of a woody trunk. How old is this tree?

summer, with the decreased availability of water, smaller xylem cells are formed. The alternating large and small cells differ in colour. This forms light and dark bands that look like rings.

Section Review

1. State two functions of stems.
2. The stems of irises, pines, maples, and cucumbers grow in different patterns. List and describe five more plants whose stems have different growth patterns.
3. a) Explain the meaning of nodes and internodes.
   b) What is a leaf axil?
   c) In which two positions are buds located on a stem?
   d) What effect does pruning have on buds in leaf axils? Why?
4. a) Why do stem tips not have a covering similar to a root cap?
   b) Name the five main tissues of a stem.
5. a) What does vascular mean?
   b) Describe the arrangement of vascular bundles in both monocot and dicot stems.
   c) What is cambium?
   d) Explain how annual rings are formed.

15.5   ACTIVITY   Stem Tissues

For Part A of this activity, your teacher will provide you with two stained prepared slides of stems. One is a cross section through a monocot stem. The other is a dicot stem. Do you remember the difference between monocots and dicots?

Problem

What do stem tissues look like?

Materials

microscope
prepared slide of the cross section of a monocot stem (such as corn, lily, or iris)
prepared slide of the cross section of a dicot stem (such as buttercup, sunflower, or geranium)
cross section of a trunk or large branch of a tree

Procedure A Monocot and Dicot Stems

a. Focus the monocot slide under low power. You should see a circular section stained red and green. Move the slide around on the stage until you have seen the whole section.
b. Locate the epidermis. How many layers of cells is the epidermis?
c. Locate the cortex inside the epidermis. Its cells are large. The large cells extend right across the stem. In the centre they form the pith. Locate the pith on your slide.
d. Scattered in the cortex and pith are vascular bundles. Note how they are arranged.
e. Make a sketch of the stem section as seen on low power. Label the epidermis, cortex, pith, and vascular bundles. Draw in a few cells of each type.
f. Look at one vascular bundle carefully. Use medium then high power. It contains xylem and phloem cells. Usually the xylem is stained red. Compare the thickness of the xylem cell walls to the cortex cell walls. Make a large labelled diagram of one vascular bundle.
g. Repeat steps (a) to (f) for a dicot stem. Note carefully the arrangement of the vascular bundles. Have they formed a continuous ring? Is cambium present?

Procedure B Woody Dicot

a. Observe the specimen of a tree trunk or branch. Count the annual rings.
b. Each annual ring represents one year's growth. Are all the rings the same width on your specimen? Suggest a reason why they might or might not be.
c. The outer covering of a woody stem is called **bark**. Observe the bark. Describe it in words.

Discussion A

1. How many layers of epidermal cells does a stem have?
2. **a)** Which type of tissue makes up most of the stem?
   b) What is the function of this tissue?
3. **a)** Which cells have the thickest walls?
   b) Why would it be important for a stem to contain some thick-walled cells?

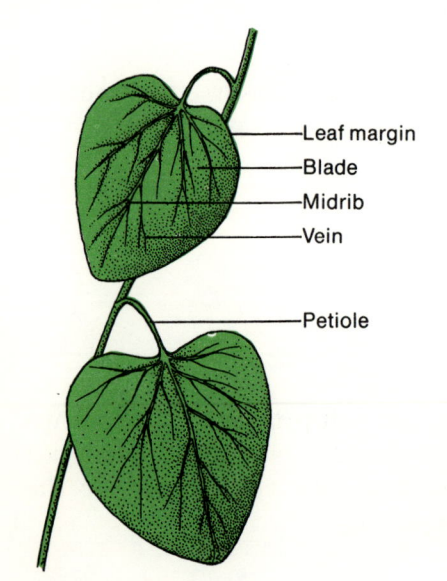

Fig. 15-14 Leaves with thin broad blades photosynthesize large amounts of food.

4. Suggest a probable function for the pith.
5. State two differences between a monocot and a dicot stem.
6. a) Where is the cambium located in a dicot stem?
   b) What is the function of the cambium?

Discussion B

1. How old is the specimen you observed?
2. a) What is the function of bark on a tree?
   b) Which layer has this function in a monocot stem? (Monocots do not produce bark.)

15.6 Leaves

Although leaves exist in thousands of shapes and sizes, they all have one purpose: photosynthesis. Most leaves have thin, broad blades that contain the chlorophyll for photosynthesis (Fig. 15-14).

The blades are supported by **veins** that contain **vascular bundles**. Recall that monocot leaves have parallel veins, and dicot leaves are net-veined. The veins carry food and water in and out of the leaf. The main vein is called the **midrib**. It extends below the blade as a stalk called a **petiole** [PET-ee-ol]. It attaches the leaf to the stem.

The blades of leaves vary in shape, from round, to oval, to oblong, to linear. Their edges may be toothed, smooth, or wavy. The blades may be thick and waxy, or thin and fragile. Figure 15-15 shows a variety of common leaf types. Many plants can be easily identified by the shape of their leaves.

Fig. 15-15 Many plants have distinctive leaf shapes that help in identification.

Section 15.6

Simple and Compound Leaves

Leaves in which the blade is all in one piece are called **simple** leaves. Maple and beech leaves are simple leaves (Fig. 15-16). If the blade is divided into several leaflets, the leaf is **compound** (Fig. 15-17). Leaflets may be arranged in two different ways. Figure 15-17 shows the **pinnately compound** leaves of walnut trees, and the **palmately compound** leaves of horse chestnuts. What is the difference in the arrangement of the leaflets?

Deciduous and Evergreen Leaves

An individual leaf may live anywhere from a few days to several years. Plants whose leaves last more than one year are called **evergreens**. Pines, cedars, and junipers are all evergreens. Each of their needlelike leaves lives for several years. The leaves fall off just a few at a time. Therefore the whole plant stays green all year round.

All the leaves of trees such as maple and oak die and fall off each autumn. These trees are said to be **deciduous** [dee-SID-you-us]. The shortened days and cold nights of autumn cause their leaves to change colour and fall off. Beautiful reds, oranges and yellows can be seen in deciduous forests at that time of year.

Leaf Organization

Leaves are organized for food production. The flat blades absorb sunlight for photosynthesis. The veins deliver the water from the roots. The carbon dioxide that is required enters through small holes in the blade, called **stomata** (singular: **stoma**). (See Figures 15-18 and 15-19).

Leaf Tissues

If a leaf were sliced from top to bottom in very thin sections, it would look like Figure 15-18 in the microscope. On both the upper and lower surfaces is a single layer of epidermal cells. The **epidermis** produces an outer waxy layer, called the **cuticle** [KYOO-ti-kel]. The cuticle protects the leaf from drying out.

The middle layers of the leaf are called **mesophyll** [ME-so-fill]. There are two types. The elongated cells near the upper epidermis are the **palisade mesophyll**. They are tightly packed together, like the logs of a palisade wall around an old fort. In fact, this is how they got their name. The palisade cells contain thousands of chloroplasts to capture sunlight.

The mesophyll cells underneath the palisade layer are called **spongy mesophyll**. They also contain chlorplasts. They vary in shape, and are separated by large air spaces. Gases such as carbon dioxide, oxygen, and water vapour diffuse from cell to cell through these spaces.

Fig. 15-16 Simple leaves

Fig. 15-17 Compound leaves

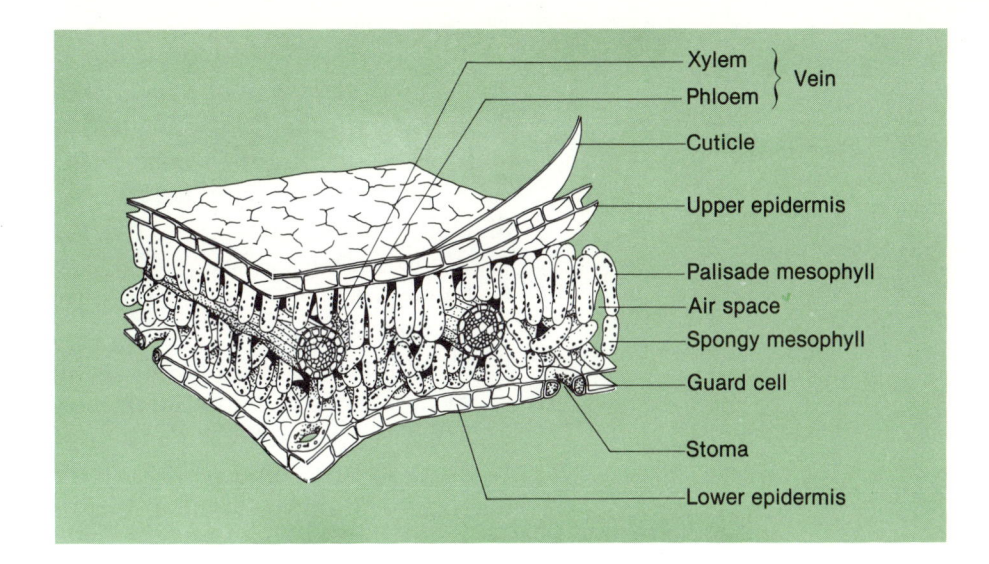

Fig. 15-18 The cross section of a leaf

Scattered through the epidermis are tiny holes, or **stomata**. There are usually more of these in the lower epidermis than the upper. Each stoma is surrounded by two curved cells called **guard cells** (Fig. 15-19). Their curved surfaces make the hole through which gases can diffuse. Guard cells are the only epidermal cells that contain chloroplasts. Their function is to open and close the stomata. The stomata are usually open during the day and closed at night. When they are open, gases are constantly diffusing into and out of the leaves. Also, a lot of water vapour evaporates from the leaf through the stomata. This helps to cool the plant off. The lost water has to be replaced by water from the soil. It is carried up to the leaves by a very efficient plumbing system: the **xylem**. Food made in the leaves is carried away to the rest of the plant by the **phloem**.

There are endless variations to this basic leaf structure. In hot, sunny climates, leaves grow thicker and smaller. The upper surfaces have no stomata and the lower surfaces have only a few. The cuticles are very thick. In shady, moist places, leaves grow large and thin. They do not need thick cuticles. But they do need to gather all the sunlight they can. For these reasons, leaves in wet tropical jungles are huge!

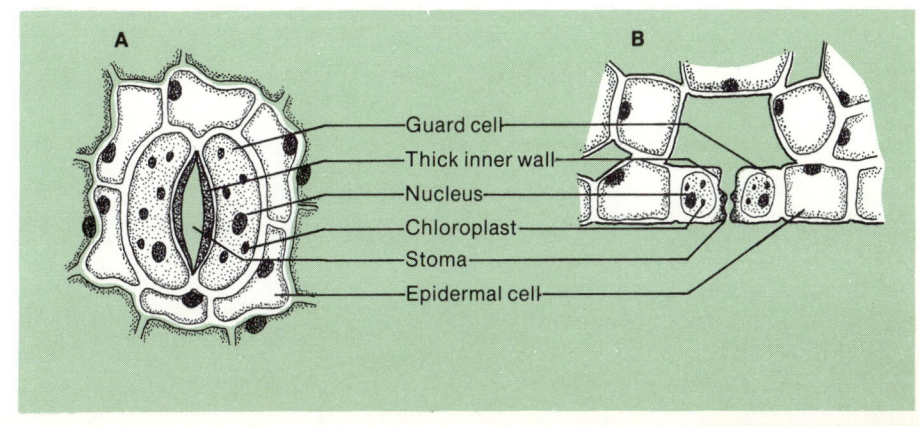

Fig. 15-19 Guard cells and stoma: surface view (A); cross section (B)

Section 15.6 209

Section Review

1. Name three parts of a leaf. State the function of each.
2. **a)** Explain the difference between simple and compound leaves.
   b) Explain the difference between pinnately compound and palmately compound leaves. Name an example of each.
3. Define deciduous. Name five Canadian deciduous trees.
4. What is a cuticle?
5. Name, describe, and state the function of the two types of mesophyll in a leaf.
6. **a)** Where in a leaf are the most air spaces?
   b) What gases would be present in the air spaces?
7. What is a stoma?
8. State two reasons why the stomata need to be open during the day.
9. Name the three types of cells in a leaf that contain chloroplasts.

15.7   ACTIVITY   The Stomata of a Leaf

In the lower epidermis of most leaves are hundreds of **stomata**. They allow gases to diffuse in and out of the leaf during the day. They are usually closed at night.

Each stoma is surrounded by two **guard cells**. How do these cells open and close the stoma? They have to change their shape. They do this by taking in or letting out water. The walls of the guard cells are very thick along the inner side, facing the stoma. The outer walls are thin. When the guard cells expand with water, the thin outer walls push outward. This pulls the thick inner walls apart, creating an opening (Fig. 15-20). When guard cells lose water and shrink, the inner walls collapse against each other. This closes the opening.

You may be wondering why water would go into or out of a guard cell. It has to do with photosynthesis. During the day, the chloroplasts in the guard cells produce sugars. These sugars cause water to enter the cells by osmosis. The cells swell, and the stomata open. This allows gases to get in and out of the leaf. At night, photosynthesis stops. The sugars produced during the day change to starch,

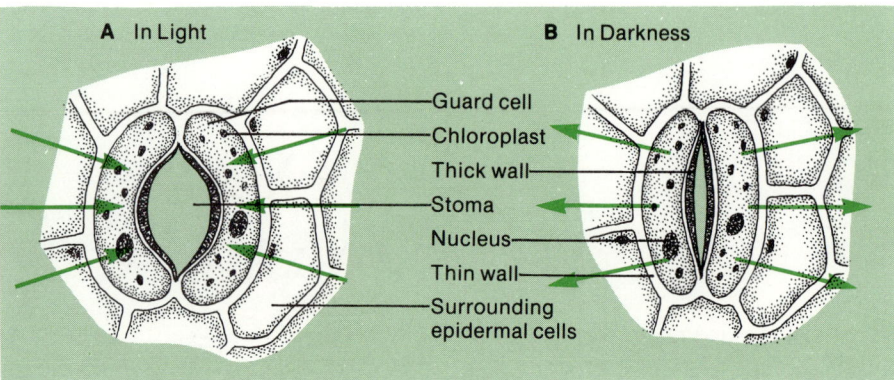

Fig. 15-20 The opening and closing of stomata. Water flows into the guard cells during the daytime (A) and out of the guard cells at night (B).

Without sugars present, water diffuses out of the cells. The guard cells shrink and the stomata close.

The guard cells are easy to see in a microscope, but only if the epidermis can be peeled off the rest of the leaf. That is what you are going to do.

Problem

Can you find and draw guard cells?

Materials

leaf of geranium, purple heart (*Setcreasea*), or Mexican hat plant (*Bryophyllum*)
forceps
microscope
slide
cover slip
iodine solution
dropper

Procedure

a. Hold the leaf with its lower surface facing up. Slowly tear it by pulling one half over the surface of the other half. This should leave a thin, clear piece of epidermal tissue sticking out from one of the torn edges.

b. Use the forceps to remove this tissue. Place it in a drop of iodine solution on a slide. Cover it with a cover slip.

c. Focus on low power. Look for tiny green football-shaped structures. Each "football" is a pair of guard cells surrounding a stoma, or hole.

d. Switch to medium power. Count the number of stomata you can see. Note how they are scattered among the epidermal cells.

e. Switch to high power. Make a large labelled diagram of a few epidermal cells and stomata. Look for chloroplasts in the cells.

Discussion

1. Iodine turns starch black. Did any of the cells contain starch.
2. **a)** What shape are the epidermal cells?
   b) Do epidermal cells contain chloroplasts?
3. **a)** What shape are the guard cells?
   b) Do guard cells contain chloroplasts?
4. Describe how guard cells open the stomata during daylight hours.

Main Ideas

1. The tissues of flowering plants are specialized for different functions.
2. Water moves from the soil by osmosis into the root hairs, across the root, and up the stem of a plant to the leaves.

3. Root and stem tissues are designed for transport and storage of materials.
4. Monocots and dicots differ in the structure of their roots, stems, and leaves. Monocots never become woody.
5. Leaves are very efficient at producing food. That's because all the materials necessary for photosynthesis are brought together in the leaf.

Glossary

cambium	KAM-bee-um	a ring of cells between xylem and phloem
cortex		storage tissue in roots and stems
cuticle	KYOO-ti-kel	waxy coating on a leaf
epidermis	ep-uh-DUR-mis	outer protective layer of cells
guard cells		two curved cells around a stoma
palisade mesophyll	PAL-i-sayd ME-so-fill	elongated leaf cells that photosynthesize
phloem	FLO-em	tissue that transports food
spongy mesophyll		photosynthetic leaf tissue with many air spaces
stoma	STO-ma	hole in the surface of a leaf
xylem	ZY-lem	water-conducting tissue

Study Questions

A. True or False

Decide whether each of the following statements is true or false. If the sentence is false, rewrite it to make it true. (Do not write in this book.)

1. Many roots store starch in the cortex tissue.
2. Flower buds form at the internodes of a stem.
3. Annual rings in tree trunks are formed by layers of phloem tissue.
4. Guard cells are the only epidermal cells of a leaf that contain chlorophyll.
5. Most Canadian deciduous trees have needlelike leaves that live for more than one season.

B. Completion

Complete each of the following sentences with a word or phrase that will make the sentence correct. (Do not write in this book.)

1. Roots have large surface areas for absorbing water because the epidermal cells produce extensions called _____.
2. As a root pushes through the soil, its tip is protected by a _____.
3. The central core of large storage cells in a stem is called the _____.
4. The two types of mesophyll cells in a leaf are called _____ and _____.
5. A compound leaf whose leaflets are attached all along the sides of the midrib is _____ compound.

C. Multiple Choice

Each of the following statements or questions is followed by four responses. Choose the correct response in each case. (Do not write in this book.)

1. The thick-walled cells that transport water in a plant are called
   a) cortex b) xylem c) phloem d) cambium
2. The vascular bundles in a dicot stem are
   a) scattered through the pith and cortex
   b) arranged in a ring
   c) located in the centre of the stem
   d) outside of the cortex under the epidermis
3. Food produced by the leaves is transported to the root by the
   a) cambium b) xylem c) phloem d) pith
4. Trees increase the diameter of their trunks by cell division in the
   a) xylem b) phloem c) cortex d) cambium
5. The leaves of desert plants are usually
   a) small, thick, and waxy
   b) large, thin, and fragile
   c) small, yellowish-green, and flattened
   d) large, thick, and deep green

D. Using Your Knowledge

1. Plants store large amounts of food in the form of sugars and starches. Name three plants that store food in roots. Name three that store food in stems.
2. Trace the path of a water molecule from the soil through a plant to the atmosphere.
3. Describe the methods by which a plant cools itself on a hot sunny day.
4. Supply evidence to support the statement "Leaves are perfectly designed for photosynthesis".
5. The equation for photosynthesis is:

   carbon dioxide + water ⟶ oxygen + glucose (sugar)

   Describe how each of these substances gets into or out of a leaf.

E. Investigations

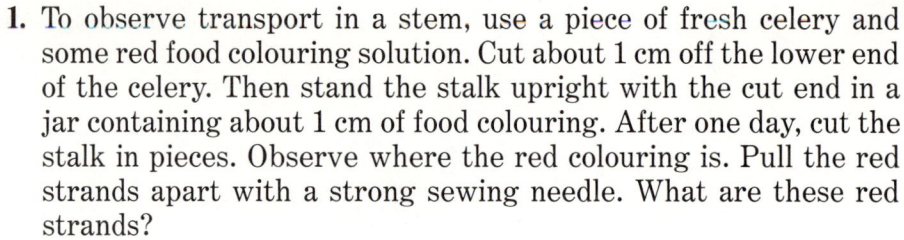

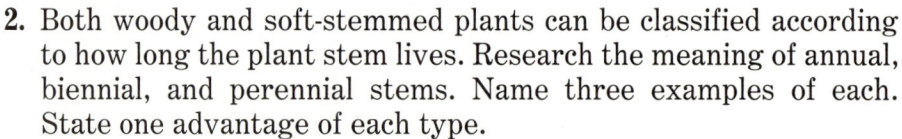

1. To observe transport in a stem, use a piece of fresh celery and some red food colouring solution. Cut about 1 cm off the lower end of the celery. Then stand the stalk upright with the cut end in a jar containing about 1 cm of food colouring. After one day, cut the stalk in pieces. Observe where the red colouring is. Pull the red strands apart with a strong sewing needle. What are these red strands?
2. Both woody and soft-stemmed plants can be classified according to how long the plant stem lives. Research the meaning of annual, biennial, and perennial stems. Name three examples of each. State one advantage of each type.
3. Design an artificial stoma using two long balloons and two pieces of masking tape. Can you make the balloons expand and contract like the guard cells of a leaf to create an opening?
4. Some plants have leaves that are modified to catch insects. Venus' flytraps, sundews, pitcher plants, and bladderworts are insectivorous (insect-eating) plants. Find out how the leaves are modified for such an unusual job! Make sketches of each type.
5. Identifying trees by their leaves is easy in the summer. But identifying trees by their bark and twigs in winter is a real challenge. Obtain a field guide to trees (such as *Native Trees of Canada*, by R. C. Hosie). Use it to identify the native trees in your neighbourhood in the winter.

16 Growing Plants

16.1 Why Grow Plants at All?
16.2 What Plants Need to Grow
16.3 Seed Germination
16.4 Activity: Growing Plants from Seed
16.5 Activity: Growth Conditions for Seeds
16.6 Activity: Grow Your Own Vegetables Indoors
16.7 Activity: New Plants without Seeds

Why do people grow plants (Fig. 16-1)? How do they do it? What information is needed to keep plants healthy? This chapter will help you to answer some of these questions. The activities will give you a chance to grow some plants of your own.

16.1 Why Grow Plants at All?

People grow plants for all sorts of reasons. Many do it to provide food. Others grow plants for the beauty they add to homes and gardens. What is a hobby for some is a life's work for others.

The science of growing plants is called **horticulture** [HOR-tuh-kul-chur]. It is a complex science. A horticulturalist must know a great deal about plant types and their requirements. Plants vary in the amount of sunlight, humidity, and soil moisture needed for best growth. Farmers, florists, and landscape gardeners learn about the needs of the plants they want to grow. They also learn about plant diseases, fertilizers, and pesticides. Then they use this information to provide the best possible growth conditions for their plants. You can do this too, consulting the library, a garden centre, or a garden club in your community.

Plants for Food

The most important use of plants is for food. Dozens of crops are cultivated by farmers, greenhouse operators and home gardeners. All human food comes originally from green plants. Photosynthesis in the leaves changes light energy into stored chemical energy. Animals that eat the plants use that stored chemical energy for growth. Some of the animals, like cattle, are eaten by humans. We may eat the plants directly or we may eat the animals that feed on them. But in both cases, we are depending on photosynthesis for our food.

Many types of food crops are grown in Canada. The most important are the grain crops, such as wheat, rye, barley, and corn (Fig. 16-2). These are milled to provide us with baked goods and cereals. Large amounts are also used to feed livestock. A second

Fig. 16-1 Can you imagine a world without flowers?

Section 16.1   215

important group of crops is the legume group. This includes peas, beans, soybeans, peanuts, clover, and alfalfa. Some of these are also used to feed farm animals.

Each summer, large quantities of fruits and vegetables are grown in orchards and market gardens all across the country. Fresh Canadian produce is sold in roadside stands, farmers' markets, and grocery stores until late October or November in most areas.

Other Canadian food products that come from plants include maple syrup, honey, vegetable oils, and many herbs and seasonings.

Other Uses of Plants

Plants are valuable to people in many other ways. Vast forests of trees provide us with lumber, pulp and paper, and resins. Some plants provide medicines and drugs to cure human diseases. Fibers from plants such as cotton and flax can be used to make clothing. The forests and fields of the countryside also provide recreational value. Thousands of people enjoy hiking or camping in the woodlands of Canada's parks and wilderness areas (Fig. 16-3).

Finally, the beauty of plants should not be overlooked when considering their importance. Imagine how bare a park, a city street, or a country lane would be without trees, shrubs, or flowers. Every spring, many people look forward to planting and caring for a flower garden, a home vegetable patch, or a balcony garden. Horticulturalists, dealers in seed, and landscape designers encourage people to enjoy plants. These people provide the seeds, plants, shrubs, and tools needed for successful growing.

Fig. 16-2 This crop of corn provides food for livestock.

Fig. 16-3 Do you enjoy camping or hiking in forests?

Section Review

1. **a)** What is horticulture?
   b) List five types of information that horticulturalists must consider to grow healthy plants.
2. Explain why humans would have no food if photosynthesis did not occur.
3. **a)** Name the major grain crops grown in Canada.
   b) Why are the grains considered to be the most important crops?
4. In addition to grains, what are the other two important categories of food grown in Canada?
5. Plants provide many jobs besides farming. List and describe at least eight occupations that involve working directly with plants.

16.2 What Plants Need to Grow

The growth and development of a plant is determined by many factors. Its genetic makeup and the environment both affect its final form. By controlling the environment, a gardener can modify the growth of a plant a great deal.

Fig. 16-4 This cactus is not watered in the winter. It is watered every 3 weeks in the summer. This watering schedule imitates the rainfall pattern of the desert.

Light

Of all the factors that affect plant growth, water and sunlight are the most important. **Light** is required for photosynthesis. Plants grown in darkness become tall, spindly, and yellowish. They soon die from lack of food. The amount of light needed each day to remain healthy varies from plant to plant. Shade-loving plants such as begonias, impatiens, and ferns die if planted in a sunny location. Sun-loving plants such as marigolds, geraniums, and tomatoes do poorly if planted in shady areas. A few plants, such as petunias, will do well in full sun or in partial shade.

Water

Water is needed by a plant for cell division, photosynthesis, and growth of cells. It is also needed to cool the plant tissues. This cooling occurs as the water evaporates from the leaves. Over 90% of the water absorbed through the roots evaporates into the air.

Some species of plants require large quantities of water to grow well. Others, such as desert plants, are adapted to survive on small amounts of water. If a cactus receives water every day, its tissues will soon rot. A person who grows plants needs to know how much water the plants require (Fig. 16-4).

Soil

Although the **soil** in which plants are growing must supply water to the roots, it must also supply air. The spaces among the particles in most soils contain both air and water. Oxygen in the air pockets is able to diffuse into the root cells of the plants. If the soil is waterlogged, no air gets to the roots. In this case, the roots turn brown and rot away. Then the plant dies. A healthy plant has firm white roots with numerous root hairs near the tips.

The quantity of air and water a soil contains depends on the amount of sand, clay, and humus in it. **Humus** is dead organic matter, such as peat moss or manure. Light sandy soils do not hold water very long. The water trickles down between the large sand particles, out of reach of most plant roots. To improve a sandy soil for gardening or farming, humus should be mixed into it. Humus helps hold rain water for a longer period of time.

Heavy clay soils are poor for plant growth for the opposite reason. They hold too much water, and not enough air. Clay particles are very tiny. They compact together and make it difficult for roots to grow. Carrots and potatoes grown in clay soils are small because the clay will not allow them to expand. In wet weather, clay absorbs water to form mud. In dry weather it shrinks and cracks. Adding large amounts of humus to clay breaks apart the particles and solves all these problems.

Of course the best soil for plant growth is a mixture of sand, clay, and humus. This is called **loam**, or **topsoil**. Nurseries and garden centres sell it by the bag or truckload. They also sell other non-soil substances to mix with soil for growing plants indoors. Two of these substances are **vermiculite** and **perlite**. They both help to lighten the soil and provide more air spaces. They are excellent for starting seeds or for rooting cuttings, because they are light weight, and free of weed seeds, fungus spores, and bacteria.

Nutrients

Soils also provide **nutrients** for plant growth. The three main nutrients needed are nitrogen, phosphorus, and potash. **Nitrogen** is needed for deep green leaves. **Phosphorus** and **potash** are needed for the development of healthy roots, stems, and flowers.

Plants remove these nutrients from the soil. Therefore gardeners replace the nutrients by adding fertilizers. Most packages of fertilizer have three numbers on them, such as 10-6-4, or 21-7-7 (Fig. 16-5). The first number shows the amount of nitrogen. The second shows the amount of phosphorus. And the third shows the amount of potash. A fertilizer with a high nitrogen number should be used on leafy plants such as grass. Root crops such as carrots and beets benefit from a fertilizer with a high phosphorus or potash number.

Fig. 16-5 This fertilizer is used to make grass grow fast in the spring. How do you know?

Temperature

One other factor that affects plant growth is **temperature**. As the temperature rises, the rates of photosynthesis and growth increase, up to a certain point. At very high temperatures (or very low ones), growth stops. For each species of plant, there is an **optimum temperature** for growth. For most plants, this best temperature is between 10°C and 40°C. For instance, peas grow best in the early spring when the temperature is below 15°C. Tomatoes and cucumbers would die at that temperature. They require hot summer days to do well.

Section Review

1. List four environmental factors that affect plant growth.
2. **a)** Describe how lack of light affects plant growth.
   b) Explain why this occurs.
3. Plant roots absorb large amounts of water. For what is this water used?
4. Why do plant roots require air in the soil?
5. **a)** Describe why neither sand nor clay are well suited for growing plants.
   b) Explain how each of these soil types can be improved.
6. Explain the meaning of the numbers 10-6-4 on a package of fertilizer.

16.3 Seed Germination

Growth of plants usually begins with a seed. Most seeds are produced in late summer or fall. They usually do not **germinate**, or sprout, until the spring. Over the winter, they remain in a resting state called **dormancy** [DOR-man-see]. During dormancy, the life processes of the embryo in the seed are very slow. The energy needed to keep the embryo alive is supplied by the stored food. If seeds are dormant for too long, all the stored food is used up. Then the embryos die. Then the seeds will not germinate. In this case, the seeds are said to be no longer **viable** [VY-a-bul].

The ability of seeds to germinate is called **viability**. When you buy packets of vegetable or flower seeds, the viability is sometimes stamped on the labels. If the packet is 80% viable, for instance, you know that 80 out of 100 seeds will probably germinate. The viability decreases as the seeds get older. Seeds of most plants remain viable for four to ten years. Storing seeds in a cool dry place helps to keep them viable for a longer time.

Conditions for Germination

After a period of dormancy, seeds will germinate if they receive oxygen, water, and warmth. The first step in germination is the absorption of water. Water softens the seed coats and makes the seed swell up. Then the young plant emerges, root first. Too much water will cause the seeds to rot because it encourages the growth of bacteria and fungi.

Oxygen is needed by the embryo for its life processes. During germination, the cells divide and grow very quickly. This requires more oxygen. Seeds that are planted too deep in the soil do not get enough oxygen to germinate.

The best temperature for germination is different for each type of seed. Tomato seeds germinate best at 26°C, while peas and spinach seeds germinate best in cold soil (15°C). Most plants require temperatures somewhere between these two.

Germination of a Bean Seed

In Chapter 14, you learned about the structure of seeds by observing beans and corn. Each of these germinates in a slightly different way. When a bean seed germinates, it absorbs water from the soil and swells in size. The seed coats soften, and the cells of the embryo begin to divide at a fast rate. A tiny root grows out of the seed. It soon branches into a mass of secondary roots. They anchor the seedling and absorb water and minerals from the soil (Fig. 16-6).

Meanwhile, the shoot begins to grow. It arches above the soil, carrying the cotyledons with it. The arch straightens out as the first two leaves unfold. The two cotyledons turn green for a short while.

Fig. 16-6 Germination of a bean seed

Section 16.3 219

They supply the seedling with food energy until the leaves begin to photosynthesize. Then they dry up and fall off.

Between the first two leaves is a bud that is able to produce more leaves and a stem. This bud allows the shoot to keep growing upward for the life of the plant.

Germination of a Corn Grain

Germination of a corn grain also begins with the absorption of water. Recall that a grain is really a fruit with a seed inside it. When the outer ovary wall and seed coats soften, the root of the embryo emerges (Fig. 16-7). A branched root system soon develops. Then the shoot pushes up through the soil. It forms the first leaves and becomes photosynthetic when it reaches the soil surface. Unlike the bean, the seed coats and cotyledons remain underground. The stored food is absorbed by the growing seedling, and the grain withers away in the soil.

Most seeds germinate in a pattern similar to either corn or bean seeds. In the activity that follows, you will be able to observe seed germination yourself.

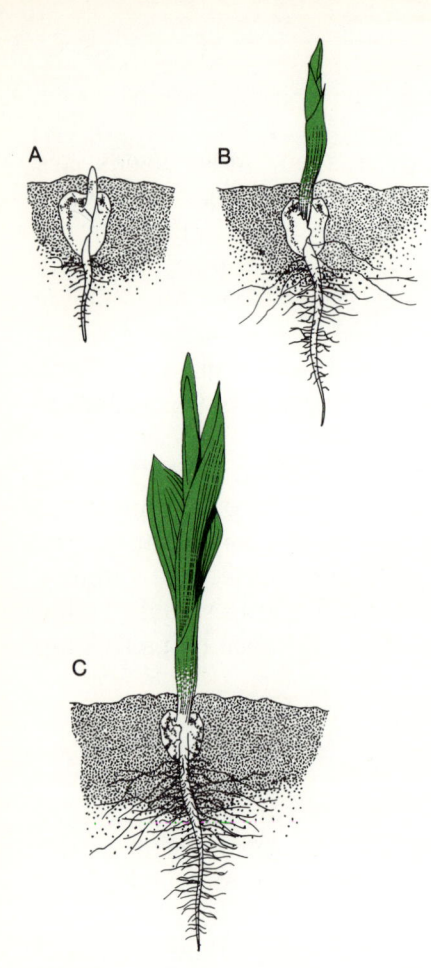

Fig. 16-7 Germination of a corn grain

Section Review

1. **a)** Define seed germination.
   b) Define dormancy.
   c) Why is dormancy an advantage to plants?
2. **a)** What does viability mean?
   b) Suppose you found some old packets of seeds in your house. How could you test their viability?
   c) What conditions would you have to provide for the best chances of germination?
3. **a)** Which part of the embryo emerges first from a germinating seed?
   b) How do the cotyledons of a bean seed get out of the soil?
   c) What purpose do the cotyledons serve? When are they no longer needed?
4. How does germination of a corn grain differ from germination of a bean seed?

16.4 ACTIVITY Growing Plants from Seed

Have you ever taken a handful of dry, apparently lifeless seeds, and watched them germinate into healthy green plants? Let's try it! Your teacher will organize the class so that some groups of students are working with beans, others with corn, and others with pea seeds.

After the seeds have developed into young plants, they will need nutrients for proper growth. You can provide the nutrients by planting them into pots of soil. This is called **transplanting**. Part B of this activity explains how to transplant seedlings.

Problem

Can you grow plants from seeds and transplant them?

Materials

6 seeds of one type (beans, peas, corn or others available)
jar (or beaker)
water
paper towel (absorbent paper)
vermiculite (or perlite or styrofoam chips)
soil mix (peat moss, vermiculite, and soil in equal amounts)
labels
8-10 cm diameter plastic pot

Procedure A Germinating Seeds

a. Soak the 6 seeds in water overnight to soften the seed coats.
b. Insert a folded paper towel into the jar so that it lines the sides. Fill the space in the centre with vermiculite (Fig. 16-8).
c. Label the outside of the jar with your name, the date, and the type of seed.
d. Between the paper towel and the sides of the jar, insert the 6 seeds. Spread them an equal distance apart. Add enough water to wet the vermiculite. But do not cover the seeds.
e. Keep the jar in a warm bright location for about two weeks. Continue to add enough water to keep the paper towel wet at all times.
f. Every second day, observe the seeds for germination and growth. Record how many days it takes each seed to germinate. Observe how the root, stem, and leaves develop. Note the colour, position, and condition of the cotyledons and the seed coats.
g. Every second day make a labelled diagram of one seedling. Title each diagram with the date. Measure and record the length of the roots and shoot on the diagram.
h. Compare your results with other groups of students who are working with different types of seed.

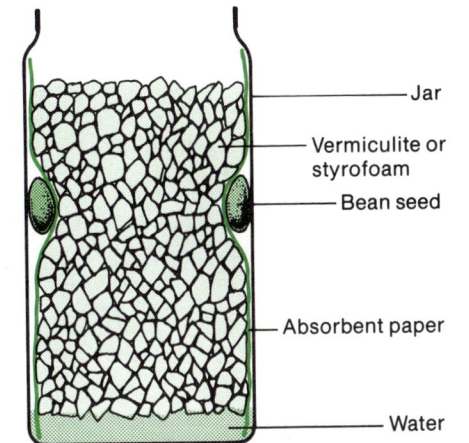

Fig. 16-8 A method of germinating seeds

Procedure B Transplanting Seedlings

a. Prepare a mixture of equal amounts of peat moss, sterilized soil, and vermiculite.
b. Place about 2 cm of soil mix in the bottom of the pot.
c. Gently remove a seedling from the beaker. Do not squeeze the stem or damage the root tips.
d. Hold the seedling by a leaf so that it is centred in the pot, with its roots touching the soil (Fig. 16-9). Add more soil and pack it down so that the seedling is firmly supported. Leave at least a 1 cm space at the top of the pot for watering.
e. Place the pot on a drainage tray. Water the plant well (until water comes out of the drainage holes in the bottom of the pot). Keep it in a sunny location. Water it whenever the top of the soil is dry.

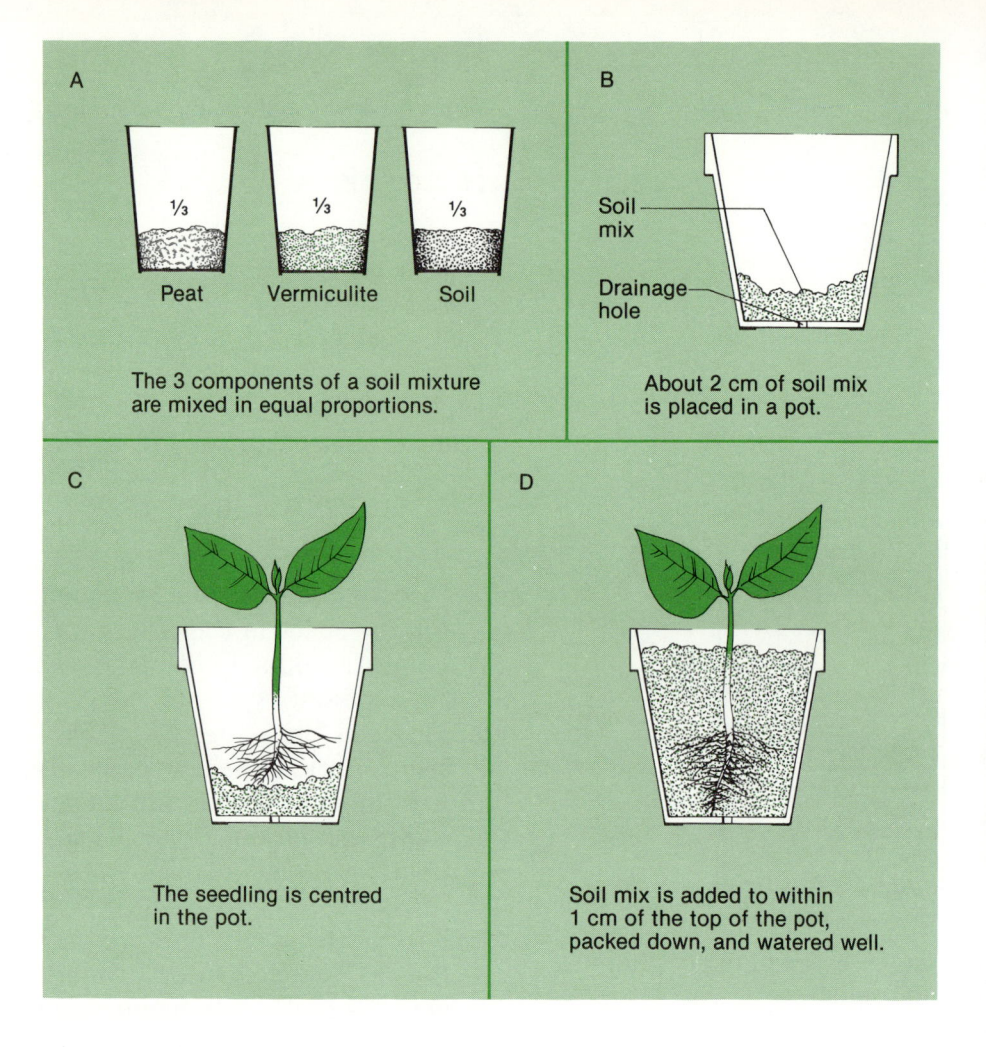

Fig. 16-9 Steps in transplanting seedlings

Discussion A

1. Compare the germination of bean, corn, and pea seeds by answering the following questions for each type:
    a) How many days did it take for any seeds of this type to germinate?
    b) How many days did it take for all 6 seeds of this type to germinate?
    c) Which part of the embryo appeared first?
    d) Which part of the seedling turned green first?
    e) What happened to the cotyledons?
    f) What happened to the seed coats?
2. a) Why would all 6 seeds of one type not necessarily germinate on the same day?
   b) What factors in the classroom might affect the length of time for germination?

Discussion B

1. Why will seedlings not thrive for very long in a beaker of water?
2. Explain what is meant by "transplanting".

16.5 ACTIVITY Growth Conditions for Seeds

You now know that temperature, moisture, oxygen, and light are all factors that affect germination. This activity tests the effects of two of these factors.

Problem
How do light and temperature affect germination?

Materials

40 tomato seeds
4 petri dishes
paper towels
masking tape
aluminum foil
labels, or a marking pen
scissors
water

Procedure

a. Cut circles out of the paper towels to fit in the bottom of the petri dishes. Put 3 layers in each dish. Soak the papers well with water. Pour off any excess water so that the papers do not float.
b. Put 10 tomato seeds into each dish. Cover and seal the dishes with masking tape to keep them from drying out.
c. Wrap two of the dishes with foil to keep out the light. Label one of them "A: dark and warm" and the other "B: dark and cold".
d. Label the remaining two dishes "C: light and warm" and "D: light and cold".
e. For the next 10 d, keep each dish in an appropriate location, as follows: Dishes A and C in a bright warm spot in the classroom; Dishes B and D in a refrigerator in which an electric light bulb has been turned on. The foil will keep the light out of dishes A and B.
f. Every second day, check the dishes for germinating seeds. Count how many have germinated in each dish. Record the number in a table like Table 16-1 in your notebook.
g. The last column in Table 16-1 is a calculation of percent germination. For example, if only 4 seeds out of 10 germinated, the percent germination would be 40%. Calculate the percent germination for each dish after 10 d, as follows:

$$\frac{\text{number of seeds germinated}}{\text{total number of seeds in dish}} \times 100$$

Record your calculations in the table.
h. Keep the germinated tomato seeds. You will need them for Activity 16.6.
i. If your teacher has other species of seeds available, such as lettuce, radish, pea, grass, or bean, repeat steps (a) to (g) for another species.

Table 16-1 Seed Germination

Type of seed _____							
Growing conditions	Number of seeds put in	\multicolumn{5}{c	}{Number of seeds germinated}	Percent germination
		2d	4d	6d	8d	10d	
A: dark; warm							
B: dark; cold							
C: light; warm							
D: light; cold							

Discussion

1. **a)** In which conditions was germination the best?
   b) In which conditions did the fewest seeds germinate?
   c) Suggest reasons for your findings.
2. **a)** What effect does light have on germinating tomato seeds?
   b) Some seeds will not germinate in the dark. Others will not germinate in the light. Do any of the seeds you used fall into these categories?
3. **a)** What effect does temperature have on germinating seeds?
   b) Would any of the species you used germinate well in the early spring in a garden? Explain.
4. Why might some seeds not germinate at all in this activity?

16.6   ACTIVITY   Grow Your Own Vegetables Indoors

In this activity you can use the seedlings germinated in the previous activity to start an indoor vegetable garden.

Problem

Can you grow tomato plants?

Materials

3 germinated tomato seeds
8-10 cm diameter plastic pot
15 cm diameter plastic pot
other seeds that your teacher
   may provide

labels
soil mix
scissors
wooden stake
pencil

Fig. 16-10 Transplanting tomato seedlings to a larger pot

Procedure

a. Prepare a mixture of equal parts peat moss, vermiculite, and sterilized soil.
b. Fill the small pot firmly with the soil mix, leaving a 1 cm space at the top.
c. Poke 3 small holes in the soil mix using a pencil. Carefully transplant the tomato seedlings into these holes. Don't damage the delicate tissues. Press the soil mix gently but firmly around each seedling.
d. Water the pot by placing it in a sink where water can soak upward through the drainage holes. After the soil mix is completely moist, remove the pot from the sink.
e. Place the pot in a sunny warm windowsill or under artificial lights. Water the seedlings whenever the top of the soil mix is dry.
f. After a week or two, either all or some of the seedlings should be growing well. Select the healthiest one to grow into an adult plant. Cut off the others at the soil line so they will not compete.
g. After several weeks, the seedling will have grown too large for its pot. Its roots will be growing out of the drainage holes. At this time you will have to transplant it into a larger pot with more soil mix. Figure 16-10 shows you how to do this.
h. Keep the plant well watered and provide it with as much light as possible. As it gets larger, support it with a wooden stake. After 4-6 weeks, watch for flowers among the leaves.
i. This activity can be expanded to include other vegetables if seeds are available. Parsley, chives, lettuce, spinach, and radishes all

Section 16.6

grow well indoors if they receive lots of light. In several months you will be able to harvest your experiment for lunch!

Discussion

1. Why is each pot planted with 3 seedlings instead of only one?
2. Describe how to transplant seedlings that have outgrown their pots.
3. If you want to move your vegetables outdoors, you have to let them adjust slowly to the new conditions. They should be put out only one hour the first day, two hours the second, and so on for about 1½ weeks. This adjustment is called **hardening off**. Describe the differences between the indoor and outdoor environments that make "hardening off" necessary.

16.7 ACTIVITY New Plants without Seeds

In Chapter 14, you learned that seeds are not necessary to produce new plants. A piece of a stem, root, or leaf can develop into a new plant under proper conditions. This process is called **vegetative propagation**.

One of the fastest ways to propagate plants is from **stem cuttings**. These are short, leafy pieces of stem that will produce their own roots. A piece of stem 10 to 15 cm long is removed from a plant using a sharp knife (Fig. 16-11). The cut is trimmed at a 45° angle just below a node (the point where the leaves arise). The roots will develop from tissues in the node. The lower leaves and all flowers are then removed. This ensures that all the energy is used to produce roots.

Rooting occurs faster if the cut end of the stem is dipped in a powder called **rooting hormone**. This powder contains a chemical that speeds up cell divisions. Since only a small amount is needed, the excess powder should be tapped off.

Once a cutting is prepared, it is transplanted firmly into a mixture of damp peat moss and vermiculite. It must be kept well watered and warm (21°-26°C) for several weeks. Covering the cutting loosely with clear plastic helps to reduce evaporation from the leaves. This gives the cutting a better chance to survive while it lacks roots for absorbing water.

Cuttings of most plants root best if kept in a well-lit location but not in bright sunlight. Rooting may take anywhere from one week to several months.

Problem

Can you grow plants from cuttings?

Fig. 16-11 Follow these steps when rooting cuttings indoors.

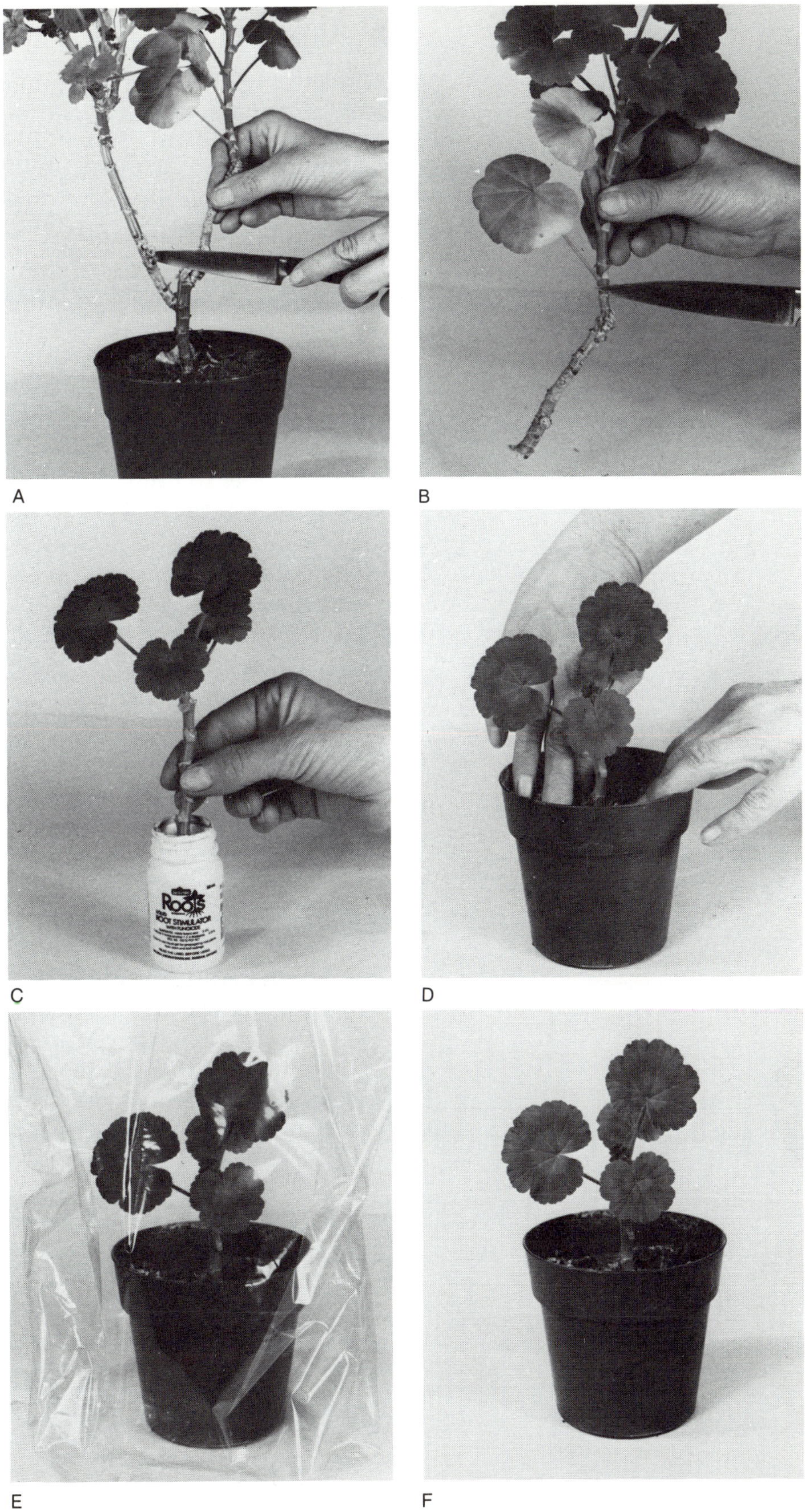

A

B

C

D

E

F

Section 16.7 227

Materials

stock plant of geranium, coleus, begonia, ivy, or other houseplant
sharp knife
rooting hormone powder
10-15 cm diameter plastic pot
mixture of peat moss and vermiculite (50:50)
clear plastic bag
water
pencil

Procedure

a. Fill a plastic pot with the peat moss-vermiculite mixture, leaving a 1 cm space at the top. Poke a hole 3-4 cm deep into the centre of it with a pencil.
b. Use the knife to prepare a stem cutting of one of the plant types available. Dip the end of the cutting in rooting hormone.
c. Insert the cutting into the hole in the vermiculite mixture. Press the mixture firmly around the stem with your fingers so that the cutting will stay upright.
d. Water the cutting well. Then cover it with a clear plastic bag.
e. Keep it in a bright warm location for several weeks. Make sure it never dries out during this time.
f. By tugging gently on the stem you will be able to tell when roots have formed. When rooting has occurred, the plastic bag can be removed. At that time, you may wish to transplant the cutting into a pot of soil mix and take it home.

Discussion

1. What is vegetative propagation?
2. Why is a stem cutting cut made just below a node?
3. Why are flowers removed when cuttings are made?
4. Explain why rooting hormone is used.
5. Under what environmental conditions will cuttings root best?

Main Ideas

1. People are interested in growing plants for many different reasons.
2. Plants require water, light, warmth, oxygen, and nutrients for successful growth.
3. Seeds will germinate from a dormant state if supplied with oxygen, water, and warmth.

4. Growing healthy plants from seed is easily accomplished with only a few simple materials.
5. Many plants can be propagated vegetatively by stem cuttings.

Glossary

dormancy	DOR-man-see	a resting state
germination	jur-muh-NAY-shun	the sprouting of seeds
horticulture	HOR-tuh-kul-chur	the science of growing plants
humus	HYOO-mus	organic matter in the soil
transplanting	TRANZ-plant-ing	moving a plant from one place to another
viable	VY-a-bul	able to germinate

Study Questions

A. True or False

Decide whether each of the following statements is true or false. If the sentence is false, rewrite it to make it true. (Do not write in this book.)
1. All human food is derived originally from green plants.
2. The first part of the embryo to emerge from a germinating seed is the stem.
3. Seeds must be buried deep in the soil before they will germinate.
4. Tomato seeds germinate best when the temperature is 15°C.
5. Stem cuttings produce roots at the nodes of the stem.

B. Completion

Complete each of the following sentences with a word or phrase that will make the sentence correct. (Do not write in this book.)
1. Dead organic material such as peat moss and decaying leaves is called _____.
2. The young embryo germinating from a bean seed receives food energy from its _____.
3. The three major nutrients needed for plant growth are _____, _____, and _____.
4. The best soil mix for transplanting seedlings is composed of _____.
5. Rooting hormone is used on stem cuttings to _____.

C. Multiple Choice

Each of the following statements or questions is followed by four responses. Choose the correct response in each case. (Do not write in this book.)

1. The two most important factors that affect plant growth are
   a) sunlight and humidity
   b) nutrients and temperature
   c) sunlight and water
   d) water and humus
2. Heavy clay soils are poor for plant growth because
   a) water soaks through them too quickly
   b) they hold too much air
   c) they lack nutrients
   d) they hold too much water
3. The ability of seeds to germinate is called
   a) dormancy   b) viability   c) hardening off   d) magic
4. The first step in the germination of a bean seed is
   a) the drying up of the cotyledons
   b) the absorption of water to soften the seed coats
   c) the curving of the stem to pull the leaves out of the seed
   d) the growth of a root system

D. Using Your Knowledge

1. The growing season in Canada is very short compared to many other parts of the world. Canadians import food from other countries during the winter. List 10 foods that are imported.
2. A gardener with a healthy indoor cactus garden decided to plant all the cacti outside under a maple tree on the May 24th weekend. By July all the cacti had died. List as many reasons for this as you can.
3. Suppose you got a part-time job working in a garden centre in the spring. List some of the tasks (at least 6) that you would likely be asked to do in this job.
4. Many houseplant cuttings will root if the stems are kept in a glass of water. Why would plants rooted this way likely be less healthy than plants rooted in a soil mix?

E. Investigations

1. Prepare stem cuttings of houseplants such as geraniums, coleus, ivy, Kalanchoe, begonia, or Wandering Jew at home. Root them in a mixture of moist peat moss and vermiculite. Then grow them in a sunny window or in a fluorescent light garden.
2. Design and try an experiment using bean seedlings to find out how much fertilizer is needed for optimum plant growth.
3. Use a library to research the growing conditions required by some of the vegetables people grow in your area. Use this information to design a vegetable patch that would feed your family.

4. **Hydroponics** is a technique of growing plants in water to which nutrients have been added. Obtain a hydroponics starter kit and set up your own hydroponic garden at home.
5. Try to propagate an African Violet using a single leaf. Cut the leaf stalk 2-3 cm long and insert it into a rooting mixture. Keep it warm, well-watered, and in good light for two to three months. One leaf may produce up to ten new plants!

Unit 5: The Animal Kingdom

CHAPTER 17
A Survey of the Animal Kingdom

CHAPTER 18
Feeding and Digestion

CHAPTER 19
Breathing Systems

CHAPTER 20
Internal Transport Systems

CHAPTER 21
Reproductive Systems

The animal kingdom includes over 1 200 000 species of living things. Countless thousands of animals still remain to be discovered. Most of them are probably insects, spiders, and other arthropods. And most of them probably live in the tropics.

A person, cow, clam, grasshopper, and sponge are all animals. That means they have much in common. Just what do they have in common? Why are they all called animals?

You will begin this unit with an overview of the animal kingdom. Then you will study certain aspects of selected animals. When you have finished, you should know what animals have in common. You will also know how some animals differ, and why.

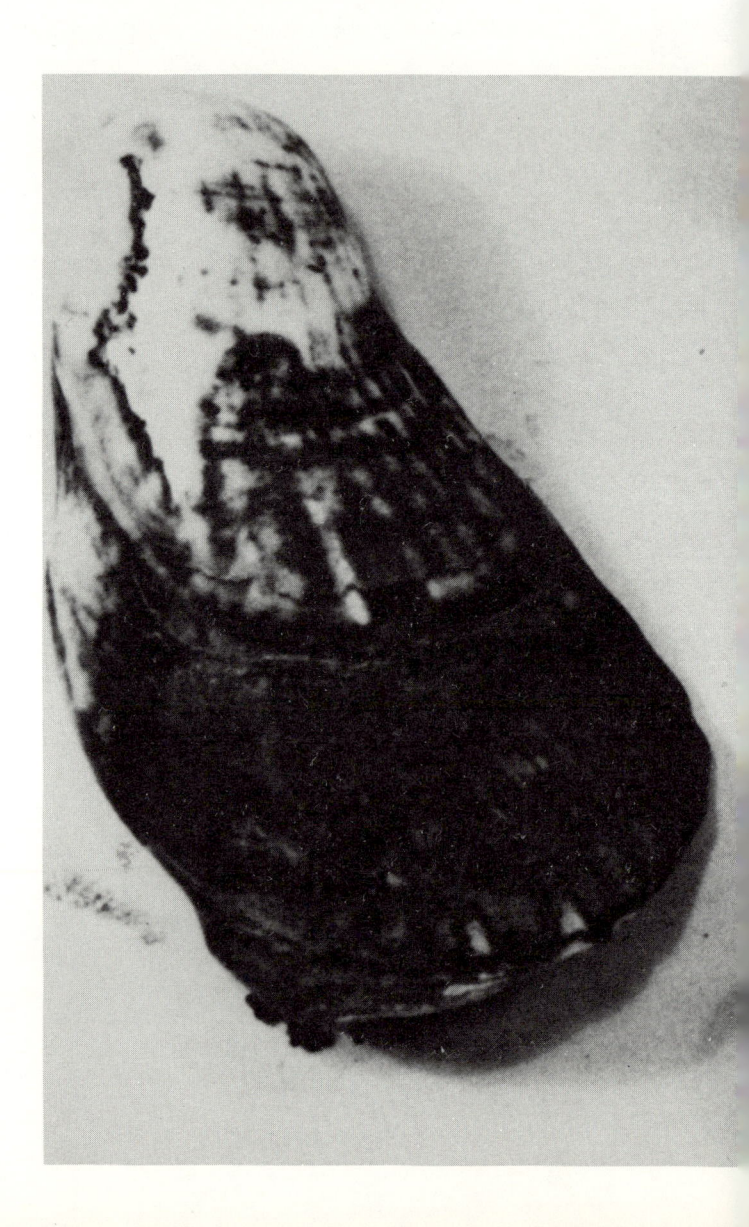

Fig. 17-0 How many kinds of animals can you see? Can you name them?

17 A Survey of the Animal Kingdom

17.1 What Is an Animal?
17.2 The Main Phyla of Animals
17.3 Activity: What Type of Animal Is It?

An earthworm is an animal; so is a horse. What do an earthworm and a horse have in common? A mosquito is an animal; so is a clam. What do a mosquito and a clam have in common? A sponge is an animal; so are you. What do the sponge and you have in common?

17.1 What Is an Animal?

How do you know an animal is an animal? Even biologists argue about what is an animal and what isn't. For example, some biologists call the amoeba an animal. But others call it a protist. All biologists agree, however, that the organisms listed in Table 17-1 are animals. This table shows just the 9 main groups (phyla) of animals.

Table 17-1 The Main Phyla of the Animal Kingdom

Common name of phylum	Approximate number of species	Example
Sponges	4 500	Bath sponge
Coelenterates	9 500	Jellyfish
Flatworms	6 500	Tapeworm
Roundworms	10 000	Hookworm
Molluscs	75 000	Clam
Segmented worms	10 000	Earthworm
Arthropods	1 000 000	Housefly
Echinoderms	5 000	Starfish
Chordates	46 000	Human

The Characteristics of Animals

What do the animals in Table 17-1 have in common? Animals have all or most of the following characteristics:

1. *Animals are made of many cells.* These cells are specialized to form different tissues. In most animals, the tissues are grouped into organs. The organs are grouped into organ systems.
2. *Animal cells contain specialized parts such as a nucleus.*
3. *Animal cells are covered by a cell membrane.* They have no cell walls. Therefore, they tend to be flexible.
4. *Animals get their energy by eating other organisms.* Some animals eat plants. They are called **herbivores** [HUR-bi-vohrs]. Others eat animals. They are called **carnivores** [KAR-ni-vohrs]. Some animals eat both plants and animals. They are called **omnivores** [OM-ni-vohrs].
5. *Most animals can move from place to place.*
6. *All animals reproduce sexually.* That is, a **male gamete** (**sperm**) unites with a **female gamete** (**egg**) to form a **zygote**. The zygote develops into an **embryo**. And the embryo eventually becomes a mature animal.

   A few animal species can also reproduce asexually. A new animal grows on the parent one. This is called **budding**.

Section Review

1. Which phylum contains the most species of animals?
2. a) How many phyla of worms are in the animal kingdom?
   b) How many species of worms are there?
3. Make a summary of the 6 characteristics of animals.

17.2 The Main Phyla of Animals

This section describes the nine phyla of animals listed in Table 17-1. As you read about each of these phyla, think of two questions:
1. How are the animals in this phylum like all other animals?
2. How are the animals in this phylum different from other animals?

Sponges

About 4500 species of animals are classified as **sponges**. These are the simplest animals. The cells of sponges are grouped together. However, they do not form true tissues. The body has many canals in it. Water flows through them.

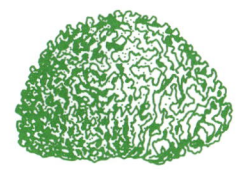

Spongia (Bath sponge)

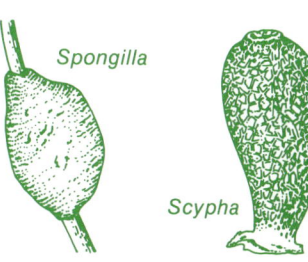

Spongilla

Scypha

Microciona

Sponges have no organs. Also, they have no movable parts or appendages. Most species have an internal skeleton made of silicates (glass-like substances), limestone, or proteins.

Sexual reproduction occurs in all species. As well, some sponges reproduce asexually by budding. Some species have both sexes in the same animal. Others have separate sexes.

Coelenterates

About 9500 species of animals are classified as **coelenterates** [suh-LEN-ter-ates]. These include hydras, jellyfish, corals, and sea anenomes.

Coelenterates have a mouth. It is surrounded by tentacles. The tentacles deliver food to the mouth. Digestion occurs in a large body cavity. Then undigested solids are expelled through the mouth.

Some coelenterates, such as jellyfish, swim or float in the water. Others, like corals, live attached to some object or to each other.

These animals have no circulatory, breathing, or excretory systems. They have a few nerve cells in the body wall, but no organized nervous system.

Reproduction may be asexual (by budding), or sexual. Some species have both sexes in the same animal. Others have separate sexes.

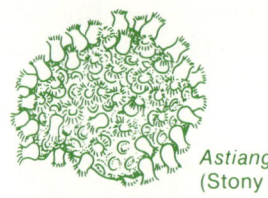

Hydra

Physalia (Portugese man-of-war)

Astiangia (Stony coral)

Aurelia (Common jellyfish)

Edwardsia (Sea anemone)

Flatworms

About 6500 species of animals are classified as **flatworms**. These include planaria, tapeworms, and flukes. Tapeworms and flukes are parasites. They feed on the tissues of other animals. Planaria are free-living. They live in ponds, lakes, and other wet environments. As the name suggests, these worms are flat. Also, they have no segments (body divisions) like an earthworm does.

Flatworms have a simple digestive system. Like coelenterates, they have a mouth but no anus. Food enters and wastes leave by the same opening. Flatworms have a simple nervous system. However, they have no skeletal, breathing, or circulatory system.

Sexual reproduction occurs in all species. Most flatworms have both sexes in the same animal.

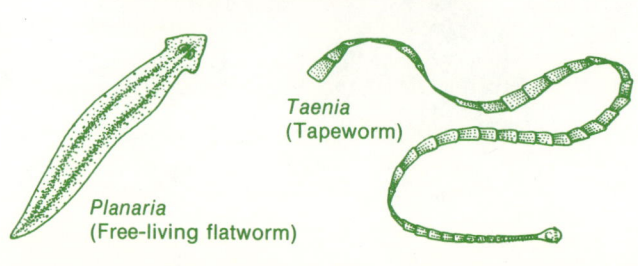

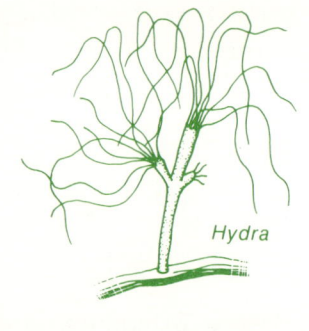

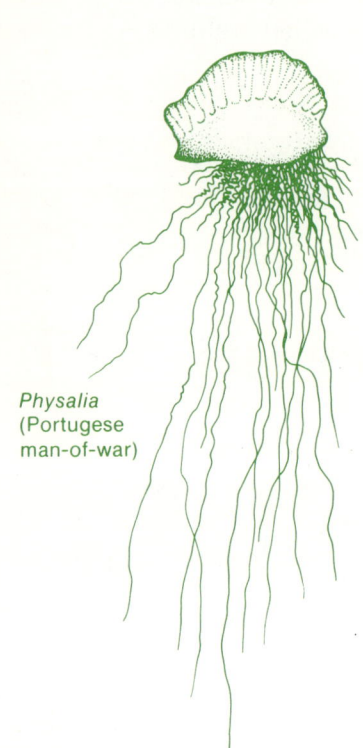

Planaria (Free-living flatworm)

Taenia (Tapeworm)

Fasciola (Liver fluke of sheep)

Roundworms

About 10 000 species of animals are classified as **roundworms**, or **nematodes**. Some of these are scavengers. This means they feed on decaying organic matter. However, most species are parasites. Some are parasites of plants. Others are parasites of animals. In fact, over 30 species can live in humans.

Roundworms are small. Most are from 0.5 mm to 1 mm long. A few are as long as 1 cm. As the name suggests, these worms are round. Also, they have no segments like an earthworm.

Roundworms have a complete digestive system. That is, it begins with a mouth and ends with an anus. Some species have simple excretory systems. Also, the nervous system is well developed.

Sexual reproduction occurs in all species. Most species have separate sexes. The male is usually smaller than the female.

Ascaris (Roundworm of pig)

Trichinella (A parasite of many animals)

Ancylostoma (Hookworm)

Molluscs

About 75 000 species of animals are classified as **molluscs** [MOL-usks]. They make up the second largest phylum of animals. Among the molluscs are clams, snails, whelks, conchs, oysters, and octopuses.

Molluscs have a soft body which is often enclosed in a shell. The shell is very small or missing in some species like the octopus.

Most species have a well-developed head region. All have a "foot" on the underside. It is used for crawling, burrowing, or swimming.

The digestive system is complete. The circulatory system has a heart and a few blood vessels. Breathing occurs through gills, the skin, or a simple lung. Excretion occurs through one or more simple "kidneys". The nervous system has organs for smell, taste, touch, and sight.

Sexual reproduction occurs in all species. Most species have separate sexes.

Anodonta (Freshwater clam)

Buccinum (Whelk)

Loliga (Squid)

A chiton

Ostrea (Oyster)

Mytilus (Mussel)

Helix (Common garden snail)

Octopus

Section 17.2

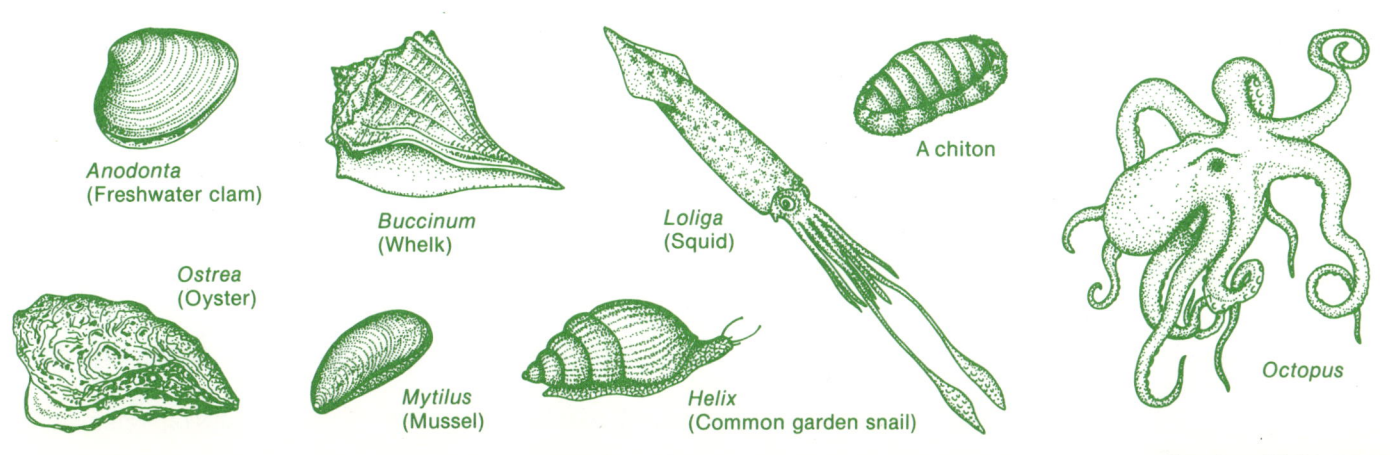

Segmented Worms

About 10 000 species of animals are classified as **segmented worms**. The most familiar segmented worms live in the soil. You have likely seen several species of earthworms. However, most segmented worms live in water. Some saltwater species reach a length of 1 m. Giant Australian earthworms are over 3 m long! As the name suggests, all these worms have their bodies divided up into segments.

The digestive system is complete. The circulatory system is well-advanced. Blood circulates through blood vessels, with the help of five pairs of "hearts". The blood contains hemoglobin to carry oxygen.

In most species, breathing occurs through a moist skin. Some marine species have gills. Excretion occurs through simple excretory organs. The nervous system is well-advanced. It is a central nervous system with a simple "brain" and a main nerve cord. It also has sensory organs for touch, taste, and detection of light.

Sexual reproduction occurs in all species. Most species have both sexes in the same animal. But a few have separate sexes.

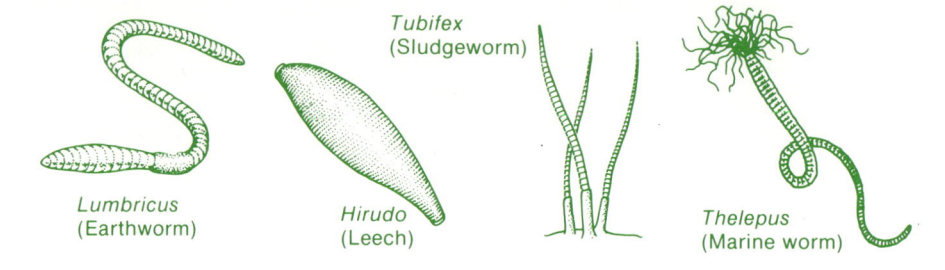

Lumbricus (Earthworm) *Hirudo* (Leech) *Tubifex* (Sludgeworm) *Thelepus* (Marine worm)

Arthropods

About 1 000 000 species of animals are classified as **arthropods** [ARE-throw-pods]. They make up the largest phylum of animals. In fact, most animals are arthropods. The name arthropod comes from two Greek words, *arthros*, meaning "joint" and *podos*, meaning "foot". Arthropods do have jointed feet. In fact, all their appendages (legs, wings, antennas, mouthparts) are jointed.

The arthropod phylum includes crustaceans such as shrimps, crayfish, lobsters, crabs, and barnacles. It also includes insects, spiders, scorpions, millipedes, centipedes, and a host of other animals.

Arthropods have three body divisions: head, thorax, and abdomen. They also have a hard **exoskeleton** (outer skeleton). Sometimes only two body divisions can be seen, since the head and thorax are joined.

The digestive system is complete. The mouth has specialized mouthparts. The circulatory system is open. That is, the blood does not travel always within blood vessels. Instead, it is pumped by a long heart through vessels into large spaces. There it bathes the organs and tissues.

Aquatic arthropods breathe by means of gills. Land species use tracheae (air tubes). Still others use their body surface. Excretion occurs through simple excretory organs. The nervous system includes a "brain" in the head and two nerve cords. Among the sensory organs are antennas, compound eyes, and simple eyes.

Sexual reproduction occurs in all species. The sexes are separate. And, the males and females often look different.

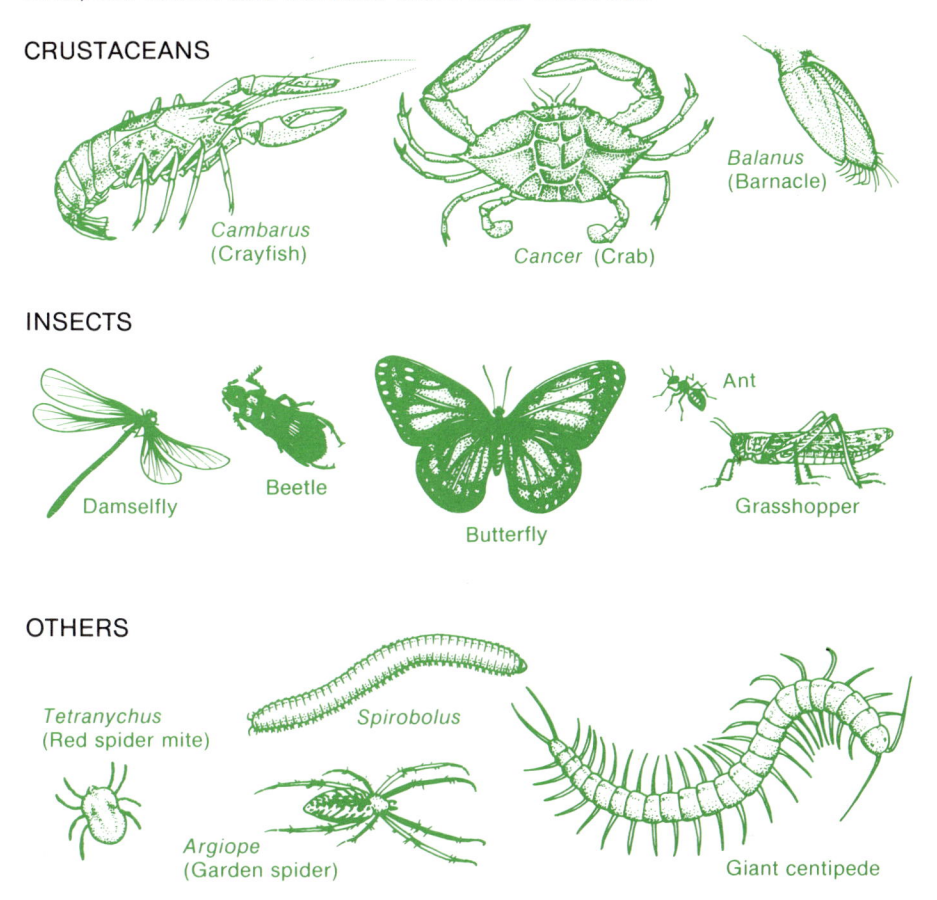

CRUSTACEANS

Cambarus (Crayfish) *Cancer* (Crab) *Balanus* (Barnacle)

INSECTS

Damselfly Beetle Butterfly Ant Grasshopper

OTHERS

Tetranychus (Red spider mite) *Spirobolus* *Argiope* (Garden spider) Giant centipede

Echinoderms

About 5000 species of animals are classified as **echinoderms** [eh-KIN-oh-derms]. Among these are starfish, brittle stars, sand dollars, sea urchins, and sea cucumbers. All echinoderms live in salt water.

Echinoderms have five-part bodies that radiate out from the centre like spokes on a wheel. Each part usually has projections

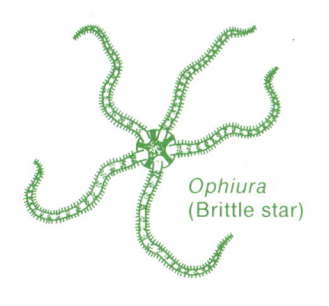

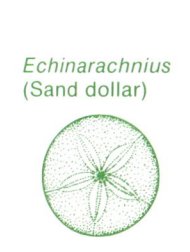

Asterias (Starfish) *Ophiura* (Brittle star) *Thyone* (Sea cucumber) *Echinarachnius* (Sand dollar) *Strongylocentrotus* (Sea urchin)

called "tube feet". These are used in locomotion, feeding, and breathing. The body is covered by a thin skin. Under it is a hard **endoskeleton** (inner skeleton).

The digestive system is complete. Breathing is by tiny gills on the body surface and by the "tube feet". The circulatory and nervous systems are simple.

Sexual reproduction occurs in all species. Most species have separate sexes.

Chordates

You belong to one of the 46 000 species that make up the **chordates** [KOR-dates]. Most chordates are **vertebrates** [VUR-teh-brates]. They have a **vertebral column** [VUR-teh-bral]. It is made up of many bones called **vertebrae** (singular: **vertebra**). Vertebrates have a **nerve cord**, or **spinal cord**, on the back of the body. It is protected by the vertebral column. The **brain** is at one end of the nerve cord.

There are six main classes of vertebrates in the chordate phylum. These are cartilaginous fishes, bony fishes, amphibians, reptiles, birds, and mammals.

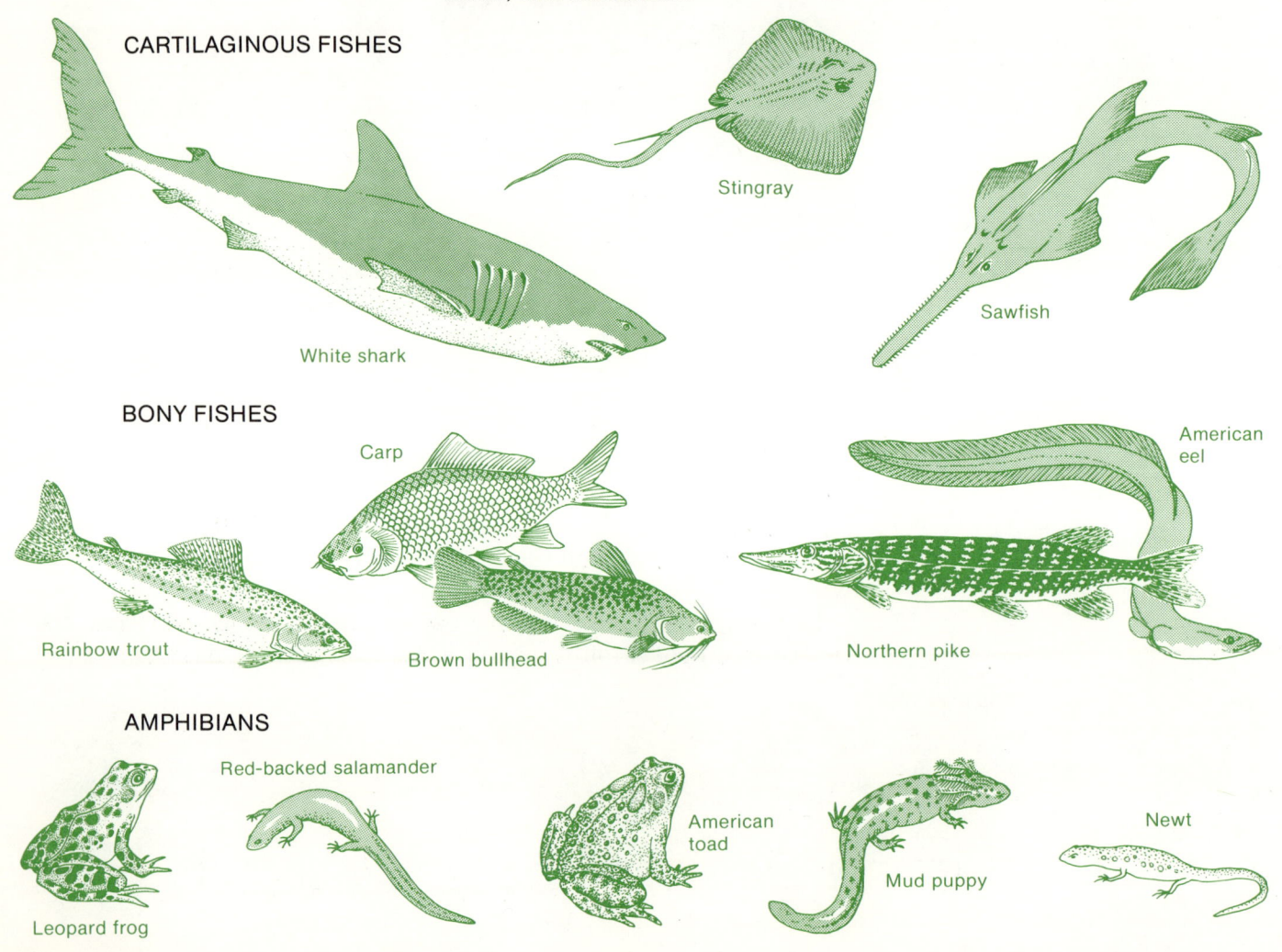

CARTILAGINOUS FISHES

White shark

Stingray

Sawfish

BONY FISHES

Rainbow trout

Carp

Brown bullhead

American eel

Northern pike

AMPHIBIANS

Leopard frog

Red-backed salamander

American toad

Mud puppy

Newt

240 Chapter 17

REPTILES
- Painted turtle
- Broad-nose crocodile
- American alligator
- Fence lizard
- Massasauga rattlesnake

BIRDS
- Red-headed woodpecker
- Canada goose
- Ostrich
- Meadow lark

MAMMALS
- Cat
- Platypus
- Seal
- Squirrel
- Elephant
- Human

Section Review

1. Make a point form summary of the key features of the animals in each of the nine phyla.

17.3 ACTIVITY What Type of Animal Is It?

Your teacher has prepared several stations around the room. At each station there is an animal that belongs to one of the nine phyla described in Section 17.2. Can you find out which phylum each animal is in?

Section 17.3 241

Problem

What type of animal is at each of the stations?

Materials

animals from several phyla

Procedure

a. Copy Table 17-2 into your notebook. Make it a page long.
b. Select a station. Write its number in your table.
c. Study the animal closely. By using Section 17.2, decide what phylum the animal is in. Record the answer in your table.
d. Record in your table the reasons why you decided the animal was in that phylum.

Discussion

The completed table is your writeup for this activity.

Table 17-2   What Type of Animal Is It?

Station number	Phylum animal is in	Your reasons for placing it in that phylum

Main Ideas

1. Most animals have six common characteristics.
2. There are nine main phyla of animals: sponges, coelenterates, flatworms, roundworms, molluscs, segmented worms, arthropods, echinoderms, and chordates.

Glossary

carnivore	KAR-ni-vohr	an animal that eats animals
endoskeleton		an inner skeleton
exoskeleton		an outer skeleton
herbivore	HUR-bi-vohr	an animal that eats plants
vertebra	VUR-teh-bra	a bone of the vertebral column
vertebrate	VUR-teh-brate	an animal with a vertebral column

Study Questions

A. True or False

Decide whether each of the following statements is true or false. If the sentence is false, rewrite it to make it true. (Do not write in this book.)
1. Most animals are arthropods.
2. There are three phyla of worms.
3. All animals can reproduce asexually.
4. Herbivores eat other animals.

B. Completion

Complete each of the following sentences with a word or phrase that will make the sentence correct. (Do not write in this book.)
1. Jellyfish and corals belong to the phylum of _____.
2. _____ make up the second largest phylum of animals.
3. The blood of segmented worms contains _____.
4. Arthropods have three body divisions, _____, _____, and _____.
5. All echinoderms live in _____ water.

C. Multiple Choice

Each of the following statements or questions is followed by four responses. Choose the correct response in each case. (Do not write in this book.)
1. An animal has an inner skeleton. This skeleton has a vertebral column. The animal is
   a) an echinoderm **b)** an arthropod **c)** a chordate **d)** a mollusc
2. An animal has a five-part body with "tube feet" on each part. The animal is
   a) an echinoderm **b)** a flatworm **c)** a mollusc **d)** a coelenterate
3. An animal has tentacles and an incomplete digestive system. The animal is
   a) a mollusc **b)** an octopus **c)** an echinoderm **d)** a coelenterate

D. Using Your Knowledge

1. How are flatworms and roundworms alike? How are they different?
2. In what three ways is an earthworm most like a human?

E. Investigations

1. Make a table like Table 17-2 in your notebook. Find at least 10 living animals outside. Complete your table for those animals.
2. There are six main classes of vertebrates. These are cartilaginous fishes, bony fishes, amphibians, reptiles, birds, and mammals. Find out the main characteristics of each of these classes.

18 Feeding and Digestion

18.1 What Is Food?
18.2 Feeding: Taking food In
18.3 Activity: How Hydra Feed
18.4 Activity: The Crayfish and How It Feeds
18.5 Digestion: Breaking Food Down
18.6 Activity: The Earthworm's Digestive System
18.7 Absorption of Nutrients

All animals need to eat (Fig. 18-1). They do so in a variety of ways. In this chapter we will find out how several animals obtain their food. We will also learn how they use this food. But first, let's find out what food is.

18.1 What Is Food?

Living Things Need Energy and Matter

In Chapter 2 you studied the characteristics of living things. You learned that all living things need a constant supply of **energy**. They

Fig. 18-1 This cow, like all animals, must eat. It eats plants such as grass but other animals, such as the wolf, eat meat.

also need a constant supply of **matter**. Without energy and matter, living things would die. Energy provides the "power" for living. Energy is needed to move about, to get rid of wastes, and to grow. In fact, all life processes require energy. Matter is needed to form new cells. New cells are needed for growth and repair of body tissues.

Nutrients

Nutrients are the raw materials that provide matter and energy for life. Carbohydrates, fats, and proteins are nutrients. Minerals and vitamins are also nutrients. Even water is considered to be a nutrient. These nutrients are contained in the food animals eat. In Chapter 22 you will find out more about these nutrients and about energy.

Digestion

Foods eaten by animals are usually not in the right form to provide energy and matter. Often the food is a whole organism. Sometimes it is just a small piece of an organism. Such foods must be broken down into very small particles before they can be useful to the body. Only food particles the size of small molecules are in the right form. This breakdown process is called **digestion**. Once digestion is complete, usable nutrients can be absorbed into cells. Finally, an animal will rid itself of any undigestible or unusable food particles.

Section Review

1. Why do living things need energy?
2. Why do living things need matter?
3. What is a nutrient?
4. What is food?
5. Why must animals digest their food?

18.2 Feeding: Taking Food In

Not all organisms require the same amounts of nutrients. Also, the form in which these nutrients are found may vary with the type of organism. And, the methods by which these nutrients are obtained may also be quite different. In this section we will study ways in which various animals feed.

Feeding is the process by which organisms obtain their food. Feeding includes the method by which food is captured as well as the method by which food is taken into an organism's body. The taking in of food is also called **ingestion**. Let us examine a few organisms to see how they obtain their food. The first two organisms are not animals. They are included here to help you understand feeding in more complex organisms.

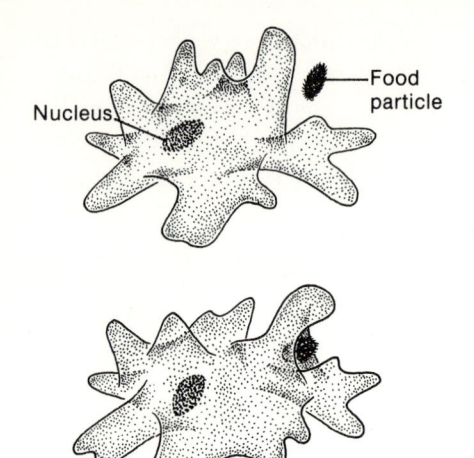

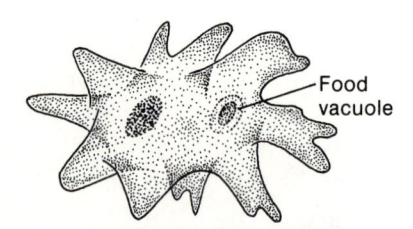

Two Simple Organisms

The **amoeba** is a single-celled organism that lives in water. Its food consists mainly of very small protists. The amoeba captures food by surrounding it (Fig. 18-2). The protist is trapped in a food vacuole. Eventually the food is digested, or broken down, in this vacuole. Required nutrients are then absorbed from the vacuole into the amoeba's body.

The **paramecium** is also a one-celled organism. It has hair-like projections covering its body. These projections are called **cilia**. They move and create a current that carries food particles into a "mouth". Once it is in the **oral groove** (mouth), the food is swept into the **gullet**. It then collects in a **food vacuole** at the end of the gullet (Fig. 18-3). The food is then absorbed into the body in the same way as in the amoeba.

The Hydra

The **hydra** is a many-celled animal. It is found in ponds, lakes, and streams. It has a simple sac-like body made of just two layers of cells.

The hydra has "stinging cells" to paralyze its prey (Fig. 18-4). They are located in special cells on its tentacles. When the prey touches a hair-like trigger, coiled filaments are discharged. They contain a poison that paralyzes the prey. Tentacles then capture the prey and carry it into the hydra's mouth. The food then enters the hydra's body cavity and is digested (Fig. 18-5).

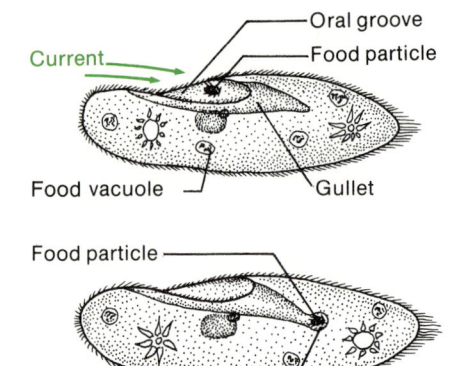

Fig. 18-2 Feeding by an amoeba. Pseudopods ("false feet") surround the food. Then the food is ingested.

Fig. 18-3 Feeding by a paramecium

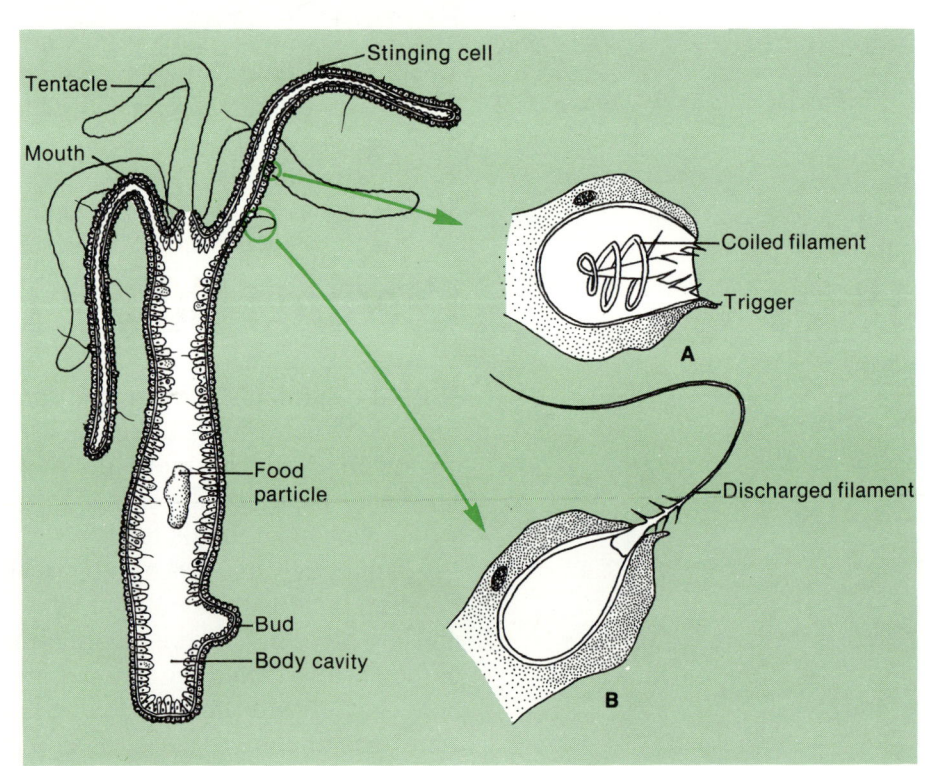

Fig. 18-4 The hydra. Note the "stinging cells" before discharge (A) and after discharge (B).

246 Chapter 18

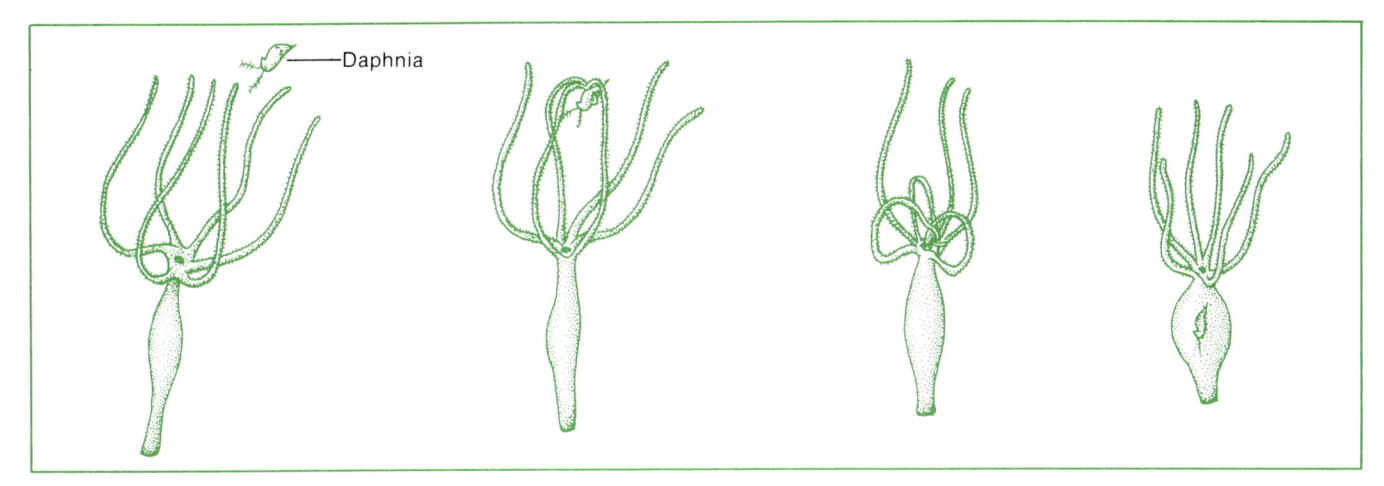

Fig. 18-5 Feeding by a hydra. Here a *Daphnia* is trapped by the tentacles. It is then drawn into the hydra's body cavity.

The Earthworm

The **earthworm** tunnels through soil. In fact, it actually eats its way through the soil. As it does this, it feeds on decaying organic matter in the soil. The food-laden dirt passes through the earthworm's mouth and into the **pharynx** [FAR-inks] (Fig. 18-6). This muscular structure forces the food into the **esophagus**. From here it travels into the **crop**, where it is temporarily stored. The earthworm has no teeth. Instead, the **gizzard** contains grit and sand. These grind the food into tiny particles. Next, the ground food passes into the **intestine**. Here it is further broken down and nutrients are absorbed.

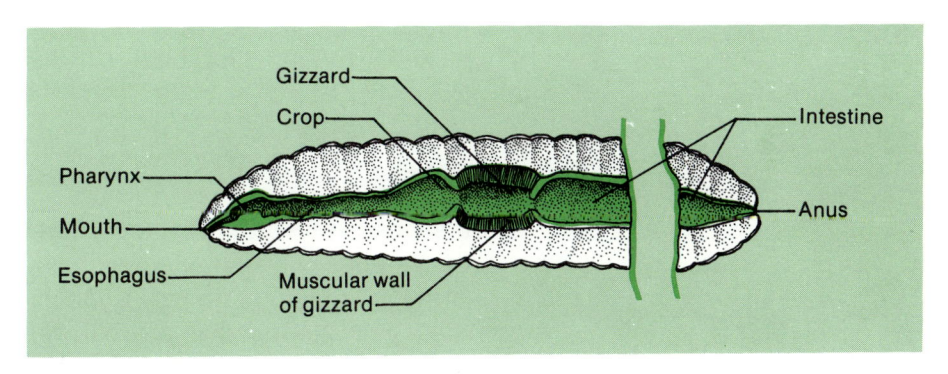

Fig. 18-6 The earthworm's digestive system

The Crayfish

The **crayfish** feeds mainly on live prey. Insect larvas, worms, snails, and small fish are its common foods. Sometimes a crayfish may eat dead fish or clams. It may even eat other crayfish. A pair of compound eyes and two pairs of "feelers" are used to sense live or dead prey.

The crayfish has five pairs of legs. The first pair have large powerful **pincers** (Fig. 18-7). These are used for defence. They are also used for grasping, killing, and tearing prey. The remaining four pairs of legs are called **walking legs**. The first two pairs also have

Section 18.2

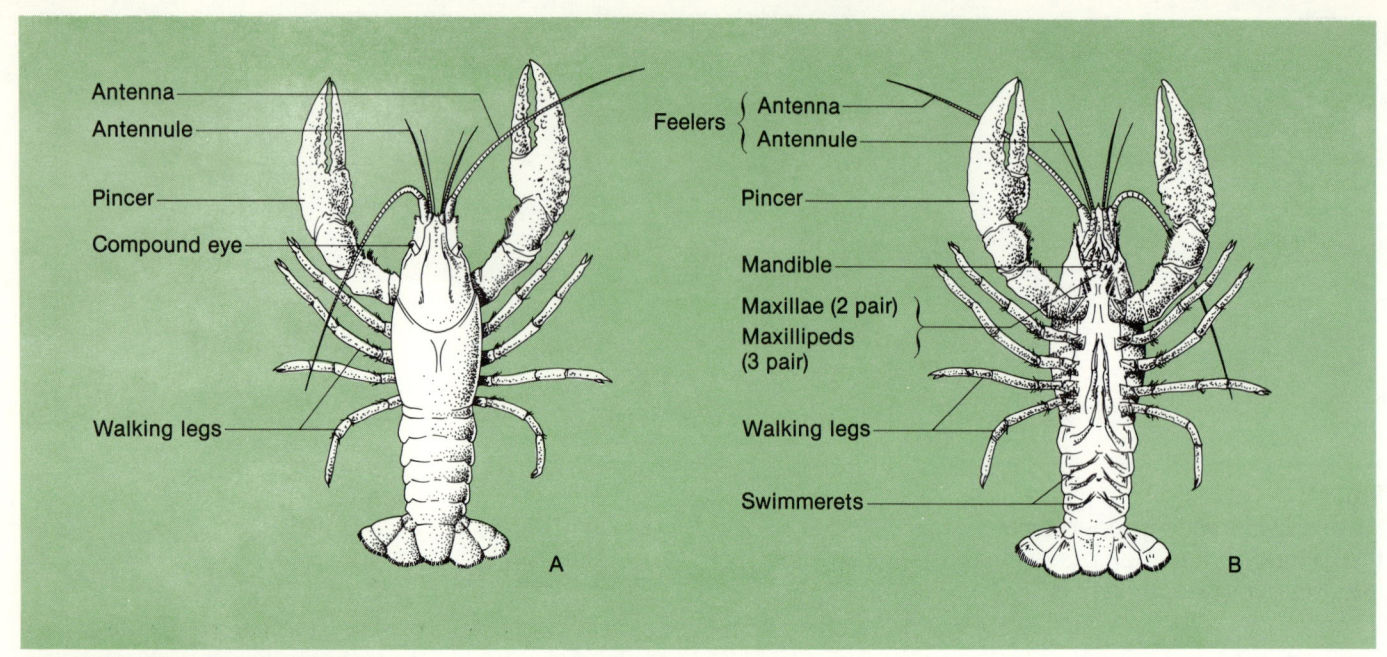

Fig. 18-7 The external features of a crayfish: top view (A); bottom view (B). Which structures are involved in feeding?

small pincers. These are used to carry small scraps of food to the mouthparts.

Just in front of the legs are three pairs of appendages called **maxillipeds** [max-ILL-i-peds]. These are handy tools for grasping food bits and holding them up to the mouth. Three pairs of mouthparts complete the feeding structures. Two pairs of jaw-like **maxillae** help chew the food. Then they pass it on to the jagged-toothed **mandibles**. The mandibles crush the food particles before they enter the mouth of the crayfish. The food then passes into the stomach and is digested.

Insects

Some insects feed by scraping, chewing, or boring into plant tissue. The grasshopper, for example, has biting and chewing mouthparts which enable it to feed on plants. Look at these mouthparts in Figure 18-8.

The **grasshopper** has a broad hinged **upper lip (labrum)**. Like the crayfish, the grasshopper also has a pair of **mandibles**. Note how these "jaws" are toothed. A pair of **lesser jaws**, or **maxillae**, have curved prongs that act like a fork and spoon. The **lower lip (labium)** has a deep notch to help direct food into the mouth. A fleshy **tongue** is located in the mouth behind the mandibles. Finally, **palps** are found on the maxillae and lower lip. These jointed appendages probably help the grasshopper tell one type of food from another.

Some insects feed on fluids or very soft tissues. Their mouthparts are very different from those of the grasshopper. Compare the biting and chewing mouth parts of the grasshopper with the mouthparts of the butterfly and housefly (Fig. 18-9).

The **mosquito** is a fluid feeder. It usually feeds on plant juices. Its mouthparts include a needle-like tube that can penetrate plant

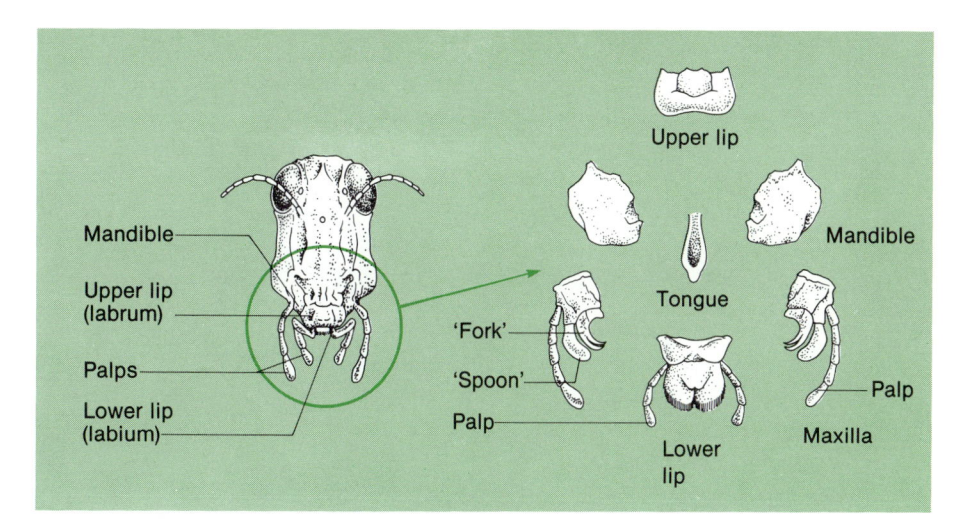

Fig. 18-8 The mouthparts of a grasshopper. These are typical of insects with biting and chewing mouthparts.

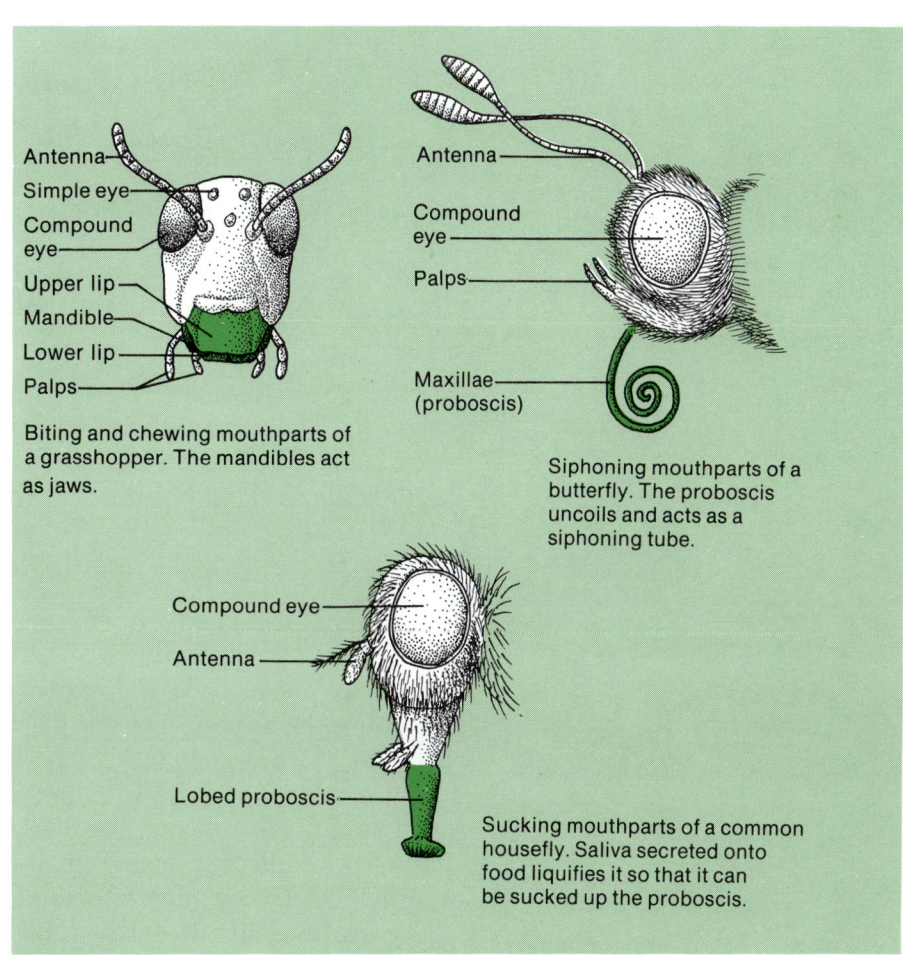

Fig. 18-9 Mouthparts of various insects. The structures are adapted for different types of feeding.

tissue. Juices containing nutrients are then sucked up the tube.

The female mosquito often gets the protein needed for producing eggs from the blood of animals. The mouthparts of the female are specially adapted for this type of feeding (Fig. 18-10). A grooved, flexible **labium** encloses the needle-like structures that pierce the skin. One of these structures, the **labrum**, is hollow. It serves as a

Section 18.2   249

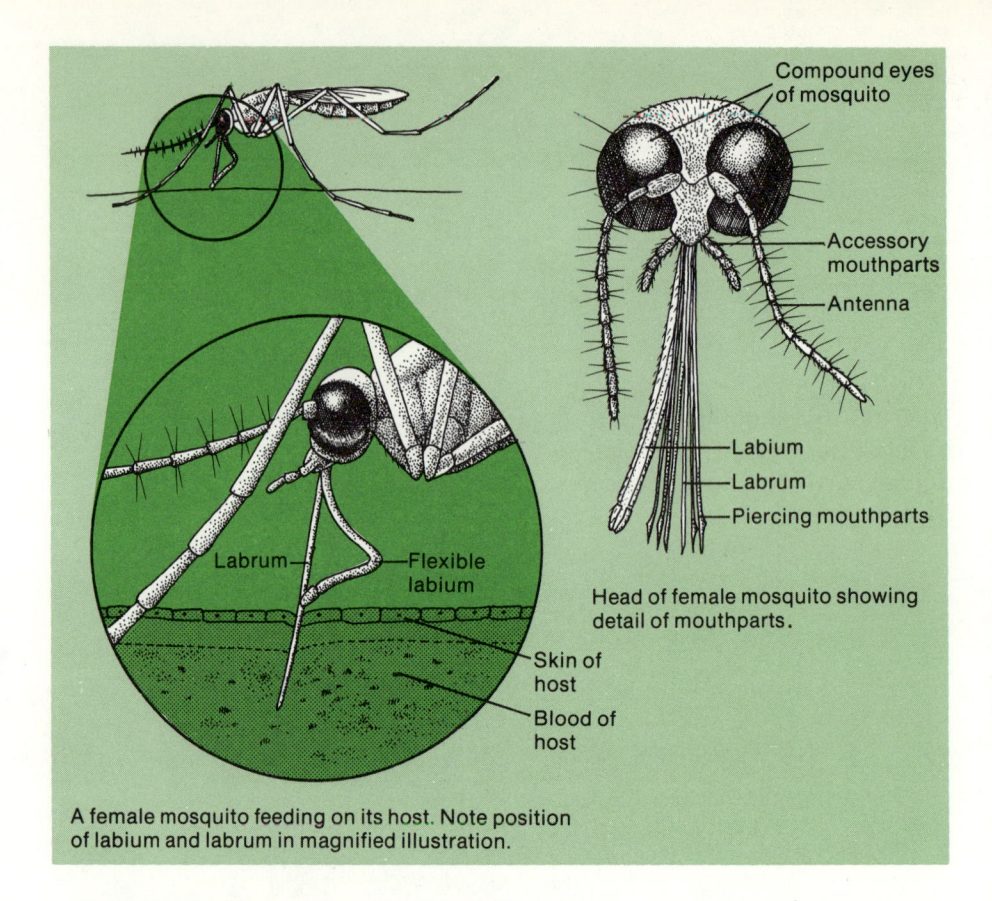

Fig. 18-10 Feeding by a female mosquito. Mouthparts pierce the skin of the prey. Then blood is pumped up the hollow labrum.

A female mosquito feeding on its host. Note position of labium and labrum in magnified illustration.

Head of female mosquito showing detail of mouthparts.

"straw" through which the victim's fluids can be drawn into the mosquito. The mosquito injects saliva which prevents the victim's blood from clotting and plugging the labrum. This saliva causes the swelling and itching of bites.

The Frog

The frog feeds mainly on insects. Sometimes snails, worms, and other small invertebrates are eaten. The frog is interested only in living, moving prey. A unique tongue enables it to capture insects. This muscular tongue is very elastic. Unlike yours, it is attached to the front of its mouth and is free behind. There are two fleshy lobes at the free end. As the tongue is flicked out at a moving insect, it picks up a sticky mucus from the roof of the mouth. The lobes and the mucus "hold" the insect until it is put into the mouth. One gulp, and the insect is swallowed whole (Fig. 18-11).

This section has described how animals obtain their food in a variety of ways. Often the food they ingest is not in a form that can be used by the animal. Further breakdown, or digestion, is usually required.

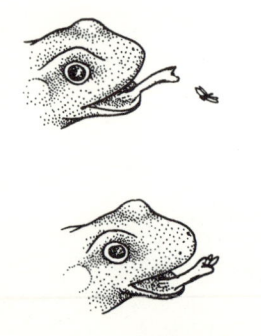

Fig. 18-11 A frog feeds with its special tongue.

Section Review

1. What is the difference between feeding and ingestion?
2. a) Describe the method by which an amoeba feeds.

    b) What are cilia? What role do they play in feeding?
3. **a)** How does the earthworm get nutrients?
    b) The earthworm has no teeth. How does it grind up its food?
4. **a)** How does the crayfish capture its prey?
    b) Describe the mouthparts of the crayfish.
    c) What role does each mouth part of a crayfish play in feeding?
5. **a)** What does a mosquito usually feed on?
    b) How is a mosquito adapted for feeding on fluids?
6. How does the frog capture its prey during feeding?

18.3   ACTIVITY   How Hydra Feed

The hydra is a simple many-celled animal. It must eat food to obtain the nutrients it requires. How do hydra obtain food (Fig. 18-12)?

Problem

How do hydra feed?

Materials

dissecting (or compound) microscope
droppers (3)
watch glass
solution of 1% acetic acid

Hydra culture
Daphnia culture
probe
cavity slide

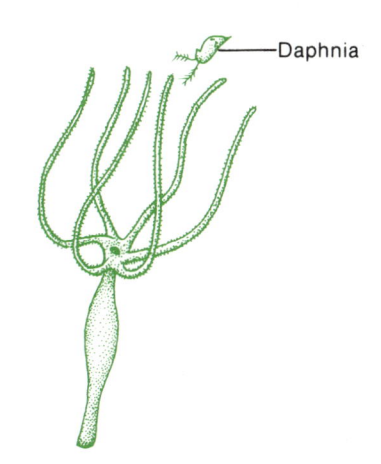

Fig. 18-12 Hydra feeding. What happens to the food once it is digested?

Procedure

a. Using the dropper, put five drops of water in the middle of a watch glass.

b. Using the dropper, remove a hydra from the culture jar. Put the hydra into the water in the watch glass.

c. Wait a minute or so for the hydra to recover from the change. Look at it with your naked eye. Can you see it move? How long is it? How many tentacles does it have? Record your observations.

d. Place the watch glass on the stage of the microscope. Focus on the hydra using low power. Observe its appearance now. Record your observations.

e. Gently touch the probe to one of the hydra's tentacles. How does the hydra react to this? Record your observations.

f. Using a second dropper, transfer a *Daphnia* (waterflea) to the watch glass containing the hydra. Observe what happens over the next few minutes. Record your observations.

g. Place another hydra into a few drops of water in the cavity of a cavity slide. Observe the hydra under medium power. Record your observations.

h. Using the third dropper, add a drop of 1% acetic acid solution to the cavity. Observe the reaction of the hydra. Record your observations.

Discussion

1. How does the hydra capture the waterflea?
2. How does the hydra eat the waterflea?
3. How will the hydra digest the waterflea?
4. In feeding, the hydra discharges structures called "stinging cells". Find out what they do.

18.4 ACTIVITY The Crayfish and How It Feeds

The crayfish is a freshwater animal. It is a relative of the lobster. You may recall that both animals are crustaceans (see Chapter 17). Have you ever observed a live crayfish? Have you ever seen one feed? How does it move? How does it obtain its food?

Problem

How does a crayfish move and feed?

Materials

live crayfish
aquarium
shallow tray of water (enough to allow crayfish to submerge)
probe
salt water
vinegar
light source
small pieces of ground beef
droppers (3)
beef bouillon

Note: Look back to Figure 18-7 as you do this activity.

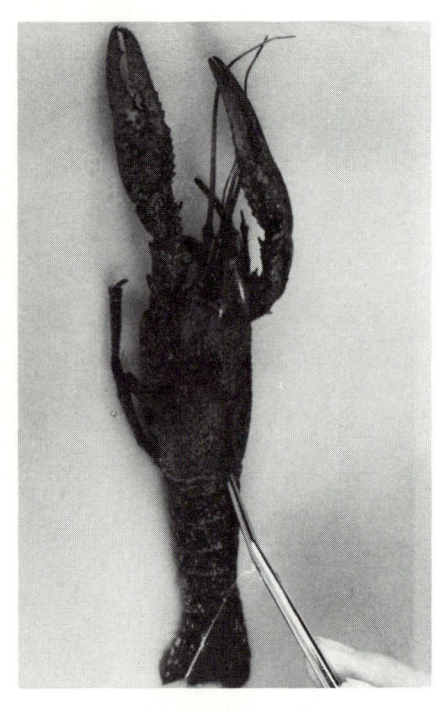

Fig. 18-13 The crayfish has a hard outer skeleton or exoskeleton. Can you think of one advantage and one disadvantage of having an outer skeleton?

Procedure A Locomotion

a. Observe the live crayfish in the aquarium. Gently touch it at the front and rear ends (Fig. 18-13). Observe and record the direction in which it moves.
b. Observe the structures used for walking. Observe how any sudden movement is made. Record your observations.

Procedure B Sensing the Environment

a. Gently touch the **antennae** (large feelers) and the **antennules** (small feelers) with the probe. Describe the reaction of the crayfish.
b. Place the crayfish in the shallow tray of water. Be sure it is submerged.
c. Place a drop of salt water in front of the crayfish. Observe and record the response of the crayfish.
d. Repeat step (c) using each of the following: a drop of vinegar, a drop of beef bouillon, a bright light.

Note: It is best to change the water each time a new substance is used.

Procedure C Feeding

a. Place a few small pieces of ground beef in a tray of fresh water.
b. Place a crayfish in the tray. Observe how the crayfish feeds.

Discussion A

1. Does the crayfish move in any particular direction? Can it walk backwards? Does it move sideways?
2. Describe the position of the pincers during the movement you observed. Of what advantage might this be to the crayfish?

Discussion B

1. What are the functions of the antennae and antennules?
2. How does the crayfish react to each substance? to the light?
3. Which senses seem to be present in a crayfish?

Discussion C

1. Describe how the crayfish feeds. How does each appendage assist in the process? Which appendages grab the food? which hold it? which tear it? which gather the scraps?
2. How does the crayfish detect the presence of food in the water?

18.5 Digestion: Breaking Food Down

What is Digestion?

You may recall that all living cells require nutrients for two main reasons. First, nutrients are needed to build new cells. Second, nutrients are used as a source of energy. Before the nutrients can be used, they must get inside the cells of an animal. But how can the food an animal eats enter its cells? Foods usually consist of very large molecules. These must be broken down into smaller molecules, or nutrients. Then these nutrients can be absorbed into living cells.

Digestion is the process by which large food molecules are broken down into smaller nutrient molecules. Chemical compounds called **enzymes** assist in digestion.

Digestive Cavities and Systems

In most animals, ingested food is broken down within a **digestive cavity** of some kind. The resulting nutrient molecules are then absorbed into the cells of the organism.

The simplest digestive cavity is the body cavity of a hydra (see Figure 18-4). Food enters the body cavity by way of the "mouth". Unused matter is expelled through the same "mouth". Most animals, however, have a one-way **digestive "tube"** or **system** (Fig. 18-14). This system has a **mouth** at one end and an **anus** at the other. Food is

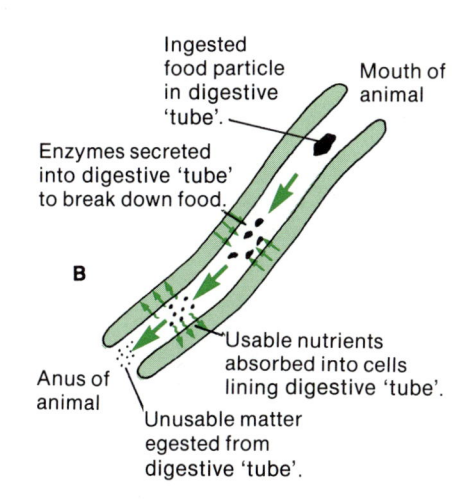

Fig. 18-14 A one-way digestive tube, or digestive system

Section 18.5

ingested through the mouth. Undigested solids are egested (given off) at the anus. Let us look at the digestive systems of a few animals to see how they work.

The Earthworm

The earthworm's digestive system has specialized sections within it (see Figure 18-6). Soil is sucked in through the mouth by the muscular **pharynx** [FAR-inks]. It then moves to the **esophagus** and into the **crop**. Here, it is temporarily stored. The food then enters the **gizzard**. When the muscles of the gizzard contract, the food is ground up. Coarse soil particles taken in by the worm help to grind up the food. The food then enters the **intestine**. Cells lining the intestine release enzymes onto the food. These enzymes break down, or digest, the food. Required nutrients can then be absorbed into the earthworm's blood system. Unused material passes out of the digestive system through the **anus**. Have you ever noticed the coiled "casts" at the entrance to an earthworm's burrow? These "casts" are made of undigested matter left by the worm.

Other Invertebrates

In the **crayfish**, food is torn apart and crushed by the mouthparts. To further aid in digestion, part of the **stomach** is lined with "teeth". They act somewhat like the earthworm's gizzard. They grind food pieces into smaller particles.

The digestive systems of insects differ yet again (Fig. 18-15). The **grasshopper** has salivary glands that produce **saliva**. The saliva, much like the saliva you produce, mixes with food in the mouth. Saliva lubricates the food so it can be easily swallowed down the esophagus. Saliva also contains enzymes to help digest the food. The crop and the gizzard act much like those of the earthworm. Do you recall what their functions are? Six pairs of **gastric pouches** produce enzymes to help digest food in the stomach. The intestine and the anus complete the digestive system.

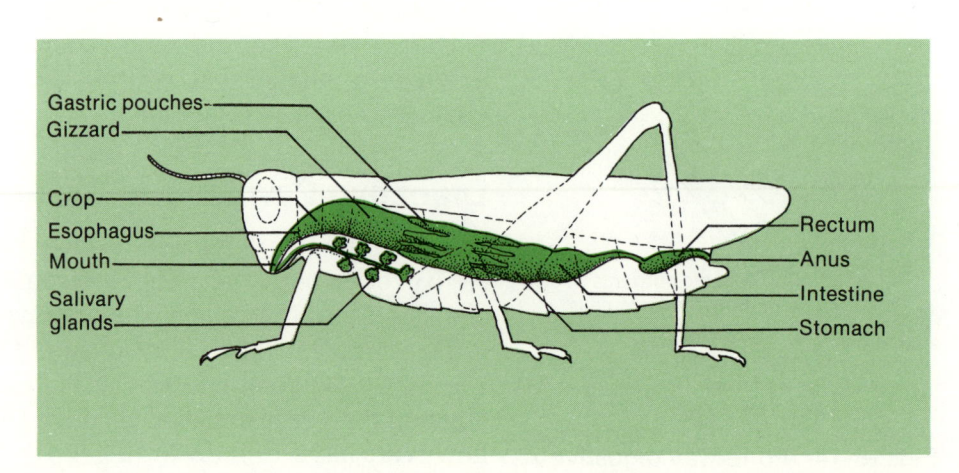

Fig. 18-15 The digestive system of a grasshopper

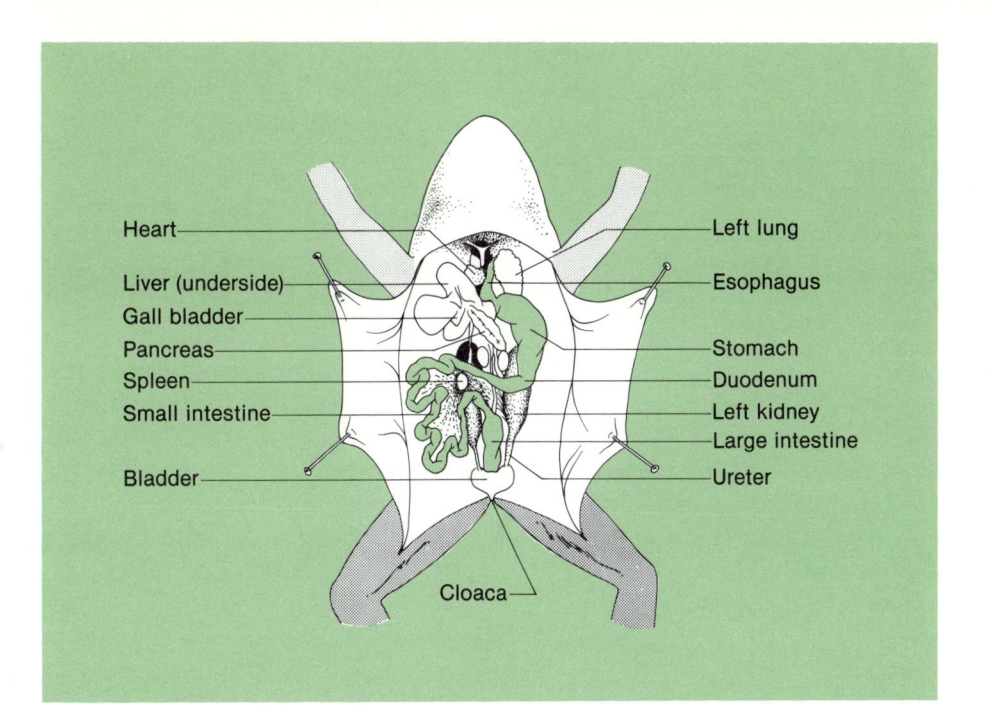

Fig. 18-16 The internal organs of the frog. Can you find the parts of the digestive system?

Vertebrate Digestive Systems

All vertebrates have the same basic design for a digestive system:
1. a mouth for ingestion;
2. an esophagus, leading to a storage area (crop) or stomach;
3. a region where digestion begins (after a stomach);
4. a region for final digestion and absorption of nutrients;
5. a region for water absorption;
6. a region for waste elimination.

The frog, for example, has an esophagus leading from the mouth to the stomach (Fig. 18-16). Initial digestion occurs in the stomach. Here the food is mixed by muscle contractions. Cells lining the stomach produce enzymes. These enzymes help digest the food. Final digestion occurs in the small intestine. Absorption of nutrients also occurs in the small intestine. The liver, gall bladder, and pancreas all assist in digestion. Water absorption occurs in the large intestine. Solid wastes are formed into **feces**. The feces are eliminated through the **cloaca** [klo-AY-ka].

Section Review

1. a) What is digestion?
   b) What is the function of an enzyme in digestion?
2. What is egestion?
3. Describe the digestive system of an earthworm. Also, state the function of each section.
4. What role do the salivary glands and gastric pouches play in the grasshopper?

Section 18.5   255

5. **a)** What are the six basic parts of a vertebrate digestive system?
   b) Using the frog as an example, describe the role played by each of these parts.

18.6 ACTIVITY The Earthworm's Digestive System

The earthworm actually eats its way through soil. How does it do this? Where is the earthworm's digestive system located? What parts does it consist of? What do they look like? What function does each part have?

Fig. 18-17 The common earthworm. Can you locate these regions on your specimen?

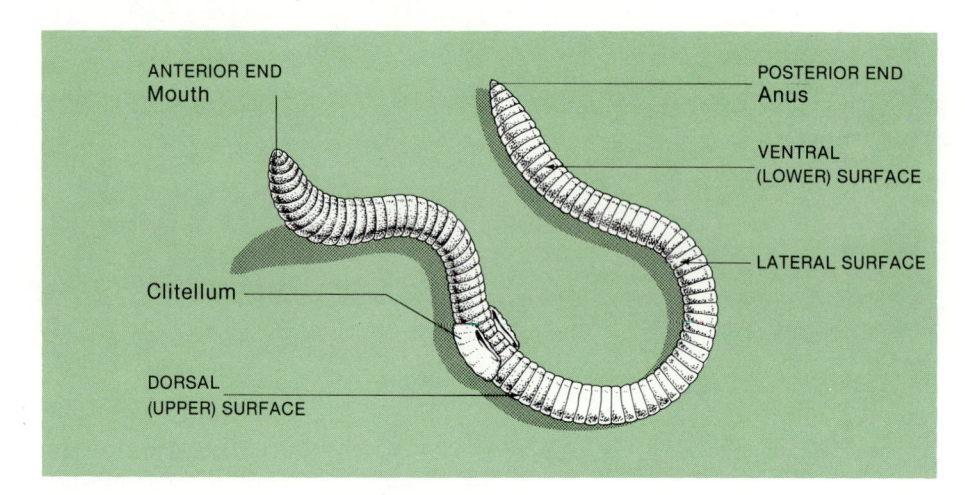

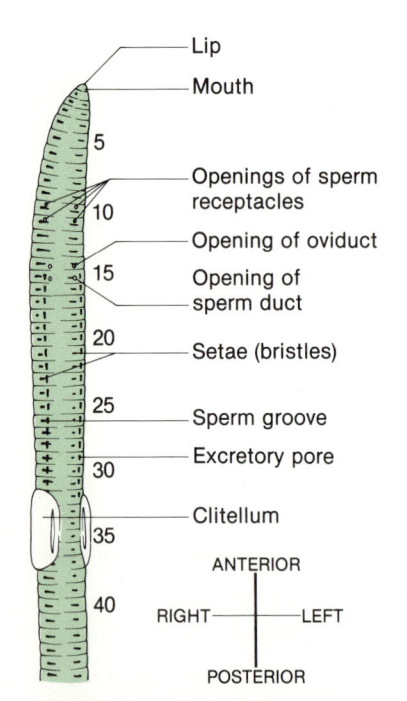

Fig. 18-18 Ventral (bottom) view of external features of an earthworm. You are looking at the anterior end.

Problem

What is the nature of the earthworm's digestive system?

Materials

freshly killed earthworm
  (or preserved specimen)
dissecting needles
dissecting scissors
forceps
straight pins
dissecting tray
hand lens

Procedure

Note: If you are using a preserved specimen, rinse the preservatives off the earthworm. Then dry it.

a. Place the earthworm on the dissecting tray with the dorsal (upper) side up. (See Figures 18-17 and 18-18).
b. Pinch up a fold of skin at the 40th segment. Snip with the scissors to make a small cut through the dorsal body wall.
c. Lay the worm across your hand. Insert the scissor point under the dorsal wall. Pull up on the scissor point. Cut slowly toward

256 Chapter 18

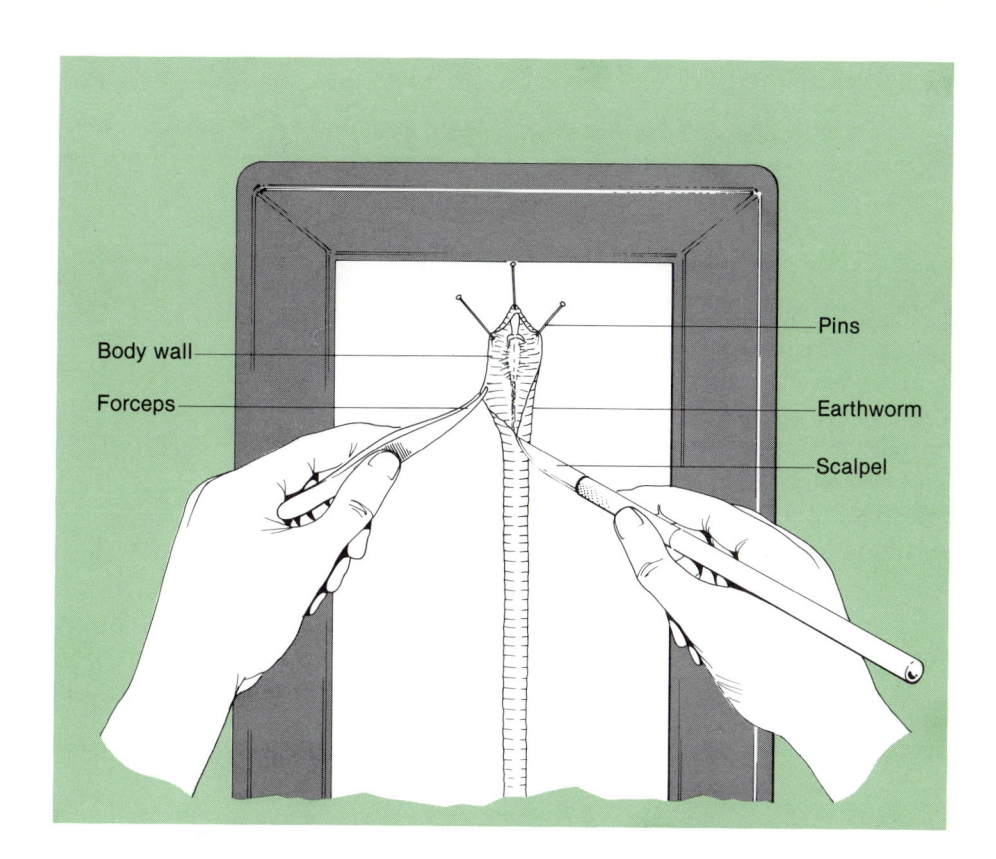

Fig. 18-19 Teasing away one tissue from another is a common dissection technique.

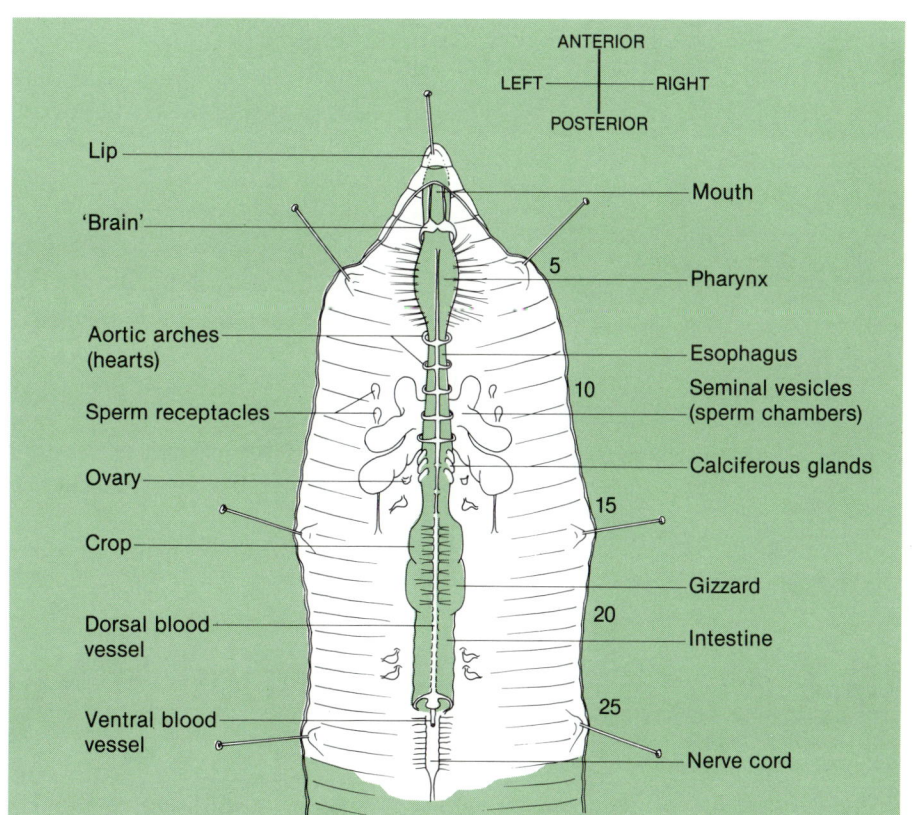

Fig. 18-20 Dissection of an earthworm: a dorsal view. You don't need to know all these parts. But you will likely see most of them in your specimen.

Section 18.6

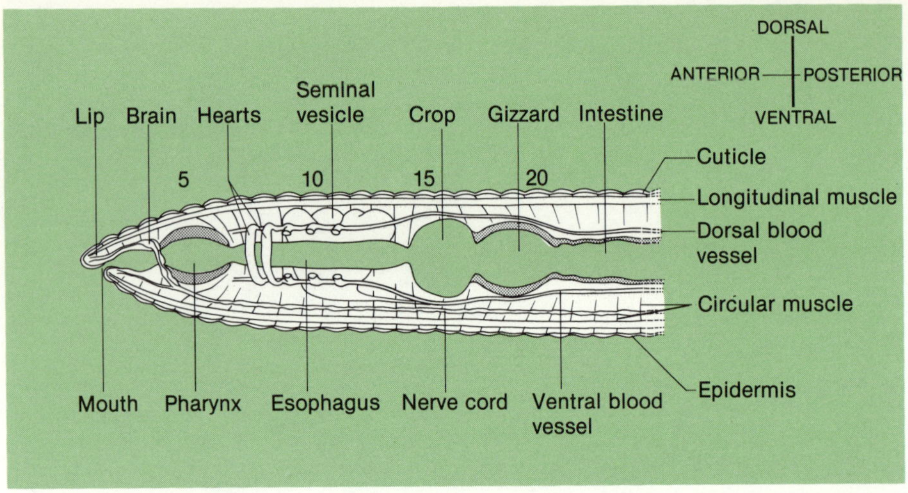

Fig. 18-21 Lateral view of an earthworm showing internal structure. Note the parts of the digestive tube.

the anterior (head) end. Do not cut too deeply. Make certain the cut is not deeper than the body wall.
d. Lay the worm, straight out, dorsal side up, on the dissecting tray. Pin the first and last segments to the wax surface.
e. Using forceps and scalpel, separate the body wall from the internal organs (Fig. 18-19). Pin the skin flaps back at 1 cm intervals.
f. Using Figure 18-20 and 18-21 as guides, find the following parts of the digestive system:
  i) the mouth
  ii) the muscular pharynx; describe its colour and texture
  iii) the esophagus; (It is hard to see. The creamy-yellow seminal vesicles (sperm containers) and aortic arches ("hearts") surround it.)
  iv) the calciferous glands
  v) the crop; describe its colour and texture
  vi) the gizzard; describe its colour and texture. Open the gizzard and describe its contents
  vii) the intestine; describe its colour

Discussion

1. Briefly describe the role each structure plays in the digestion of food.
2. Calculate what fraction of the length of the digestive system is intestine.

18.7 Absorption of Nutrients

Digestion breaks down large food particles into smaller molecules. Most of these simple basic nutrients can easily dissolve in water. This occurs in the small intestine. Once dissolved, the nutrients can

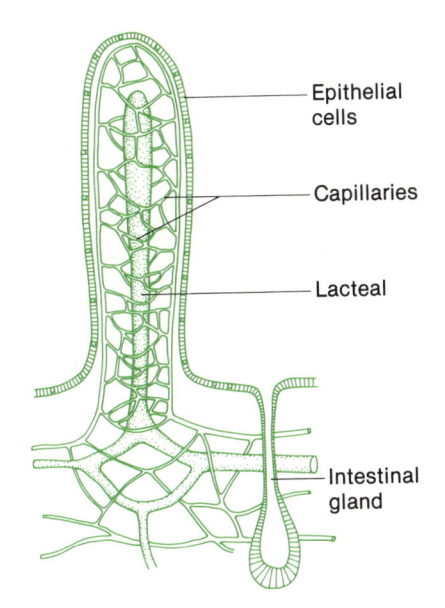

Fig. 18-22 Absorption by the villus. Fats are absorbed into the lacteal. Other basic nutrients are absorbed by the capillaries. What happens to the nutrients after they are absorbed?

pass through the membrane that lines the small intestine. This process is called **absorption**. In complex animals the absorbed nutrients enter the blood stream. The blood stream transports the nutrients to cells in all parts of the body.

Villi

If you examined the inner surface of the small intestine you would notice that it is not smooth. Instead it appears to be ridged and wrinkled. It also appears to look like a tufted brush. Under a microscope you would notice that the tufts are very small finger-like projections. These finger-like projections are called **villi** (singular: **villus**). The wrinkles or ridges are covered with villi. In fact, the entire inner surface of the small intestine is covered with villi. There are millions of them within the small intestine.

The ridges and the villi increase the surface area available for absorption. The villi are also capable of vigorous movement. This wriggling motion helps mix the food and enzymes to assist digestion. It also helps speed up absorption of nutrients.

Each villus has a very thin lining. It is one cell layer thick (Fig. 18-22). Within each villus is a network of very small blood vessels. These vessels are called **capillaries**. During absorption, nutrients diffuse through the one-celled lining of the villus and into the capillaries. The blood then carries the nutrients to other parts of the body. Also within each villus is a **lacteal** or **lymph vessel**. Digested fat products are absorbed into the lacteals. The lacteals form part of the **lymphatic system**. It is separate from the blood system. You will learn more about these two systems in Chapter 25.

Section Review

1. Define absorption.
2. In what two ways is surface area increased in the small intestine?
3. Describe the structure of a villus.
4. What is absorbed into the capillaries? into the lacteals?

Main Ideas

1. Nutrients are the raw materials which provide the matter and energy for living organisms.
2. Animals obtain nutrients from food they eat.
3. Most foods eaten by animals must be broken down or digested.
4. Digestion of food in animals occurs within a digestive "tube" or system.
5. The digestive systems of animals have specialized sections. Each of these sections has a specific function in digestion.

Glossary

digestion		the breaking down of large food particles into smaller basic nutrients
enzymes	EN-zymz	the chemical substances formed by living cells to speed up reactions
esophagus	e-SOF-a-gus	a "food tube" that connects the pharynx with the stomach
nutrients	NEW-tree-ents	the raw materials that provide energy and matter for living things
pharynx	FAR-inks	the "back of the mouth" region which connects the mouth to the esophagus
villus	VIL-us	a small finger-like projection in the small intestine that increases the surface area for absorption of nutrients

Study Questions

A. True or False

Decide whether each of the following statements is true or false. If the sentence is false, rewrite it to make it true. (Do not write in this book.)

1. Living organisms require food just for energy.
2. Enzymes are chemical compounds that assist in digestion.
3. Absorption of nutrients occurs mainly in the stomach.
4. Fat nutrients are absorbed into the lacteals of the small intestine.

B. Completion

Complete each of the following sentences with a word or phrase that will make the sentence correct. (Do not write in this book.)

1. Living organisms require a constant supply of ▓▓▓▓ and ▓▓▓▓.
2. The breakdown of large food particles into smaller nutrient particles is called ▓▓▓▓.
3. ▓▓▓▓ is the process by which organisms obtain their food.
4. The ▓▓▓▓ is a temporary storage structure in the earthworm's digestive tract.

C. Multiple Choice

Each of the following statements or questions is followed by four responses. Choose the correct response in each case. (Do not write in this book.)

260   Chapter 18

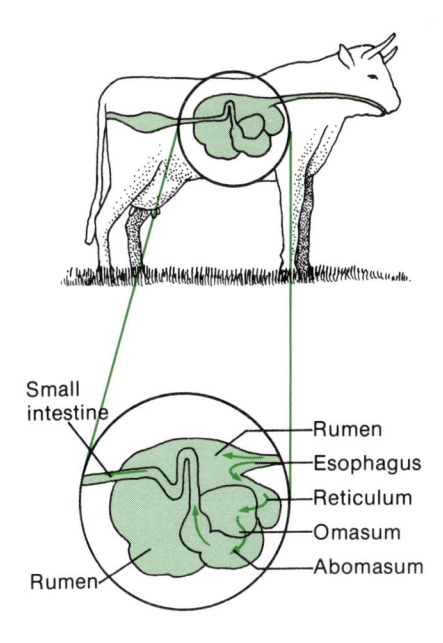

Fig. 18-23 The digestive system of a cow. Note how the stomach is divided into four compartments. The arrows show how food travels through the stomach.

1. In the small intestine, the villi
   a) absorb only fats
   b) increase the surface area to speed up absorption of nutrients
   c) slow down the digestive process so that enzymes can act
   d) absorb only water
2. The grit and sand that earthworms swallow with their food
   a) help move food along the digestive tube
   b) help grind the food into smaller particles
   c) clean out the digestive tract
   d) are used in the formation of teeth
3. The region of the vertebrate digestive system in which water is absorbed is called the
   a) large intestine   b) small intestine   c) esophagus   d) gizzard

D. Using Your Knowledge

1. Briefly explain what nutrients are and why animals require them.
2. In most animals, food must be digested before it is of use. Explain why this is so.
3. Using an example, describe how the structure of the digestive system in vertebrates is adapted for specific types of food eaten.
4. Birds, like earthworms, have a gizzard. Why is it required?

E. Investigations

1. Figure 18-23 is a diagram of the digestive system of a cow. Using a library or resource centre, prepare a report on how it works.
2. "Crustaceans have been a favourite food of gourmets for centuries." Provide proof of the accuracy of this statement.

19 Breathing Systems

19.1 Gas Exchange: Why It Is Needed
19.2 Gas Exchange in Water
19.3 Activity: How Water Temperature Affects the Breathing Rate of Fish
19.4 Activity: The Crayfish and How It Breathes
19.5 Gas Exchange on Land

In order to live, animals need oxygen. During life processes they also produce carbon dioxide. Too much carbon dioxide, however, can be harmful to animals. Therefore animals must get rid of it. Both oxygen and carbon dioxide are gases. In this chapter we will study how animals exchange oxygen and carbon dioxide with their environments (Fig. 19-1).

Air and water are two very different environments. Each presents problems that complicate gas exchange. Let us see how land and water animals have solved these problems.

19.1 Gas Exchange: Why It Is Needed

The Need for Gas Exchange

Animals need a constant supply of energy in order to live. Nutrients that animals eat provide this energy. However, animals also need oxygen to release the energy from these nutrients. This process of

Fig. 19-1 The membrane allows gases to pass through but not water. The mouse lives normally within the glass cage. How is this possible?

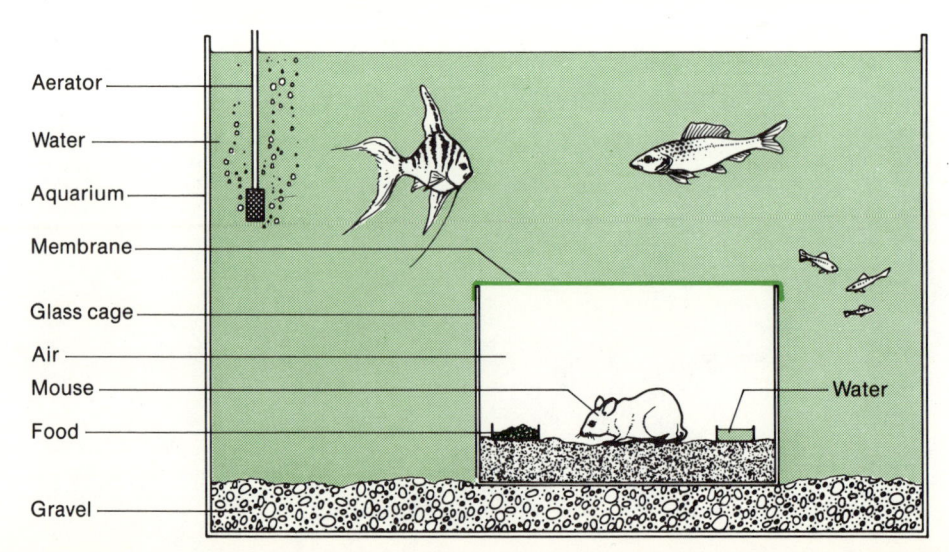

energy release is called **respiration** [res-puh-RAY-shun]. It occurs in all living cells. During respiration, carbon dioxide is released. If left in cells, this gas is harmful.

Animals, therefore, must have a way of obtaining oxygen. They must also be able to get rid of carbon dioxide. The movement of oxygen and carbon dioxide between living organisms and their environment is called **gas exchange**. In many animals this is also called **breathing**.

The Problems of Gas Exchange

In small many-celled animals, all cells are very close to their external environment. Each cell can easily exchange gases with the environment. But for larger animals, gas exchange is a problem. They have many cells deep within their bodies, far from the external environment. Therefore, in most larger animals gas exchange takes place at a special region of the body called a **gas exchange surface**.

In order for gases to pass into or out of cells, they must first be dissolved in a liquid (usually water). They can then diffuse across cell membranes. Moisture, therefore, is required for gas exchange to occur. Obviously this is not a problem for animals which live in water. For land animals, however, this presents a serious problem. Air is a dry environment. But gas exchange surfaces must be kept moist. Land animals must have ways to solve this problem. You will see how they do this later.

Many-celled animals also face an additional problem. The gas exchange surface must be large enough so that enough gas can be exchanged for all the cells of the organism. Large animals use a variety of devices to solve the problems of gas exchange. However, all these devices are designed to meet the same general needs.

Requirements for Gas Exchange in Animals

1. The gas exchange surface must be large enough.
2. The gas exchange surface must be thin and kept moist to allow diffusion to occur.
3. There must be some means of transporting the gases between the exchange surface and every body cell.

Keep these requirements in mind as we study the gas exchange systems of various animals. See if you can determine how each system meets these requirements.

Section Review

1. Why do animals need oxygen?
2. Why must animals get rid of carbon dioxide?
3. What is gas exchange?
4. Why does a land environment make the process of gas exchange more complicated for animals?
5. What general requirements must be met for gas exchange to occur in large many-celled animals?

19.2 Gas Exchange in Water

Gas exchange can only occur across a moist membrane (Fig. 19-2). Aquatic animals are completely surrounded by water. Therefore, their gas exchange surfaces are constantly kept moist. As a result, diffusion of gases can easily occur across these surfaces. Let us examine a few animals that live in water to see how they meet their needs of gas exchange.

Simple Animals

Many simple animals that live in water do not have special gas exchange surfaces. Also, they usually do not have special systems for transporting oxygen and carbon dioxide to and from every body cell. In the **hydra**, for example, all body cells are in direct contact with water (see Figure 18-4, page 246). Oxygen, dissolved in the water can diffuse into each cell and carbon dioxide can diffuse into the surrounding water from each cell.

Another example is the **planarian** (Fig. 19-3). It has a flattened shape. As a result, no cell is very far from the body surface. Therefore, oxygen and carbon dioxide can be exchanged between all cells and the surrounding water by diffusion. There is no need for a special transporting system.

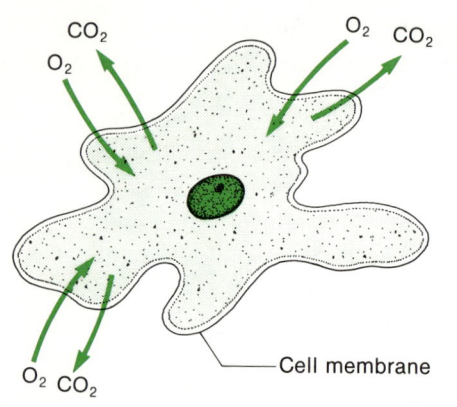

Fig. 19-2 Gas exchange in an amoeba. The entire body surface of this single-celled organism is in contact with water. Therefore, oxygen and carbon dioxide can easily be exchanged across any part of the cell membrane.

Use of Gills

In most large aquatic animals, gas exchange cannot occur directly between all body cells and the surrounding water. Most cells in these animals are far from the body surface. Diffusion from cell to cell occurs too slowly to keep the cells alive. As a result, these animals have special gas exchange surfaces. They also have special systems to transport gas between the gas exchange surface and all body cells. The special gas exchange surfaces are called **gills**. The special transport system is called a **circulatory system**.

Gills are thin walled and finely divided structures. They provide a very large surface area for gas exchange. Also, they have a rich supply of blood vessels. Dissolved oxygen and carbon dioxide can easily be exchanged across the thin membrane that separates the blood vessels from the surrounding water. The blood vessels contain the fluid, **blood**. It transports the dissolved gases to and from all body cells.

Gills may differ somewhat from animal to animal (Fig. 19-4). However, all gills usually have the same characteristics: They provide a large surface area for gas exchange. Also, they contain blood vessels (part of the circulatory system) to transport the gases.

The Crayfish

The crayfish "breathes" by means of gills. These gills have a feathery structure. They occur in rows in a chamber on both sides of

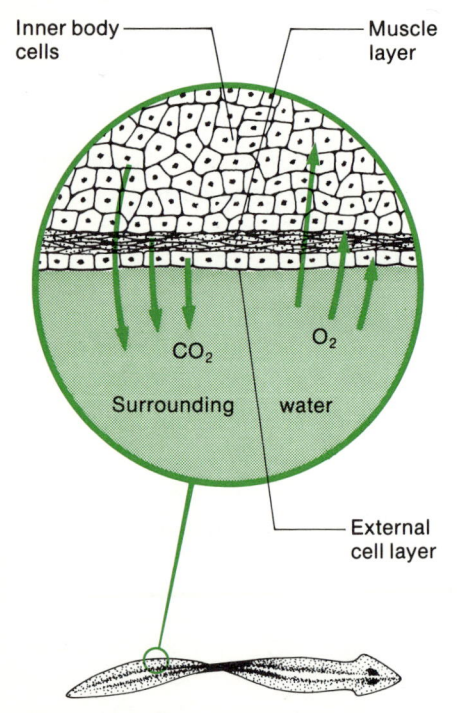

Fig. 19-3 Gas exchange in a planarian (flatworm). All cells in this flattened animal are very close to the surrounding water. This lets all cells exchange gases by simple diffusion.

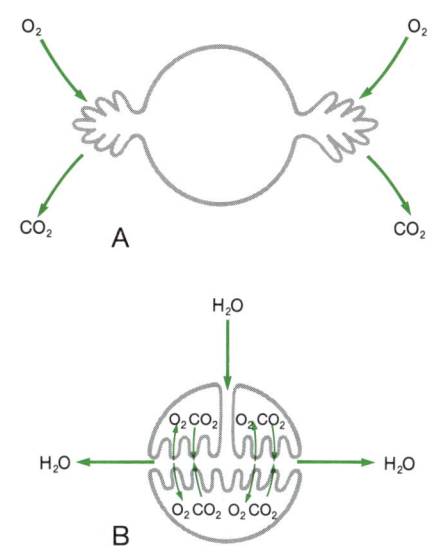

Fig. 19-4 Gills are a means of increasing surface area for gas exchange. The gills of many invertebrates are external (A). In fish, water passes in one opening (the mouth), moves over the gills where gases are exchanged, and out through different openings (B).

the crayfish. They are just beneath part of its exoskeleton called the **carapace** [ka-rah-PAYS] (see Fig. 19-6). The chamber is open at both ends and along the lower edge. Water enters the gill chamber near the walking legs. It passes upwards and forwards over the gills, and leaves the chamber near the mouth.

When the crayfish is not moving, water is circulated over the gills in currents created by **gill bailers**. The gill bailers are fan-like structures attached to the maxillae at the forward end of the gill chamber. Waving **swimmerets** assist by pushing water forward toward the gills. As water passes over the **gill filaments**, oxygen diffuses from the water into the blood vessels. Also, carbon dioxide diffuses out of the blood and into the surrounding water. Oxygen-rich blood is transported to all parts of the crayfish's body through its circulatory system.

The problem of gas exchange in aquatic animals is complicated by the fact that very little oxygen is dissolved in water. Also, as the temperature of the water increases, the amount of dissolved oxygen decreases. Many aquatic animals must constantly move water over their gills to get enough oxygen.

Section Review

1. Describe how gas exchange occurs in the hydra.
2. **a)** What are gills?
   b) What characteristics do all gills have in common?
3. Describe the breathing process in the crayfish.
4. What effect does water temperature have on the amount of oxygen dissolved in the water?

19.3 ACTIVITY How Water Temperature Affects the Breathing Rate of Fish

Like most animals, fish need oxygen in order to live. Since fish live in water, the oxygen must come from the water. Fish cannot break down H_2O (water molecules) to remove oxygen. Instead, they remove *dissolved* oxygen from the water. Also, carbon dioxide must be given off into the water. Fish use gills to exchange these gases.

Problem

How do fish gills operate? What effect does an increase in water temperature have on their action?

Materials

goldfish in an aquarium
beaker (1000 mL)
dip net

crushed ice
hot water
thermometer

Section 19.3   265

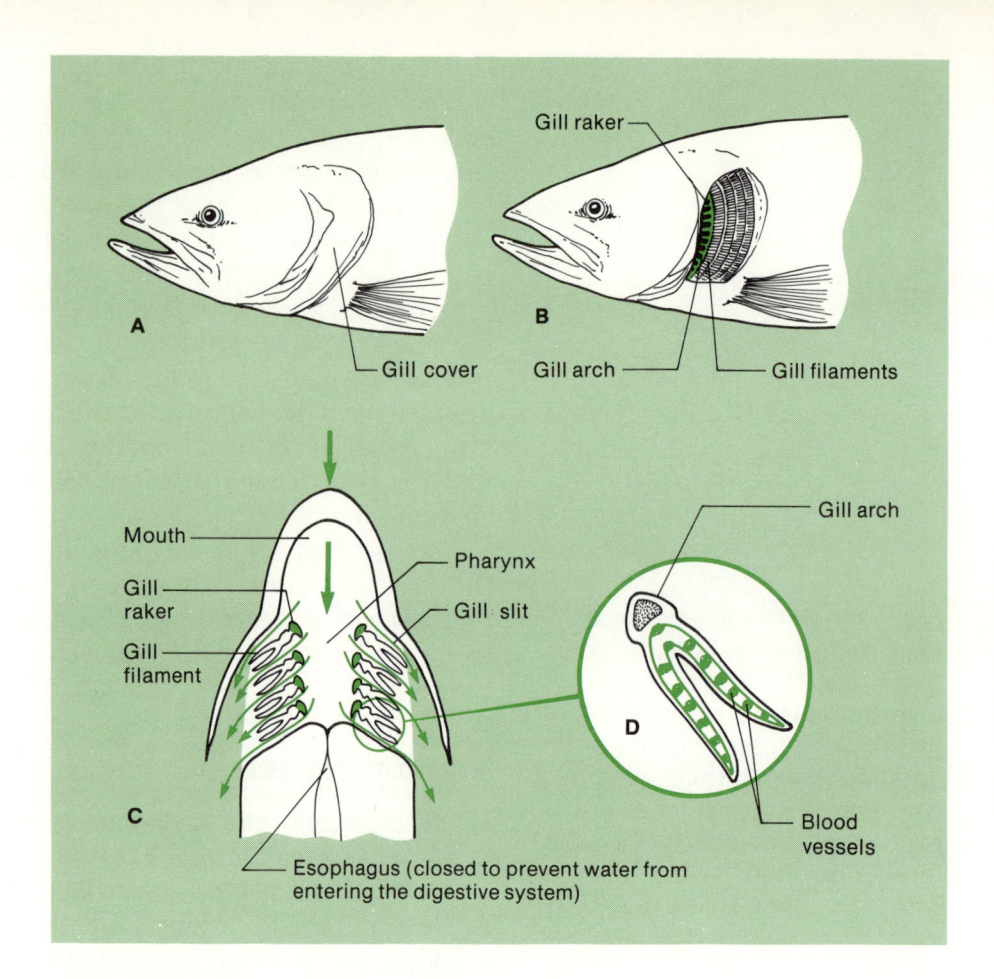

Fig. 19-5 Gills of a fish: gills covered for protection (A); gill cover cut away to show a gill (B); path of water through gills (C); cross section through a gill filament (D)

Procedure

Note: Before doing this activity, study Figure 19-5. Note that the gills are found on the sides of the head beneath the **gill covers**. The gills consist of feathery **gill filaments** attached to a supporting **gill arch**. Projections called **gill rakers** strain solid particles from the water before it passes over the gills. This helps to protect the delicate gill filaments.

a. Copy Table 19-1 into your notebook.
b. Observe the fish in the aquarium. Look at its head region for a few minutes. Describe the movement of its mouth and gill covers.
c. Fill the beaker three-quarters full of aquarium water. Then use the dip net to transfer the fish to the beaker.
d. After a few minutes, find the fish's breathing rate. You can do this by counting the number of times the gill covers open and close in one minute.
e. Take the temperature of the water.
f. Record the breathing rate and the temperature in your table.
g. Add crushed ice to the beaker a little at a time. Observe the water temperature. Each time it drops by 5°C record the fish's breathing rate and the temperature. (*Note:* As you add ice, pour enough water out of the beaker to keep the water level constant.)

h. When the temperature is close to 0°C, start adding hot water in small portions. You do not want to harm the fish. Therefore, pour the hot water into an area of the beaker away from the fish. Stir the water gently to equalize the temperature throughout. Find the breathing rate with each 5°C change in temperature until the original water temperature is reached. Record your results.
i. Use the dip net to return the fish to the aquarium.

Table 19-1 Temperature and Breathing Rate

Water temperature	Breathing rate (gill cover movements per minute)

Discussion

1. Describe what happens to the breathing rate of the goldfish as the water temperature changes. Why does this happen?
2. Table 19-2 shows the amount of dissolved oxygen that water is able to hold at various temperatures. If an industry or nuclear power plant pollutes the water with heat, what effect will this have on the fish? Use both the data given in Table 19-2 and the results of this activity to answer the question.

Table 19-2 Temperature and Dissolved Oxygen

Temperature (°C)	Amount of dissolved oxygen water can hold (µg/g)
0	14.6
5	12.7
10	11.3
15	10.1
20	9.1
25	8.3
30	7.5

19.4   ACTIVITY   The Crayfish and How It Breathes

The crayfish "breathes" by means of gills. These gills are located on both sides of the crayfish. They are within a chamber beneath the carapace or gill cover (Figure 19-6).

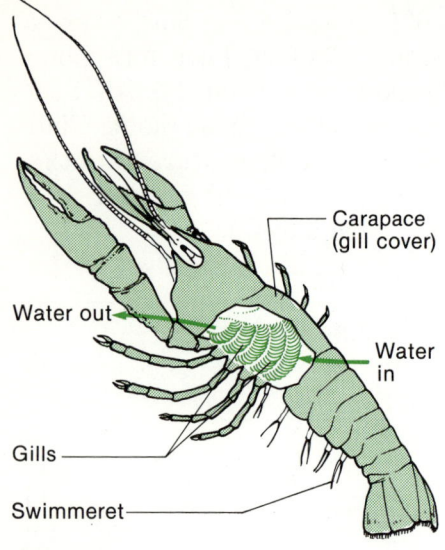

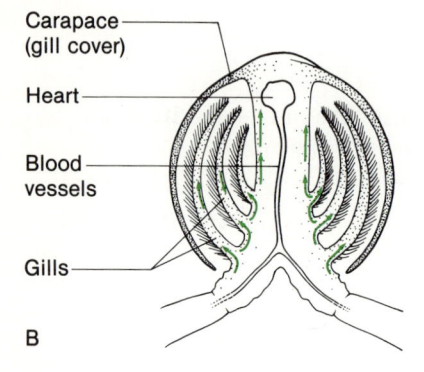

Fig. 19-6 Gills of a crayfish: part of carapace removed to show the feather-like gills. Note the direction in which the water moves (A). Cross section of a crayfish showing gill arrangement (B)

Problem

What is the structure of a gill? How does water pass through the gill chamber of a crayfish?

Materials

Part A

live crayfish
small aquarium or container of fresh water
dropper
India ink

Part B

preserved crayfish
dissecting equipment
dissecting tray
hand lens
watch glass

Procedure A Breathing Movements In a Live Crayfish

a. Transfer the live crayfish to a container of fresh water. Hold the container so that you can examine the ventral surface of the crayfish.
b. Describe the movements of the tiny appendages (swimmerets) attached to the abdomen. These structures assist in moving water. In what direction is the water moved?
c. Locate the gill bailers at the sides of the mouth. Describe any movements the gill bailers make. What does the name of these structures tell you about their function?
d. Carefully place a couple of drops of India ink into the water at the posterior end of the gill cover. Try not to disturb the crayfish. Observe the movement of the ink.
e. Remove the crayfish from the container and place it back in its original tank.

Procedure B Dissection of Breathing Organs

a. Using the scissors, cut along the mid line of the carapace (dorsal surface). Go from the posterior end to the groove that divides it from the head region.
b. Cut along the groove from the right side up to the first cut.
c. Using the forceps remove the right side of the carapace to reveal the gills.
d. Observe the number and arrangement of the gills. Record your observations.
e. Remove a walking leg with the gills attached.
f. Cut a gill from the leg. Place it in a watch glass with water. Allow the water to spread the gill filaments apart. Observe them carefully. Record your observations.
g. Remove a maxilla with a gill bailer. Examine it with a hand lens. Describe its shape.
h. Examine the swimmerets. Describe their shape.

Discussion

1. How is water circulated over the gills of the crayfish?
2. How does gas exchange occur in the crayfish?
3. Why must fresh water be circulated past the gills?

19.5 Gas Exchange on Land

You might think that gas exchange would not be a problem for animals which live on land. After all, compared to water, the land environment is quite rich in oxygen (see Table 19-3). Remember, though, that a gas exchange surface must be kept moist. Air is a very dry environment. Land animals, therefore, must have some means of keeping their breathing surfaces from drying out. Let us examine a few land animals to see how they solve this problem.

Table 19-3 The Main Gases in Air

Gas	Percent by volume
Nitrogen	78.9
Oxygen	20.94
Argon	0.93
Carbon dioxide	0.034

The Earthworm: A Simple Breathing System

The earthworm breathes through its skin. Its skin consists of a very thin membrane. This membrane is kept moist by a secretion of **mucus** [MYOO-kus]. In addition, the earthworm lives in moist soil.

The skin of the earthworm also has a rich supply of blood vessels. Oxygen can easily dissolve in the film of moisture covering the skin and diffuse into the blood vessels. Carbon dioxide diffuses outward from the blood into the soil (Fig. 19-7). This process occurs all over the body. Therefore, the gas exchange surface of the earthworm is large enough to supply all the oxygen the animal needs.

Have you ever noticed dead earthworms on the ground after a rain? As water filled their burrows, they were forced to the surface. These worms, unable to get back into the soil, probably died from lack of oxygen once their moist skins dried out.

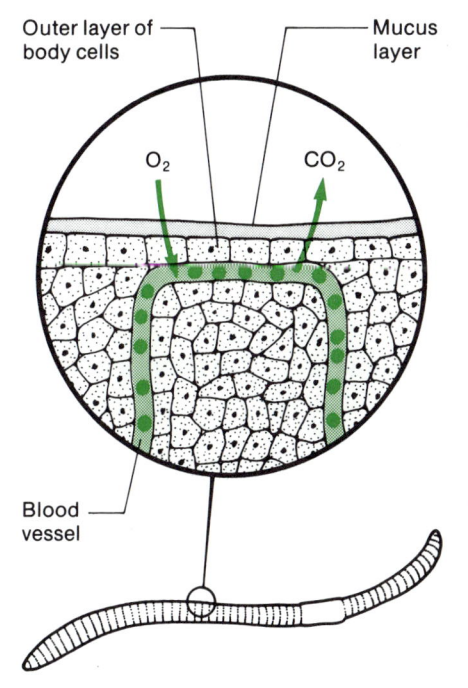

Fig. 19-7 The breathing system of an earthworm. Its moist skin is a special gas exchange surface. Blood vessels carry gases to and from this surface.

Insects: Tracheal Breathing Systems

Insects solve the problem of gas exchange in a land environment in a unique way. Their breathing systems consist of a network of branched tubes called **tracheae** [TRAY-kee-ay]. Let us look at the grasshopper to see how this **tracheal system** [TRAY-kee-uhl] works.

The grasshopper has a series of small openings called **spiracles** [SPIR-ih-kuhlz] along the sides of its body (Fig. 19-8). These openings

Section 19.5   269

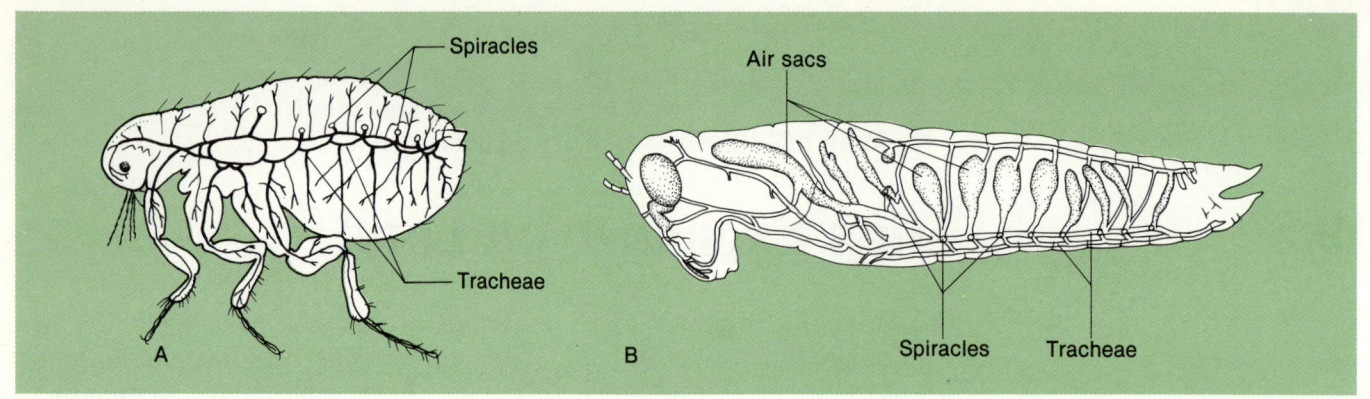

Fig. 19-8 Insect tracheal systems: the flea (A); the grasshopper (B). A network of air filled tubes or tracheae carry gases to and from all parts of the body. Air enters and leaves the tracheae through openings called spiracles.

lead into a branched system of tracheae. The tracheae subdivide into smaller and smaller tubes called **tracheoles** which reach all the cells in the insect's body (Fig. 19-9). The inner end of the smallest tubes contain a fluid. This fluid provides the moist surface needed for gas exchange by diffusion. In addition, a double row of **air sacs**, for air storage, is located opposite the spiracle openings.

As the lower part of the grasshopper's abdomen pulses up and down, air enters and leaves the spiracles. It then travels directly to the body cells by the system of tracheal tubes. Oxygen and carbon dioxide are easily exchanged across the moist membranes of the cells.

The Frog: A Lung Breathing System

The breathing system of an adult frog is quite different from that of the earthworm and insects. The frog can actually exchange gases in three ways. In the water, a frog hibernates in the mud at the bottom of a pond. During this time, oxygen and carbon dioxide are exchanged through the frog's thin moist **skin**. Blood vessels beneath the skin transport these gases to and from all body cells. In fact, at all times of the year, the skin is the major breathing surface when the frog is under water.

Gas exchange can also occur across the lining of the frog's mouth. This thin membrane is covered with a moist mucus coating. Also, it has a large supply of blood vessels beneath its surface. This is usually referred to as **mouth breathing**.

The frog also has a pair of special gas exchange structures called **lungs** (Fig. 19-10). These lungs resemble thin-walled sacs. A network of small blood vessels surrounds each lung. As the floor of the mouth is lowered air enters the frog's mouth cavity (Fig. 19-11). The nostrils are then closed and the mouth floor raised. This forces air into the lungs. Oxygen is absorbed into the blood vessels and carbon dioxide is released into the lungs. Air is forced out of the lungs by contraction of muscles in the body wall.

A few fish, most amphibians, and all reptiles, birds and mammals have lung breathing systems. Although they may vary slightly in structure, they are all designed to solve the problem of gas exchange for land animals. We, too, have a lung breathing system. You will see how our gas exchange system works in Chapter 24.

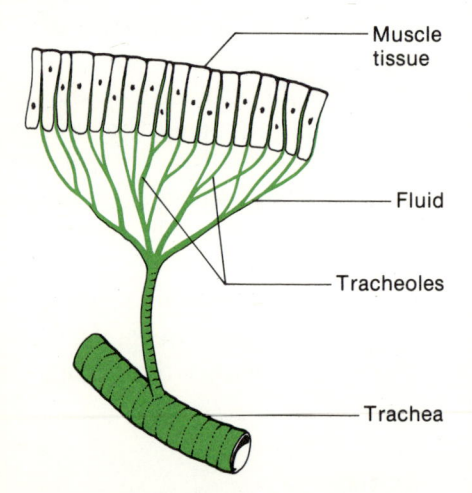

Fig. 19-9 The breathing system of an insect carries oxygen to all cells (here, muscle tissue) through a system of tracheal tubes. Why would muscle cells require such an extensive tracheal system?

Fig. 19-10 The breathing system of a frog. An adult frog has lungs: as seen from the side (A); as seen from the top (B). In the early tadpole stage, external gills are the gas exchange surfaces (C).

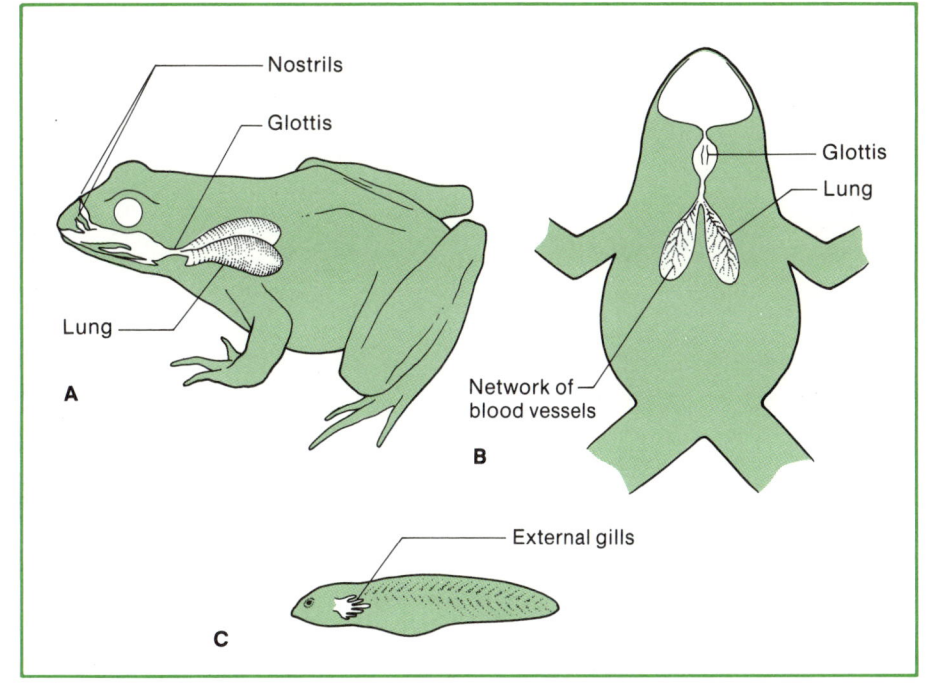

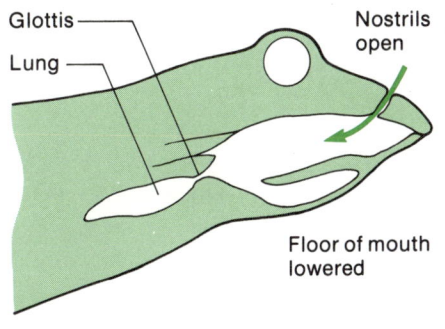

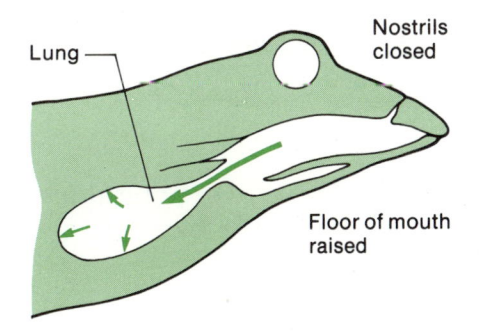

Fig. 19-11 Breathing movements of a frog. Air enters the mouth cavity as the nostrils are opened and the floor of the mouth is lowered (A). Air is forced into the lungs as the nostrils are closed and the floor of the mouth is raised. The elastic lungs expand (B).

Section Review

1. Why is gas exchange a problem for land dwelling animals?
2. How does gas exchange occur in the earthworm?
3. How does the tracheal system solve the problem of gas exchange for the grasshopper?
4. **a)** Describe three ways an adult frog can obtain oxygen.
   b) Which method do you think can supply oxygen at the fastest rate? Why?
   c) A frog that is resting on the edge of a pond may use only mouth breathing. In contrast, a frog that is hopping must use lung breathing as well. Why is this so?

Main Ideas

1. Most animals need oxygen in order to release the energy from nutrients they eat.
2. An animal must have some means of ridding itself of carbon dioxide.
3. The exchange of oxygen and carbon dioxide is carried out in a variety of ways in animals.
4. The gas exchange surface of an animal must be moist. It must also be large enough to allow for adequate gas exchange.
5. Temperature has an effect on the amount of oxygen an animal requires.
6. Large land animals usually have lung systems working together with blood transport systems.

Glossary

carapace	ka-rah-PAYS	the hard exoskeleton covering the gill chambers of the crayfish
gills		special gas exchange structures or breathing organs of certain animals
mucus	MYOO-kus	a protective fluid found on the body surface of fish, frogs and earthworms. It also lines the inside of our breathing system
respiration	res-puh-RAY-shun	the process by which energy is released in living organisms
tracheae	TRAY-kee-ay	branched tubes through which air is transported in the breathing system of insects and some other animals

Study Questions

A. True or False

Decide whether each of the following statements is true or false. If the sentence is false, rewrite it to make it true. (Do not write in this book.)

1. Oxygen is required to release energy from nutrients in most animals.
2. Carbon dioxide is a harmless by-product of respiration.
3. In order to obtain oxygen, a cell simply has to be exposed to air.
4. In the frog, gas exchange can occur across the skin and mouth as well as the lungs.

B. Completion

Complete each of the following sentences with a word or phrase that will make the sentence correct. (Do not write in this book.)

1. Most animals require _____ for respiration and release _____ _____ in the process.
2. The movement of oxygen and carbon dioxide between living organisms and their environment is called _____.
3. The gas exchange surfaces of a crayfish are called _____.
4. Air enters the tracheae of insects through openings called _____.

C. Multiple Choice

Each of the following statements or questions is followed by four responses. Choose the correct response in each case. (Do not write in this book.)

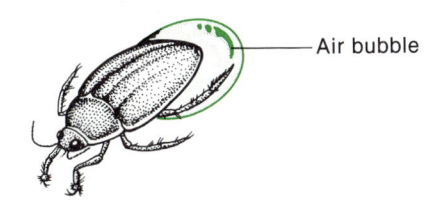

Fig. 19-12 The water beetle, *Dytiscus*, traps a bubble of air with its hind legs.

1. Earthworms breath through
   a) spiracles b) moist skin c) gills d) lungs
2. As the temperature of water increases
   a) the breathing rate of fish decreases
   b) the breathing rate of fish increases
   c) the breathing rate of fish remains the same
   d) the amount of oxygen in the water increases
3. To increase diffusion rate, breathing systems have
   a) low surface to volume ratios
   b) well dried gas exchange systems
   c) large gas exchange surface areas
   d) thick walls surrounding the gas exchange surface
4. A characteristic of all breathing systems is
   a) tracheal tubes b) gills c) lungs d) thin, moist membranes

D. Using Your Knowledge

1. Describe the structure and operation of the gills of a crayfish. How is this gill breathing system related to the blood transport system?
2. Describe how gas exchange takes place in a grasshopper. Why does its breathing system not have to be linked to a blood transport system?
3. One requirement for gas exchange surfaces is that they be of adequate size. How does the earthworm meet this requirement?
4. Air contains far more oxygen than can be absorbed by water. Why, then, do fish suffocate on dry land?
5. If you place a fish in boiled water (after it has cooled, of course) the fish will become very inactive. It may even die. Explain why.
6. The hydra picks up O_2 and gets rid of CO_2 directly through its body wall. Why doesn't it need special gas exchange structures like lungs or gills?

E. Investigations

1. Figure 19-12 is a diagram of the water beetle *Dytiscus*. The trapped air bubble is actually used as a lung. Find out how it works.
2. Using the library or resource centre, prepare a report on the breathing system of a bird. (Refer to Figure 19-13.) Be sure to indicate how it enables a bird to fly great distances without stopping.
3. Look at Table 19-3. The most abundant gas in air is nitrogen. Animals need nitrogen. Find out what for. Also, find out if animals can use nitrogen gas from the air.

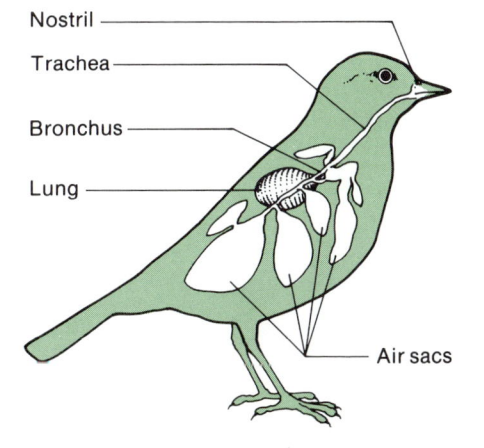

Fig. 19-13 The breathing system of a bird. Gas exchange takes place in the lungs. Air sacs provide a continuous supply of oxygen-rich air that is demanded during flight.

20 Internal Transport Systems

20.1 Why Transport Systems Are Necessary
20.2 Some Simple Transport Systems
20.3 Circulatory Systems in Animals
20.4 Activity: Circulation in the Fish: The Role of the Capillaries
20.5 Activity: How Temperature Affects Circulation in the Earthworm

In Chapter 18 you saw how animals get the nutrients they need for their life processes. In Chapter 19 you saw some methods by which animals exchange oxygen and carbon dioxide with their environments. But how are these substances transported within an animal? Let us look at the reasons why internal transport systems are needed in living organisms. Also, let us examine some of the methods by which materials are transported within animals.

20.1 Why Transport Systems Are Necessary

Transport Systems Carry Raw Materials and Wastes

In order to remain alive, every cell must have raw materials such as food and oxygen. Also, harmful by-products of life functions, such as carbon dioxide, must be gotten rid of by every cell. This is true for single-celled organisms. It is also true for *every cell* of a many-celled organism. As a result, *all* cells of an organism must be in contact with the medium from which they can get required raw materials. Also, all cells must be in contact with a medium into which they can get rid of the harmful by-products of life activities (sometimes called wastes).

For single-celled organisms living in water, there is no problem in this respect. They are completely surrounded by a medium, namely water. They can get needed raw materials from it. And they can dispose of their wastes in it. Even in some simple many-celled aquatic animals, every cell is either surrounded by, or very close to, a water medium.

Some larger and more complex animals, however, do have a problem in this respect. Many cells may be far removed from an external medium. In such animals a **division of labour** occurs. This means that certain groups of cells specialize to perform certain life functions. For example, gas exchange may take place in lungs. Or digestion may occur in a special digestive system. Materials must be carried between these special structures and all cells of the animal. An **internal transport system** is required to do this.

Transport Systems Carry Hormones

In some animals, special cells produce important substances that must be delivered to every cell. A **hormone** is an example of such a substance. It is a chemical messenger which helps control and coordinate cell activities. A hormone may be produced by special cells in one part of an organism. Yet it may control the actions of some cells in another part of an organism. Obviously, there is a need in such an organism for an internal system to transport these hormones. You will study the role of hormones in more detail in Chapter 28.

Section Review

1. What kinds of materials must a living cell exchange with its environment?
2. Why are internal transport systems necessary in larger many-celled organisms?

20.2 Some Simple Transport Systems

Transport in Simple Organisms

In single-celled organisms and some small many-celled animals, substances are transported by **diffusion**. Diffusion carries materials from where they enter or are found in these organisms to where they are used or disposed of by the organism. But diffusion is a very

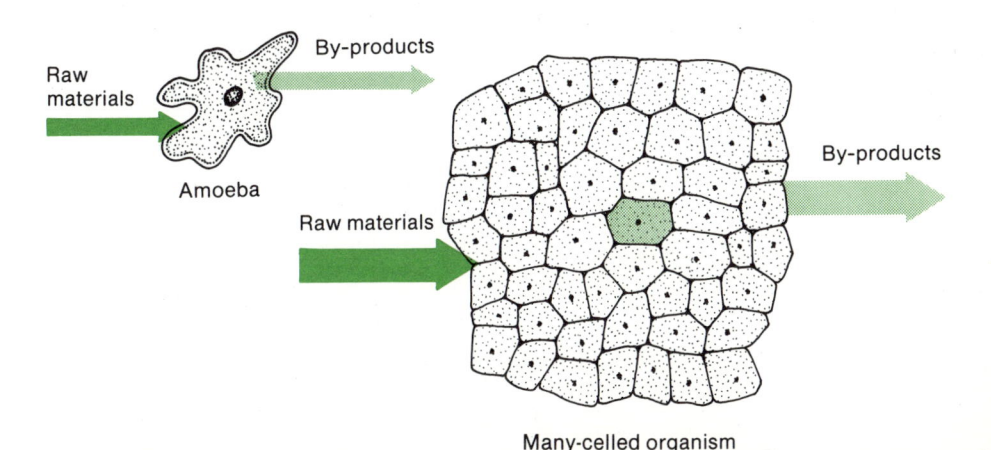

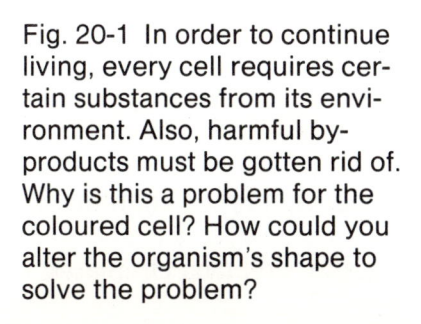

Fig. 20-1 In order to continue living, every cell requires certain substances from its environment. Also, harmful by-products must be gotten rid of. Why is this a problem for the coloured cell? How could you alter the organism's shape to solve the problem?

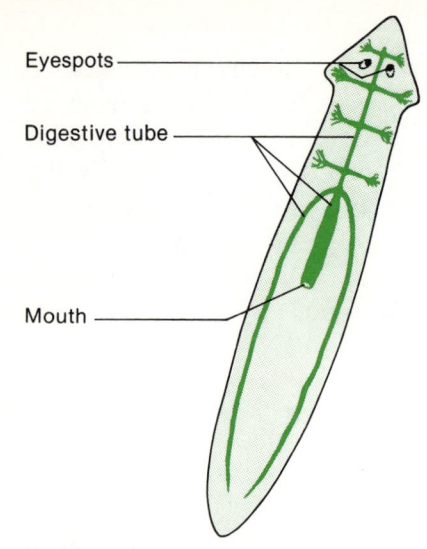

Fig. 20-2 The transport system of a planarian. Its digestive tube also serves as its transport system.

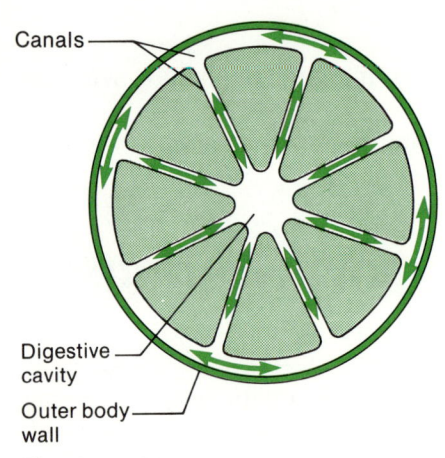

Fig. 20-3 The transport system of a jellyfish. A series of canals transports materials to and from all cells.

slow means of transport. Some organisms require that materials be transported much more quickly (Fig. 20-1). Therefore, in some organisms the cytoplasm within the cell moves. Materials in the cytoplasm get carried along with it. No one knows for sure how this **streaming of cytoplasm** works. However, it is known that energy is required in the process. Materials can be transported by this means much faster than by diffusion.

Transport in Hydra

The **hydra** has a body shape that brings all cells close to its external environment (see Figure 18-4, page 246). The inner layer of cells is in direct contact with water because of the presence of a body cavity. Special cells with **flagella** [fluh-GEL-uh] provide a sweeping action that creates a current in the water. In addition, special muscle cells enable the hydra body wall to contract. These contractions and currents cause circulation of water in the cavity. This ensures that a constant supply of water, rich in raw materials, enters the cavity. Furthermore, waste-laden water can be forced out of the cavity at the mouth opening.

Each cell of the hydra exchanges oxygen and carbon dioxide directly with its water environment. Not all cells take in food. However, since no cell is very far from the special ones that do, this food is easily transported to all cells by diffusion.

Many other simple animals also have body plans that ensure adequate transportation of materials to and from all cells. The **planarian** has a tube system which branches into all parts of its body (Fig. 20-2). The **jellyfish** has a series of canals which connect the digestive cavity to all body regions. These canals form a simple transport system (Fig. 20-3).

Section Review

1. By what two means do single-celled organisms and small many-celled animals transport materials?
2. **a)** Describe the body plan of a hydra.
   b) How is the hydra structurally adapted to ensure that all cells are in contact with water?

20.3 Circulatory Systems in Animals

Blood Circulatory Systems

Most many-celled animals have special internal transport systems. Materials are carried in a body fluid, usually **blood**. Blood moves round and round through the body. Such a system is called a **blood circulatory system**. Blood circulatory systems generally contain:
1. A fluid (blood) to carry materials;
2. A pumping device called the heart to make the blood circulate;

3. Valves to control the direction of blood flow;
4. Tubes or vessels through which blood travels to all parts of the body and then back to the heart.

All blood circulatory systems perform the same basic function. They ensure that the materials which they transport can be easily exchanged with any cell in the body. Let us examine the blood circulatory systems of a few animals.

The Earthworm

In the earthworm, blood flows within tubes or vessels throughout its entire circulatory system. This is called a **closed transport system** (Fig. 20-4). A large tube called a **dorsal blood vessel** runs along the upper surface of the earthworm, just beneath its skin. This tube carries red blood forward to the front end of the earthworm. Blood returns to the hind end through a large **lower blood vessel**. Tiny thin-walled tubes connect the upper and lower vessels throughout the earthworm. These small tubes are called **capillaries** [KAP-uh-ler-ees]. Their thin walls allow materials to be exchanged between the blood and surrounding body cells.

The earthworm has five "**hearts**". These are muscular tubes which form five loops connecting the upper and lower blood vessels. The hearts pump blood from the dorsal (upper) to the lower vessel. The blood carries food and oxygen to all parts of the earthworm's body. It also transports harmful waste products away from cells to organs that will excrete them.

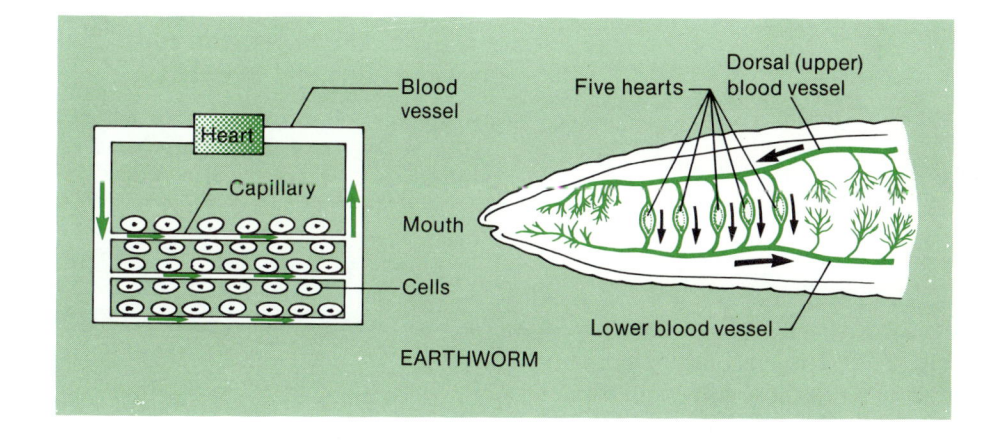

Fig. 20-4 The circulatory system of an earthworm. The blood circulates only in blood vessels (tubes). This is called a closed circulatory system. Tiny vessels (capillaries) connect the upper and lower blood vessels. Because of their small size they are not shown in the earthworm diagram.

The Grasshopper

Unlike the earthworm, blood in the grasshopper's circulatory system does not always remain in blood vessels. Instead, during its circuit, blood passes out of a blood vessel into large open spaces between body organs. This is called an **open circulatory system** (Fig. 20-5).

A single blood vessel is located in the upper region of the grasshopper's body. The rear portion of this vessel acts as a series of "**hearts**". A colourless blood is pumped forward towards the head

Fig. 20-5 The circulatory system of a grasshopper. Blood flows through a vessel and then into open spaces (sinuses) between body organs. This is called an open circulatory system.

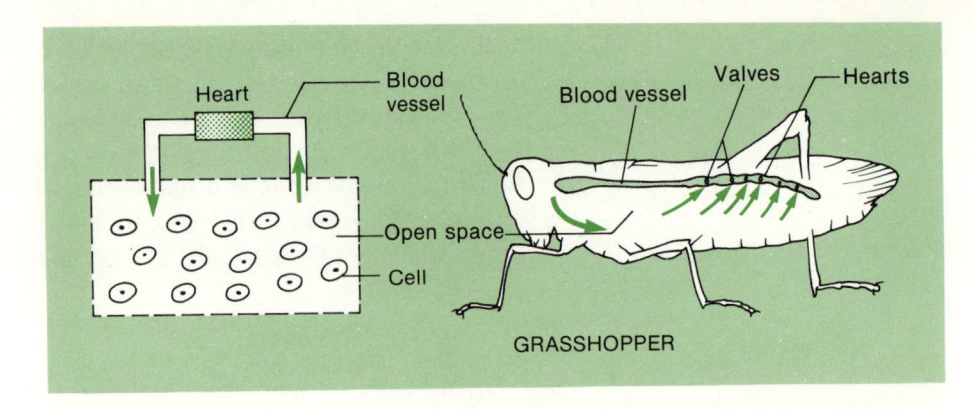

and out into the large open body spaces. There, blood bathes the body organs. Nutrients and waste products are exchanged between the blood and all body cells. The blood eventually re-enters the blood vessel near the "heart" region through a system of one-way openings or valves.

You may recall from Chapter 19 that the circulatory system of the grasshopper does not transport oxygen and carbon dioxide. This role is carried out by a system of **tracheal tubes**.

The Crayfish

The crayfish also has an open circulatory system. However, unlike the grasshopper, its blood does carry oxygen and carbon dioxide. Tiny capillaries extend into the **gills** of a crayfish. Here oxygen and carbon dioxide are exchanged with the water. Blood flows between a series of vessels and open body spaces (Fig. 20-6). In the body spaces, blood bathes the cells. This allows materials to be exchanged between cells and blood.

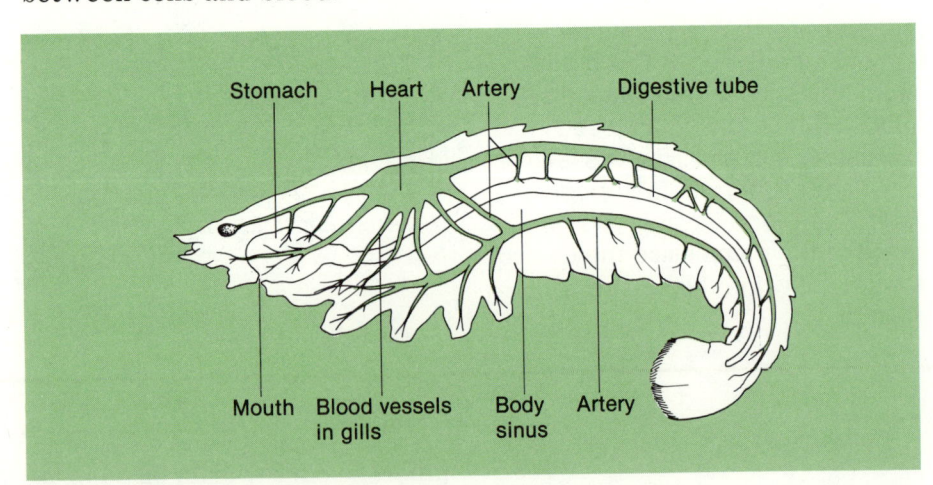

Fig. 20-6 The circulatory system of a crayfish. Although it contains more blood vessels than that of a grasshopper, it is still an open circulatory system. Arrows indicate the direction of blood flow.

Vertebrate Transport Systems

Transportation occurs much more rapidly in a closed circulatory system than in an open system. Therefore, very active animals

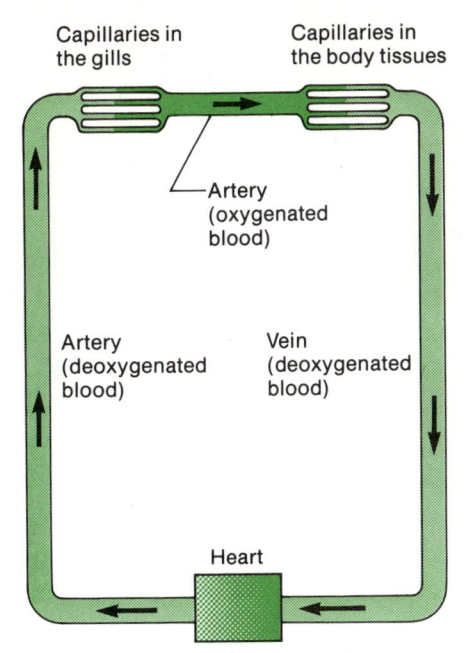

Fig. 20-7 A diagram of a single circulatory system. Blood passes through the heart only once during each circuit.

usually have a closed system. All vertebrates have a closed circulatory system.

Besides a heart, vertebrate circulatory systems contain three types of blood vessels: arteries, veins, and capillaries. **Arteries** [AR-tuh-rees] carry blood away from the heart. **Veins** transport blood toward the heart. Tiny **capillaries** [KAP-uh-ler-ees] connect the arteries and veins.

In some animals, blood passes through the heart only once each time it completes a circuit. Such an arrangement is called a **single circulatory system** (Fig. 20-7). This type of system is characteristic of fish. In birds and mammals, blood passes through the heart twice each time it completes a circuit. This is called a **double circulatory system** (Fig. 20-8).

The frog also has a double circulatory system. Its heart contains three chambers (Fig. 20-9,B). The **left atrium** [AY-tree-um] receives oxygen-rich (oxygenated) blood from the lungs. The **right atrium** receives oxygen-poor (deoxygenated) blood returning from body tissues. As the two atria pump together, blood from each is forced into a muscular **ventricle** [VEN-truh-kul]. This mixture is then pumped to all body cells. Of course, a system of arteries, veins, and capillaries completes the circuit in this closed system.

Heart structure varies in the other vertebrates. A comparison of the hearts of each vertebrate type is shown in Figure 20-9.

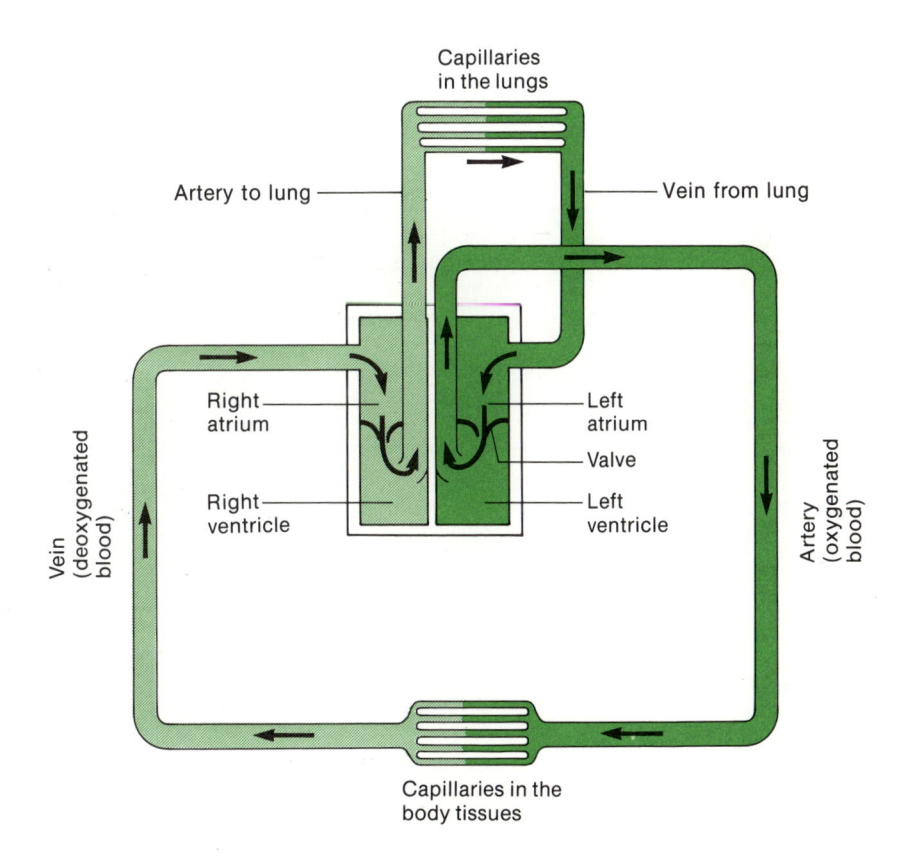

Fig. 20-8 Diagram of a double circulatory system. Blood passes through the heart twice during each circuit.

Section 20.3

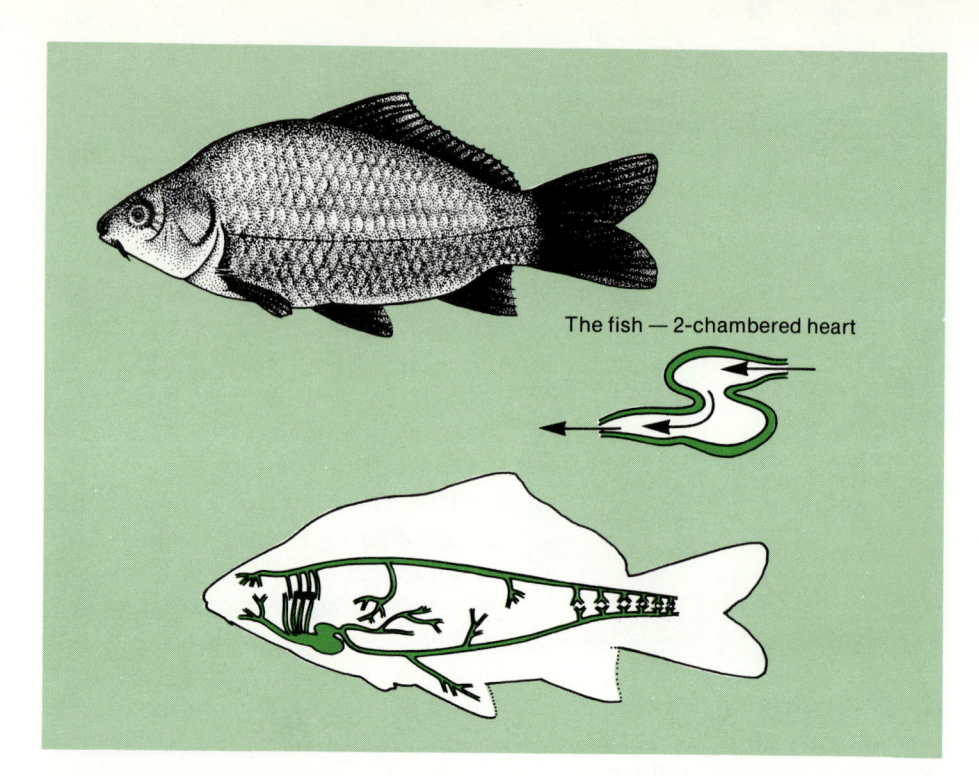

Fig. 20-9 A comparison of the hearts and circulatory systems of the five classes of vertebrates. Why is the mammal's heart considered to be the most efficient? Heart structure of a fish (A); of an amphibian (B); of a reptile (C); of a bird (D); of a mammal (E)

The fish — 2-chambered heart

A. Heart structure of a fish

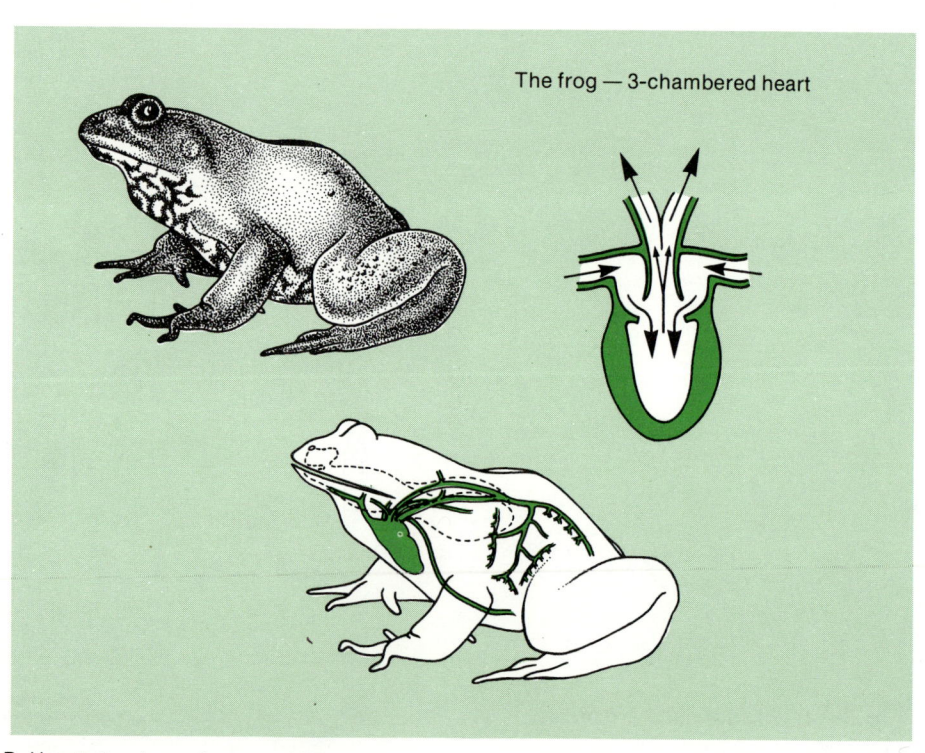

The frog — 3-chambered heart

B. Heart structure of an amphibian

280 **Chapter 20**

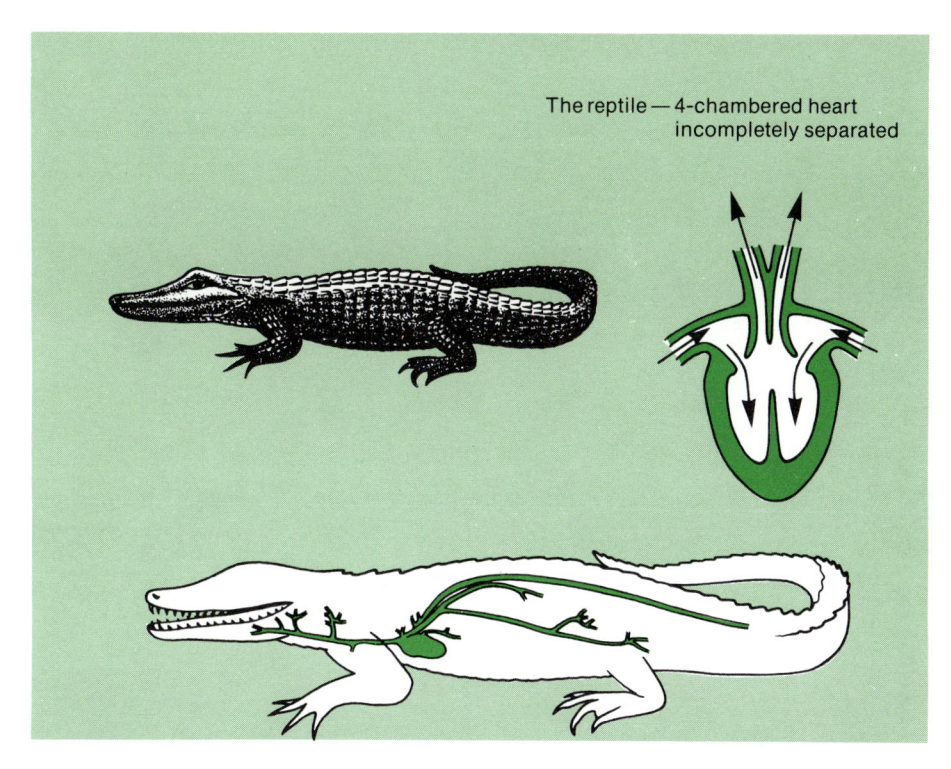

C. Heart structure of a reptile

D. Heart structure of a bird

Section 20.3

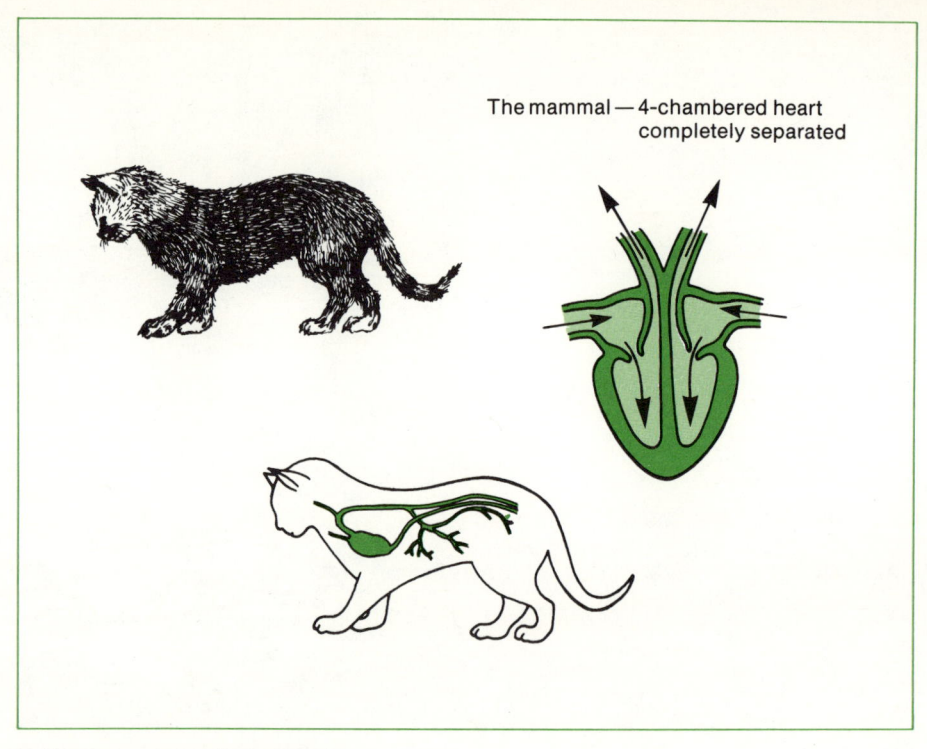

The mammal — 4-chambered heart completely separated

E. Heart structure of a mammal

Section Review

1. a) What are the parts of a blood circulatory system?
   b) What is the function of a blood circulatory system?
2. How does an open transport system differ from a closed transport system?
3. Describe the circulatory system of an earthworm.
4. a) Describe the circulatory system of a grasshopper.
   b) Why is it not necessary for the grasshopper's blood to carry oxygen?
5. Is the crayfish circulatory system more like that of the earthworm or the grasshopper? Explain your answer.
6. Distinguish between arteries, veins, and capillaries.
7. Use examples to distinguish between single and double circulatory systems.

20.4 ACTIVITY Circulation in the Fish: The Role of the Capillaries

Background

The pattern of blood circulation, as we know it today, was not discovered until the 17th century. At that time, an English doctor named William Harvey studied the pumping action of various animal hearts. He noticed the presence of vessels carrying blood to the heart. Today, these vessels are called **veins** [VAYNS]. He also observed

282 Chapter 20

the presence of vessels carrying blood away from the heart. We now refer to these vessels as **arteries** [AR-tuh-rees]. Harvey's observations led him to hypothesize that blood travelled in a complete closed circuit (Fig. 20-10).

Of course, without the aid of a microscope, Harvey never saw the fine blood vessels connecting the arteries and veins. But just after Harvey's death, Marcella Malpighi, an Italian anatomist, did observe these fine connecting vessels. Using a microscope, Malpighi was able to see the tiny vessels connecting the arteries and veins in the lungs of a frog. He called these tiny vessels **capillaries** [KAP-uh-ler-ees].

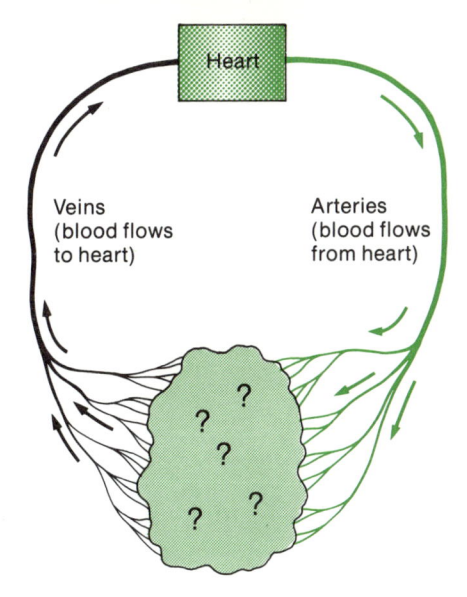

What is the structure that joins the arteries and veins to complete the circuit?

Fig. 20-10 Harvey hypothesized that blood was transported in vessels in a complete circuit. However, he could not see the structure that completes the circuit. What is this structure?

Problem

What do capillaries look like? How does blood flow through them? In this activity you will observe the flow of blood through the capillaries in the tail of a fish.

Materials

goldfish in aquarium
dip net
petri dish (one half only)
cotton
microscope
dropper

Procedure

a. Using a dip net, obtain a small goldfish from the aquarium. If you follow the procedures quickly and carefully, the fish will not be harmed. **After 5 min it must be returned to the aquarium.**

b. Place the goldfish in a petri dish which has been moistened with water. Try not to touch the fish with your fingers. Handling a fish may remove its mucus coating. This coating protects the fish from fungal and bacterial infections (Fig. 20-11).

c. Wrap the fish in wet cotton. Be sure to leave the mouth and tail exposed.

d. Place the petri dish on the microscope stage. Focus on the tail fin with low power.

e. Observe the movement of blood through the vessels. The smallest vessels you see are capillaries.

f. Switch to medium power to see the blood cells flowing through the capillaries.

g. Compare the rate and method of blood flow in the arteries, capillaries, and veins. Record your observations in a table similar to Table 20-1.

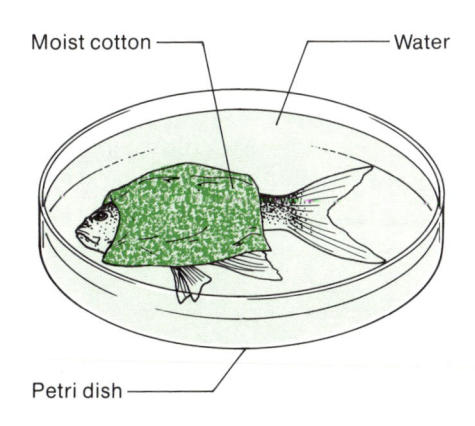

Fig. 20-11 Preparation of a goldfish for examination of capillaries in the tailfin

Table 20-1 Comparison of Blood Flow

Type of vessel	Rate of blood flow	Method of blood flow

h. Compare the diameters of the arteries, capillaries, and veins.
i. Note the shape of the blood cells.
j. Sketch an outline of the fish tail. Then draw in and label the various vessels you observed. Indicate the direction of blood flow.
k. Return the goldfish to the aquarium **(no later than 5 min from the time it was removed)**.

Discussion

1. In which vessel does the blood flow most rapidly?
2. Capillary walls are very thin in comparison to the walls of arteries or veins. What is the reason for this?
3. How is the shape of the red blood cell adapted for movement through the capillaries?
4. Does blood always flow in the same direction through a particular capillary? Or, does the flow change directions?
5. Would you expect to find a capillary network in the gills of the goldfish? Why?

20.5 ACTIVITY How Temperature Affects Circulation in the Earthworm

Birds and mammals are warm-blooded animals. That is, they maintain a constant body temperature, despite changes in the temperature of their surroundings. Their body processes are also kept at a high level. An efficient internal transport system helps these animals to maintain high body temperatures and high metabolic rates.

In contrast, some animals, like the earthworm, are cold-blooded. What happens to the body processes of the earthworm as the external temperature changes? In this activity we will examine the effect of changing the external temperature on an earthworm's circulatory system.

Problem

How does temperature affect circulation in the earthworm?

Materials

live earthworms (2)
thermometer
hand lens
paper towels
timer
dissecting tray
refrigerator

Note: One earthworm should be placed in a refrigerator for at least 2 h before beginning this activity. The other earthworm is to be left at room temperature.

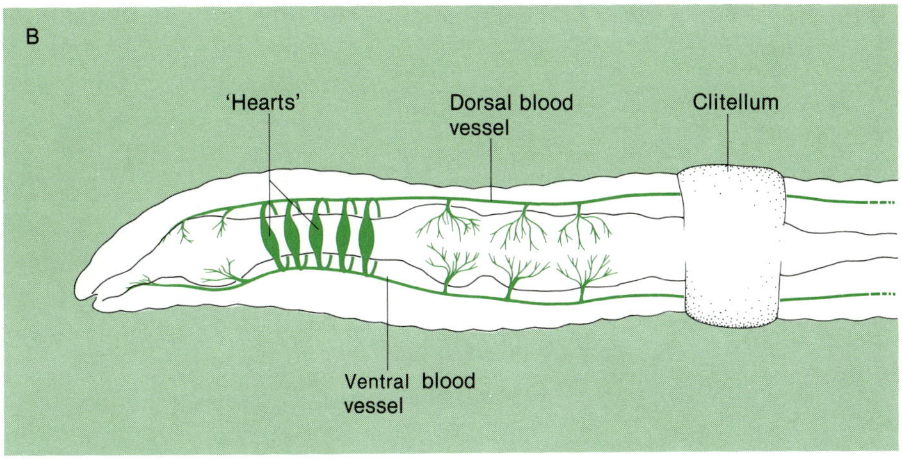

Fig. 20-12 The Earthworm: a drawing showing the location of the dorsal blood vessel. Arrows indicate the direction of blood flow.

Procedure

a. Place the worm at room temperature on the dissecting tray lined with wet paper towels.

b. Using the hand lens, locate the earthworm's dorsal blood vessel. It is easiest to see in the anterior end near the ring shaped clitellum (Fig. 20-12). You should notice it pulsating in a steady rhythm. The vessel becomes thinner as it contracts and may even seem to disappear.

c. Measure and record the room temperature. Use a table similar to Table 20-2.

d. Count the contractions of the dorsal vessel each time they occur. Find the time taken for 10 contractions. Repeat this three times. Then calculate the average rate. Record your results in your table.

e. Repeat step (d) using the refrigerated earthworm. Also, record the temperature of the refrigerator.

f. Allow the refrigerated earthworm to remain on your dissecting tray for about 10 min. Then repeat step (d).

Table 20-2 Temperature and Circulation

External temperature (°C)	Average time for 10 contractions (s)

Discussion

1. What effect did the change in external temperature have on the circulation rate of the earthworm's dorsal vessel? Why?
2. After the refrigerated earthworm was allowed to remain at room temperature for 10 min, what happened to the contraction rate? Explain why.

3. As the environmental temperature decreases, would a cold-blooded animal require more or less food? Explain why. Why is this an advantage?

Main Ideas

1. Animals must have a means of transporting required materials and harmful waste products to and from all cells.
2. Simple animals transport materials by diffusion and movement of the cytoplasm.
3. More complex animals have circulatory systems.
4. Blood circulatory systems generally consist of a fluid (blood), a pumping device (heart), and a network of tubes (blood vessels).
5. An open circulatory system is less efficient than a closed circulatory system.
6. Circulation in blood systems moves blood through arteries, capillaries, and veins.

Glossary

artery	AR-tuh-ree	a blood vessel that carries blood *away from* the heart
atrium	AY-tree-uhm	a chamber of the heart that receives blood from the veins
blood		a fluid in transport systems in which materials are carried
capillaries	KAP-uh-ler-ees	the tiny thin walled blood vessels which connect arteries and veins
vein		a blood vessel that carries blood *to* the heart
ventricle	VEN-truh-kul	a chamber of the heart which pumps blood into the arteries

Study Questions

A. True or False

Decide whether each of the following statements is true or false. If the sentence is false, rewrite it to make it true. (Do not write in this book.)

1. Diffusion and cytoplasmic movement provide adequate transport in large complex animals such as the frog.

2. Blood circulatory systems generally consist of just arteries and veins.
3. The earthworm and the grasshopper both have closed circulatory systems.
4. The circulatory system of a grasshopper transports carbon dioxide and oxygen.
5. A double circulatory system is characteristic of mammals.

B. Completion

Complete each of the following sentences with a word or phrase that will make the sentence correct. (Do not write in this book.)
1. A _____ is a chemical messenger carried by the blood transport system.
2. Single-celled organisms rely on _____ and _____ to transport materials within their "bodies".
3. In most many-celled animals, a fluid called _____ carries materials within the transport system.
4. _____ carry blood away from the heart in circulatory systems.

C. Multiple Choice

Each of the following statements or questions is followed by four responses. Choose the correct response in each case. (Do not write in this book.)
1. The smallest blood vessels (tubes) in a closed circulatory system are the
   a) arteries   b) veins   c) capillaries   d) sinuses
2. An organism without a special transport system is the
   a) hydra   b) earthworm   c) grasshopper   d) crayfish
3. The exchange of materials between the cells of the body and the blood occurs at
   a) the heart   b) capillaries   c) arteries   d) veins
4. An open circulatory system means that
   a) the blood is open to the air or water surrounding an animal
   b) arteries and veins are separated by open body spaces
   c) the circulatory system is easily seen
   d) there is more than one heart in the circulatory system

D. Using Your Knowledge

1. Why doesn't a goldfish die when it is taken out of the water to have its tail examined?
2. The grasshopper does not transport oxygen and carbon dioxide in its circulatory system. Explain how this is possible.
3. Compare the structure of the mammalian heart with that of the frog. Why is the mammalian heart considered to be more efficient?

Study Questions 287

E. Investigations

1. William Harvey helped to explain how circulatory systems work. Using the library or resource centre, write a report on Harvey's findings.
2. Find out how cold-blooded animals avoid freezing in the water.
3. Find out what hypothermia is. Explain your findings using knowledge you have gained in this chapter.

21 Reproductive Systems

21.1 Asexual Reproduction: Only One Parent Needed
21.2 Sexual Reproduction: Two Parents Involved
21.3 Sexual Reproduction: Not without Its Problems
21.4 Reproductive Patterns in Animals
21.5 Activity: Reproductive Systems of Some Animals

Reproduction is the process by which organisms produce new organisms of their own kind. It helps ensure the survival of the species, not the individual. All organisms have a limited life span. They will eventually die from such causes as accidents, disease, or old age. Therefore, without reproduction, a species would become extinct.

In this chapter you will study the reproduction of animals. But first, you will review reproduction of some simple organisms.

21.1 Asexual Reproduction: Only One Parent Needed

There are many different methods by which organisms reproduce. However, they can all be grouped into two categories: **asexual reproduction** [ay-SEK-shoo-ul] and **sexual reproduction**.

Asexual reproduction is the formation of one or more offspring from only one parent. The parent organism, by the process of mitosis (see Chapter 8), produces a new cell or cells which eventually develop into offspring. The new individuals produced have traits identical to those of the parent. Mitosis does not allow for any variation in the offspring.

Asexual reproduction ensures that the hereditary traits remain the same, generation after generation. It is also a very fast method of reproduction. However, there is a disadvantage to this method. If the environment changes, the offspring produced may not be able to adapt to the new conditions. Why is this?

Asexual reproduction can occur in a variety of ways. Let us examine a few of these methods.

Fig. 21-1 Cell splitting or fission is a common means of asexual reproduction in simple organisms such as the amoeba.

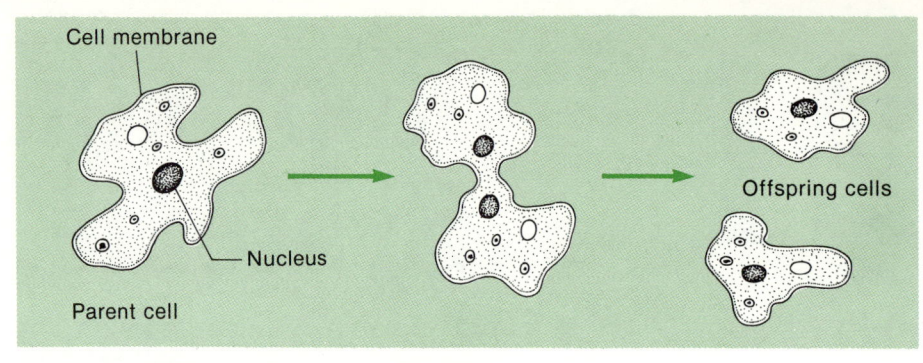

Cell Splitting (Binary Fission)

Many one-celled organisms reproduce asexually by dividing into two. The splitting of the cell occurs following nuclear division by mitosis. Examine Figure 21-1. A single amoeba undergoes mitosis and divides into two new offspring. Mitosis ensures that the hereditary material is duplicated. Therefore the two offspring will have traits identical to those of the parent. Bacteria also reproduce in this manner (see Chapter 9). This process of cell splitting is called **binary fission** [FISH-un]. The new cells produced in this process usually separate and live independently.

Budding

Budding is another type of asexual reproduction. In this process, bud-like growths form on the parent organism. The bud grows and eventually breaks away from the parent. It then develops into a new individual, identical to the parent organism.

The single-celled yeast organism reproduces by budding (Fig. 21-2). Some simple many-celled animals, such as the hydra, also reproduce by budding. Reproduction of the hydra will be discussed in Section 21.4.

Fragmentation

Fragmentation is another form of asexual reproduction. It involves the breaking off of part of the organism. The separated fragment develops into a new complete organism. The sponge, for example, can be cut up into small pieces, or fragments. Each piece can develop into a new sponge.

Fig. 21-2 Budding in a yeast cell. This type of asexual reproduction may develop into a chain of buds. Each bud is a new offspring.

Figure 21-3 shows fragmentation in the planarium. Each fragment is capable of **regeneration** [re-jen-uh-RAY-shun], or regrowth, of any missing parts.

Section Review

1. What is asexual reproduction?
2. List three advantages of asexual reproduction.
3. Compare binary fission to budding. How are these two processes alike? How are they different?
4. What is meant by the term fragmentation?

21.2 Sexual Reproduction: Two Parents Involved

Many animals can reproduce asexually. However, very few depend entirely on this method. Most animals also reproduce sexually. In fact, in animals, sexual reproduction is more common than asexual reproduction. How does sexual reproduction differ from asexual reproduction? Of what advantage is sexual reproduction? Let us find out the answers to these questions.

Special Reproductive Cells

In **sexual reproduction**, two parents are involved. Each parent produces special reproductive cells called **gametes** [GAM-eets]. In most animals, the gametes of each parent are different. For example, the gamete produced by the male parent is called a **sperm cell**. A female gamete is called an **egg cell**. A sperm cell unites with an egg cell to form one new cell. The combining process is called **fertilization** [FUR-tuh-ly-ZAY-shun]. The new cell which results is called a **zygote** [ZY-goat] or fertilized egg. The zygote then divides repeatedly by mitosis. Eventually it develops into a new complete organism.

Each gamete contains half the hereditary information of each of its parents. (Chapter 31 explains why this is so.) Therefore, the new individual will have hereditary information from both parents. Examine Figure 21-4. The result of asexual reproduction is quite different from that of sexual reproduction.

The Reproductive Organs

In most sexually reproducing animals, gametes are formed in special reproductive organs. The special organs which produce sperm cells are called **testes** [TES-teez], or male reproductive organs. Those that produce eggs are called **ovaries** [OHV-uh-rees], or female reproductive organs.

In most animal species, testes and ovaries are found in separate parents. However, in some animals, they are found within the same

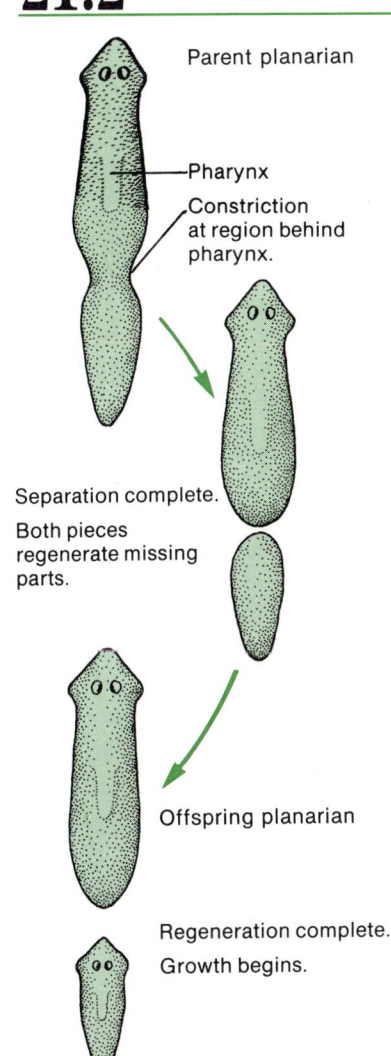

Fig. 21-3 Fragmentation in the planarian. In this form of asexual reproduction each fragment regenerates to form two complete planarians.

Parent planarian
Pharynx
Constriction at region behind pharynx.
Separation complete. Both pieces regenerate missing parts.
Offspring planarian
Regeneration complete. Growth begins.

Section 21.2

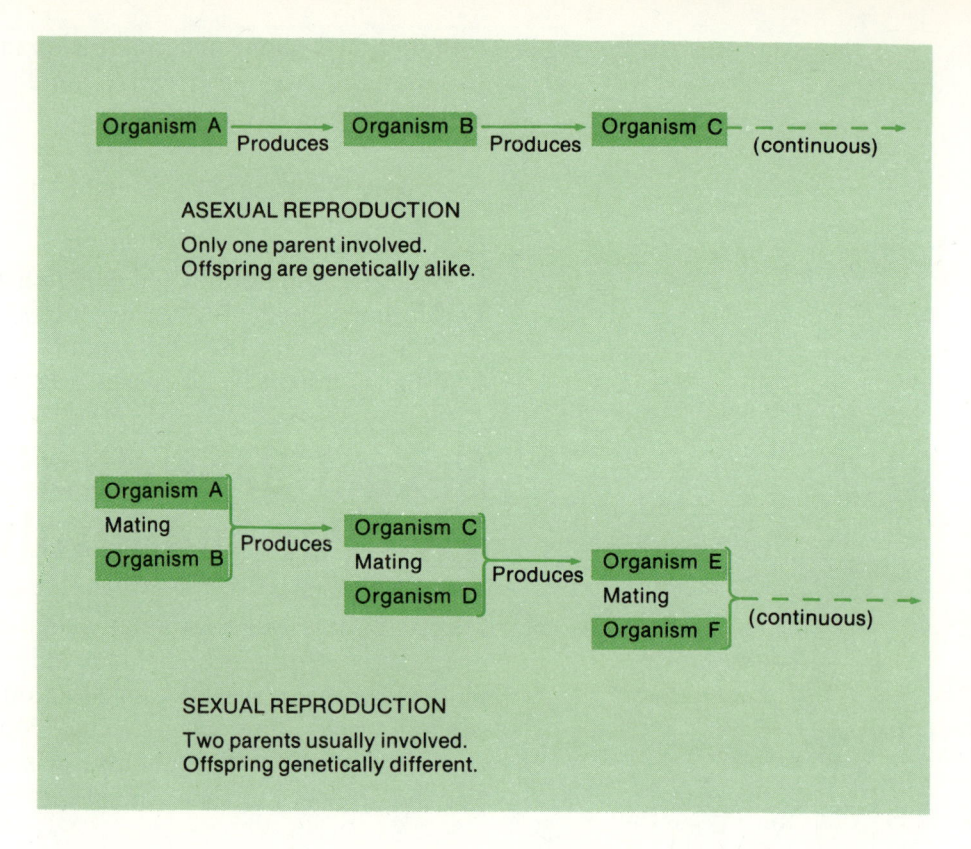

Fig. 21-4 A comparison of asexual and sexual reproduction. Unlike sexual reproduction, the new individuals resulting from sexual reproduction differ from their parents.

parent. Such animals are called **hermaphrodites** [hur-MAF-ruh-dyts]. Each individual is part male and part female. In Section 21.4 we will discuss reproduction in two hermaphrodites, the hydra and the earthworm.

External and Internal Fertilization

During reproduction, many aquatic animals release their gametes into the surrounding water. Fertilization occurs outside either parent's body. This is referred to as **external fertilization**. In many fish and in amphibians such as the frog, fertilization is usually external.

In most land animals, however, fertilization occurs within the female parent's body. This is called **internal fertilization**. Fertilization is internal in animals such as insects, reptiles, birds, and mammals. It also occurs in a few types of fish and amphibians.

Once fertilization occurs, the resulting zygote undergoes repeated cell division (mitosis). It becomes a small rapidly developing organism called an **embryo** [EM-bree-oh]. Then it finally becomes a fully functioning animal.

Section Review

1. **a)** What is a gamete?
   b) Name the two types of gametes produced by most animals.
2. How is a zygote formed?
3. What is a hermaphrodite?

4. a) Describe the difference between internal and external fertilization.
   b) Which type of fertilization is more likely to be characteristic of land animals? Explain why.

21.3 Sexual Reproduction: Not without Its Problems

Among animals, sexual reproduction is more common than asexual reproduction. However, for sexual reproduction to be successful, two conditions are necessary. First, fertilization must take place. That is, a sperm and an egg must unite to form a zygote. Second, the zygote must develop into an embryo and then into a mature organism. This may sound simple. But animals must overcome a number of problems before these two conditions are met. What are some of these problems? How do different animals solve them? In this section we will find out the answers to these questions.

Getting Reproductive Cells Together: Problems of Fertilization

There are three problems involving fertilization:
1. Sperm and egg need a liquid medium in which to move.
2. The release of sperm and egg must be timed to ensure that they meet one another.
3. In some way, sperm and egg must be protected before fertilization.

Now let's look at each of these problems briefly.

A Liquid for Movement

An egg has no means of locomotion. That is, it cannot move from place to place on its own. However, a sperm can. It has a tail-like **flagellum** which propels it. However, this type of locomotion only works if a sperm is in a liquid medium. Animals that live in or near water simply release eggs and sperm into the water. The sperm swim toward the eggs and fertilization takes place in the water. This, of course, is external fertilization.

In most land animals, fertilization is internal. It occurs within the female's reproduction tract. Sperm are released by the male into the female reproductive tract. They propel themselves along the tract in a liquid released by the testes of the male. Once egg and sperm meet, fertilization takes place.

The chances of sperm and egg meeting are much greater in the confined space of the female reproductive tract than if they were simply released into water. As a result, animals that undergo internal fertilization produce fewer eggs.

Section 21.3

A Matter of Timing

Sperm usually do not have a food supply. This means that they have a limited amount of energy. Therefore, once released, they can only live for a short period of time. And, they can only travel short distances. Eggs can be fertilized for only a short period of time after their release. To ensure that fertilization can occur, sperm and eggs must be released about the same time. They must also be released in the same general location.

Often, proper timing is achieved through various **courting** or **mating behaviour** patterns. In animals that undergo external fertilization, such behaviour patterns stimulate each sex to release gametes into the water in the same place and at the same time. In Section 21.4 we will look at the mating behaviour pattern of the frog. In animals in which fertilization is internal, mating behaviour patterns may, in part, serve as an advertisement or signal to attract male and female when they are ready to mate.

Chemical substances called **hormones** [HOR-moanz] appear to play an important role in controlling mating behaviour patterns. The role of hormones in human reproduction is described in Chapter 28.

Protection of Egg and Sperm before Fertilization

In animals that undergo internal fertilization, egg and sperm are contained within the female reproductive tract after mating. This provides them with protection from the external environment.

As we have seen, though, many animals simply release gametes into a water environment. In this case, the sperm and egg are subject to many dangers prior to fertilization. Predators may eat them. Also, the availability of dissolved oxygen and the temperature of the water may present some problems.

With the exception of a jelly-like coating around some types of eggs, the gametes generally do not have any physical protection.

However, animals that carry out external fertilization usually produce and release large numbers of gametes. Though this may seem rather wasteful, it does increase the chances that fertilization will occur.

Problems after Fertilization

After fertilization has taken place, the zygote begins to divide by mitosis. Soon it becomes an embryo that will eventually develop into a mature individual. A developing embryo that has resulted from external fertilization usually has little or no protection. Most of them will not survive. This is another reason why animals which reproduce by external fertilization release large numbers of gametes.

A few animals, such as the frog, release eggs surrounded by a jelly-like coating. The coating provides some protection for the eggs before and after fertilization. The eggs may also be coloured in such a way as to hide them from predators.

Fig. 21-5 Reproductive behaviour of the ten-spined stickleback. After a complex courting behaviour, the female sheds her eggs into a nest previously prepared by the male. Once the eggs are released, the male enters the nest and releases sperm over them. The male then guards the developing embryos and young offspring until they are capable of living on their own.

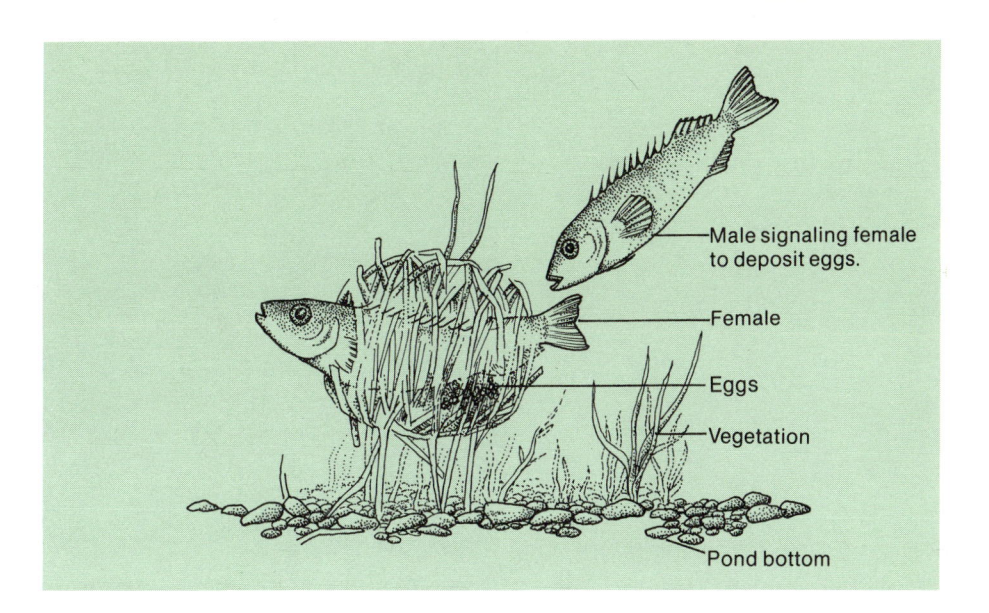

Some fish are **mouth breeders**. A parent places fertilized eggs in its mouth until they are fully developed. Other fish, such as the stickleback, are **nest breeders** (Fig. 21-5). After fertilization, the embryo develop within the tunnel-like nest. The male stands guard and also creates currents that supply the embryos with dissolved oxygen.

Animals that undergo internal fertilization provide much more protection to the developing embryo. They also usually supply the embryo with plenty of food. The developing embryo of a reptile or bird is enclosed in a **shell** (Fig. 21-6). Encased within the shell is a **yolk sac**. The yolk is a food supply for the developing embryo. The **embryo sac** surrounds the embryo. It contains a fluid which keeps the embryo moist and protects it from shock. A supply of blood vessels allows for gas exchange.

In most mammals, the embryo develops within a female's body. It is aided by a special structure called the **placenta** [pluh-SEN-tuh]. This structure consists of a network of tiny blood vessels. It is located

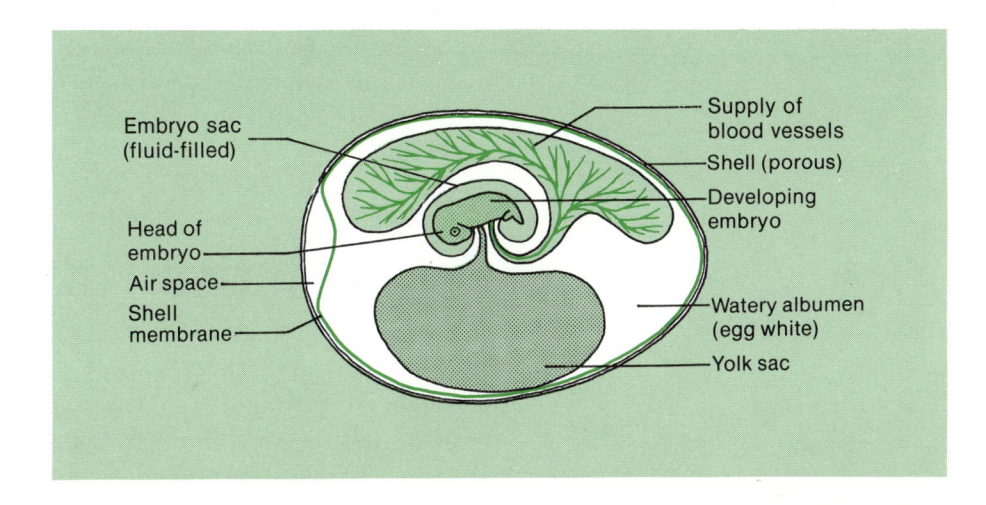

Fig. 21-6 A developing chicken embryo is surrounded by a protective shell. The embryo is also provided with a food supply (yolk).

Section 21.3

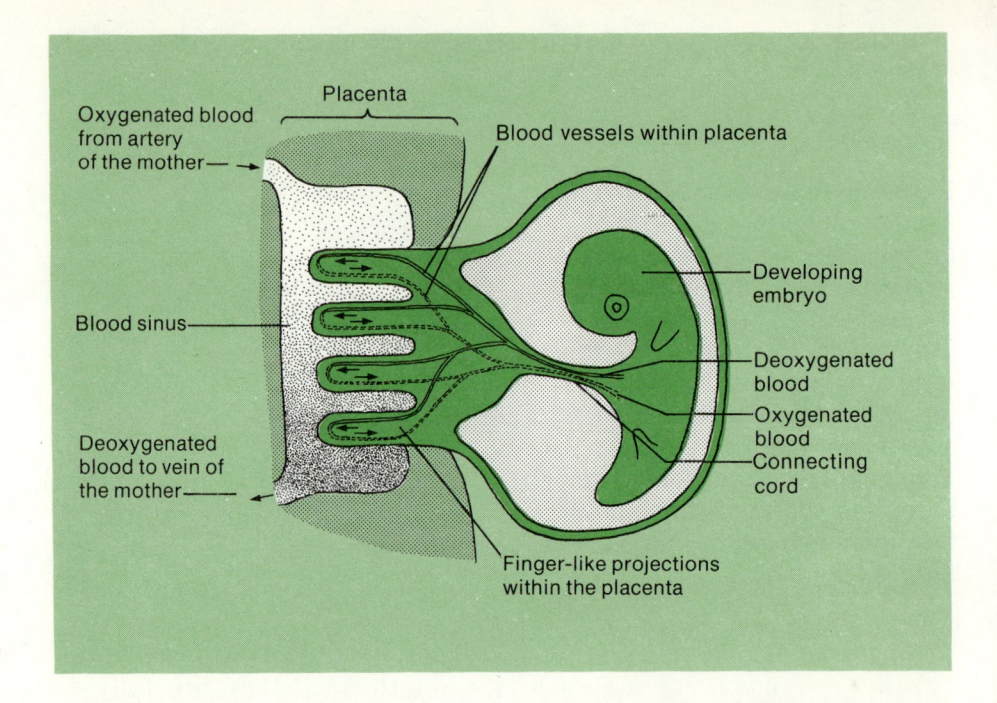

Fig. 21-7 The developing embryo of a mammal. The placenta allows for an exchange of nutrients and water between the mother and the embryo.

between the developing embryo and the mother's body. Nutrients and waste are exchanged across the network (Fig. 21-7). Once development is complete, the offspring is born as a miniature adult. It is usually fed milk from the female's mammary glands until it can find its own food. Humans are placental mammals. So are cats, dogs, horses, cows, and elephants.

Section Review

1. What are the two conditions necessary for sexual reproduction to be successful?
2. What three problems must animals deal with to make sure that fertilization does occur?
3. Why is the timing of the release of gametes so important for successful fertilization?
4. How do aquatic animals generally solve the problem of protection of gametes before fertilization?
5. Describe how the egg of a bird provides protection and nourishment for the developing embryo.
6. a) What is a placenta?
   b) What role does it play in the development of an embryo?

21.4 Reproductive Patterns in Animals

Before you read this section, review the problems that were described in Section 21.3. Keep these problems in mind and try to determine how each of the following animals solves these problems.

The Hydra

The hydra can reproduce both asexually and sexually. Most often it reproduces by **budding** (Fig. 21-8). A number of body cells divide to form a **bud** on the body wall. In 2 to 3 d, the bud resembles the parent hydra. It is complete with tentacles and mouth opening. As the bud develops, it obtains food from the parent. Once it is fully developed, the bud pinches off and begins its life as an independent animal.

Sexual reproduction usually takes place in the fall (Fig. 21-9). When hydra are ready to mate, reproductive organs develop. Each hydra has both a **testes** and an **ovary**. However, each must mate with another hydra. Once the sperm are released, they swim toward an ovary. One sperm enters the ovary. Then fertilization takes place, internally, within the ovary. The zygote divides many times to form an embryo. A thick protective coat forms around the embryo. The encased embryo eventually breaks away from the ovary and drops to the bottom of the pond. It may remain dormant through the winter. Then, when conditions are favourable, the embryo completes its development into a new hydra.

The Earthworm

You may recall that earthworms are **hermaphrodites**. However, an individual earthworm is not capable of fertilizing its eggs with its own sperm. Two worms must get together.

When two earthworms mate, sperm are released by each. They travel along grooves to the **sperm storage sacs** of the other worm

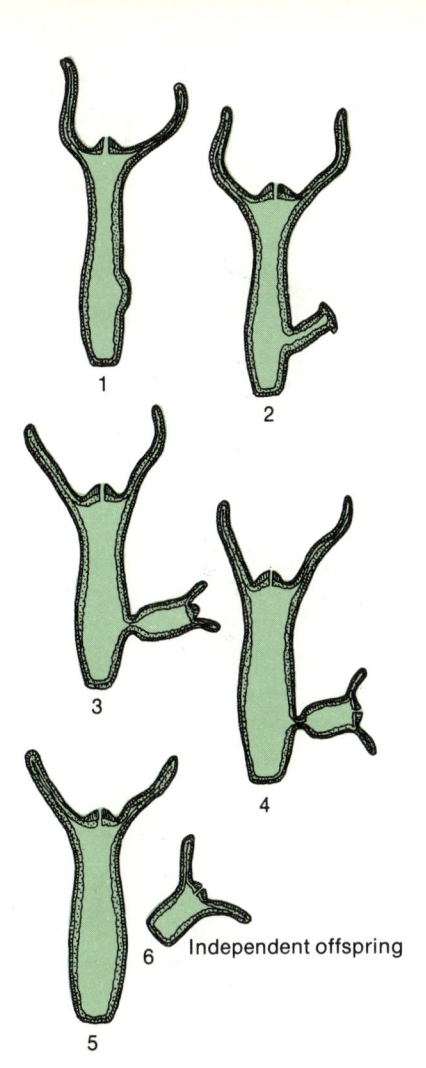

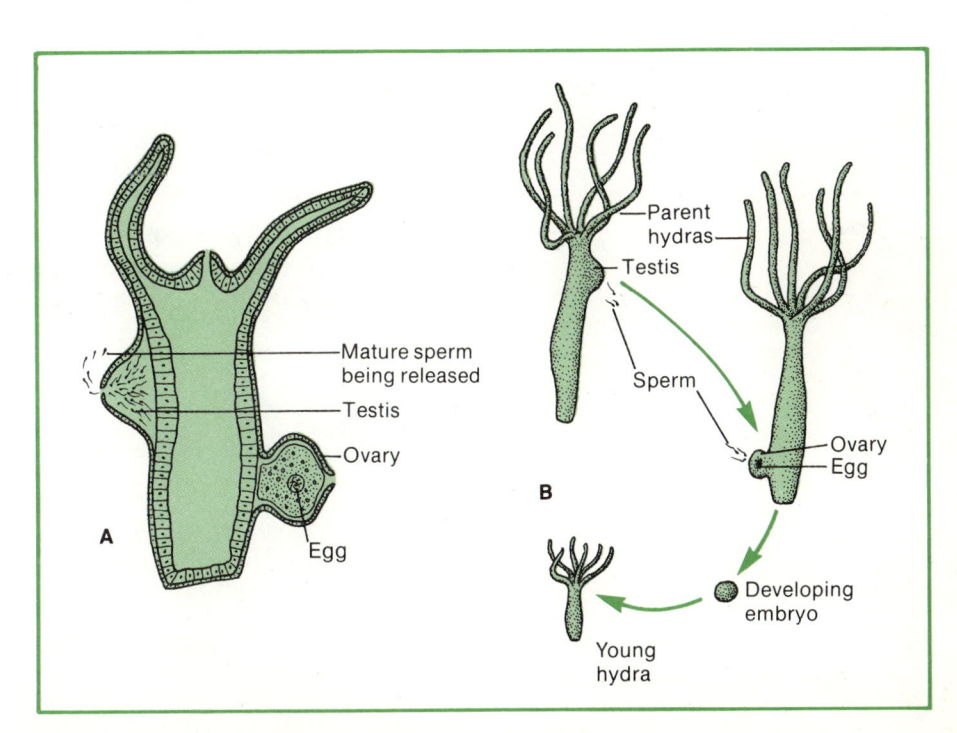

Fig. 21-8 Longitudinal section of hydra showing stages of developing bud

Fig. 21-9 The hydra can also reproduce sexually: a longitudinal section showing sex organs (A); the pattern of reproduction in the hydra (B)

Section 21.4 297

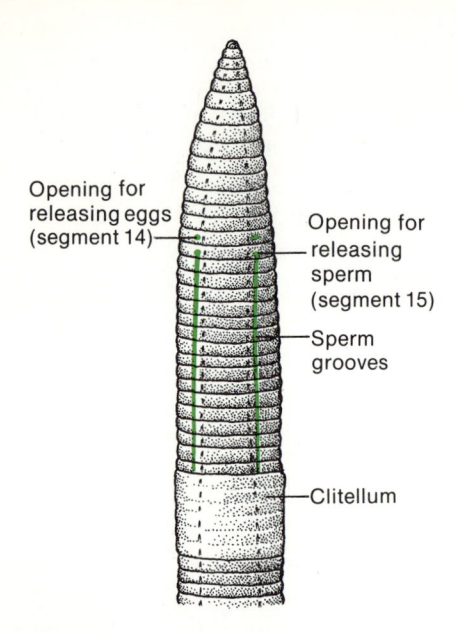

Fig. 21-10 External features of the earthworm's reproductive system. This is a vertical view of the front end of an earthworm.

(Fig. 21-10 and 21-11). After exchanging sperm, the two earthworms separate. The band-like **clitellum** secretes a mucus sleeve. A food reserve is also deposited in the sleeve to later feed the developing embryo. The sleeve slips forward toward the front of the earthworm. As it passes segment 14, it receives eggs released through special openings on the earthworm's lower surface. Sperm are released into the sleeve as it continues forward past segments 9 and 10.

Fertilization takes place within the mucus sleeve. Once the sleeve slips off the earthworm, its ends seal, forming a capsule called a **cocoon**. Only one zygote will develop within a cocoon. In a few weeks the embryo is fully developed. A young earthworm then escapes from the cocoon.

The Crayfish

During mating, the female is held on her back. Sperm are released by the male into storage sacs on the female's abdomen. They remain there until eggs have fully developed. Then the female releases her eggs and the stored sperm. Fertilization takes place. The resulting embryo become attached to the female's **swimmerets**. The attached embryos resemble a bunch of berries. The female is said to be "in berry" (Fig. 21-12,A). The swimmerets wave back and forth constantly. This supplies the developing embryos with fresh, oxygen-rich water.

Full development of the embryos to young crayfish takes about five to eight weeks. When they hatch, the young crayfish remain attached to the mother's swimmerets until they are able to survive on their own (Fig. 21-12,B).

The Frog

Although adapted for life partly on land, the adult frog must return to the water to reproduce. Separate sexes, external fertilization, and development outside the mother are all characteristics of its reproductive pattern. **Metamorphosis** [MET-uh-MOR-fuh-sis] is also a characteristic of the frog. That is, during development from fertilized egg to adult, a series of changes in form take place.

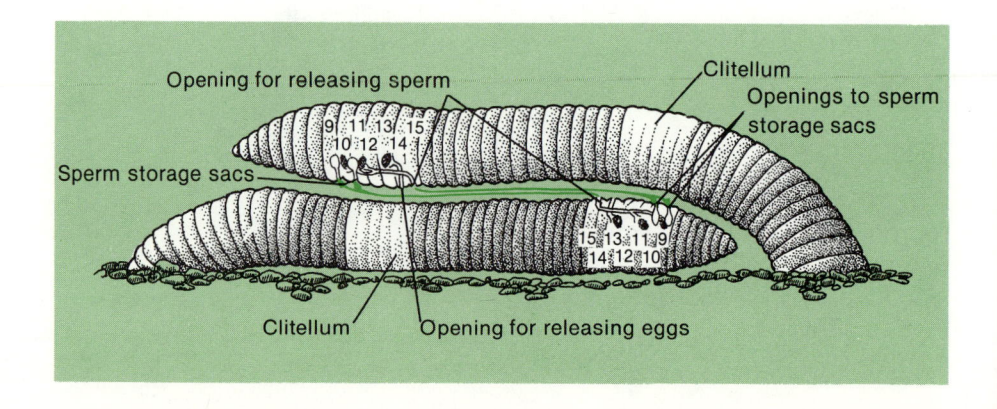

Fig. 21-11 Diagram shows the pathway of exchange of sperm between two mating earthworms.

Frogs mate in the spring and usually at night. The female, body swollen with eggs, seeks out a male. The male perches on the back of the female and clasps his front legs around her body (Fig. 21-13). As eggs are released into the water, the male releases sperm over them and fertilization takes place.

A jelly-like coating around the eggs swells on contact with the water. This forms a large egg mass. The mass may float freely or become attached to vegetation. Each egg contains a **yolk** (white portion) and a dark coloured **zygote**. This colour pattern serves to "hide" the developing embryos from predators.

A **tadpole** develops from the embryo in two to three weeks. The tadpole has a tail for swimming and external gills for breathing. Gradually, the tadpole undergoes metamorphosis. In this process the external gills are replaced by internal gills, then by lungs. The mouth broadens, teeth develop, and legs appear. The tail is gradually absorbed. It serves as a nutrient source for the developing embryo. In about three months, development is completed. The adult frog is now capable of living on land.

Section Review

1. Describe the pattern of asexual reproduction in the hydra.
2. a) Describe the pattern of sexual reproduction in the earthworm.
   b) Refer to the problems outlined in Section 21.3. How does this pattern overcome these problems?

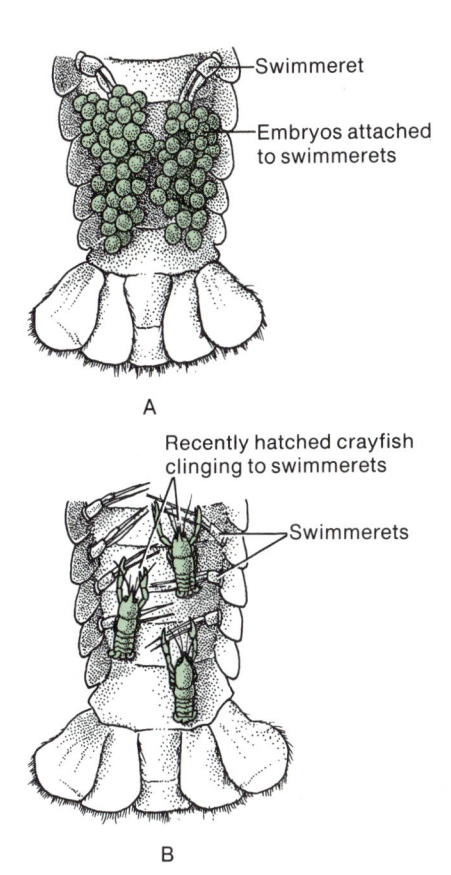

Fig. 21-12 Undersurface of a female crayfish showing development stages. Clusters of eggs; the female is "in berry" (A); "hatchlings" cling to mother for protection (B)

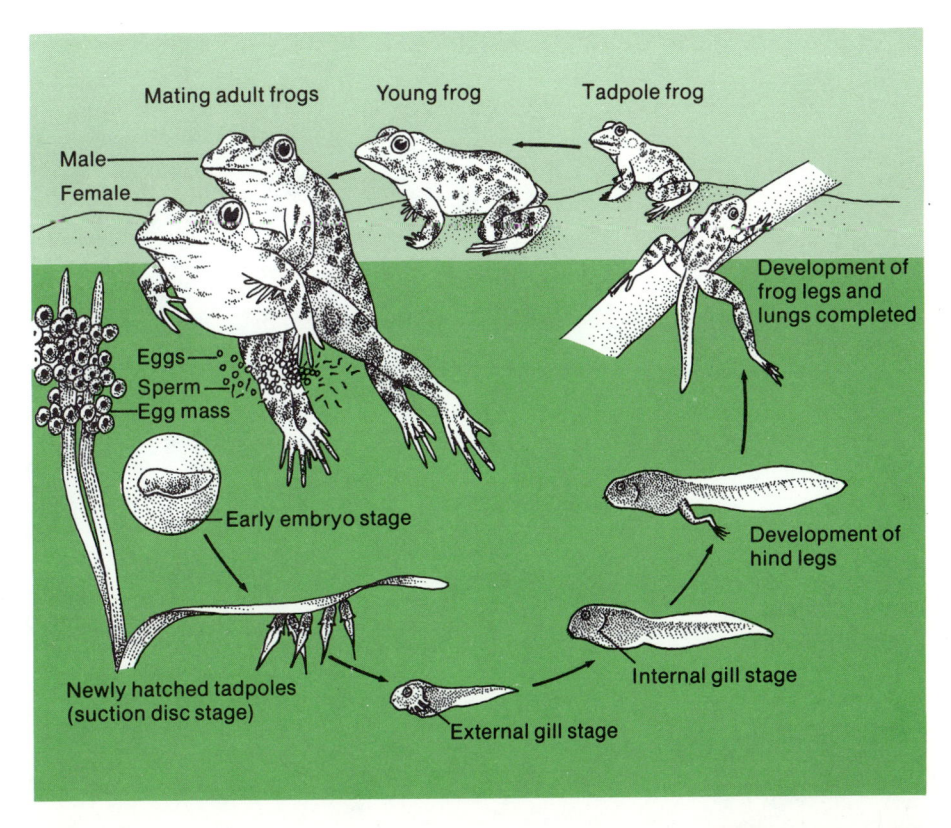

Fig. 21-13 The life cycle of a frog

3. What features of the reproductive pattern of the crayfish ensure a high survival rate of the zygote?
4. Describe the changes that occur in the metamorphosis from fertilized egg to adult in the frog.
5. What features of the frog's reproductive pattern enable it to overcome the problems referred to in Section 21.3?

21.5　ACTIVITY　The Reproductive Systems of Some Animals

In this chapter, we have discussed the reproductive patterns of a number of the animals. In this activity, you will observe and compare the reproductive systems of the earthworm, crayfish, and frog. Keep in mind the problems of reproduction mentioned in this chapter. Then try to determine how each system is designed to solve the problems.

Problem

How are the reproductive systems of earthworms, crayfish, and frogs alike? How are they different?

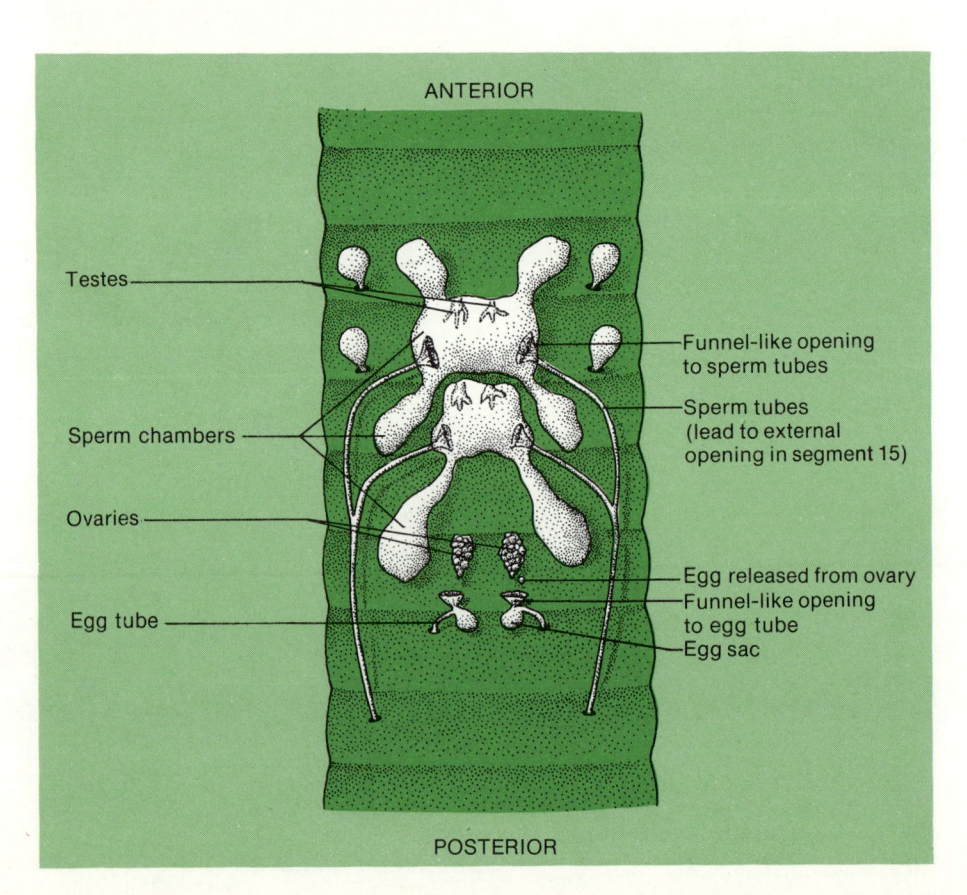

Fig. 21-14 Internal features of the reproductive system of an earthworm

Materials

For each group of six to eight students, a hand lens and the following freshly dissected animals are to be provided.

an earthworm
a female crayfish
a male crayfish
a female frog
a male frog

Procedure A The Earthworm

The earthworm has two reproductive systems. It is a hermaphrodite. Refer to Figures 21-10 and 21-14 as you proceed through this portion of the activity.

a. Locate the **sperm chambers** surrounding the esophagus. The sperm-producing **testes** are found inside these chambers.

b. Locate the opening of the **sperm tube** on the 15th segment. Using a hand lens, find the sperm tube leading from the sperm chambers to the opening.

c. Locate the egg-producing **ovaries** in segment 13 beneath the sperm chambers. You may also be able to notice the **egg tubes** with their funnel-like ends. The egg tubes lead to two small pores in segment 14.

d. Find the **sperm storage sacs** on segments 9 and 10. Sperm from another worm are stored in them until the eggs are ready for fertilization.

e. Finally, locate the **clitellum**. It is about 31 segments from the front end. It produces the cocoon in which egg and sperm are collected and in which the developing young are protected.

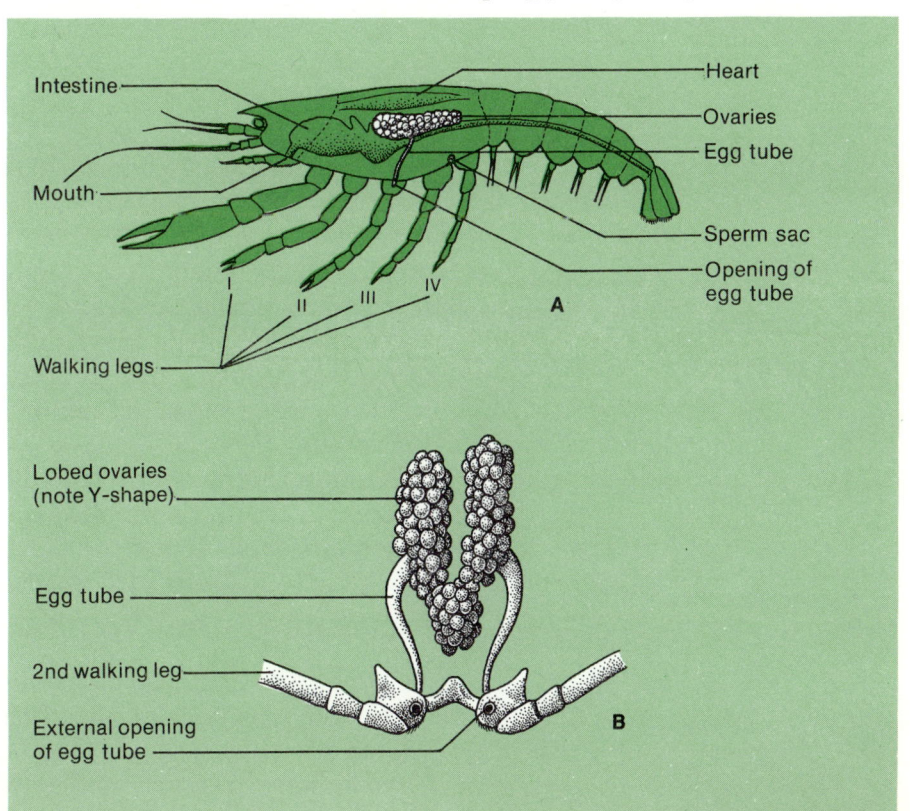

Fig. 21-15 Reproductive organs of a female crayfish: view of left side showing relative position (A); top view showing details of structure (B)

Section 21.5   301

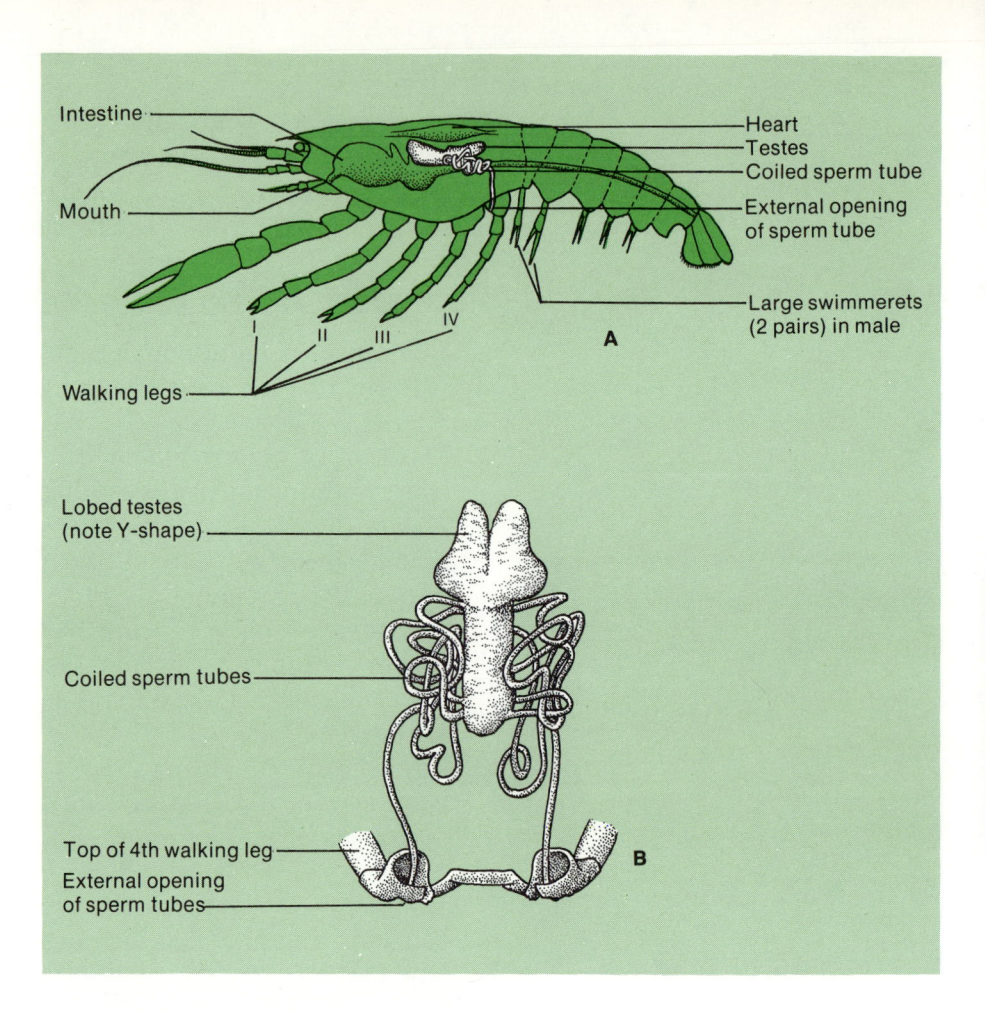

Fig. 21-16 Reproductive organs of a male crayfish: view of left side showing relative position (A); top view showing details of structure (B)

Procedure B The Crayfish

Refer to Figures 21-15 and 21-16 as you proceed through this part of the activity. The male crayfish is identified by four large, stiff, sharp pointed swimmerets. These are the first two of six pairs located beneath the abdomen. In the female, the first two pairs of swimmerets are much smaller in size.

a. Examine the female crayfish. Locate the **ovaries**. They consist of three large lobes arranged much like the shape of the letter Y. Trace the egg tubes to their openings at the top of the second pair of walking legs.

b. Examine the male crayfish. The **testes** are also shaped like the letter Y. Find them. Trace the coiled **sperm tubes** to their openings at the top of the fourth pair of walking legs.

Procedure C The Frog

Refer to Figure 21-17 as you proceed through this part of the activity.

a. Examine the female frog. Locate the **ovaries**. They are large, lobed structures. One is found on the lower surface of each of the two kidneys. You may notice the black and white **eggs**.

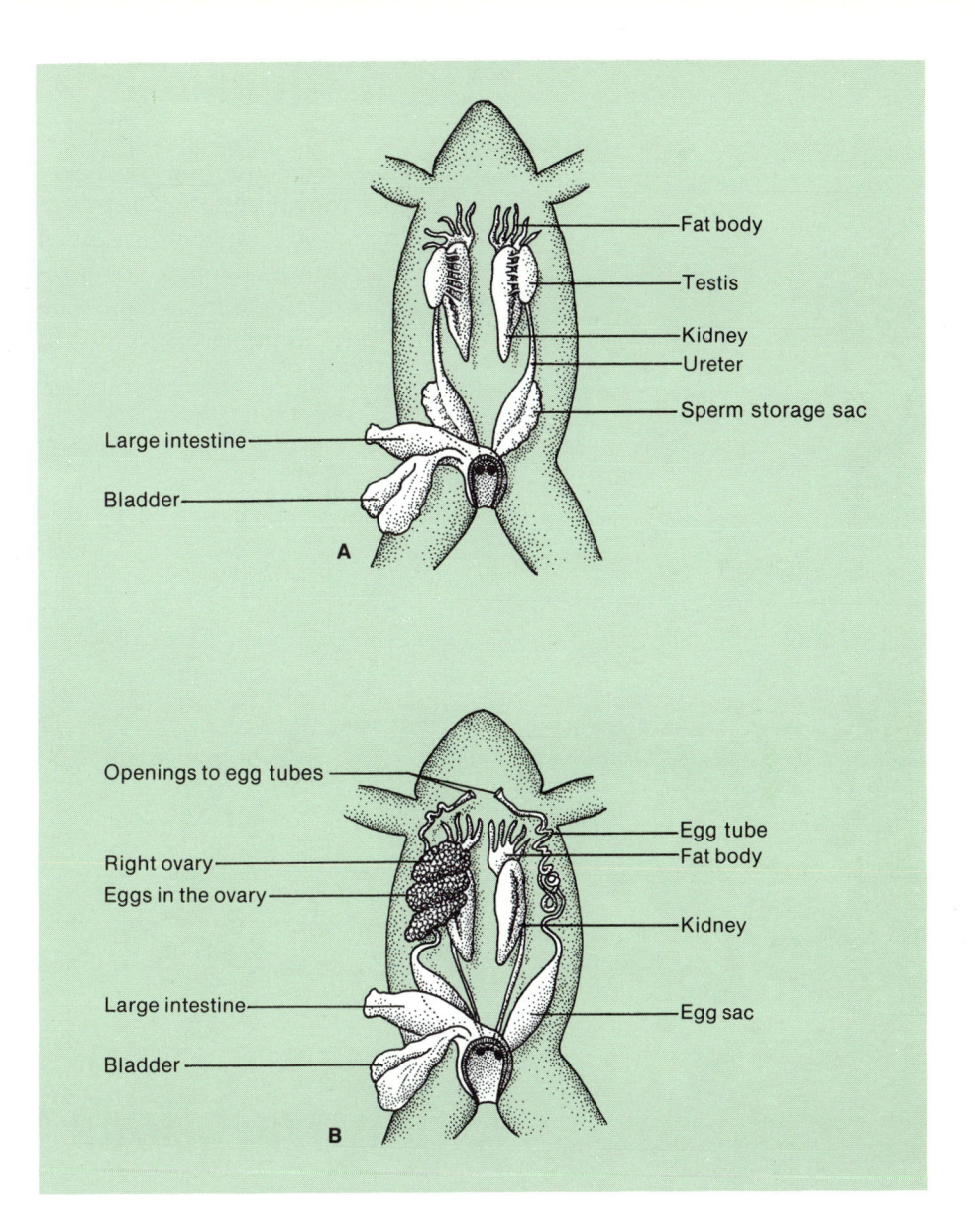

Fig. 21-17 Reproductive system of the frog: a male frog (A); a female frog (B) (Note: left ovary not shown)

b. Next, locate the **egg tubes** and **egg sacs**. Eggs, released by the ovary, are swept into the egg tubes and coated with a jelly-like substance. Finally, they are stored in the egg sacs until mating occurs.

c. Examine the male frog. Find the **testes** on the surface of each kidney. They are oval shaped and creamy-white in colour. You should be able to spot the **sperm tubes**, **ureters**, and **sperm chambers**. Sperm produced in the testes are released into the tubes. Then they travel down the ureters and are stored in the sperm chambers until mating occurs.

Discussion

There is no writeup for this activity. Its function is to help you understand Section 21.4.

Main Ideas

1. Asexual reproduction requires only one parent.
2. Sexual reproduction requires two parents.
3. Male gametes (sperm) are produced in testes.
4. Female gametes (eggs) are produced in ovaries.
5. Many aquatic animals reproduce by external fertilization.
6. Most land animals reproduce by internal fertilization.
7. Fertilization requires a liquid medium, proper timing, and protection of the gametes.

Glossary

hermaphrodite	hur-MAF-ruh-dyt	an animal that has both testes and ovaries on the same individual
ovaries	OHV-uh-rees	the female reproductive organs; they produce eggs
placenta	pluh-SEN-tuh	a group of membranes through which food, oxygen, and wastes are exchanged between the blood supplies of the mother and the developing embryo
testes	TES-teez	the male reproductive organs; they produce sperm
zygote	ZY-goat	a fertilized egg cell

Study Questions

A. True or False

Decide whether each of the following statements is true or false. If the sentence is false, rewrite it to make it true. (Do not write in this book.)

1. Reproduction is concerned with the survival of the individual.
2. Budding is a form of sexual reproduction.
3. Land animals, as a rule, undergo internal fertilization.
4. The hydra is a hermaphrodite.

B. Completion

Complete each of the following sentences with a word or phrase that will make the sentence correct. (Do not write in this book.)

1. The methods by which animals reproduce can be grouped into two categories called ▨▨▨ and ▨▨▨ reproduction.
2. Bacteria reproduce asexually by a process called ▨▨▨ .

Fig. 21-18 The wallaby (a type of kangaroo) is a "pouched mammal". Mammary glands within the female's "pouch" supply food to the developing young.

3. A fragment from a sponge can develop into an entire new sponge by a process called _____.
4. _____ cells are male gametes and _____ are female gametes.

C. Multiple Choice

Each of the following statements or questions is followed by four responses. Choose the correct response in each case. (Do not write in this book.)

1. Internal fertilization occurs
    a) only in mammals
    b) only in land animals
    c) in many land and aquatic animals
    d) in all animals at some time in their life
2. Reproduction differs from all other functions in an organism because it is not
    a) an energy requiring process
    b) necessary for the individual to live
    c) a normal body function
    d) important to the species to which the animal belongs
3. In animals, sperm are produced in organs called
    a) testes   b) ovaries   c) kidneys   d) buds
4. Which one of the following includes all the others?
    a) regeneration   b) asexual reproduction   c) budding   d) fission

D. Using Your Knowledge

1. Would asexual reproduction be an advantage or a disadvantage to an animal in a constantly changing environment? Explain your answer.
2. What do you think would happen to a developing embryo within a shell if the shell was coated with wax? Explain.
3. The hydra is capable of reproducing both asexually and sexually. What are the possible advantages and disadvantages for the hydra to be able to reproduce in two ways?
4. Describe how the physical features of the various stages during metamorphosis adapt the frog to the environment in which it lives.

E. Investigations

1. The platypus is a rather unusual animal. Find out about its pattern of reproduction. How does the platypus solve the problems related to reproduction that were discussed in this chapter?
2. The kangaroo is a mammal. It is also referred to as a "pouched" animal (Fig. 21-18). Using the library or resource centre, investigate the reproductive pattern of this animal.
3. Animals vary in their ability to regenerate parts of themselves. Research the topic of regeneration. What damaged parts can humans regenerate?
4. Design and perform an experiment that might produce a planarian with two heads attached to the front end.

Unit 6

Human Biology

This unit is about your body — how it works and how you can keep it working properly. We think this is an important unit for you. After all, you have only one body. And it has to last you a long time!

CHAPTER 22
Food and Nutrition

CHAPTER 23
The Digestive System

CHAPTER 24
The Breathing System

CHAPTER 25
The Circulatory System

CHAPTER 26
The Excretory and Reproductive Systems

CHAPTER 27
Structural Systems

CHAPTER 28
Control Systems

Fig. 22-0 Your body is made of about sixty million million cells (60 000 000 000 000). They work together to make you what you are. How do they do this? Why do they sometimes fail to do their jobs properly?

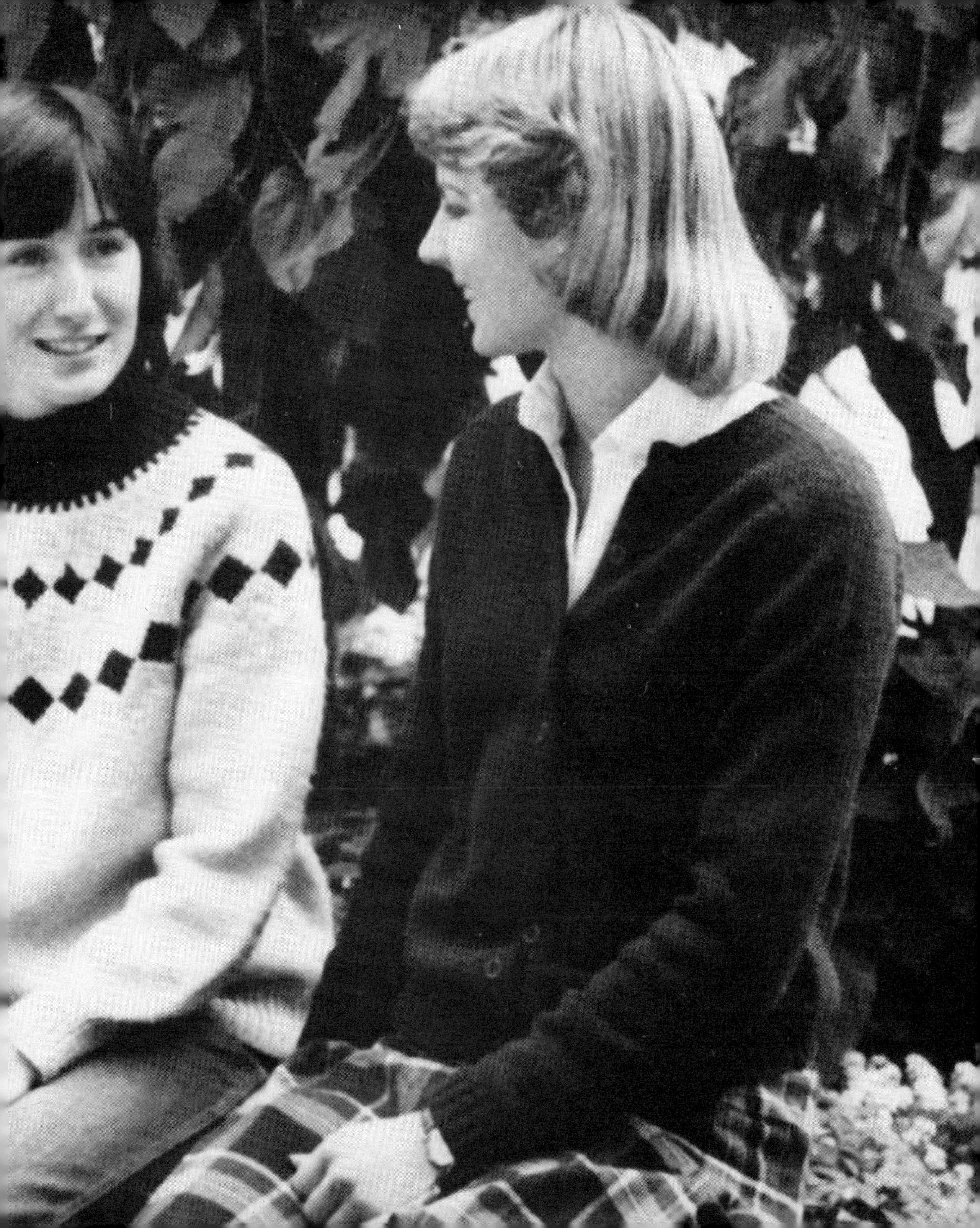

22 Food and Nutrition

22.1   Food, Nutrients, and Energy
22.2   Carbohydrates
22.3   Activity: Testing for Carbohydrates
22.4   Fats and Oils
22.5   Activity: Testing for Fats and Oils
22.6   Proteins
22.7   Activity: Testing for Proteins
22.8   Vitamins
22.9   Activity: Testing for Vitamin C
22.10   Water
22.11   Minerals
22.12   Activity: Testing for Minerals
22.13   Food for a Healthy Body

You know that you need food to live. But do you know why? Just what does the food do for you? Does your body use all the food you eat? Do you really need vitamins? What foods should an athlete eat? What is a proper diet?

22.1 Food, Nutrients, and Energy

Food provides the energy you need to live. Food also helps you grow. You are constantly repairing and replacing your body cells. For example, your body makes nearly two million new blood cells every second! But to do such things, your body needs the right nutrients (Fig. 22-1).

Nutrients

You eat many kinds of food. Parts of this food cannot be used. You release these parts as waste. The useful parts of food are called **nutrients** [NEW-tree-ents]. Some nutrients supply energy. Others are needed for growth or repair of damaged cells (Fig. 22-2).

   Some nutrients are **organic**. They contain carbon atoms. **Carbohydrates**, **fats and oils**, **proteins**, and **vitamins** are examples of organic nutrients. **Water** and **minerals** are **inorganic** nutrients. These come from the non-living environment. Figure 22-3 shows the six basic groups of nutrients.

Cells and Energy

Your living cells use energy constantly. The more active you are, the more energy they use. Also, humans are warm-blooded. This means that energy is needed just to maintain your high body temperature. How do your cells obtain this energy?

Respiration occurs in all living cells. During respiration, certain nutrients are slowly broken down (oxidized). Heat and other forms of energy are released. (You may wish to review respiration in Chapter 7.)

Carbohydrates and fats are oxidized most easily. But proteins can also provide energy. Without food, you soon use up the supply of carbohydrates in your body. Then your stored fat is used. This is why you gradually lose mass if you eat less. Suppose you continued to fast, or starve. Then the proteins in your muscles and other body tissues would be broken down for energy. Your starving body would actually feed upon itself! This is why victims of famine look like walking skeletons.

Measuring Food Energy

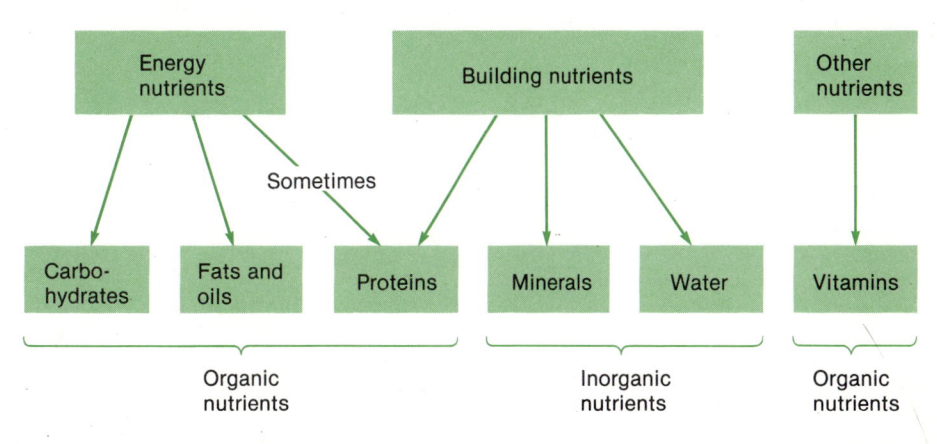

The energy content of different foods can be found. A certain mass of each food is burned. Then the resulting heat energy is measured. Food energy is measured in **kilojoules (kJ)**. Table 22-1 shows a few examples. It gives the energy in kilojoules (kJ) in 1 kg of some foods. What foods would you take on a hike on a cold day?

Section Review

1. What are nutrients?
2. What do nutrients do for your body?
3. How do your cells get energy?
4. Why are nuts good foods to take on a long hike?

Fig. 22-1 What nutrients do these foods give to your body?

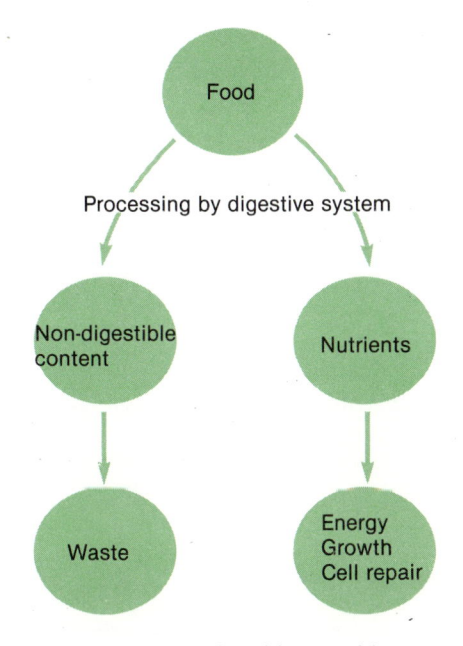

Fig. 22-2 How food is used by your body

Fig. 22-3 The basic groups of nutrients

Section 22.1

Table 22-1 Energy Content of Some Foods

Food	Energy content per kilogram (kJ/kg)	Food	Energy content per kilogram (kJ/kg)
Bacon, raw	27 900	Eggs, fried	9 100
Beef, T-bone steak	16 700	Eggs, hard boiled	6 800
Bread, white	11 300	Peanuts	24 600
Bread, whole wheat	10 200	Potatoes, boiled	3 200
Carrot, raw	1 800	Potato chips	23 900
Cereal, shredded wheat	14 900	Walnuts	27 300

22.2 Carbohydrates

Carbohydrates are your main source of energy. These compounds contain carbon, hydrogen, and oxygen. Carbohydrates include **sugars**, **starches**, and **cellulose**. Table 22-2 lists the four major food groups that provide carbohydrates.

Table 22-2 Food Groups Providing Carbohydrates

Food group	Description
Cereal group	Foods made from corn, wheat, oats, and rice provide starch and cellulose.
Vegetable group	Potatoes, squash, carrots, beets, beans, and peas provide starch and cellulose.
Fruit group	Peaches, plums, and grapes provide sugar. Bananas provide sugar and starch.
Concentrated sweets group	Jams, honey, sugar, candies, and syrups provide sugar.

Sugars

There are many different sugars. **Glucose** [GLOO-kohs] is one of the **simple sugars**. It is made during photosynthesis.

You do not have to digest simple sugars. They can diffuse from the digestive tract to the blood. Glucose is often injected directly into the blood of hospital patients. This procedure is called **intravenous feeding** [in-tra-VEE-nus].

Sucrose (ordinary table sugar) is one of the **compound sugars**. These sugars consist of two simple sugar units linked together. Such

310 Chapter 22

sugars must be digested. They are broken down into simple sugar molecules. Then these can diffuse into the blood.

Starches

Plants often make more sugar than they need. They store this extra sugar in the form of **starches**. Starches are just many simple sugar molecules linked together. During digestion starch molecules are broken down into simple sugar molecules. These can then diffuse into the blood and be used for energy. Many plant foods are rich in starch. Potatoes and bread are good examples.

Cellulose

Plants also make **cellulose** [SEL-yoo-lohs]. A cellulose molecule is a very long chain of simple sugar molecules. Plants use cellulose to make cell walls. The rigid cellulose strengthens and supports plant cells. Wood is mainly cellulose.

Plant-eating mammals such as horses and cows can digest cellulose molecules. But many other organisms cannot. Humans cannot digest cellulose. This is why you cannot eat wood or graze on your front lawn. You would starve on a diet of grass or hay. The cellulose in plant foods such as celery passes through your body undigested. But this cellulose is very important. It acts as **roughage** in your digestive system. Roughage helps other foods pass through your body.

North Americans eat many refined foods such as white bread. These foods lack cellulose (fibre). Cancers of the colon are more common in North America. Many doctors blame this on the lack of roughage in our diets.

Section Review

1. Name three kinds of carbohydrates.
2. Why are carbohydrates important?
3. Name a simple sugar. What is its source?
4. How do you obtain energy from starches?
5. Why is cellulose important in your diet?

22.3　ACTIVITY　Testing for Carbohydrates

We can use chemical tests to identify some nutrients. In this activity you will test for sugars and starches.

PART A TEST FOR SIMPLE SUGARS

Problem

How can we test for simple sugars?

Materials

10% glucose solution
Benedict's solution
test tubes (3)
10% sucrose solution
25 mL graduated cylinder
Bunsen burner
adjustable clamp

CAUTION: Wear safety goggles during this activity.

Procedure

a. Pour 10 mL of the 10% glucose solution into a test tube.
b. Add 5 mL of Benedict's solution.
c. Heat the test tube carefully until the mixture is boiling (Fig. 22-4). Boil it for 2-3 min.
CAUTION: Do not point the test tube at anyone. Heat the mixture as directed by your teacher.
d. Record any changes that you see.
e. Repeat steps (a) to (d) using 10% sucrose solution.

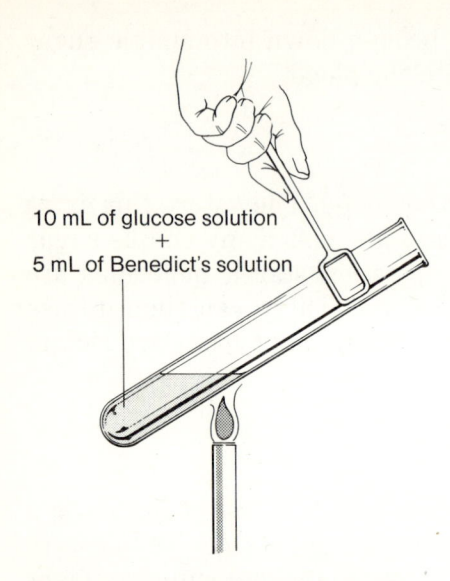

Fig. 22-4 Heat carefully. Move the test tube around. Heat at the top of the liquid. Don't point the test tube at anyone.

Discussion

1. Benedict's solution reacts with a simple sugar to produce a reddish-orange colour. Is glucose a simple sugar?
2. Why did sucrose not give a reddish-orange colour with Benedict's solution?

PART B TEST FOR STARCH

Problem

How can we test for starch?

Materials

corn starch
iodine solution
cracker
piece of bread
slice of potato
piece of apple
petri dish

CAUTION: Wear safety goggles during this activity.

Procedure

a. Place a pinch of corn starch in the petri dish.
b. Add 4-5 drops of iodine solution (Fig. 22-5).
c. Record any changes that occur. *This is the test for starch.*
d. Place a piece of cracker in the petri dish. Add 4-5 drops of iodine solution. Does the cracker contain starch?
e. Repeat step (d) using bread, potato, and apple.

Fig. 22-5 Add 4-5 drops of iodine solution. Do not get it on your hands.

312   Chapter 22

Discussion

1. *Iodine reacts with starch to form a blue colour.* (Sometimes the colour may be deep purple or almost black.) Which foods contained starch?

22.4 Fats and Oils

Many Canadians fight a constant "battle of the bulge." Over-eating is a major pastime in our society. As a result, so is dieting. Fat is regarded as a public enemy. Too much fat can cause many problems. Yet a certain amount is essential for a healthy body.

Fats are solids at room temperature. **Oils** are liquids at room temperature. Otherwise they are much the same. We will refer to both as "fats".

The Role of Fats

Many people eat more carbohydrates than they need. The body changes extra carbohydrates into fats. Fats are a highly concentrated form of food energy. Some Inuit and other northern people use fats as their chief source of energy.

Like carbohydrates, fats are made of carbon, hydrogen, and oxygen. But a gram of fat can give twice as much energy as a gram of carbohydrate. However, fats are harder for the body to use. And, once fats are in your body, they are hard to get rid of. That's why overweight people find it hard to reduce.

Fats do more than supply energy. They also build tissues. Many parts of your cells contain fat molecules. Cells could not function without fats. Also, certain vitamins do not dissolve in water. But they can dissolve in fats. Without fats, your cells could not use these vitamins.

Unused fats are stored in your tissues. Fats give your body contours. They insulate your body from the cold. Fats protect certain vital organs. For example, fats cushion your hands, feet, arms, legs — and even your eyeballs! So you see, fats have many benefits. Fatty foods tend to satisfy your hunger better. Just remember — a little goes a long way. And too much can be harmful to your health. Table 22-3 lists some foods which supply fats.

Table 22-3 Some Foods Containing Fats and Oils

Butter	Cooking oil	Sausage
Margarine	Fried foods	Side bacon
Lard	Peanuts	Cream

Section Review

1. Which has the most energy per gram, fats or carbohydrates?
2. List five functions of fats.

22.5 ACTIVITY Testing for Fats and Oils

In this activity you will test cooking oil to see how the test for fats and oils works. Then you will test some foods to see if they contain fats and oils.

Problem

How can we test for fats and oils?

Materials

cooking oil	whole milk	potato
margarine	dropper	peanut
filter paper		

Procedure

a. Rub a small quantity of margarine on a piece of filter paper.
b. Wait 2 min. Then hold the paper up to a light (Fig. 22-6). Record your observations.
c. Repeat steps (a) and (b) using 3-4 drops of cooking oil.
d. Continue to use steps (a) and (b) for all the other foods.

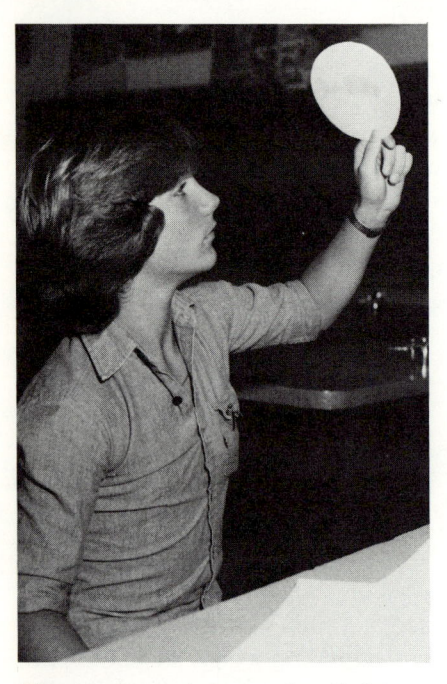

Fig. 22-6 Look towards a light. What has the food done to the filter paper?

Discussion

1. If the food contains a fat or an oil, a semi-transparent spot forms on the filter paper. Which foods contain a fat or an oil?

22.6 Proteins

We need fats and carbohydrates for energy. Proteins also supply energy. In addition, they build new cell protoplasm. In fact, they make up about 20% of our body tissues, including muscles and skin.

The Nature of Proteins

Proteins are large complex molecules. They are made of smaller molecules called **amino acids**. Living things need about 20 different kinds of amino acids. Amino acids are like the letters of an alphabet. They link together in different numbers and arrangements to form giant "words" — the protein molecules. Even small protein molecules contain about 100 amino acids. Larger proteins have thousands. The hemoglobin in your blood is a protein. A hemoglobin molecule consists of 547 amino acid molecules.

The Role of Proteins

You have hundreds of different proteins in your body. Each one has a special job. Proteins build parts of your cells. They form your different tissues. Your muscles, bones, skin, hair, and nails are all made of proteins.

Some proteins are **enzymes** [EN-zimes]. They control different chemical activities in your body. For example, an enzyme in your saliva helps digest starch.

Proteins can also provide energy. If you fast, you use up your stores of fat and carbohydrates. Then the proteins in your muscles supply energy.

Sources of Protein

The proteins in your body are all formed from about 20 different amino acids. Your body can make 12 of these amino acids. But at least 8 other amino acids must be provided in your diet. These are called **essential** amino acids. During digestion, proteins in your food are broken down. This releases their amino acids. These amino acids are absorbed and then used to build the proteins needed by your body.

The best sources of proteins are animal foods. So vegetarians must choose their diets carefully. They must eat a wide variety of plant foods to obtain all of the essential amino acids. Table 22-4 lists some of the foods which are rich in protein.

Table 22-4 Good Sources of Proteins

Milk	Fish	Nuts
Eggs	Beef	Beans
Cheese	Pork	Soybeans
Poultry	Cereals	Peas

Protein Deficiency

Millions of people all over the world suffer from a lack of protein. Many poor people exist on starchy foods which only supply carbohydrates. They have no milk, meat, or vegetables from which to obtain essential amino acids.

Kwashiorkor [Kwash-ee-or-kor] is a disease caused by protein deficiency. Its effects can be seen in the child shown in Figure 22-7. The texture and colour of the hair changes. it becomes a reddish-orange colour. In fact, "Kwashiorkor" means "little red boy". The liver swells and the stomach becomes bloated with water. Arms and

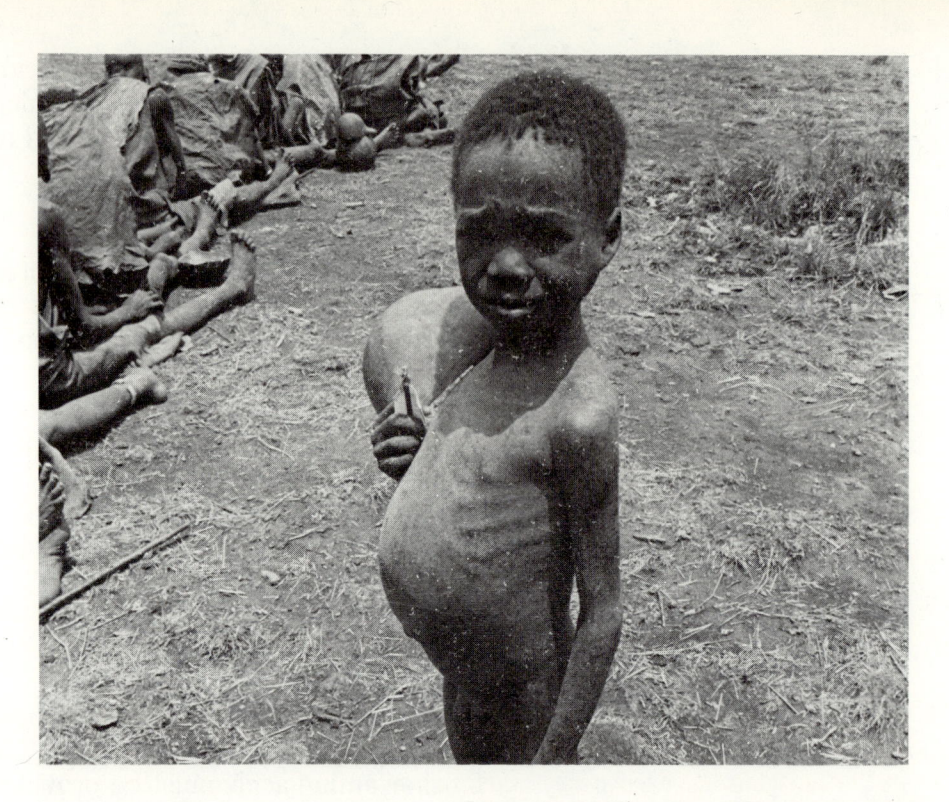

Fig. 22-7 This child does not get enough protein. Can we do anything about this?

legs have little muscle. They look like matchsticks. The body stops growing. And even worse, brain development is retarded.

Children deprived of protein from birth suffer permanent damage. They have no energy — no hope. Without help, they simply die. Yet just one cup of whole milk or a serving of beans daily could prevent this suffering.

Section Review

1. How does your body make the proteins it needs?
2. State 3 functions of proteins in your body.
3. State 3 effects of protein deficiency.

22.7   ACTIVITY   Testing for Proteins

Gelatin is a protein. In this activity you will use it to see how the test for proteins works. Then you will test milk and egg white. Do they contain proteins?

Problem

How do we test for proteins?

Materials

Biuret reagent	egg white	test tubes (3)
gelatin	milk	25 mL graduated cylinder

316   Chapter 22

CAUTION: Biuret reagent is very caustic. If you spill any on your hands, wash them immediately and tell your teacher. Wear safety goggles during this activity.

Procedure

a. Put a pinch of gelatin in a test tube.
b. Add 5 mL of water to the test tube.
c. Add 5 mL of Biuret reagent (Fig. 22-8).
d. Turn the test tube several times to mix the contents.
e. Record your observations.
f. Repeat steps (a) to (e) using a few pieces of egg white instead of gelatin.
g. Repeat steps (a) to (e) using milk instead of gelatin and water.

Discussion

1. **Biuret reagent produces a faint violet colour with proteins.** Is gelatin a protein?
2. Do egg white and milk contain proteins? How do you know?

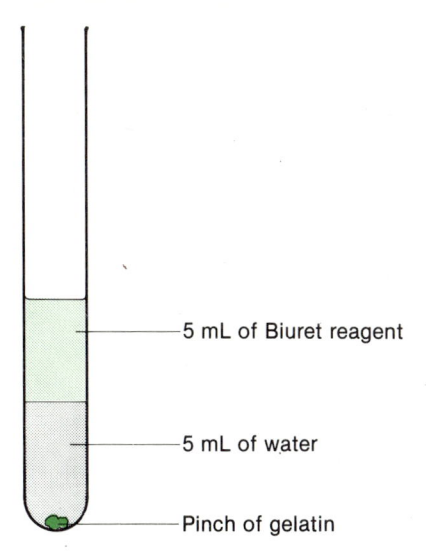

Fig. 22-8 How does Biuret reagent affect gelatin? This is the test for proteins.

22.8 Vitamins

Vitamins are remarkable organic compounds. They do not have to be broken down before they can be used. Also, you normally need very small amounts. In fact, your daily requirement of some vitamins would fit on the head of a pin! Yet your body cannot operate properly without them.

Why Do You Need Vitamins?

Vitamins are not used directly for energy or for building cells. Your body needs vitamins to form enzymes. **Enzymes** are molecules that control the chemical activities of your cells. So vitamins help your body grow and function properly.

Vitamins also prevent **deficiency diseases** [di-FISH-un-see]. These diseases result from a lack of some essential nutrient in the body. A well-balanced diet normally supplies all the vitamins you need.

An excess of certain vitamins can be very harmful. Vitamin A is a good example. Lack of vitamin A can retard growth. It can also cause night blindness and poor tooth formation. But too much vitamin A can cause weak bones, an enlarged liver, and hair loss.

Sources and Functions of Vitamins

Table 22-5 lists some important vitamins. It also gives sources of those vitamins and their main functions.

Table 22-5 Sources and Functions of Vitamins

Vitamin	Sources	Functions
A	Leafy and yellow vegetables, egg yolk, milk, liver, butter, margarine	Aids growth; prevents night blindness; helps form healthy hair and skin
B_1	Whole grain cereals, yeast, milk, green vegetables, egg yolk, liver, lean meat, fish	Aids nerve function; aids growth, aids muscle strength
B_2	Lean meat, wheat germ, yeast, milk, cheese, eggs, liver, yeast, bread, leafy green vegetables	Helps form healthy skin; helps eyes adapt to light; helps form healthy tissues
B_{12}	Liver, lean meats, milk, eggs, kidney, fish	Prevents anemia; helps nervous system; aids growth
C	Citrus fruits, berries, green vegetables, tomatoes	Prevents scurvy, aids tooth and bone formation, makes strong blood vessels
D	Eggs, vitamin-enriched milk, fresh fish, egg yolk	Strong teeth and bones, helps get calcium and phosphorus from digestive tract into blood
E	Vegetable oils, leafy vegetables, wheat germ	Aids fertility, maintains cell membranes
K	Leafy vegetables, soybeans, pork liver	Aids blood clotting

Section Review

1. Give two reasons why you need vitamins.
2. What is a deficiency disease?
3. Make a list of the vitamins that were in the foods you have eaten so far today.

22.9 ACTIVITY Testing for Vitamin C

Vitamin C aids tooth and bone formation. It also helps make strong blood vessels. Without it you will get a disease called scurvy. Some people say that large doses of vitamin C will prevent colds. But *never* take large doses of vitamins without a doctor's advice. They can harm you.

In this activity you will test some drinks to see if they contain vitamin C.

Problem

Which of the following contain vitamin C: fresh orange juice, C-plus orange drink, and orange soda pop?

Materials

indophenol solution
0.5% ascorbic acid solution (vitamin C)
fresh orange juice
C-plus orange drink
orange soda pop
dropper
test tubes (3)
25 mL graduated cylinder

Procedure

a. Put 10 drops of indophenol solution in a test tube.
b. Add ascorbic acid solution (vitamin C) one drop at a time. Shake after each drop. Continue until a colour change occurs (Fig. 22-9).
c. Record the colour change. Also, record how many drops were needed.
d. Repeat steps (a) to (c), first using fresh orange juice. Then use C-plus orange drink and, finally, orange soda pop.

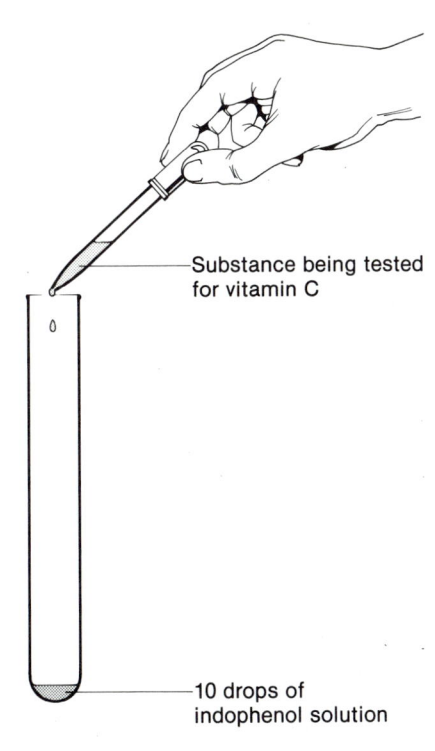

Fig. 22-9 Testing for vitamin C. Shake after each drop. Count the drops.

Discussion

1. **Vitamin C (ascorbic acid) turns indophenol solution from blue to colourless.** This is the test for vitamin C. Which of the three drinks contain vitamin C?
2. Which drink contained the most vitamin C? How do you know?

22.10 Water

A large part of all living matter is water. In fact, about 70% of your body mass is water. You need water for several reasons. Therefore it is classed as a nutrient.

The main reasons you need water are:
1. It aids in the digestion (breaking down) of food.
2. Water in the blood carries substances throughout the body. These include minerals, vitamins, hormones, and glucose.
3. It lubricates your joints.
4. It helps you to excrete wastes such as urea.
5. It helps control your body temperature. Sweating cools the body.

You get much of the water you need by drinking water and other liquids. But you also get some from your foods. Most foods contain some water. Lettuce, for example, is about 95% water.

Section Review

1. State 5 reasons why your body needs water.

22.11 Minerals

The main elements in your body are carbon, hydrogen, oxygen, and nitrogen. However, your body needs small amounts of many other elements. These elements are usually called **minerals**.

The **major minerals** the body needs are calcium, phosphorus, potassium, sodium, magnesium, chlorine and sulfur. The **trace minerals** (those needed in *very* small amounts) include iron, iodine, zinc, copper, and manganese. Your body needs at least 14 trace minerals.

The functions of some minerals are well-known. Table 22-6 gives the functions and sources of three minerals. Do your foods contain enough of these minerals?

Table 22-6   Three Important Minerals

Mineral	Sources	Functions
Calcium	Milk, eggs, cheese, beans, peas, asparagus	Makes strong bones and teeth; aids muscle and nerve function
Phosphorus	Milk, eggs, liver, grains	Makes strong bones and teeth; helps body processes
Iron	Liver, heart, whole wheat, raisins, spinach, lettuce	Helps make hemoglobin; prevents anemia

Section Review

1. What are minerals?
2. How do trace minerals differ from major minerals?
3. Name 7 major minerals.
4. Name 5 trace minerals.

22.12   ACTIVITY   Testing for Minerals

A food can contain as many as six nutrients. These are carbohydrates, fats and oils, proteins, vitamins, water, and minerals. If a food is burned, the first five disappear as gases. But minerals do not burn. They stay behind as an ash.

In this activity you do not test for specific minerals. Instead, you simply burn some foods to see if they contain minerals.

Problem

Which foods contain minerals?

Materials

deflagrating spoon	oatmeal	sugar cube
Bunsen burner	potato	apple

CAUTION: Wear safety goggles during this activity.

Procedure

a. Fill the deflagrating spoon with oatmeal.
b. Heat the spoon and oatmeal strongly (Fig. 22-10). Continue until no further changes occur.
c. Record the results.
d. Repeat steps (a) to (c) using the sugar cube.
e. If time permits, try 1 cm³ of apple, then 1 cm³ of potato.

Discussion

1. *Any white ash that remains after heating is mineral.* Which foods contained mineral?
2. What other nutrients are in oatmeal?
3. What other nutrients are in glucose?

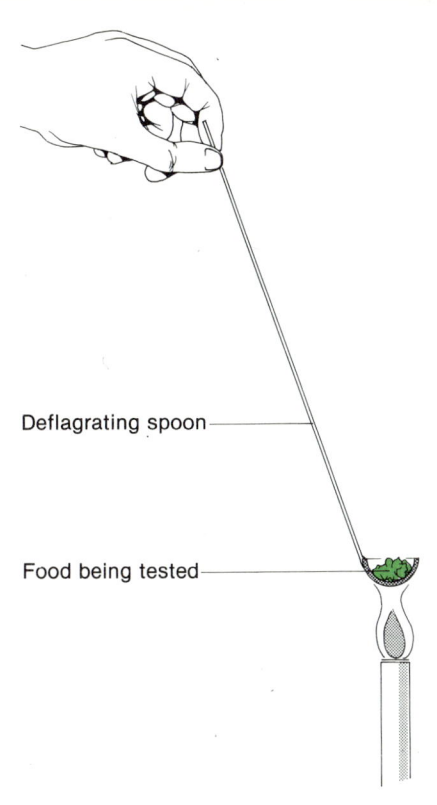

Fig. 22-10 Testing for minerals. Minerals do not burn. They stay behind as an ash.

22.13 Food For a Healthy Body

Food and Energy

A 50 kg boy at the age of 13-15 needs about 13 000 kJ of energy a day. A 50 kg girl at the same age needs 11 000 kJ. The energy is used for growth and activity.

Look back to Table 22-1 on page 310. It tells you how many kilojoules (kJ) of energy there are in 1 kg of several foods. You can see that 1 kg of white bread would give a 50 kg girl all the energy she needs for a day. It would also give her a stomach ache! That's because 1 kg of bread is about 2 loaves! Also, this bread would not provide all the minerals, vitamins, and other nutrients that are needed.

We usually get our energy needs from several foods, not just one. Table 22-7 shows the energy content of a dinner. How many more kilojoules would a boy have to add in his breakfast and lunch to give 13 000 kJ?

Section 22.13 321

Table 22-7 Energy Content of a Dinner

Food	Energy content (kJ)
85 g sirloin roast	1050
1 medium sized baked potato	750
115 cm³ cooked carrots	80
115 cm³ cooked broccoli	85
60 mL white sauce	440
1 large roll (white)	480
9 cm wedge of apple pie (doubled-crusted)	2000
1 glass whole milk	700
Total	5585

Food and Exercise

The number of kilojoules you need per day depends, in part, on how active you are. It also depends on the kinds of activities you do. Table 22-8 shows how fast a 70 kg person uses energy in different activities.

Table 22-8 Activity and Energy Use

Activity (70 kg person)	Rate of energy use (kJ/min)
Lying down	6
Walking fast	22
Bicycling	35
Swimming	47
Running	82

Table 22-9 Time Taken to Burn up One Chocolate Bar

Activity	Time (min)
Lying down	190
Walking fast	52
Bicycling	33
Swimming	24
Running	14

Some people have a tendency to become overweight. If you are one of those, you can control your mass two ways. First, you can eat less and cut back on energy-rich foods. Second, you can exercise.

Table 22-8 shows what exercise can do to that surplus energy in your body. Figure 22-11 shows how long you must do various activities to "burn" up one slice of bread.

A medium-sized chocolate bar has about 1150 kJ of energy in it. Study Table 22-9 before you eat your next chocolate bar!

A Balanced Diet

A **balanced diet** is one that gives your body the right amounts of all six nutrients. If you eat a wide variety of foods, your diet will likely be balanced. But to make sure, you can use a book called *Canada's Food Guide*. You can get it from the Department of Health and Welfare in Ottawa or from your school nurse.

Canada's Food Guide divides foods into **four food groups**. Table 22-10 shows you the foods that are in each group. You need foods from all four groups each day. Otherwise, your diet will not be balanced.

Fig. 22-11 Exercise uses up energy. But diet control is a better way to keep your mass down. You have to walk 15 min just to use up one slice of bread.

Section 22.13 323

Table 22-10 The Four Food Groups

Group	Foods in the group
Milk group — 3 glasses of milk a day — occasionally eat other foods in the milk group	Milk, yoghurt, cheese, ice cream, butter
Meat group (includes vegetable substitutes for meat) — 2 servings a day	Beef, pork, chicken, turkey, fish, lamb, seafood, eggs, beans, peas, peanut butter
Vegetable-fruit group — 4 to 5 servings a day — including a Vitamin C and a Vitamin A food each day	Oranges, grapefruits, berries, tomatoes, salad greens, spinach, carrots, peas, apples, mushrooms, cabbage
Bread-cereal group — 3 to 5 or more servings each day	Whole grain cereals, whole grain bread, macaroni, spaghetti, crackers, noodles

Section Review

1. Which food provides more energy, a medium sized baked potato or a large white roll?
2. Which activity uses more energy per minute, bicycling or running?
3. Name the four food groups.

Main Ideas

1. The six nutrients are carbohydrates, fats and oils, proteins, vitamins, water, and minerals.
2. Sugars and starches are our main source of energy.
3. Fats are a concentrated form of food energy.
4. Proteins build your tissues.
5. Vitamins help make enzymes.
6. A balanced diet contains food from all four food groups.

Glossary

amino acid	ah-MEEN-o	a building block of a protein
enzyme	EN-zime	a protein that controls a chemical activity in your body
nutrient	NEW-tree-ent	a useful part of a food

Study Questions

A. True or False

Decide whether each of the following statements is true or false. If the sentence is false, rewrite it to make it true. (Do not write in this book.)
1. All nutrients supply energy for your body.
2. Cellulose acts as roughage in your digestive tract.
3. A soda cracker contains starch.
4. Fats and oils are carbohydrates.
5. Your body can make all the amino acids it needs.
6. Minerals can be burned.

B. Completion

Complete each of the following sentences with a word or phrase that will make the sentence correct. (Do not write in this book.)
1. Food energy is measured in units called _____.
2. Glucose is a _____ sugar.
3. Vitamins help prevent _____ diseases.
4. Running uses about _____ as much energy as walking for the same time.
5. You would have to walk _____ min to burn up one chocolate bar.

C. Multiple Choice

Each of the following statements or questions is followed by four responses. Choose the correct response in each case. (Do not write in this book.)
1. The nutrients that are oxidized most easily are
    a) carbohydrates and fats
    b) carbohydrates and proteins
    c) minerals and vitamins
    d) minerals and proteins
2. The best food to take on a long hike on a cold day would be
    a) raw carrots
    b) whole wheat bread
    c) peanuts
    d) boiled potatoes
3. Humans cannot digest
    a) proteins
    b) cellulose
    c) starches
    d) sucrose
4. A food that provides Vitamins A, B_1, B_2, B_{12}, and D is
    a) orange juice
    b) green vegetables
    c) margarine
    d) milk
5. You have just eaten a large piece of pie. Its energy content is 2500 kJ. To burn up the pie you would have to run for about
    a) 30 min
    b) 10 min
    c) 150 min
    d) 5 min

D. Using Your Knowledge

1. Why is glucose used in hospitals for intravenous feeding?
2. Why is it best that a person have just the right amount of a vitamin?

3. What is meant by a balanced diet?
4. People who have lost blood often need to be given iron. Why?
5. Why might potato chips be called "junk food" in one person's diet but not in another person's diet?

E. Investigations

1. Bring to class a food of your choice. Test it for sugar, protein, and fats.
2. Select any mineral mentioned in Section 22.11. Find out why your body needs that mineral. Then list several foods that contain it.
3. Make up a test to show that foods contain water. Write out your method. Check it with your teacher. Then try it on 2 or 3 foods. Include a food, like crackers, that looks dry.
4. Find out what "food additives" are. Prepare a report on the food additives in one food that you eat. Discuss the nature of the additives and the reasons they are put in the food.
5. Keep track of what you eat each day for a week. Use *Canada's Food Guide* to find the number of kilojoules you ate each day. Also, see if you had a balanced diet each day. Write a report on your findings.

23 The Digestive System

23.1 Overview of the Digestive System
23.2 Activity: Why Is Chemical Digestion Necessary?
23.3 Before the Stomach
23.4 Activity: Effect of Amylase on Starch
23.5 At the Stomach
23.6 Activity: Effect of Pepsin on Protein
23.7 After the Stomach
23.8 Care of Your Digestive System

Your **digestive system** has one main job. It **digests** food, or *breaks it up* into tiny particles so they may diffuse into your blood to be used by your body.

The main purpose of this chapter is to show you how the digestive system does this job.

23.1 Overview of the Digestive System

Parts

Your digestive system is made up of two main parts: the digestive tract and the assisting organs (Fig. 23-1).

The **digestive tract** is a tube about 10 m long. It begins at the mouth and ends at the anus. Most of it is curled up in your abdomen. As Figure 23-1 shows, it has six main parts. Each of these parts plays a role in the breaking down, or digestion, of food.

Now, look at Figure 23-2. Follow an imaginary piece of food from the mouth to the anus. Try to guess what happens to the food in each of the six main parts. You will see later how good your guesses were.

The **assisting organs** are also listed in Figure 23-1. Do you know what any of them do? (Hint: Note where they connect to the digestive tract.)

Function

Food is no good to you if it stays in your digestive tract. It must move from the digestive tract into the blood. Then your blood carries it to the cells of your body. There it is used for life processes such as growth and respiration.

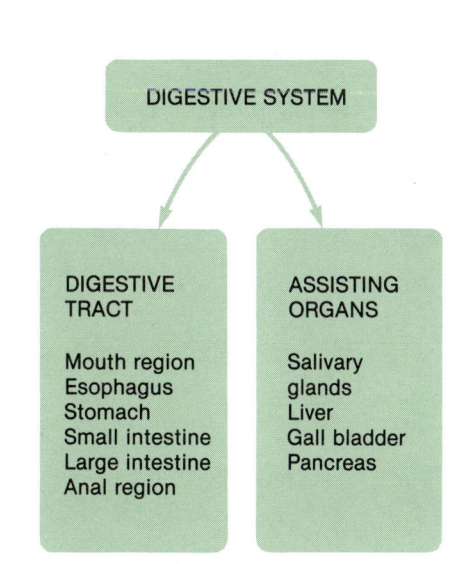

```
              DIGESTIVE SYSTEM
                  /      \
        DIGESTIVE       ASSISTING
        TRACT           ORGANS

        Mouth region    Salivary
        Esophagus         glands
        Stomach         Liver
        Small intestine Gall bladder
        Large intestine Pancreas
        Anal region
```

Fig. 23-1 Parts of the digestive system

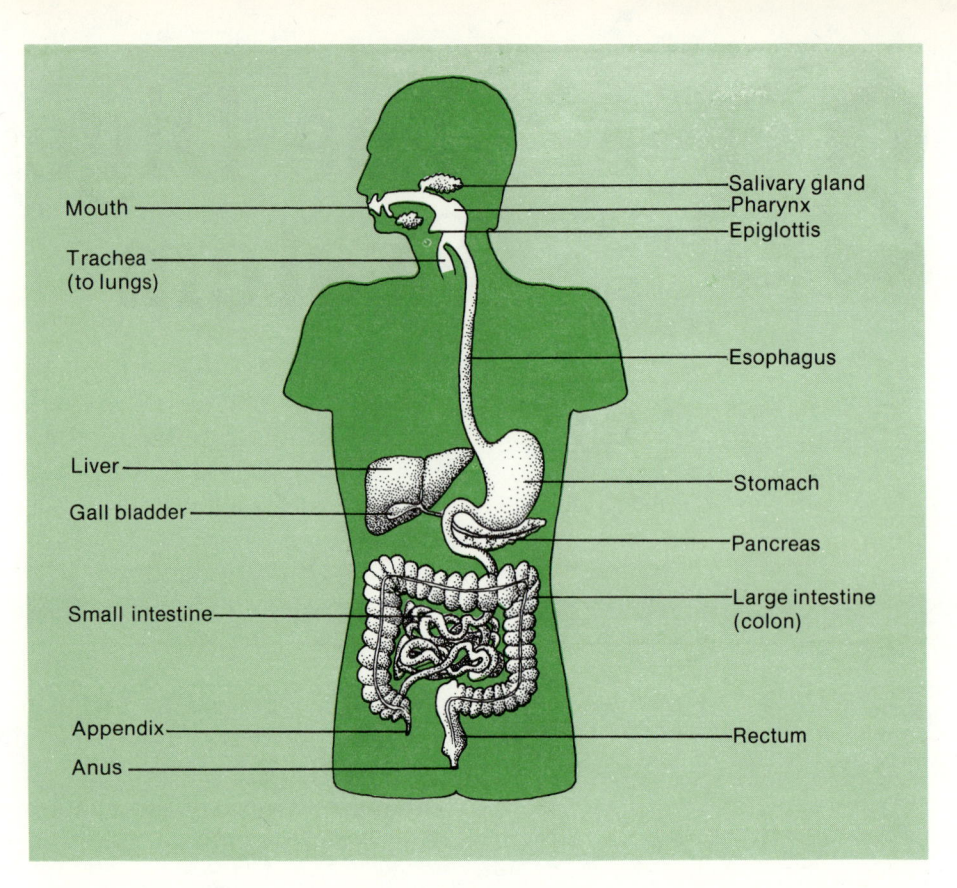

Fig. 23-2 The human digestive system

To get out of your digestive tract, the food must pass through cell membranes. (You will see how later.) This means that the food must be in the form of small molecules. Your digestive system, then, has to break food down into small molecules. We call this **digestion**. It begins this job by crushing, grinding, and mixing the food. This breaks the food into smaller pieces. We call this **mechanical digestion**. The food now consists of molecules of water, minerals, vitamins, fats, proteins, and carbohydrates.

Water, minerals, and vitamins are already small molecules. They do not need any further treatment. But the fat, protein, and carbohydrate molecules are too large. They have to be broken down into smaller molecules before they can get to the blood. We call this **chemical digestion**. Chemical digestion is caused by **enzymes** in your digestive system. The rest of this chapter explains how your digestive system carries out mechanical and chemical digestion.

Section Review

1. Cover up the labels in Figure 23-2. Then practice until you know all the parts.
2. Why is digestion necessary?
3. What is mechanical digestion?
4. What is chemical digestion?

23.2 ACTIVITY Why is Chemical Digestion Necessary?

Some molecules need chemical digestion. Others don't. Do you remember why?

In this activity you will put two kinds of molecules in a sac made of dialysis tubing. Some of the molecules are salt, a mineral. The rest are starch, a carbohydrate. The tubing behaves much like your intestine. It lets some molecules through but not others. Which molecules do you think will get through the tubing?

Problem

Can salt and starch molecules get through dialysis tubing?

Materials

20 cm of dialysis tubing
starch suspension
salt solution
warm distilled water
iodine
silver nitrate solution

250 mL beaker
wooden stick
string
cavity slide
dropper

Procedure

a. Copy Table 23-1 into your notebook.
b. Wet the dialysis tubing to soften it. Then roll one end between your finger and thumb to open it. Tie one end to make a bag.
c. Add to the bag about 4 mL of starch suspension and 4 mL of salt solution.
d. Tie the open end tightly.
e. Rinse the outer surface of the bag with water.
f. Hang the bag in 50 mL of warm (not hot) distilled water as shown in Figure 23-3.
g. Wait 5 min. Then do the 2 tests outlined in steps (h) and (i). Record your results in the table.

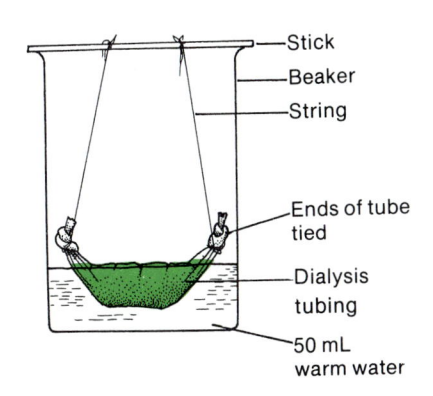

Fig. 23-3 The materials should be set up like this. Keep the ends of the tubing out of the water.

Table 23-1 Why Is Chemical Digestion Needed?

Test Solution	Reaction with water from beaker		
	After 5 min	After 20 min	After 1 d
Iodine solution			
Silver nitrate solution			

Section 23.2 329

h. Test the water in the beaker for the presence of starch as follows. Place 1 drop of the water in the cavity slide. Then add 1 drop of iodine solution. *(A blue or purple colour will form if starch is present in the water.)*
i. Test the water in the beaker for the presence of salt as follows. Wash the cavity slide. Then dry it. Place one drop of the water in the cavity slide. Then add 1 drop of silver nitrate solution. *(A white solid will form if salt is present in the water.)*
j. Repeat the 2 tests after 20 min.
k. If time permits, set the beakers aside for 1 d in a warm place. Repeat the 2 tests.

Discussion

1. Can salt molecules pass through dialysis tubing? How do you know?
2. Can starch molecules pass through dialysis tubing? How do you know?
3. A starch molecule is thousands of times larger than a salt molecule. Explain the results you have summarized in questions 1 and 2.

23.3   Before the Stomach

Before food gets to the stomach, it passes through the **mouth region** and **esophagus** [ih-SOF-uh-gus]. Let's see what they do to the food.

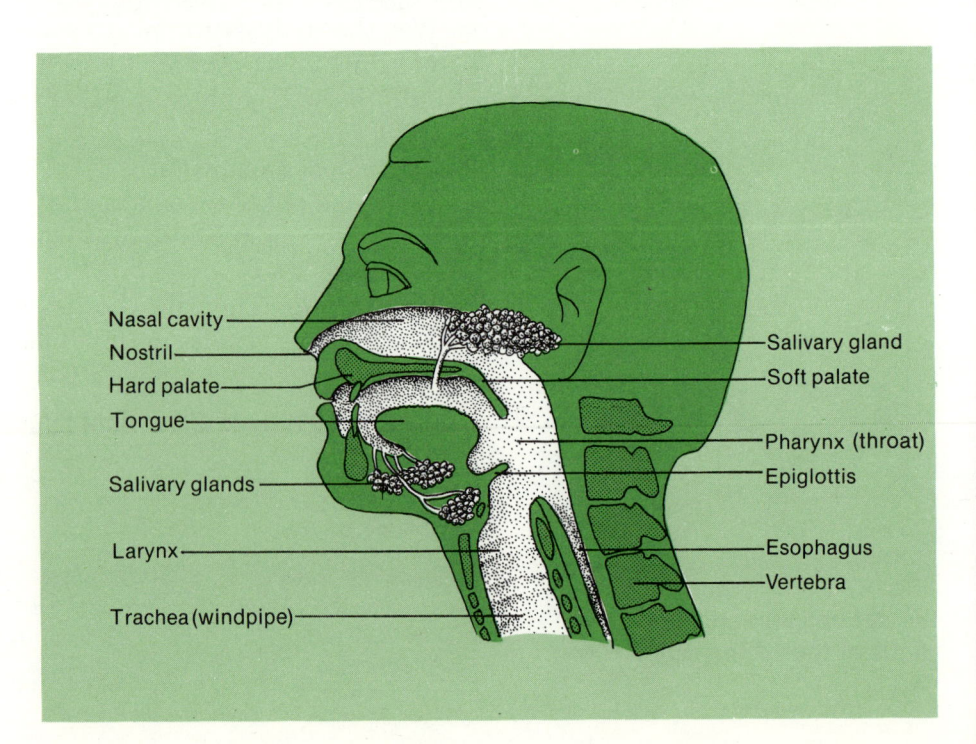

Fig. 23-4 The mouth region of the digestive tract. Also shown are the upper parts of the breathing system.

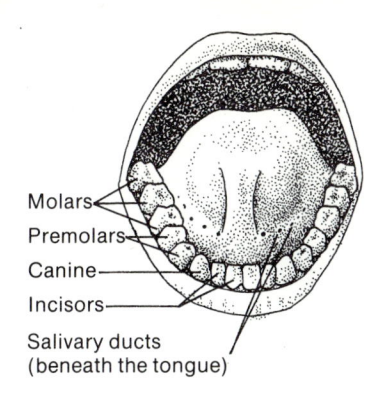

Fig. 23-5 Types of teeth in the lower jaw. The upper jaw has the same types.

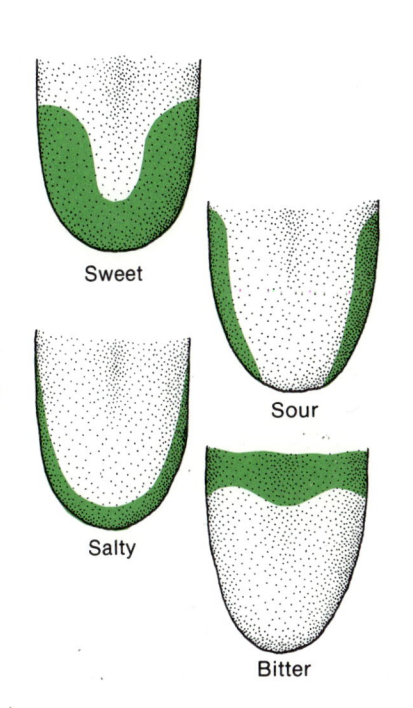

Fig. 23-6 The green areas indicate where the tongue is most sensitive to the taste. If you want to prove this, try Investigation 4 on page 341.

The Mouth Region

The mouth region includes lips, teeth, jaws, tongue, salivary glands, and hard and soft palates (Fig. 23-4). This region carries out three main functions:
1. It crushes and grinds the food (mechanical digestion);
2. It mixes the food with saliva (some chemical digestion);
3. It monitors taste and temperature.

Crushing and Grinding Look into a mirror. Find the following teeth on your lower jaw. (Use Figure 23-5 as a guide.)
1. Four flattened, sharp **incisors**. They cut food.
2. Two single-pointed **canines**. They pierce and tear food.
3. Four **premolars**. They crush food.
4. Four **molars** (six if you have "wisdom teeth"). They grind food.

Your upper jaw has the same type and arrangement of teeth. The total, then, is 28 teeth, or 32 if you have "wisdom teeth".

All teeth have a **crown** and one or more **roots**. The crown is the part you can see. It is covered with a hard substance called **enamel**.

The teeth, with the help of the tongue, carry out mechanical digestion.

Mixing with Saliva **Salivary glands** in your mouth make up to 1.5 L of **saliva** every day. Saliva is made when food is being tasted or chewed. It's even made when you think about food. (Try thinking about something you'd like to eat right now!) Saliva contains three things: **water**, the enzyme **amylase**, and **mucin**.

The water and mucin lubricate the food. Then it moves more easily down the esophagus. Amylase breaks some of the starch molecules into sugar molecules. Starch molecules are large. They cannot get through the digestive tract into the bloodstream. But the sugar molecules are small. They can get through.

Monitoring Temperature and Taste Special nerve endings in the lips, tongue, and **hard palate** (roof of your mouth) monitor temperature. They tell you if the food is too hot.

Your tongue does the tasting in your mouth. The tongue is covered by tiny areas called **taste buds**. Each taste bud is most sensitive to one taste: salty, sweet, sour, or bitter. Different areas of your tongue detect different tastes (Fig. 23-6).

The Esophagus

Swallowing Swallowing moves the food to the esophagus. First, the tongue pushes the food back to the **pharynx** [FAR-inks]. The pharynx is also called the throat. It's where the nasal cavity and mouth cavity meet. As the food moves back, it pushes the **soft palate** up (Fig. 23-7). This closes the passage to the nasal cavity. Now food cannot get into the nasal cavity.

Section 23.3 331

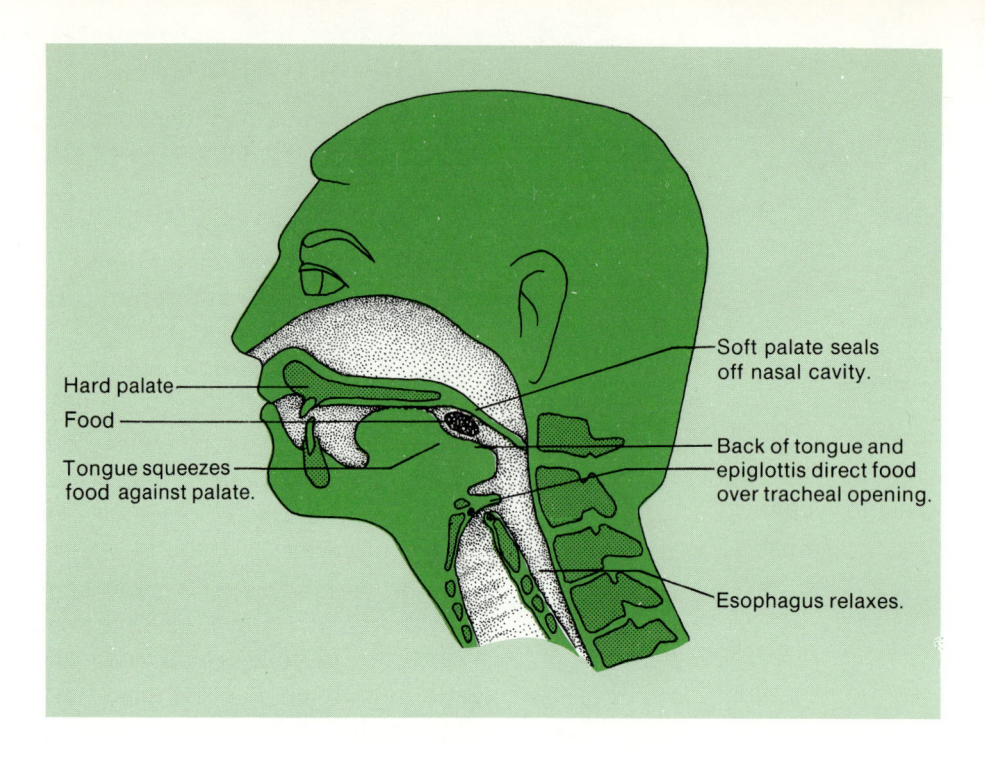

Fig. 23-7 How food is swallowed

At the same time, the **epiglottis** covers the **trachea** (windpipe). Now food cannot get into the trachea and choke you. It has only one path left — it slides into the esophagus.

How Food Moves Down the Esophagus Food is moved down the esophagus by an action called **peristalsis** [per-ih-STALL-sis]. Circular muscles behind the food contract. Those in front of the food relax. The food is squeezed forward by this action (Fig. 23-8). Longitudinal muscles (those going the length of the esophagus) contract as well. This shortens the esophagus and helps move the food along. This action is something like squeezing toothpaste from a tube.

Peristalsis explains why you can swallow when upside down. Your food is forced down the esophagus. It does not simply slide down due to gravity. Peristalsis occurs in the stomach and intestines as well.

Fig. 23-8 Peristalsis moves food down the esophagus.

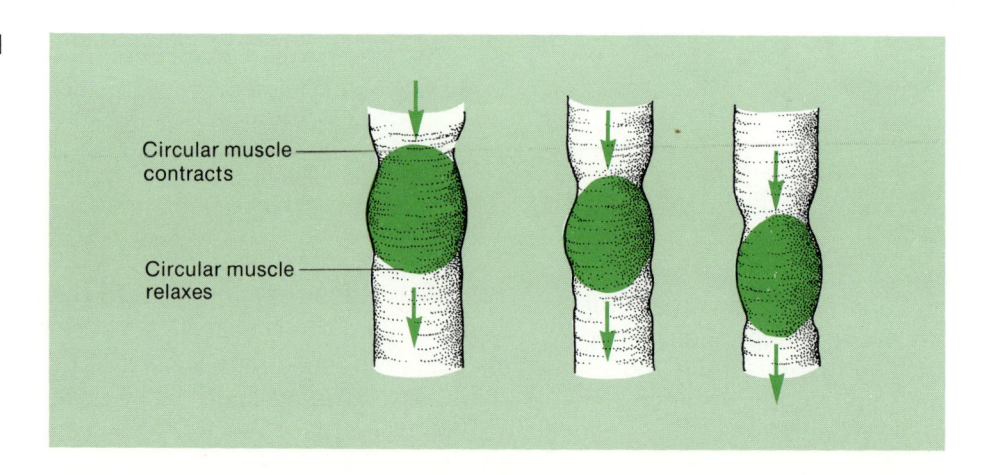

Section Review

1. What are the 3 functions of the mouth region?
2. Name the 4 types of teeth and state their functions.
3. What is saliva?
4. What does amylase do?
5. How are temperature and taste monitored?
6. How is food swallowed?
7. Explain how peristalsis moves food down the esophagus.

23.4 ACTIVITY Effect of Amylase on Starch

Starch molecules are too large to move from the digestive tract into the blood. Therefore, they must be digested, or broken down into smaller molecules. Starch can be broken down into sugar molecules such as maltose and glucose. They are small enough to move into the blood. In this activity you will find out if the enzyme amylase in your saliva can do this job.

Problem

Can the amylase in your saliva digest starch to form sugars?

Materials

dilute starch suspension
saliva (contains amylase)
clean, new rubber band
new drinking straw
thermometer
iodine solution
Benedict's solution
marking pen
water bath (stand, ring, clamp, wire gauze, Bunsen burner, 250 mL beaker 2/3 full of water)
test tubes in test tube rack (4)
250 mL beaker with 150 mL of water at about 37°C

CAUTION: Wear safety goggles during this activity.

Procedure A Preparing for the Experiment

a. Set up the water bath. Get the water boiling.
b. Label 4 test tubes CI, SI, CB, SB. Table 23-2 explains what these labels mean.
c. Put 10 mL of starch suspension into each of the 4 test tubes.

Table 23-2 Code for Test Tube Labels

Code	Meaning
CI	Control; only iodine is added
SI	Saliva and iodine are added
CB	Control; only Benedict's solution is added
SB	Saliva and Benedict's solution are added

Table 23-3   Effect of Amylase on Starch

Test tube	Initial colour	Final colour	What results mean
CI			
SI			
CB			
SB			

Procedure B Does Amylase Break Down Starch?

a. Chew the rubber band to stimulate saliva production. Using the straw, put 2 mL of saliva into SI.
b. Put 2 drops of iodine solution into CI and SI.
c. Sit CI and SI in the 37°C water bath. Observe what happens over the next 10 min.
d. Copy Table 23-3 into your notebook.
e. Record your observations from step (c) in the table.

Note: Iodine reacts with starch to form a blue or purple colour. The more starch there is, the darker the colour will be.

Procedure C Are Sugars Formed When Starch Breaks Down?

a. Put 2 mL of saliva into SB.
b. Add 5 drops of Benedict's solution to both CB and SB.
c. Sit CB and SB in the 37°C water bath for 10 min. Now place them in the boiling water bath. Note any colour change after 5 min. Record your observations in your table.

Note: Benedict's solution turns orange-red if glucose (a sugar) is present.

Discussion

1. What do the results of Procedure B prove?
2. What do the results of Procedure C prove?
3. What does the amylase in saliva do to starch?
4. Why are the test tubes CI and CB needed?

23.5 At the Stomach

Peristalsis moves the food down the esophagus to the **stomach** When empty, the stomach appears to be no more than a thick section of the esophagus. When full, however, it holds up to 2 L. And, it takes on a J-shape.

A thick ring of circular muscle forms a **valve** at each end of the stomach (Fig. 23-9). **Valve 1** opens to let food enter from the esophagus. **Valve 2** opens to let food pass on to the **duodenum**. When food is being digested in the stomach, these valves are closed most of

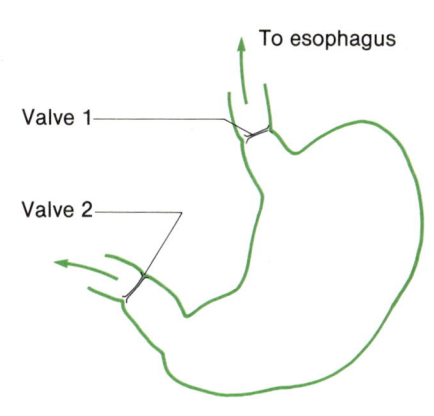

Fig. 23-9 The J-shaped stomach has a valve at each end.

the time. Peristalsis churns the food back and forth in the stomach. Sometimes peristalsis happens even when you have no food in your stomach. Then you have "hunger pains".

Glands in the stomach secrete many digestive juices. These include hydrochloric acid, pepsin, and mucin.

The **hydrochloric acid** aids in the digestion of starch and protein. It also has many other functions. Sometimes hydrochloric acid bubbles up through valve 1 into your esophagus. This causes some discomfort. It is called "heartburn". But, it has nothing to do with your heart.

The **pepsin** is an enzyme that digests (breaks down) proteins. Hydrochloric acid helps the pepsin do this. The proteins become smaller molecules. But they need to be broken down further. This happens in the small intestine.

The **mucin** coats the lining of the stomach. It prevents the acid from dissolving the lining. Sometimes not enough mucin is produced and the acid dissolves the lining and even muscles under it. This painful condition is called an **ulcer**.

Section Review

1. Describe the shape of a full stomach.
2. What are the functions of the two valves?
3. Explain what each of the following does: hydrochloric acid, pepsin, mucin.

23.6 ACTIVITY Effect of Pepsin on Protein

Your stomach juices contain pepsin and hydrochloric acid. In this activity you will see how proteins (egg white) are digested by these chemicals.

Problem

What does pepsin do to proteins? Is hydrochloric acid needed?

Materials

5% pepsin
0.2% hydrochloric acid
hard boiled egg
graduated cylinder
marking pen
400 mL beaker
test tubes (4)
incubator (optional)
stoppers (4)

Procedure

a. Label the test tubes A, B, C, and D.
b. Put a piece of egg white from a hard boiled egg in each test tube. Try to make the pieces the same size and shape. For example each could be a cube that is 1 cm on each side.

c. Add the following substances to the test tubes:
   Tube A — 10 mL of water
   Tube B — 10 mL of 0.2% hydrochloric acid
   Tube C — 10 mL of 5% pepsin
   Tube D — 10 mL of 5% pepsin plus 2 drops of 0.2% hydrochloric acid
d. Put stoppers in the test tubes. Stand the test tubes in a beaker (Fig. 23-10).
e. Put the set-up in an incubator at 37°C. Leave if there for at least 24 h. (If you do not have an incubator, let the set-up sit at room temperature for 48 h.)
f. Look at the egg whites closely. Note any evidence of digestion (breaking down).

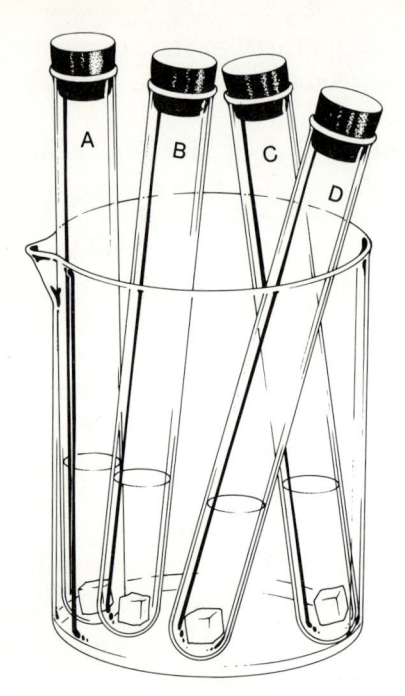

Fig. 23-10 Digestion of proteins

Discussion

1. In which tube did digestion occur?
2. What substances are needed for the digestion of protein?
3. **a)** Are these substances present in the stomach juices?
   b) Your stomach is made of protein. Why is it not digested by its own juices?
4. Why were the test tubes incubated at 37°C?

23.7 After the Stomach

Food is partly digested in the stomach. Then it moves on to the **small intestine**, **large intestine**, and, finally, the **anal region**. Look back to Figure 23-2 to make sure you know where these parts are. Now let's see what they do in the digestion of food.

The Small Intestine

The **small intestine** is about 7 m long. It's called "small" because of its diameter, not its length. Its diameter is about 3 cm. That of the large intestine is about 7 cm. The small intestine is coiled and folded inside your abdomen. Food moves through it by peristalsis.

Intestinal Glands The lining of the small intestine has many **intestinal glands**. They secrete enzymes that complete the digestion of proteins. They also digest large sugar molecules into smaller ones.

Liver The **liver** makes no enzymes. Instead, it makes a green liquid called **bile**. Bile can be stored in the **gall bladder** until needed. When it is needed, it passes down a tube to the **duodenum** (the first part of the small intestine). In the intestine it helps break up droplets of fats and oils into smaller droplets.

Pancreas The **pancreas** [PAN-cree-us] makes several enzymes. It makes amylase that continues the digestion of starches. It makes an enzyme that digests the fats and oils broken down by the bile. And it makes an enzyme that completes the digestion of proteins. These enzymes enter the small intestine through a tube from the pancreas to the duodenum.

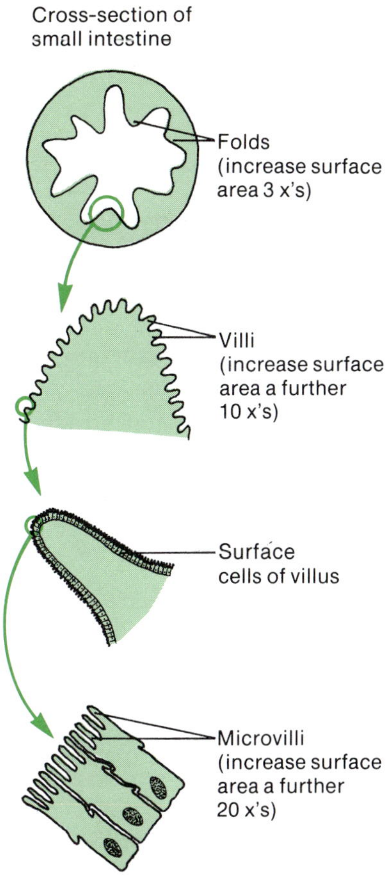

Absorption of Nutrients The small intestine is an important digestive organ. But it has another important use. It is the main organ that transfers digested food from the digestive tract to the blood. Let's see how this happens.

The lining of the small intestine is not smooth. Instead, it is gathered in large **folds** (Fig. 23-11). These folds are covered with millions of projections called **villi** [VILL-ee]. Each one is called a **villus**. The surface of each villus contains hundreds of cells. And each cell has projections called **microvilli**.

The folds, villi, and microvilli give the small intestine 600 times more surface area than it would have if it had a smooth lining. Why is this important? The villi are full of tiny blood vessels and lymph vessels. Digested foods are absorbed by the villi and go into these vessels. That's how food gets to the blood.

The long length of the intestine is also important in absorption of nutrients. It, together with the folds, villi, and microvilli, gives the intestine a surface area of over 200 m². That's more than the area of the floor in your classroom!

Fig. 23-11 The small intestine is not a tube like a garden hose. Its lining has folds, villi, and microvilli. These increase its surface area by 600 times.

The Large Intestine

The **large intestine**, or **colon**, is about 1.5 m long. It absorbs water from the food wastes and recycles it into your bloodstream. If it did not do this, you would always have diarrhea. You would also have to drink much more water than you do. Your digestive juices, alone, put about 5.5 L of water into your digestive tract every day. The large intestine recycles all of this but 0.2 L. If it didn't, you would have to drink the 5.5 L of water. That's a lot of water!

Your large intestine is home for many bacteria. Some of these make vitamin K and some of the B vitamins. Others, called **coliform** bacteria, help the intestine recycle water.

As the food wastes move through the large intestine, it becomes **feces**, or **fecal matter**. About half of the feces consists of dead bacteria. The rest is undigested food and water.

The Anal Region

Feces are stored for about a day in the **rectum** which is about 20 cm long. Then they are eliminated through the **anus**.

Section Review

1. What three parts of the digestive tract follow the stomach?
2. State the role in digestion of each of the following: intestinal glands, liver, pancreas.

Section 23.7

3. a) Describe the structure of the lining of the small intestine.
   b) Explain why the lining has this structure.
4. What functions does the large intestine perform?
5. What functions does the anal region perform?

23.8 Care of Your Digestive System

My stomach's sick. I have heartburn. I have acid stomach. I have gas. I have cramps in my stomach. I brought up. I have diarrhea. I'm constipated.

How many of these problems have you had? Probably all of them. All of us have problems with our digestive systems from time to time. But these problems usually solve themselves. Often they were caused by too much food, the wrong foods, too little food, lack of sleep, worry, or tension.

Four rules help keep the digestive system working properly and prevent many problems:

1. Eat well-balanced meals. Eat regularly. Include fibre in every meal (Examples: whole wheat bread, bran cereals, bran muffins, raw celery, raw carrots).
2. Exercise regularly.
3. Get regular and adequate sleep.
4. Avoid the use of drugs. Even legal drugs such as caffeine, nicotine, and alcohol can upset the digestive system.

Some Common Problems

Acid conditions such as **heartburn** and **acid stomach** can often be cured best with a glass of warm milk. Many antacid remedies are available. But use them with care. They often make the problem worse. If acid conditions persist, **ulcers** can form. They commonly form in the stomach and duodenum.

Many serious diseases affect the digestive system. **Food poisoning** is one. You can avoid it by following the precautions given in Chapter 10. **Influenza** (the 'flu) is another. Many forms of this virus cause upsets in the digestive system.

Appendicitis is another common illness. The appendix is located in your lower right abdomen (see Figure 23-2). Food sometimes gets caught in the appendix. Then the appendix becomes infected. It can even burst, dumping poisons into your body cavity. An infected appendix can cause pains much like gas pains or indigestion. And they may occur far from the appendix. **Don't ignore any persistent pains!**

Cancer

Cancer, of course, is the most serious disease of the digestive system. The mouth, esophagus, pancreas, stomach, and large intestine are common sites of cancer. Many cancers are caused by lifestyle. Alcoholic beverages and smoking cause cancer of the mouth and esophagus. Smoking is a major cause of cancer of the pancreas. The lack of fibre in the diet is thought to contribute to cancer of the large intestine.

The cure rate for most cancers is getting higher all the time. Early detection increases the chances of survival.

Four warning signals of cancer of the digestive tract are:
1. changes in bowel habits
2. unusual bleeding or discharge
3. indigestion or difficulty in swallowing
4. a thickening or lump

See a doctor if you have a warning signal.

Two Things to Remember

1. Many problems of the digestive tract are caused by lifestyle. What you do today can cause ulcers or cancer tomorrow. Are you taking any risks?
2. Always see a doctor for any pains or problems that don't go away quickly.

Section Review

1. Give four rules for keeping the digestive system working properly.
2. What organs are common sites of cancer?
3. What do we mean when we say some diseases are caused by lifestyle?
4. Give four warning signals of cancer of the digestive system.

Main Ideas

1. The mouth carries out mechanical digestion and some chemical digestion.
2. Food moves through the digestive system by peristalsis.
3. The stomach helps digest starch and proteins.
4. The small intestine digests sugars, fats and oils, and proteins.
5. Foods are absorbed from the small intestine into the bloodstream.
6. The large intestine recycles water.
7. Many problems of the digestive system are caused by lifestyle.

Glossary

amylase	AM-uh-lays	an enzyme that digest starch
feces	FEE-sees	the waste material from digestion
mucin	MYOO-sin	a protective coating in the digestive system

peristalsis	per-ih-STALL-sis	the method by which food is moved through the digestive tract
villus	VILL-us	the absorbing area of the intestine

Study Questions

A. True or False

Decide whether each of the following statements is true or false. If the sentence is false, rewrite it to make it true. (Do not write in this book.)
1. The duodenum is part of the large intestine.
2. A glass of warm milk is a good cure for acid stomach.
3. Smoking is a major cause of cancer of the pancreas.
4. Appendicitis pains always occur in the lower right abdomen.

B. Completion

Complete each of the following sentences with a word or phrase that will make the sentence correct. (Do not write in this book.)
1. Food is moved through the digestive tract by an action called _____ .
2. Bile is stored in the _____ until needed.
3. The large intestine absorbs _____ from food wastes.
4. A change in bowel habits can be a warning signal of _____ of the digestive tract.

C. Multiple Choice

Each of the following statements or questions is followed by four responses. Choose the correct response in each case. (Do not write in this book.)
1. The part of the digestive tract just before the stomach is the
   a) epiglottis   b) esophagus   c) small intestine   d) trachea
2. If your stomach doesn't make enough mucin, you may get
   a) heartburn   b) acid stomach   c) ulcers   d) hunger pains
3. You can help avoid cancer of the esophagus by
   a) eating foods high in fibre   c) drinking warm milk
   b) exercising regularly   d) avoiding smoking

D. Using Your Knowledge

1. Explain why you can swallow when you are upside down.
2. A dog's saliva contains no amylase. Why not?
3. A cheese sandwich is mainly starch, protein, and fat. Describe the changes in the sandwich as it passes through the digestive tract.
4. What are the advantages of chewing food well?

E. Investigations

1. Find out the latest information on the care of teeth. Write a paper of about 300 words on your findings.
2. Find out why fluorides are put into some toothpastes and drinking water.
3. Chew an unsalted soda cracker for 5 min. Don't swallow. What evidence do you have that starch is digested in the mouth? (The cracker contains starch.)
4. Do an experiment to verify the diagrams in Figure 23-6. Hint: Use a cotton swab.
5. Interview 5 smokers. Find out what they know about the effects of smoking. Find out why they smoke. Write a report on your findings.

24 The Breathing System

24.1 Your Breathing System
24.2 Activity: How Breathing Occurs
24.3 Breathing
24.4 Activity: What Is Your Breathing Rate?
24.5 Activity: What Is Your Useable Lung Capacity
24.6 Artificial Respiration
24.7 Care of Your Breathing System

You can live for days without water and food. But you can live only a few minutes without breathing. Like other animals, you need oxygen. Your cells use it during **respiration** (see Chapter 7). Respiration provides energy for life processes. But it also produces carbon dioxide. Too much of this gas can be harmful. Therefore you must get rid of it.

Breathing gets oxygen into your blood. And, it gets carbon dioxide out of your blood. This chapter shows how your body does these two things.

24.1 Your Breathing System

Your lungs and the tubes that lead to them make up your **breathing system** (Fig. 24-1). Let's follow some air as it enters your body and goes to your lungs.

The Air Passage

Air usually enters your breathing system through the **nostrils**. Hairs in the nostrils filter dust particles from the air. The air then moves to the **nasal cavity**. Here the moist lining traps small particles that got past the hairs in the nostrils. This lining also moistens and warms the air.

The air now enters the **pharynx** [FAR-inks]. The pharynx is also called the **throat**. Here the air may join with air inhaled by the mouth. You can breathe faster through your mouth, but the air is not warmed as much. Nor is it as well filtered and moistened.

Next, the air goes down the **trachea** [TRAKE-ee-a], or **windpipe**. The trachea is about 10 cm long. It is prevented from collapsing by U-shaped pieces of cartilage (Fig. 24-2). The trachea is lined with

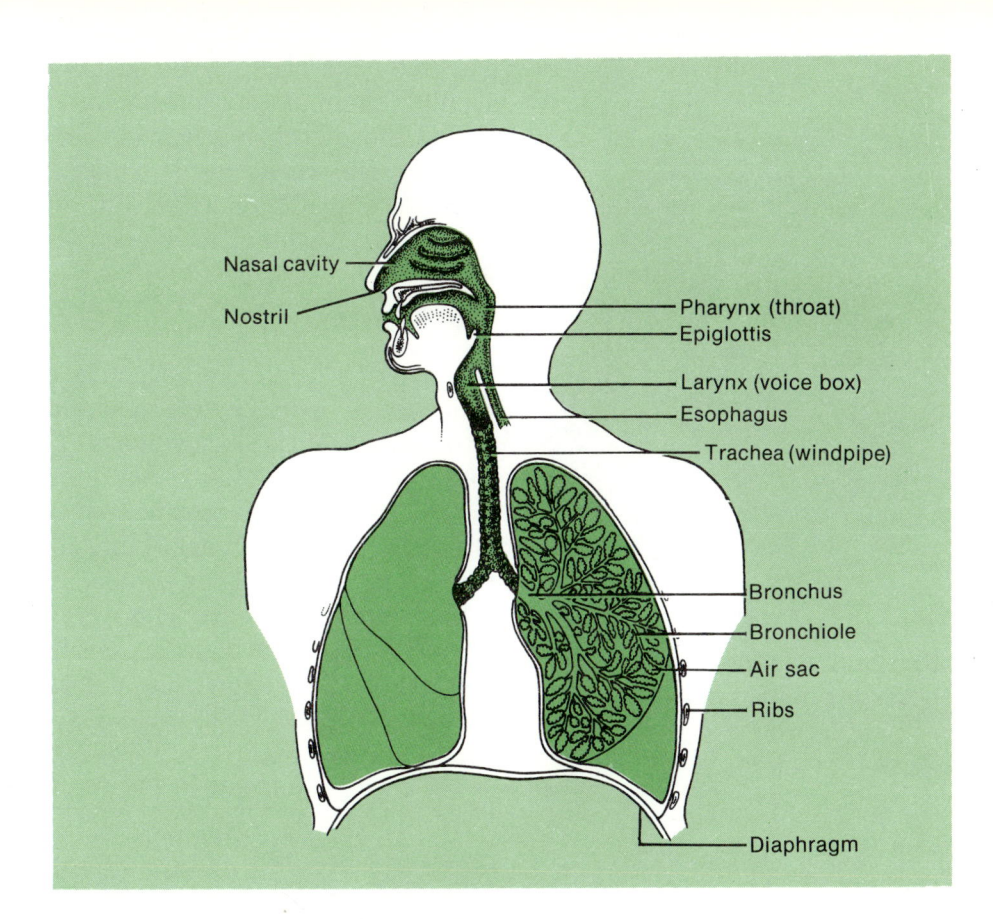

Fig. 24-1 The human breathing system

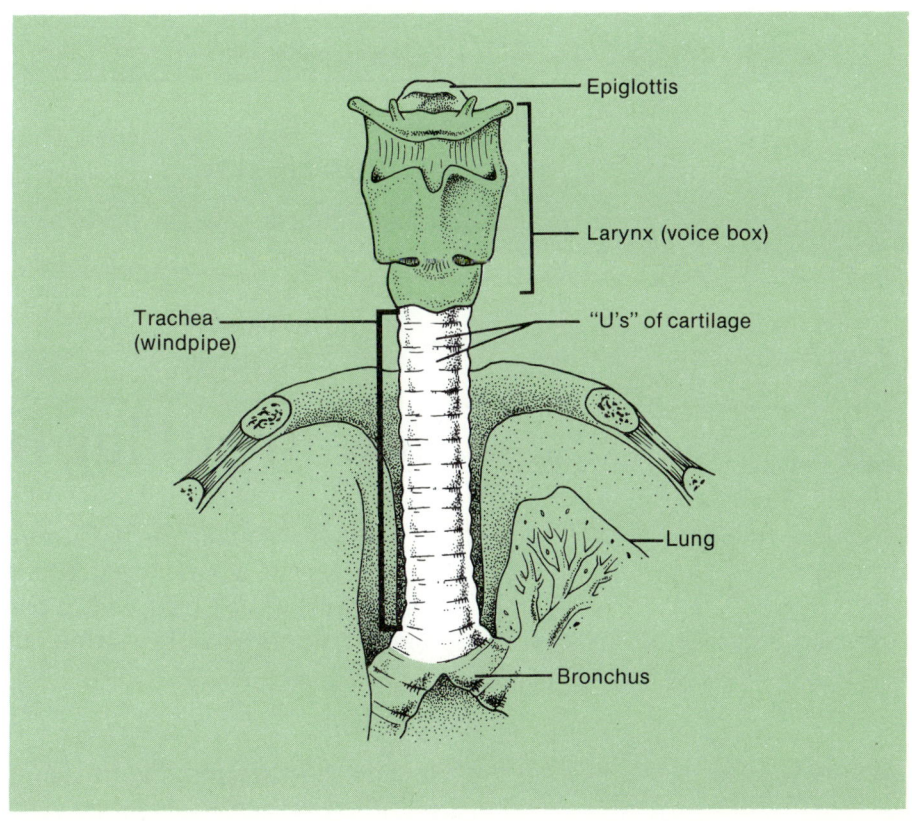

Fig. 24-2 The trachea

Section 24.1 343

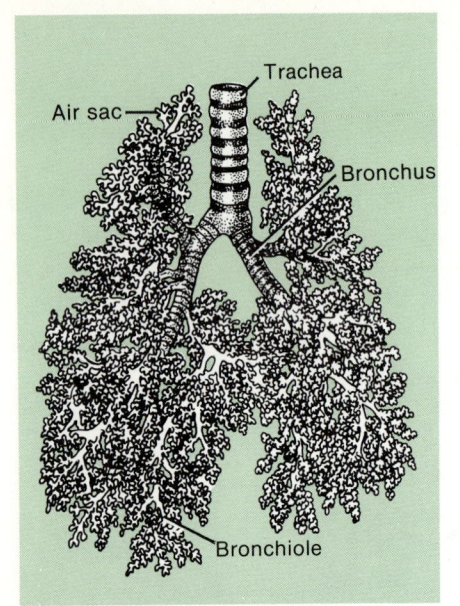

Fig. 24-3 A model of the air passages of the lungs

hair-like **cilia**. They sweep particles up to the pharynx. Then you can swallow them or cough them out.

The top of the trachea is the **larynx** [LAR-inks] or **voice box**. It contains the **vocal cords**. They vibrate to make sounds when you speak or sing. The bottom of the trachea branches into two **bronchi** [BRON-key]. Each **bronchus** [BRON-cuss] leads to a **lung**.

The Lungs

Within each lung, the bronchus branches into smaller tubes called **bronchioles** [BRON-key-oles]. The bronchioles divide again and again into millions of tiny bronchioles. The air goes down these bronchioles to **air sacs** (Fig. 24-3). Each lung has over three hundred million (300 000 000) of these grape-like air sacs (Fig. 24-4).

The air sacs have very thin walls. These walls are moist and full of tiny blood vessels called **capillaries**. Blood flows through the capillaries as shown in Figure 24-4. Oxygen diffuses from the air sacs into the blood. And carbon dioxide diffuses from the blood into the air sacs. Table 24-1 shows the changes that take place in the air you breathe.

Table 24-1   Changes in the Air You Breathe

	Percent oxygen	Percent carbon dioxide
Inhaled air	21	0.034
Exhaled air	16	4

Section Review

1. List, in order, all the parts of your breathing system from nostrils to air sacs.
2. How does oxygen get into the blood? How does carbon dioxide get out of the blood?

24.2   ACTIVITY   How Breathing Occurs

Your lungs are attached to the rest of your body only by the bronchi. They hang in a cavity surrounded by your ribs and diaphragm [DIE-a-fram]. (Look back to Figure 24-1). You breathe by moving your ribs and diaphragm. Let's find out how this works.

Problem

How do the ribs and diaphragm move during breathing?

Vein leaving the lung. Blood is *high* in oxygen and *low* in carbon dioxide.

Artery coming into the lung. Blood is *low* in oxygen and *high* in carbon dioxide.

Bronchiole

Capillaries

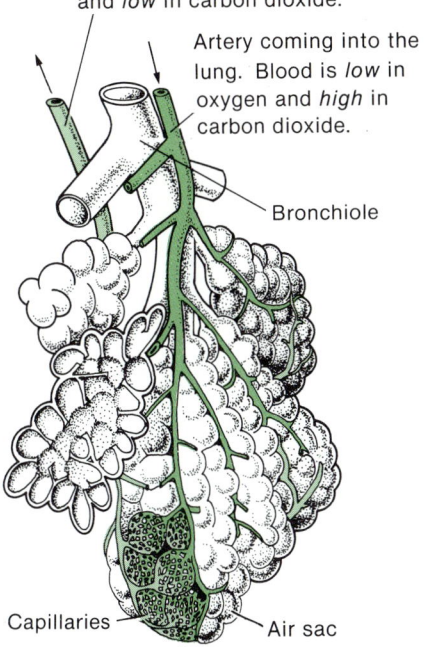

Air sac

Fig. 24-4 The bronchioles lead to grape-like air sacs.

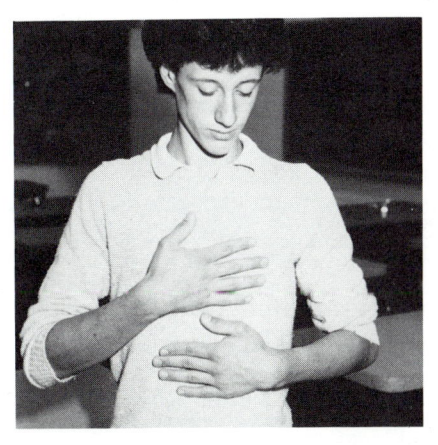

Fig. 24-5 How do the ribs and diaphragm move during breathing?

Materials

your body

Procedure

a. Copy Table 24-2 into your notebook.
b. Place one hand on your chest and the other on your abdomen (Fig. 24-5).
c. Try to inhale (breathe in) and exhale (breathe out) a few times without moving your ribs. What happened to your abdomen? Did your diaphragm move up or down? (Your diaphragm moves down to push your abdomen out. It moves up to pull your abdomen in.) Record your results in your table.
d. Try to inhale and exhale a few times moving only your ribs. Do not move your abdomen. How do your ribs move when you inhale? How do your ribs move when you exhale? Record your results in your table.

Table 24-2 How Breathing Occurs

	Movement of diaphragm	Movement of ribs	Change in size of chest cavity
Inhale		None	
Exhale		None	
Inhale	None		
Exhale	None		

Discussion

1. a) How do your ribs move when you inhale? How does your diaphragm move when you inhale?
   b) What does this do to the size of the chest cavity?
2. a) How do your ribs and diaphragm move when you exhale?
   b) What does this do to the size of the chest cavity?
3. What makes air go in and out of your lungs? Try to explain what happens.

24.3 Breathing

You learned in Activity 24.2 that the ribs and diaphragm move when breathing occurs. Let's look at this process more closely.

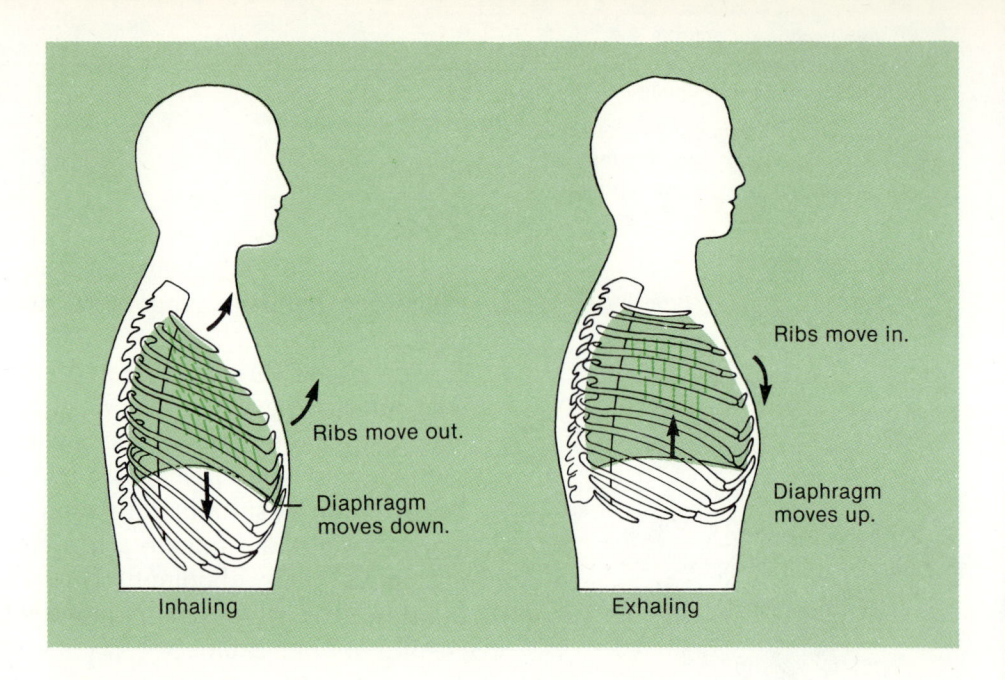

Fig. 24-6 How breathing occurs

Inhaling

A control centre in your brain monitors the level of carbon dioxide in your blood. When the level gets to a certain point, the control centre sends a message to the rib muscles and diaphragm. It directs the rib muscles to pull the ribs *out*. It also directs the diaphragm muscles to pull the diaphragm *down* (Fig. 24-6).

The ribs move out; the diaphragm moves down. This makes the chest cavity larger in size. As a result, the air pressure drops in the air sacs of the lungs. The outside air pressure is higher. This pushes air into the air sacs.

Exhaling

To exhale, your rib and diaphragm muscles simply relax. The ribs move *in* and the diaphragm moves *up*. This makes the chest cavity smaller in size. The air pressure in the air sacs increases. And air is forced out of the air sacs.

In healthy lungs, the air sacs are elastic. Thus they contract to help force air out.

Make yourself cough, or exhale rapidly. Do you know what's happening in your body? You are forcing your stomach and other abdominal contents up against the diaphragm. This pushes the air quickly out of the lungs.

Section Review

1. Explain how inhaling occurs.
2. Explain how exhaling occurs.
3. How does your body exhale quickly?

24.4 ACTIVITY What Is Your Breathing Rate?

Breathing is automatic. It speeds up and slows down as needed by your body. You don't have to think about it. In this activity you will find your breathing rate when resting. Then you will see how it changes when you become active.

Problem

What is your breathing rate when resting and when active?

Materials

your body
watch or timer that reads in seconds

Procedure

a. Sit quietly for 2 or 3 min. Breathe normally. Now count the number of breaths you take in 1 min. Record your results.
b. Repeat step (a) two times. Record your results.
c. Calculate the average number of breaths you take per minute while resting. Record your result.
d. Run on the spot for 2 min. Then count the number of breaths you take in 1 min. Record your results.
e. Repeat step (d) two times. Record your results. Do not average these figures.

Discussion

1. What is your breathing rate while resting?
2. a) Compare your rate with those of several classmates.
   b) Why are there some differences?
3. a) Why did your breathing rate increase when you ran?
   b) Compare your rates after running with those of several classmates.
   c) Why are there some differences?

24.5 ACTIVITY What Is Your Useable Lung Capacity?

Breathe as deeply as you can. Now exhale all the air you can. The volume of air you exhaled is your useable lung capacity. Some air stays in your lungs. Therefore your lungs actually hold more air than this activity suggests. Why is the term "useable" used here?

Problem

What is your useable lung capacity?

Materials

large bottle (at least 2 L)
100 mL graduated cylinder
overflow tray and sink
tubing
glass plate

Procedure

a. Fill the large bottle with water.
b. Invert it in the overflow tray as shown in Figure 24-7. Use the glass plate so that you do not lose any water from the jar.
c. Shove the tubing well up into the bottle.
d. Inhale as deeply as you can. Then exhale as much of your breath as you can through the tubing.
e. Remove the tubing. Then cover the mouth of the bottle with the glass plate.
f. Remove the bottle from the tray. Set it upright on the desk.
g. Add water with the graduated cylinder until the jar is full. Keep track of how much you add. This is the volume of air that was in the bottle.
h. Repeat steps (b) to (g) two times.
i. Average your three results.

Discussion

1. What is the average of your three results? This is your useable lung capacity.
2. Compare your useable lung capacity to that of several classmates.
3. What factors affect the useable lung capacity?

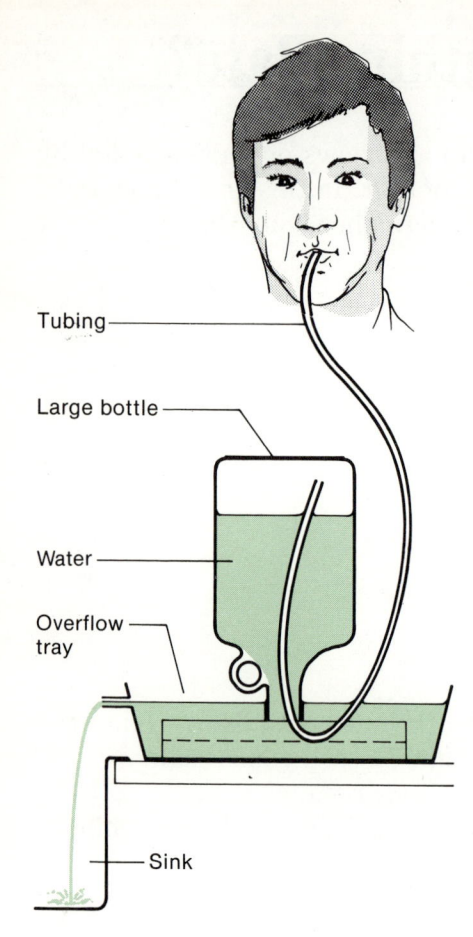

Fig. 24-7 Measuring lung capacity

24.6 Artificial Respiration

Everyone should know how to perform artificial respiration. A person can live only 5-6 min without air. Therefore, if you find someone who is not breathing, you do not have time to get help. You must perform artificial respiration. Otherwise the person will die.

348 Chapter 24

Victims of drowning and severe electrical shock usually need artificial respiration. The best method to use is **mouth-to-mouth resuscitation** [ree-sus-i-TAY-shun].

Procedure for Mouth-to-Mouth Resuscitation

CAUTION: Before you try this procedure, you should have instruction from a qualified instructor.

If the victim is unconscious and not breathing follow these steps:

1. Make sure the air passage is clear. This is best done by lifting the victims neck and tilting the head back by pressing on the forehead. The chin should now point up (Fig. 24-8,A).
2. Keep the victim's head in this position. Then the air passage will stay clear. Otherwise the tongue could block the air passage.
3. Pinch the victim's nose. Use the same hand to press on the forehead. This will keep the victim's head in the proper position (Fig. 24-8,B).
4. Inhale deeply. Open your mouth wide. Now place it over the victim's mouth. Make sure there is a tight seal between your mouth and the victim's.
5. Blow your entire breath into the victim's mouth. Do so in one breath. Don't use several small puffs. Watch out of the corner of your eye to see if the victim's chest rises (Fig. 24-8,C).
6. Remove your mouth from the victim's mouth. Watch and listen to see if the victim exhales (Fig. 24-8,D).
7. Keep the victim's head in the same position. Repeat steps 4, 5, and 6 every 5 s. (This is about your normal breathing rate.)
8. Send the next person on the scene to get medical help.
9. Continue this procedure until breathing resumes or until you are relieved by a medical person.

Note: For very young children, place your mouth over the victim's mouth and nose. Also, use short, faster breaths.

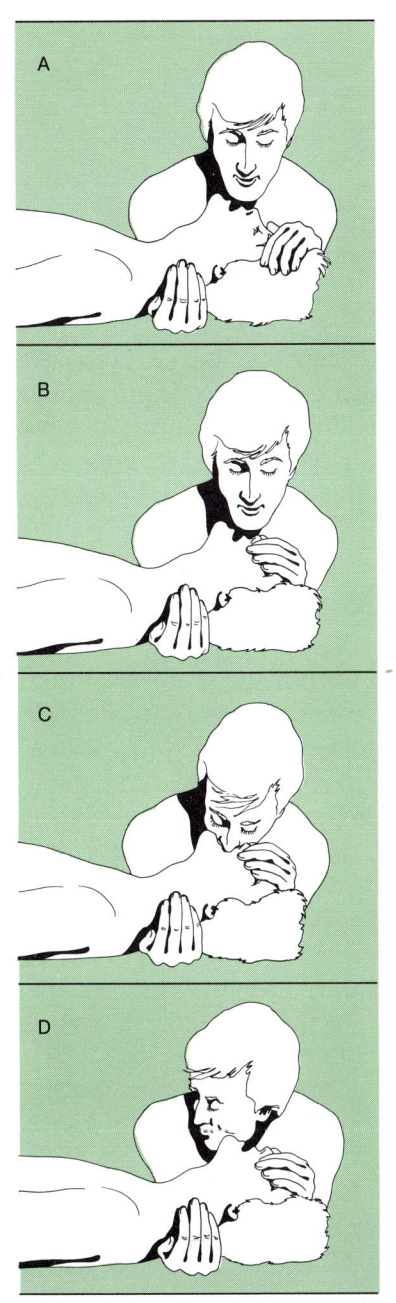

Fig. 24-8 Mouth-to-mouth artificial respiration

Section Review

1. Make a summary of the steps in mouth-to-mouth resuscitation.
2. If possible practise this method on a dummy. You could attend a class offered by your school, the Red Cross, St. John Ambulance, or another group.

24.7 Care of Your Breathing System

Even the cleanest "country air" contains about 40 000 particles in every breath. But the **cilia** (tiny hairs) in your nasal cavity, trachea, bronchi, and bronchioles stop most of these (Fig. 24-9). Only the smallest particles of dust, bacteria, and viruses reach the air sacs of your lungs.

Section 24.7 349

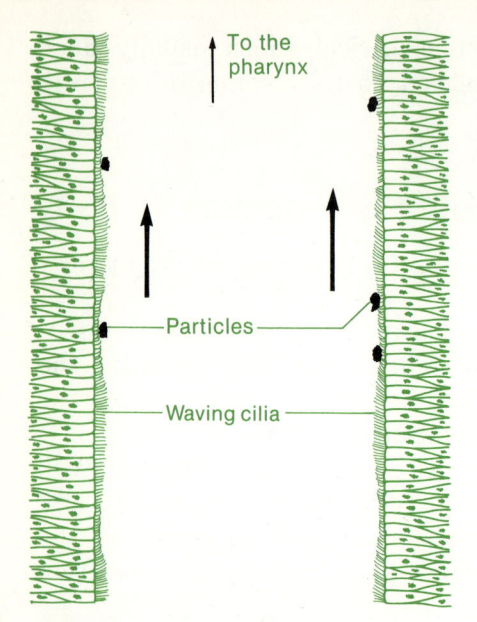

Fig. 24-9 Cilia sweep particles to the pharynx.

Fig. 24-10 This worker is wearing a filter mask to protect his air passage and lungs from dust.

Bacterial and Viral Diseases

The air passage and lungs are moist and warm. They are an ideal habitat for some bacteria and viruses. When these bacteria and viruses multiply, you get diseases. Colds, flu, and pneumonia are examples of diseases of the breathing system. Proper exercise, diet, and rest are the best protection against such diseases.

Occupational Diseases

Many people must work in areas where the air is polluted. Asbestos fibres, coal dust, or dust from sand blasting can cause lung cancer. Dust from farming operations can also damage the lungs. Fumes from pesticides and other chemicals can also damage the lungs.

You should *never* breathe air that is polluted with such things. You should *always* wear a proper filter mask (Fig. 24-10).

Smoking

The best advice for the care of your breathing system is simple: **Don't smoke!** The second best advice for most people is also simple: When someone asks "Do you mind if I smoke?" be honest. Say "Yes". Second-hand smoke is also harmful to your breathing system.

Less than one-third of Canadians smoke. Even so, over 30 000 Canadians will die of smoking-related diseases this year! Lung cancer alone will kill about 9000 people. At least 80% of lung cancer is caused by smoking. Also, 30% of all cancers are linked to smoking. This includes cancer of the mouth, larynx, and esophagus of the air passage. It also includes cancer of the bladder, pancreas, and kidneys. Heart attacks, strokes, high blood pressure, emphysema,

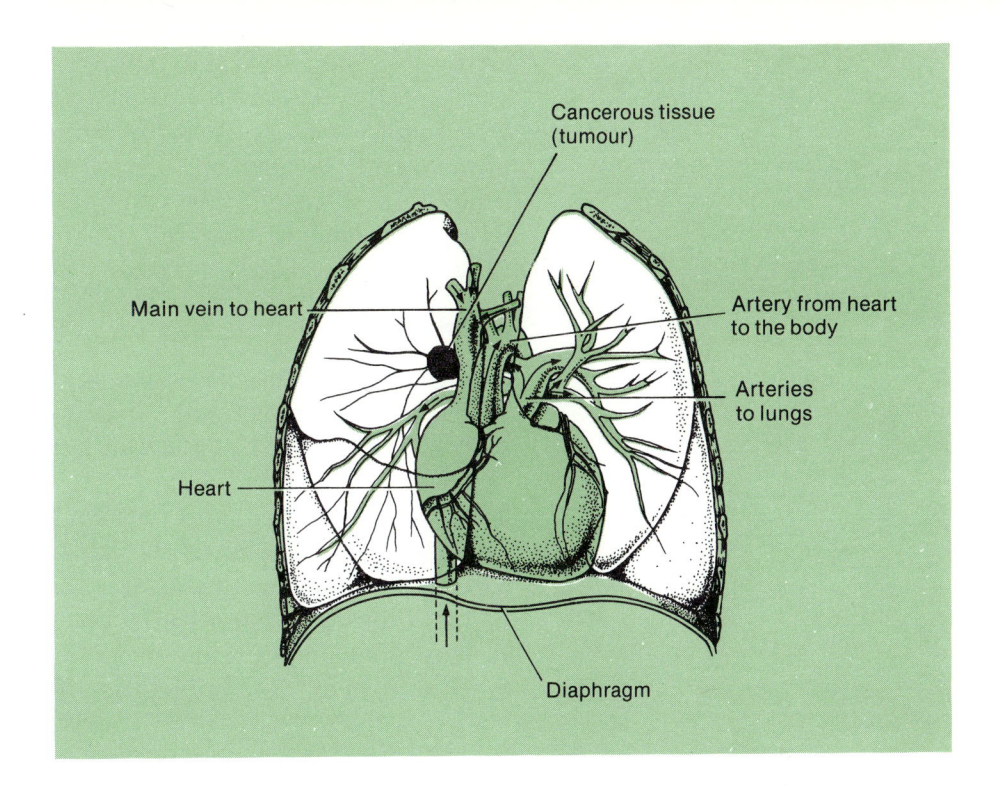

Fig. 24-11 A tumour in the lung can block the flow of blood. It can also block the flow of air.

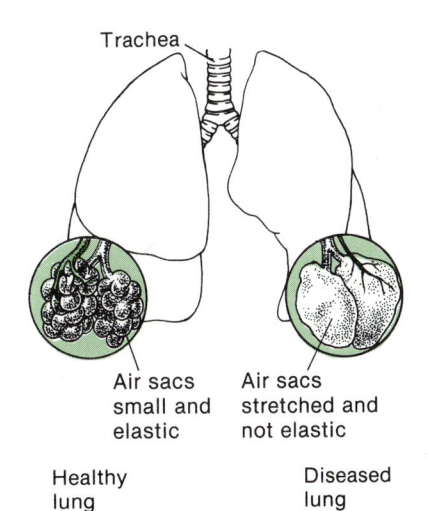

Fig. 24-12 Emphysema. Note how the air sacs change.

and chronic bronchitis also occur much more frequently among smokers.

Smoking is our most serious health problem. Let's look more closely at three common diseases of smokers: lung cancer, emphysema, and chronic bronchitis.

Lung Cancer A heavy smoker can inhale 2 L of tar per year! Tobacco tar "gums" up the cilia in the air passage. Then they can no longer sweep particles up and out of the passage. Thus dangerous chemicals stay in the lungs. Some of these chemicals cause **lung cancer** (Fig. 24-11).

Lung cancer is hard to detect. Often the tumour is not found until it is too late to operate successfully. Then it may block circulation of the blood through the lung. Or, it may block the passage of air. Also, the cancer can spread from the lung to the rest of the body.

Lung cancer is the third-ranked cause of death in Canada. Yet most lung cancers can be prevented. That's because over 80% are caused by smoking.

Emphysema Most people who get **emphysema** [em-fi-SEE-ma] are smokers. Chemicals in the smoke damage the walls of the air sacs. The sacs lose their elasticity and they stretch (Fig. 24-12). Symptoms include wheezing and breathlessness. Breathing becomes very difficult.

In the advanced stages of emphysema, the person often cannot blow out a match! This disease cannot be cured. The victim often dies because of a lack of sufficient air.

Section 24.7

Chronic Bronchitis Like emphysema, **chronic bronchitis** [KRON-ik bron-KI-tis] tends to be a disease of smokers. In fact, it often occurs with emphysema in the same person. Chemicals in smoke cause the bronchi and bronchioles to swell up. Also, mucus builds up in these tubes. As a result, air cannot move easily through the tubes. Symptoms include coughing, wheezing, and breathlessness.

Section Review

1. What is the best protection against bacterial and viral diseases of the breathing system?
2. What is the best protection against occupational diseases of the breathing system?
3. What is Canada's most serious health problem?
4. Name 10 serious diseases that can be caused by smoking.
5. Explain how smoking affects the cilia of the air passage.
6. How does lung cancer affect the body?
7. How does emphysema affect the body?
8. How does chronic bronchitis affect the body?

Main Ideas

1. Air follows this path to the lungs: nasal cavity, pharynx, larynx, trachea, bronchi, bronchioles, air sacs.
2. In the air sacs oxygen diffuses into the blood; carbon dioxide diffuses out of the blood.
3. The ribs and diaphragm move during breathing.
4. Everyone should be able to perform mouth-to-mouth resuscitation.
5. Smoking is our most serious health problem.

Glossary

bronchiole	BRON-key-ole	a tiny air tube in a lung
bronchus	BRON-cuss	a tube leading from the trachea to a lung
chronic bronchitis	KRON-ik bron-KI-tis	a disease common in smokers in which air tubes become swollen and clogged
emphysema	em-fi-SEE-ma	a disease, mainly of smokers, in which air sacs are damaged
larynx	LAR-inks	the voice box
pharynx	FAR-inks	the throat
trachea	TRAKE-ee-a	the main windpipe

Study Questions

A. True or False

Decide whether each of the following statements is true or false. If the sentence is false, rewrite it to make it true. (Do not write in this book.)
1. The esophagus is part of the breathing system.
2. Bronchioles are smaller than bronchi.
3. The diaphragm moves down when you are inhaling.
4. Most people who get lung cancer are smokers.

B. Completion

Complete each of the following sentences with a word or phrase that will make the sentence correct. (Do not write in this book.)
1. The bronchi join the _____ to the lungs.
2. The pharynx is also called the _____ .
3. The trachea is also called the _____ .
4. Three lung diseases that may be caused by smoking are _____ , _____ , and _____ .

C. Multiple Choice

Each of the following statements or questions is followed by four responses. Choose the correct response in each case. (Do not write in this book.)
1. The vocal cords are in the
    a) larynx   b) pharynx   c) bronchi   d) bronchioles
2. Your lungs are attached to the rest of your body by
    a) the ribs only   c) the bronchi
    b) the ribs and diaphragm   d) the diaphragm only
3. What happens when you inhale?
    a) The diaphragm moves up and the ribs move in.
    b) The diaphragm moves down and the ribs move out.
    c) The diaphragm moves up and the ribs move out.
    d) The diaphragm moves down and the ribs move in.
4. Two diseases that occur mainly in smokers are
    a) high blood pressure and strokes
    b) lung cancer and heart attacks
    c) strokes and chronic bronchitis
    d) lung cancer and emphysema

D. Using Your Knowledge

1. What is the function of the breathing system?
2. What is the difference between breathing and respiration?

Put a flame here from time to time.

Hold the test tube at this angle.

Break the cigarette into 4 or 5 pieces.

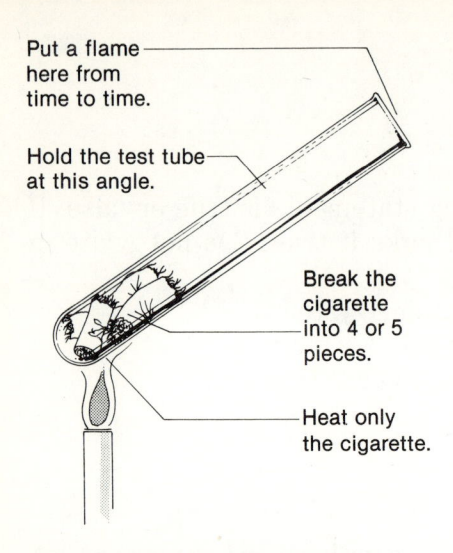

Heat only the cigarette.

Fig. 24-13 What happens to tobacco when it is heated?

3. Mouth-to-mouth resuscitation is usually called artificial respiration. It might be better to call it artificial breathing. Why?
4. Suppose you found a person unconscious and not breathing at an accident scene. This person's mouth is badly injured. Describe how you would conduct artificial respiration.

E. Investigations

1. Design and do an experiment to show that cigarette smoke contains tars.
2. Should smokers be allowed to smoke in public places? Prepare to take one side in a debate on this question.
3. Collect at least 10 cigarette ads from magazines. Look at them carefully. What methods are used to encourage people to smoke?
4. The government makes over $800 000 000 a year from tobacco tax. But it pays out over $2 000 000 000 in health care for smokers. Find out from your member of parliament why smokers don't have to pay the full cost of their health care.
5. Heat a cigarette in a test tube for 3-4 min as shown in Figure 24-13. Discuss the results with a smoker. Write a report on the smoker's reaction.

25 The Circulatory System

25.1 Parts of the Circulatory System
25.2 Path of Blood through the Body
25.3 The Lymphatic System
25.4 The Heart
25.5 Activity: Listening to Your Heart
25.6 Activity: Studying Your Pulse
25.7 Activity: Finding Your Heart Performance Score (HPS)
25.8 Blood
25.9 Activity: Examining Your Blood Cells
25.10 Blood Types and the Rh Factor
25.11 Activity: Finding Your Blood Type
25.12 Care of Your Circulatory System

Your body has about 60 000 000 000 000 (sixty million million) cells. Each one of those cells needs food, water, and oxygen. Also, each one needs to get rid of wastes such as carbon dioxide. How are these needs met? They are met by your circulatory system. This chapter deals with the circulatory system and how it carries out these tasks.

25.1 Parts of the Circulatory System

Your circulatory system is a **closed system**. This means that the blood is always inside blood vessels. Such a system is needed in complex animals like humans and other vertebrates (Fig. 25-1).

The human circulatory system has four main parts:
1. A fluid (**blood** and **lymph**);
2. Tubes to carry blood and lymph (**arteries**, **veins**, and **lymph ducts**);
3. Tiny tubes where the exchange of materials between the blood and cells occurs (**capillaries**);
4. A pump to circulate the blood (**heart**).

The Blood and Lymph

The **blood** and **lymph** [limf] carry the food, oxygen, and water that cells need. They also carry wastes away from cells. And they carry out many other functions. You will study these later.

Fig. 25-1 The circulatory system. Only the main arteries and veins are shown. The arteries are *dark green*; the veins are *light green*.

Heart

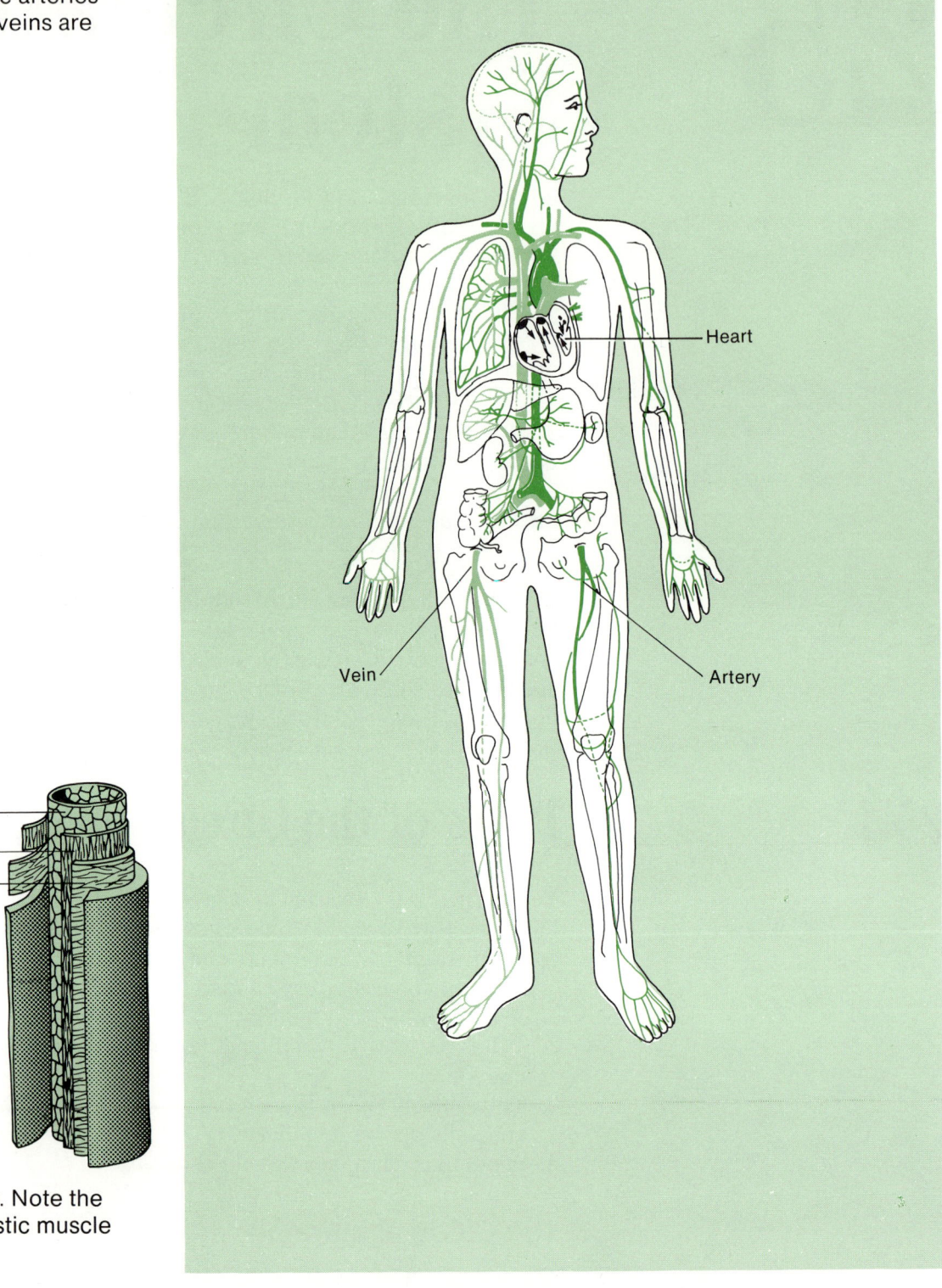

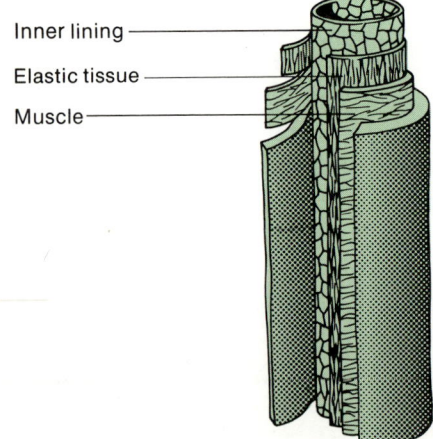

Vein

Artery

Inner lining
Elastic tissue
Muscle

Fig. 25-2 An artery. Note the thick walls with elastic muscle tissue.

356   Chapter 25

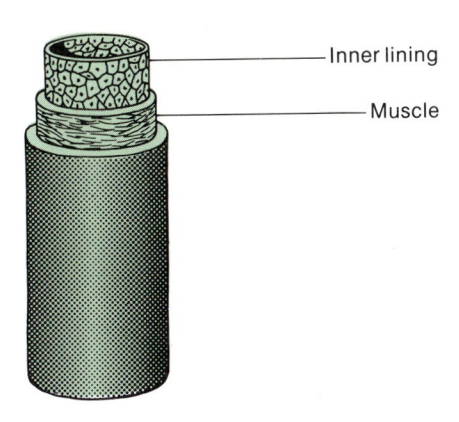

Fig. 25-3 A vein. Why do veins not need walls as thick and elastic as those of arteries?

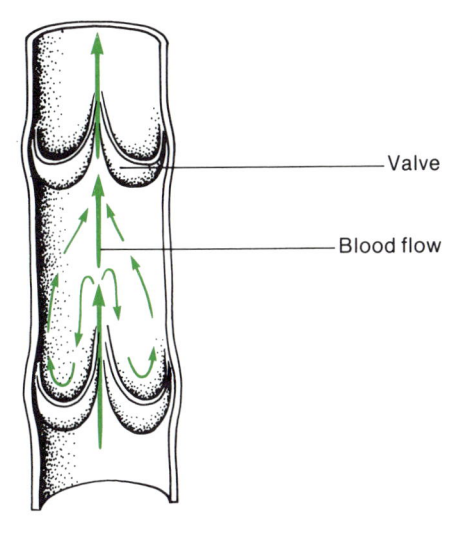

Fig. 25-4 Valves in veins. These ensure that returning blood flows in one direction only.

Fig. 25-5 Capillaries. These thin-walled tubes run through all tissues in the body.

Heart

The **heart** pumps the blood through the tubes of your circulatory system. It forces the blood along by creating a high pressure. You will see how it works later.

Arteries

Arteries carry blood *away from* the heart. Arteries carry blood under the highest pressures. Therefore they must have strong walls. Otherwise they would burst. Arteries are thick-walled tubes that are wrapped with elastic muscle tissue (Fig. 25-2). Small arteries are called **arterioles** [are-TEER-ee-oles].

Veins

Veins carry blood *toward* the heart. The blood in veins is under lower pressure than the blood in arteries. Therefore the walls of veins do not need to be as thick or elastic as those of arteries (Fig. 25-3).

Some veins in the legs and arms have one-way valves (Fig. 25-4). These ensure that the blood flows only toward the heart. Without valves, the low pressure would let the blood flow the other way.

Leaky valves can make **varicose veins**. Blood builds up in veins and stretches them. Varicose veins near the skin surface bulge out and can be easily seen.

Small veins are called **venules**.

Capillaries

Capillaries are a finely divided network of tiny tubes (Fig. 25-5). The walls of capillaries are just one cell thick. Therefore nutrients and oxygen can diffuse through the walls from the blood to the tissues. Also, wastes can diffuse from the tissues into the blood.

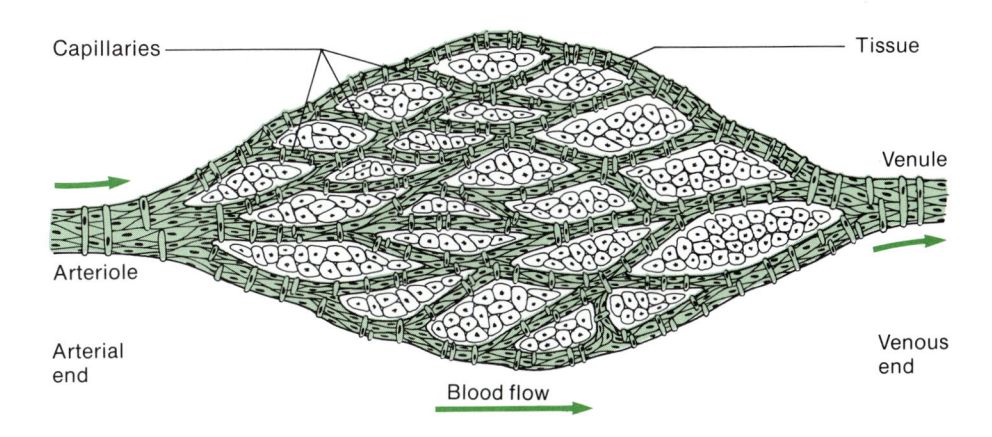

Section 25.1

Section Review

1. What is a closed circulatory system?
2. Name the four main parts of the circulatory system.
3. What are the differences between arteries and veins?
4. What are arterioles? venules?
5. The walls of capillaries are only one cell thick. Why is this important?

25.2 Path of Blood through the Body

Blood always follows the same general path through the body. It moves from the heart through arteries to a capillary. Then it moves from the capillary through veins back to the heart (Fig. 25-6). The path ends up where it started. Therefore it is called a **circuit**.

Your body has three main blood circuits. They are the **lung circuit**, the **systemic circuit**, and the **coronary circuit**. Follow Figure 25-7 as you read about these circuits.

Fig. 25-6 The path followed by blood

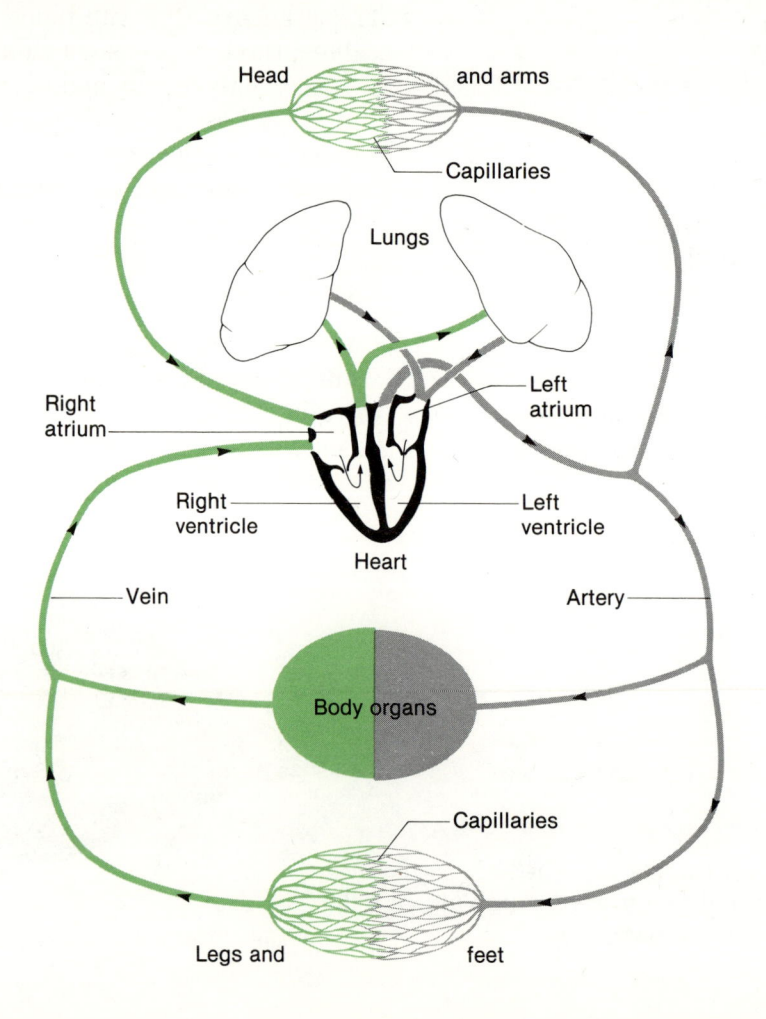

Fig. 25-7 A diagram of the blood circuits

358 Chapter 25

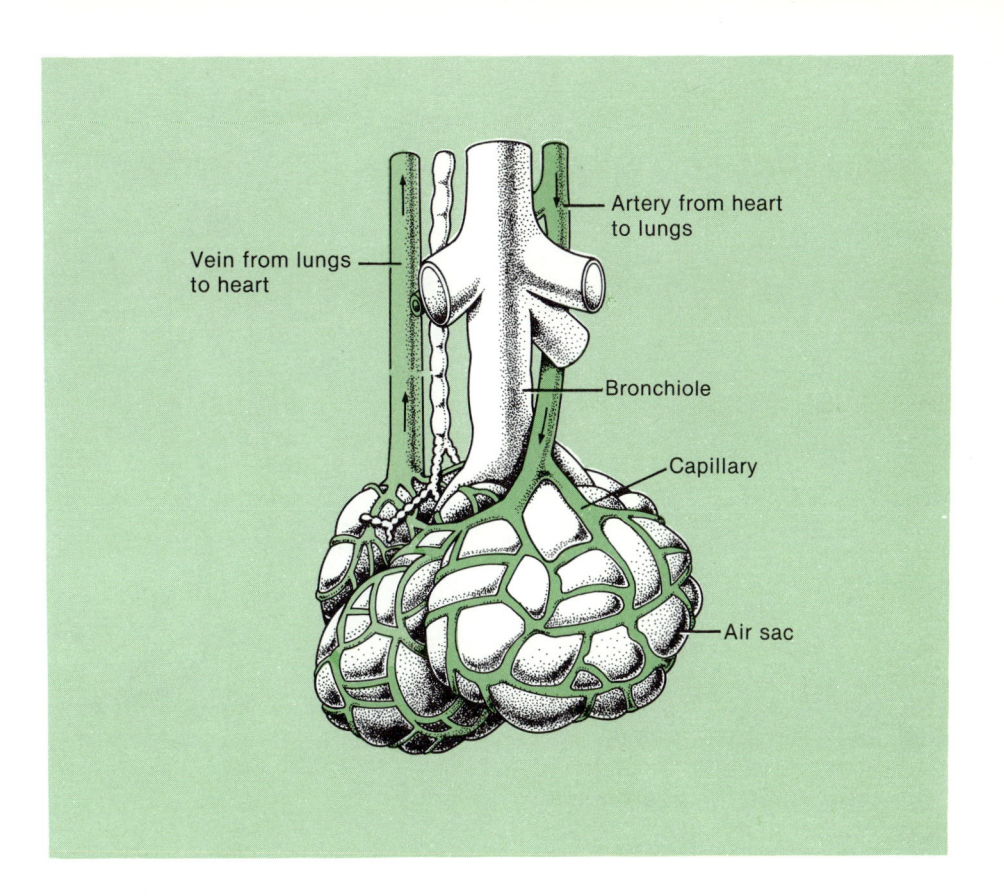

Fig. 25-8 Gas exchange takes place between the air sacs of the lungs and the blood in the capillaries.

Lung Circuit

The **lung circuit** puts oxygen into the blood. Blood returning to the heart from the legs, arms, head, and abdomen has little oxygen left in it. Also, it is rich in carbon dioxide. This blood returns to the top right chamber of the heart. It is called the **right atrium** (see also Figure 25-11). Then it moves through a one-way valve to the **right ventricle**. This chamber pumps the blood through arteries to the lungs.

In the lungs the arteries divide into **arterioles**. Then these divide into **capillaries**. The capillaries cover the small **air sacs** of the lungs (Fig. 25-8).

Carbon dioxide diffuses from the blood in the capillaries into the air sacs. Oxygen diffuses from the air sacs into the blood. The blood is now low in carbon dioxide and rich in oxygen.

The blood now moves through **venules** and into veins. These veins take the blood back to the top left chamber of the heart. It is called the **left atrium**.

Systemic Circuit

Blood from the lung circuit is now ready to be sent on a circuit through the body. This circuit is called the **systemic circuit**. The blood moves from the **left atrium** through a one-way valve into the **left ventricle**. This chamber pumps the blood into a large artery

Fig. 25-9 The green colour shows the main arteries and veins that serve the heart muscles.

called the **aorta**. Branches from the aorta take blood to all parts of the body as shown in Figure 25-7.

As in the lungs, arteries divide into arterioles. These, in turn, divide into capillaries. The capillaries supply cells with nutrients and oxygen, and pick up wastes from the cells. Then the capillaries unite to form venules. These, in turn, unite to form veins. The blood follows these veins back to the **right atrium**. The blood now begins the lung circuit again.

Coronary Circuit

The first arteries to branch off the aorta are the **coronary arteries**. They take blood to the muscles of the heart itself (Fig. 25-9). Sometimes these arteries get blocked. Then the person has a **"coronary" heart attack**. Sometimes they get partly blocked. Then the person has bad chest pains, particularly when physically active.

Doctors can now perform **coronary bypass operations**. These help people with blocked or partly blocked coronary arteries. A long vein is taken out of the patient's leg. It is used to replace the damaged coronary arteries.

Section Review

1. What is the function of the lung circuit?
2. Describe the path of blood in the lung circuit.

3. What is the function of the systemic circuit?
4. Describe the path of blood in the systemic circuit.
5. What are coronary arteries?
6. What is a "coronary" heart attack?
7. What is a coronary bypass operation?

25.3 The Lymphatic System

Your body has another system of tubes besides the blood circulatory system. It is called the **lymphatic system** [lim-FAT-ik] (Fig. 25-10).

As blood is pushed through capillaries, the liquid portion is squeezed out of the capillaries. This liquid is called **plasma**. Some of this plasma diffuses back into the capillaries. The rest builds up among the body cells. Tubes called **lymph ducts** pick up this plasma. They return it to the blood stream near your left and right shoulders.

The fluid in the lymph ducts is called **lymph** [limf]. It is mainly plasma. However, it also contains white blood cells. These cells help

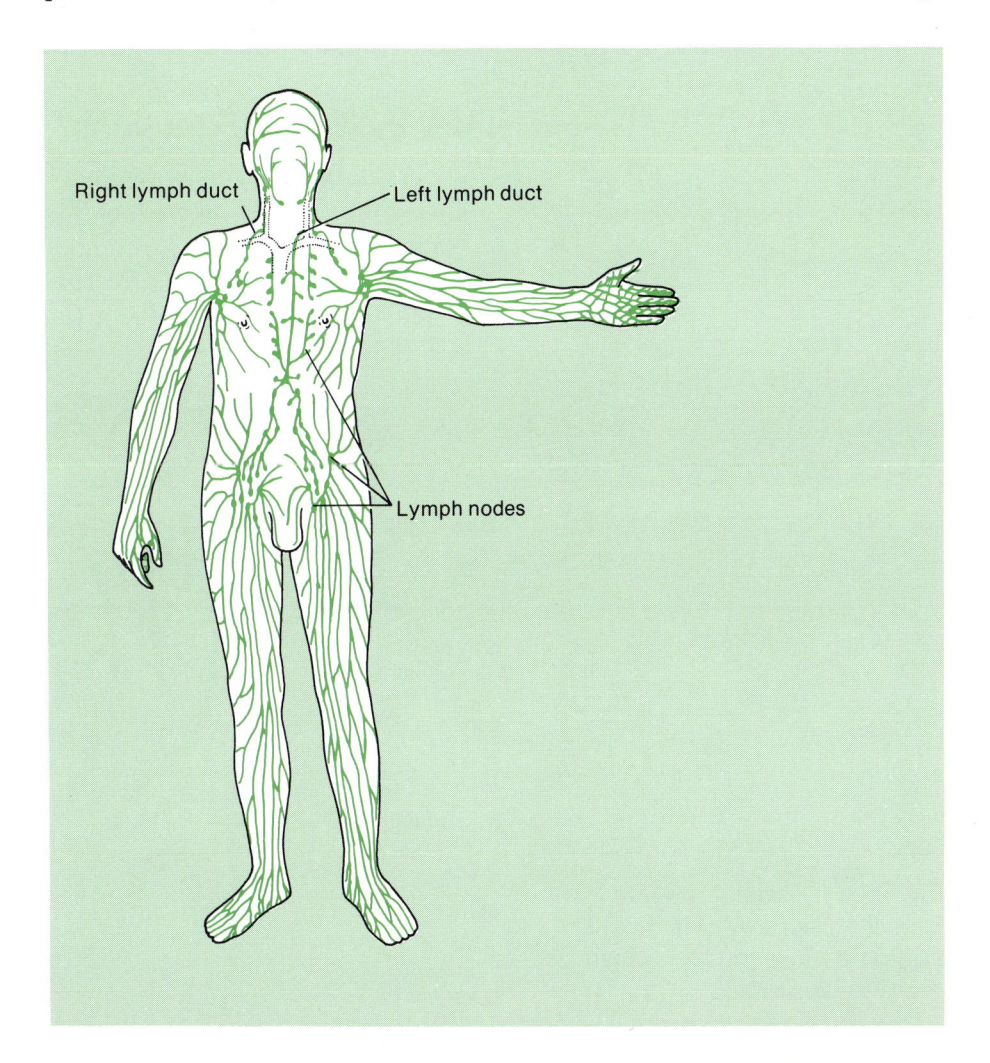

Fig. 25-10 The lymphatic system. The green lines show the main lymph ducts.

Section 25.3　361

your body fight infections. You have likely seen lymph. It is the watery substance that oozes from old wounds, scrapes, and blisters.

Sections of lymph ducts are enlarged to form **lymph nodes**. These destroy bacteria and many other infectious organisms. Figure 25-10 shows the main lymph nodes. Among these are the tonsils. Sometimes lymph nodes swell up. They may even hurt. We call this condition **"swollen glands"**.

Section Review

1. What is plasma?
2. What do lymph ducts do?
3. What is lymph?
4. What are lymph nodes?
5. What are swollen glands?

25.4 The Heart

Your heart pumps about 130 mL of blood for each beat. This amounts to about 7 L in 1 min. How does the heart pump blood?

How the Heart Functions

You read a little about how your heart pumps blood in Section 25.2. Follow Figure 25-11 as we look at this more closely.

Your heart is really two pumps in one. One pump is for the lung circuit. The other pump is for the systemic circuit. The right side of the heart receives blood from the body and pumps it to the lungs. The left side receives blood from the lungs and pumps it to the body.

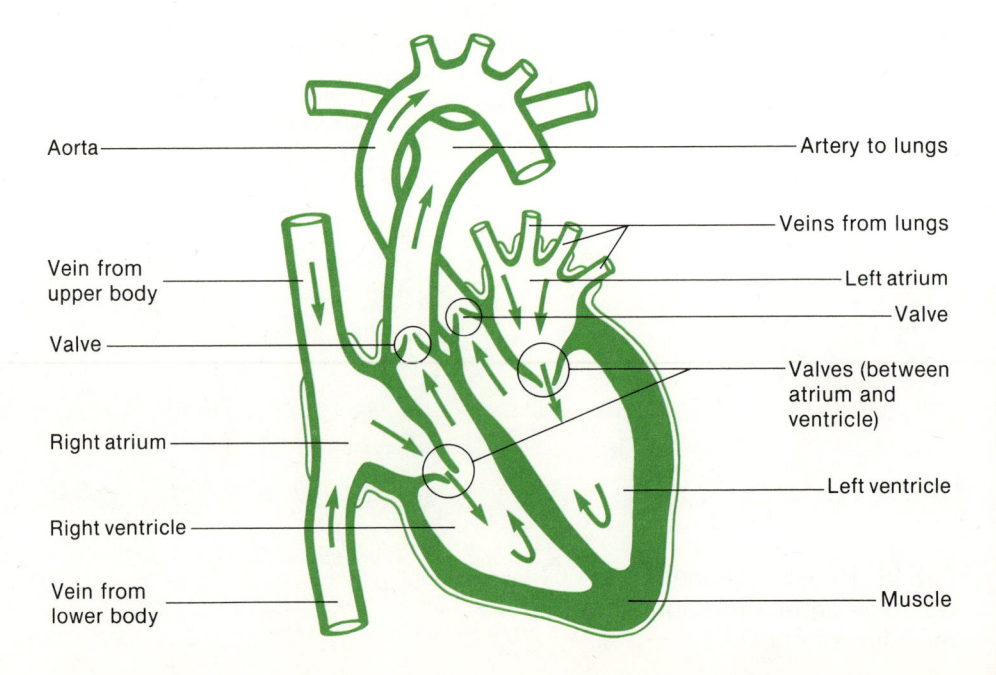

Fig. 25-11 The heart. Note the four chambers. Why is the left ventricle more muscular than the right ventricle?

A

First heart sound as these slap shut

B

Second heart sound as these slap shut

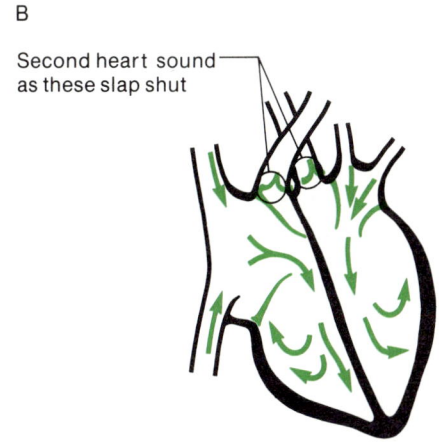

Fig. 25-12 The heartbeat. Why do the valves close?

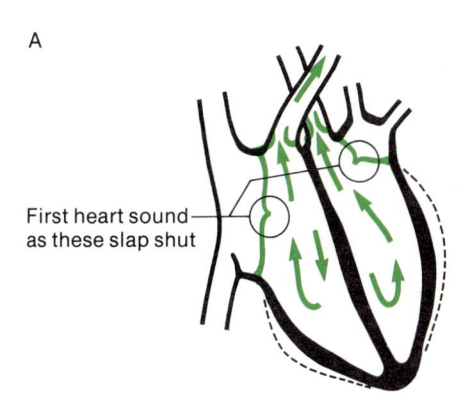

Blood returns from the body to the **right atrium**. Then the right atrium contracts. This forces the blood through a one-way valve into the **right ventricle**. Now the right ventricle contracts. This forces the blood through another valve into an artery that goes to the lungs.

The blood returns from the lungs to the **left atrium**. Then the left atrium contracts. This forces the blood through a one-way valve into the **left ventricle**. Now the left ventricle contracts. This forces the blood through another valve into the **aorta**. The aorta, you may recall, is the main artery that takes blood to the body.

The Heartbeat

"Lub-dub . . . lub-dub . . . lub-dub . . . " Have you ever wondered what causes that familiar sound?

The "lub" sound is caused by the closing of the valves between the atria and ventricles. They close when blood is being forced from the ventricles into the arteries (Fig. 25-12,A). The "dub" sound is caused by the closing of the valves between the ventricles and arteries. They close when blood is being forced from the atria into the ventricles (Fig. 25-12,B). These valves prevent the blood that was just forced out of the ventricles from flowing back to where it came from.

For adults, the average resting heart rate is 72 beats/min. For young children it is as high as 100 beats/min. During exercise, the rate can go over 170 beats/min. A long distance runner may have a resting heart rate as low as 40 beats/min.

Section Review

1. Your heart is really two pumps in one. What does this mean?
2. Explain how the heart functions.
3. What causes the "lub-dub" sound of the heartbeat?

25.5 ACTIVITY Listening to Your Heart

You can hear the valves of your heart close by using a stethoscope. By moving it to different places on the chest, you can hear the valves separately. The heart, lungs, and bones all muffle and deflect the sounds. Therefore the best places for listening are not directly over the actual valves. Figure 25-13 shows where you should listen.

Problem

Can you find the sound of each valve in your heart?

Materials

stethoscope

Fig. 25-13 Finding the heart sounds

- Actual location of valve
- Area where valve is best heard

Valve between left ventricle and aorta

Valve between right atrium and ventricle

Valve between right ventricle and artery to lungs

Valve between left atrium and ventricle

Fig. 25-14 Use the stethoscope as shown here.

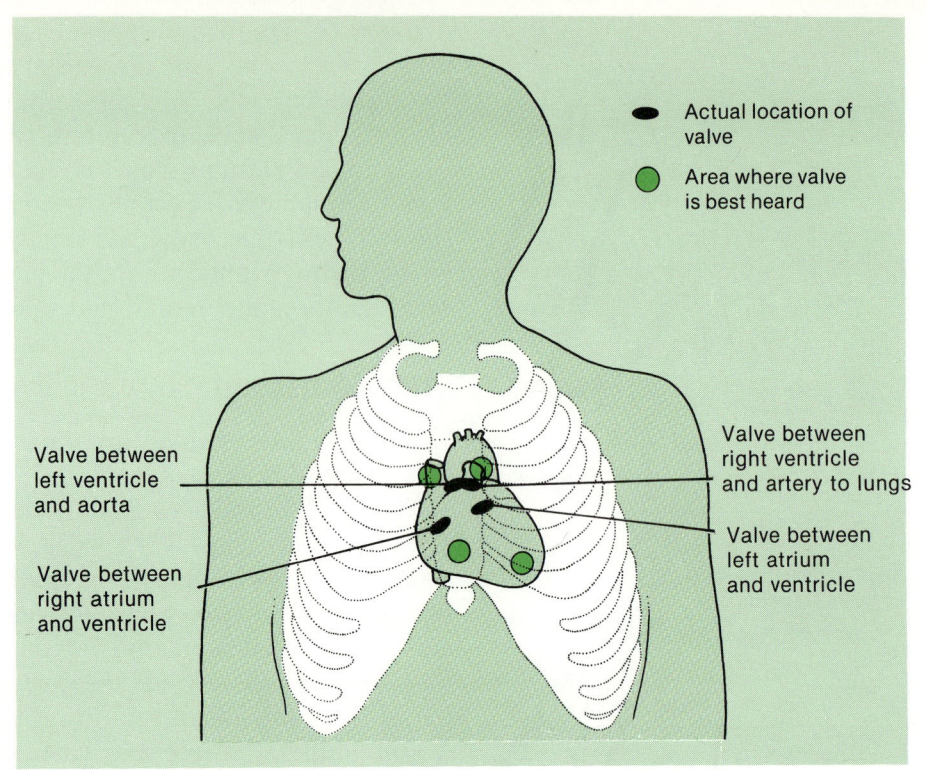

Procedure

a. Place the stethoscope in the centre of your chest (Fig. 25-14).
b. Try other locations around the centre of your chest.
c. Now try to listen to one of the four valves separately. Pick a valve in Figure 25-13. Then find on your body the area where the valve is best heard. Listen there with the stethoscope.
d. Repeat step (c) for each of the 3 other valves.
e. Run on the spot for about 30 s. Repeat step (c).
f. See if you can prove that the first sound ("lub") is caused by the closing of the valves between the atria and ventricles.

Discussion

1. Describe the sound made by each of the 4 valves.
2. How did the sounds change after you ran?
3. Describe what you did in step (f).

25.6 ACTIVITY Studying Your Pulse

Each time your ventricles contract, the blood pressure is raised for a moment in your arteries. This causes your **pulse**. You can feel this pulse almost anywhere an artery can be squeezed by a finger.

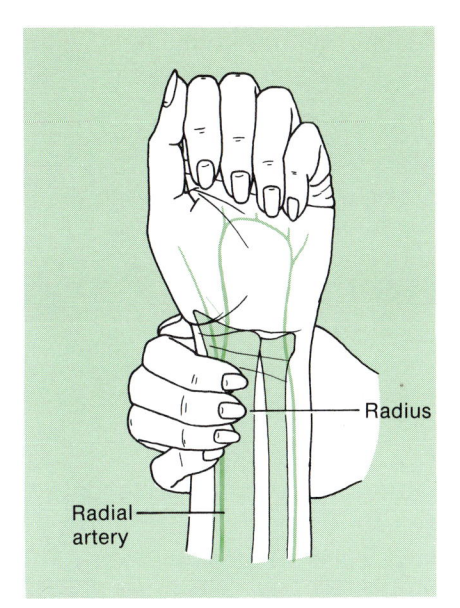

Fig. 25-15 Finding your pulse. This pulse is called the radial pulse. That's because it occurs in the radial artery that runs along the radius, a bone in your arm.

Problem

Can you find your pulse? What factors change your pulse?

Materials

stethoscope   watch or timer that reads in seconds

Procedure

a. Find your **radial pulse** as shown in Figure 25-15. Do *not* use your thumb. Its pulse can often be felt.
b. Find the radial pulse of your other arm.
c. Check for a pulse at your temple, jaw, neck, arm, and ankle. Use Figure 25-16 to help you find the locations. To find the pulse at your ankle, cross one leg over the other. Also, try changing the angle of the foot.
d. Find your ankle pulse with one hand. At the same time, find your neck pulse with the other hand. Do they occur at the same time?
e. Listen to your heartbeat with the stethoscope. At the same time, find your radial pulse. Does your pulse correspond to the first or second heart sound (the "lub" or the "dub")?
f. Find out if your radial pulse is exactly in time with your heartbeat.
g. Repeat step (f) for your ankle pulse.
h. Find your **pulse rate**. Do this as follows: Take your radial pulse for 20 s. Then multiply by 3. This gives your pulse rate in beats per minute.

Discussion

1. a) Which occurs first, the ankle or neck pulse?
   b) Which pulse location is closest to the heart?
   c) Why are the ankle and neck pulses not in time with each other?
2. a) What is a pulse?
   b) What is your pulse rate when resting?
   c) Is your pulse rate the same as your heartbeat rate?
   d) Why can your pulse be used to measure heartbeat rate?

25.7 ACTIVITY Finding Your Heart Performance Score (HPS)

When you exercise, your heartbeat rate, or pulse rate, increases. The amount that it increases depends on how fit you are. A strong heart pumps blood more efficiently than a weak one. Therefore the pulse rate of a fit person will not increase as much with exercise as will that of an unfit person.

How fit is your heart? You can find out by finding your **Performance Score (HPS)**.

Problem

What is your heart performance score? What is your general fitness level?

Materials

watch or timer that reads in seconds

Procedure

Note: Find all pulse rates as follows: Take the radial pulse for 20 s; then multiply by 3. This gives the pulse rate (in beats per minute).

a. Copy Table 25-1 into your notebook. Record all measurements in this table. Sample measurements have been placed in our table.

Table 25-1 Finding Your HPS

Pulse rate while lying down	70
Pulse rate while standing	85
Difference between the above	15
Pulse rate after exercising	130
Pulse rate after resting	105
Your HPS (add the 5 numerals)	405

b. Lie down for 2-3 min. Find your pulse rate while lying down.
c. Stand up and wait for 10 s. Then find your pulse rate.
d. Calculate the difference between the two pulse rates you have measured.
e. Exercise as follows: Obtain a chair about 0.5 m high. Have your partner hold it securely. Then, raise your left leg and step up on the chair (Fig. 25-17). Stand right up on it with both feet. Then get off the chair using your right leg. Switch legs after every 5 complete steps. (A complete step is an up and down motion.) Continue this exercise for 60 s. You must do about 15 complete steps in the 60 s.
f. Immediately find your pulse rate.
g. Rest for 60 s. Then find your pulse rate again.
h. Calculate your HPS by adding the 5 numerals you have put in your table.

Discussion

1. Find your general fitness level from Table 25-2.
2. What is the average HPS for your class?
3. What is the average HPS for the boys? the girls?

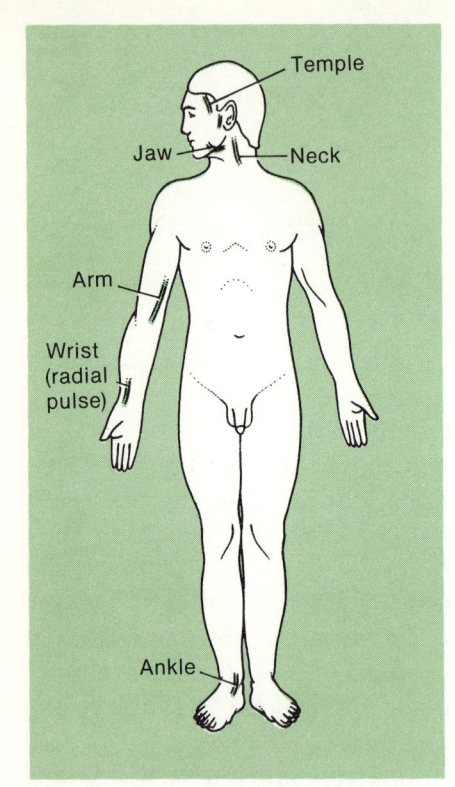

Fig. 25-16 Locations for finding your pulse

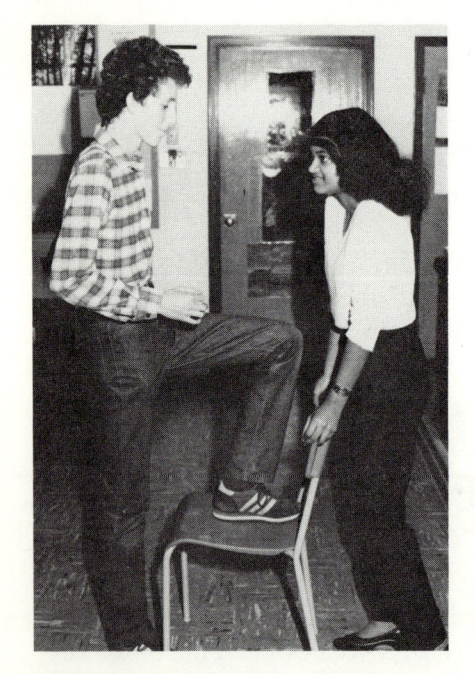

Fig. 25-17 Hold the chair securely. Step up as described in step (e).

4. Compare your HPS to the class average. Try to account for any difference.
5. Compare your HPS to the average for your sex. Try to account for any differences.
6. Why does the heartbeat rate increase during exercise?

Table 25-2 HPS and Fitness Level

HPS	General fitness level
200-250	Endurance athlete
250-300	Athletic
300-325	Very good
325-350	Good
350-375	Fairly good
375-400	Fair
400-450	Poor
450-500	Very poor

25.8 Blood

You have at least 4 L of blood in your body. Part of it is a liquid called **plasma**. You read about plasma in Section 25.3. The rest of the blood is a solid (Fig. 25-18). This solid consists of **blood cells**.

Blood Cells

There are three types of blood cells: red blood cells, white blood cells, and platelets (Fig. 25-19).

Red Blood Cells The mature **red blood cells** of mammals, including humans, have no nuclei. Yet each cell functions well for almost four

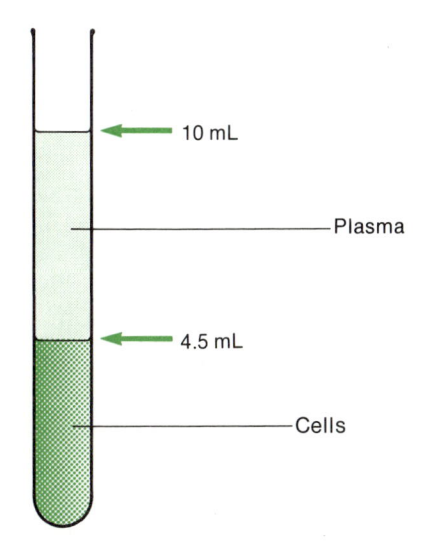

Fig. 25-18 Every 10 mL of blood contains 4.5 mL of cells (red cells, white cells, and platelets). Plasma is largely water. It contains dissolved proteins and minerals.

Fig. 25-19 Types of human blood cells

Type of cell	Size and shape	Number in 4L of human blood
Red blood cell		2000×10^{10}
White blood cell		3×10^{10}
Platelet	7 μm	120×10^{10}

Section 25.8

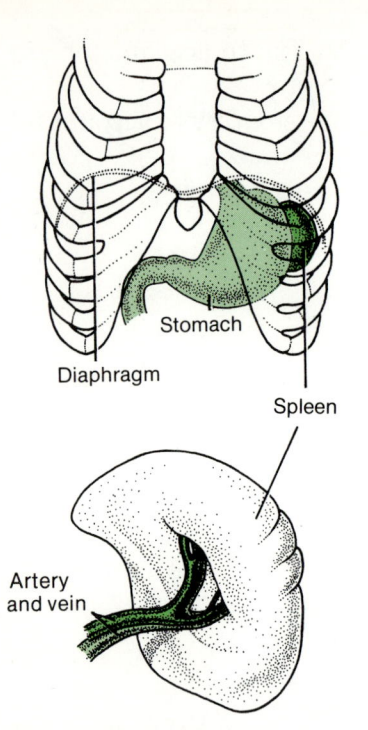

Fig. 25-20 The spleen. It stores red blood cells. It also destroys old red blood cells.

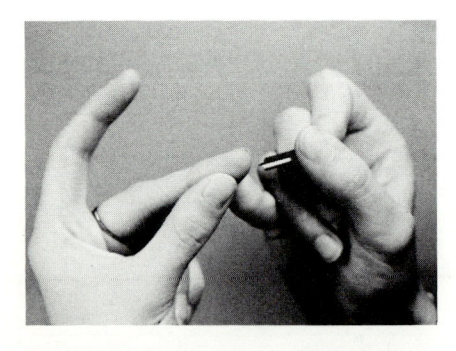

Fig. 25-21 How to use a lancet. Prop the wrists of both hands on the edge of the desk. The lancing hand should dart firmly toward the middle finger of the other hand. *Use the lancet only once!*

months. Red blood cells are shaped like a donut with no hole. They contain molecules of **hemoglobin** [HE-mow-glow-bin]. These molecules carry oxygen throughout the body.

Hemoglobin is dull red in colour. When the blood reaches the lungs, the hemoglobin picks up oxygen. This turns the hemoglobin bright red. Thus blood in your arteries is bright red. In the tissues, oxygen leaves the hemoglobin. Now the blood turns dull red again. Thus blood in veins is dull red. (Layers of skin make the veins look blue.)

Hemoglobin molecules contain **iron** atoms. Therefore your body cannot make hemoglobin without iron. People without enough iron are often pale and tired. Do you know why?

Spare red blood cells are stored in your **spleen** (Fig. 25-20). If you lose blood, your spleen pours extra cells into your blood. Old cells are destroyed by the spleen and liver. New red blood cells are made in the **bone marrow**.

White Blood Cells **White blood cells** are similar to amoebas. They have nuclei; they are colourless; they can change shape; they can move in any direction. There are several types of white blood cells. Some stay outside the capillaries in the plasma between the body cells. Some are in the lymph of the lymph vessels and nodes.

Like amoebas, white blood cells engulf particles such as bacteria. In this way they protect the body from disease. You have likely seen **pus** near a cut. This pus is white blood cells that died while fighting an infection.

Platelets Strictly speaking, **platelets** are not true cells. They have no nuclei. And, they are only 2 or 3 μm in diameter. They speed up the clotting process in blood.

Plasma

Plasma is over 90% water. The rest is dissolved substances such as nutrients, wastes, special proteins, hormones, and minerals. You may have seen this clear, straw-coloured liquid oozing from a reopened cut or scrape.

Nutrients are taken to the body cells by the plasma. Then cell **wastes** such as urea and ammonia are taken from the cells by the plasma.

Many special **proteins** are in the plasma. Some of these help the blood to clot. Others, called **antibodies**, help protect the body against disease (see Chapter 10).

The plasma also contains **hormones**. These chemicals control the body processes. **Adrenalin** is a hormone that you may have heard of. It causes the reactions of fright: cold sweat, racing heartbeat, and fast shallow breathing.

Most of the **minerals** you need are carried to your cells in the plasma. These include sodium, chloride, calcium, and potassium. You need such elements to function properly. For example, calcium is needed for proper blood clotting and bone growth.

Section Review

1. Name the two main parts of blood.
2. Describe the function of red blood cells.
3. Why is it important for you to have iron in your diet?
4. What roles do your spleen and bone marrow play in your body?
5. What is the function of white blood cells?
6. What is the function of platelets?
7. List the 5 kinds of dissolved substances in plasma.

25.9 ACTIVITY Examining Your Blood Cells

It is easy to see red blood cells under a microscope. However, platelets and white blood cells are hard to see. They must be stained first.

Problem

Can you find blood cells with a microscope?

Materials

microscope slides (2)
Wright's stain
sterile, disposable blood lancet
rubbing alcohol
cotton batting
dropper
microscope
distilled water

CAUTION: Use the lancet once only. Discard it after using. Never share it with another person. You could spread blood diseases.

Procedure

a. Clean and dry the 2 microscope slides. Handle them by the edges only.
b. Wash your hands well. Now wipe the end of a middle finger with the alcohol.
c. Move blood towards your fingertips by shaking your hand for 10 s.
d. Make a small puncture in the fleshy part of the fingertip with the lancet (Fig. 25-21). Discard the lancet after using.
e. Place a drop of blood near the end of one slide. Then cover the puncture with a cotton ball soaked in alcohol.
f. Use the other slide to make an even smear across the first slide as shown in Figure 25-22.
g. Let the smear dry completely to a dull finish.
h. Cover the entire smear with a few drops of Wright's stain.
i. After 2 min, add drops of distilled water until a golden sheen appears.
j. Let the slide sit undisturbed for 13 min.
k. While you are waiting, study the prepared slides as described in steps (m) to (o).

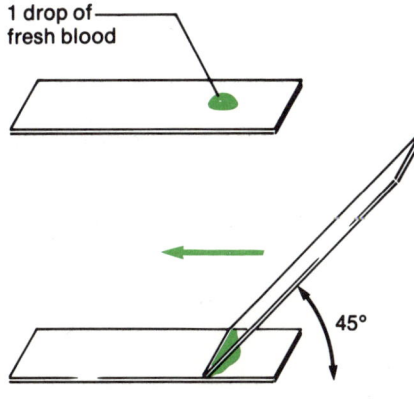

Fig. 25-22 Making a smear. Hold the slide at 45°. Push it *away* from the blood.

l. After the 13 min, gently rinse the slide with distilled water. Wipe the bottom of the slide dry. Let the top air dry.
m. Look at the slide under low, medium, then high power. (You do not need a coverslip.)
n. Find and describe red blood cells and white blood cells. You should see at least two types of white blood cells. They differ in nuclear shape and amount of cytoplasm. Look for platelets between the red and white cells.
o. Make a sketch of a red blood cell and two kinds of white blood cells.

Discussion

1. What is the purpose of the stain?
2. Why is your finger wiped with alcohol?
3. Why do you not share your lancet with someone else?
4. Why are the centres of red cells so much lighter in colour?
5. How do white blood cells differ from red blood cells?
6. Why are platelets so hard to see?

25.10 Blood Types and the Rh Factor

Blood Types

Your blood cells may look like those of other people. However, all human blood is not exactly the same. In fact, there are four main types of blood. They are called **Type A, B, AB, and O**.

Do you know your blood type? Your doctor needs to know it before an operation. When some blood types are mixed, a clumping of red blood cells occurs. Suppose you were given the wrong type of blood during an operation. Large clots will form, and you could die. Table 25-3 shows which types can be safely mixed.

Type O blood can be given to people with the other three blood types. But a person with Type A blood can be given only Type O or Type A blood.

People with Type AB blood can receive blood of all four types.

Table 25-3 Mixing Blood Types

Your blood type	You can be given blood of this type	You can give your blood to a person of this type
A	A and O	A and AB
B	B and O	B and AB
AB	A, B, AB, and O	AB
O	O	A, B, AB, and O

The Rh Factor

Certain proteins in the blood give the blood what is called the **Rh Factor**. About 85% of Canadians have **Rh positive (Rh+)** blood. The other 15% have **Rh negative (Rh−)** blood. Rh+ blood has certain proteins. Rh− blood lacks those proteins.

One in every 300 babies born in Canada has an Rh− mother and a Rh+ father. If the baby inherits the father's Rh+ blood, its life may be in danger. When an Rh− mother carries a Rh+ baby, antibodies from the mother can destroy the baby's red blood cells. The baby must have a transfusion at birth. This problem can also be averted before the birth of the baby by injections given by a doctor.

Section Review

1. What are the 4 main blood types?
2. What can happen if the wrong blood types are mixed?
3. Why is the Rh factor important?

25.11   ACTIVITY   Finding Your Blood Type

In this activity you will find out what your blood type is.

Problem

Is your blood Type A, B, AB, or O?

Materials

microscope slides (2)
anti-A and anti-B sera
   (singular: scrum)
disposable, sterile blood lancet

rubbing alcohol
cotton batting
marking pen
toothpicks

CAUTION: Use the lancet once only. Discard it after using. Never share it with another person. You could spread blood diseases.

Procedure

a. Clean and dry both slides. Label one as "1" and the other as "2".
b. Wash your hands well. Now wipe the end of a middle finger with alcohol.
c. Move blood towards your fingertip by shaking your hand for 10 s.
d. Make a small puncture in the fleshy part of the fingertip with the lancet (see Figure 25-20). Discard it after using!
e. Discard the first drop of blood. Then place one drop in the centre of each slide. Then cover the puncture with a cotton ball soaked in alcohol.

Section 25.11   371

f. Add 1 drop of anti-A serum beside the blood on slide 1.
g. Add 1 drop of anti-B serum beside the blood on slide 2.
h. Use a toothpick to mix the two drops on slide 1.
i. Use a new toothpick to mix the two drops on slide 2.
j. Wait 2 min for reactions to occur.
k. Record your observations in a table like Table 25-4.
l. Wash the slides.

Table 25-4 Blood Typing

Slide	Anti-serum used	Sketch after 2 min
1	Anti-A	
2	Anti-B	

Discussion

1. Find your blood type using Table 25-5. Share your results with the class.
2. How do your class results compare with the average for Canada?
   46% have Type O blood
   42% have Type A blood
   9% have Type B blood
   3% have Type AB blood

Table 25-5 Blood Type Determination

If the reaction with anti-A produced:	and	If the reaction with anti-B produced:	Your blood type is:
No change		No change	O
No change		Black dots	B
Black dots		No change	A
Black dots		Black dots	AB

25.12 Care of Your Circulatory System

Do you know anyone who has had a stroke or heart attack? Or, do you know anyone with high blood pressure? You probably do. Diseases of the circulatory system are very common. In fact, *heart disease is the number one killer in Canada*. Heart disease and other diseases of the circulatory system cause over half of all deaths in Canada. Some people inherit diseases of the circulatory system. But most of these diseases seem to be related to lifestyle — diet, exercise, rest, smoking, and tension.

Fig. 25-23 Exercises that make you puff for 15-20 min are best for your circulatory system.

Diet

A proper diet is essential for a healthy circulatory system. You need iron, proteins, and minerals to build red blood cells. You need calcium for flexible arteries. But you don't need much salt. It causes **high blood pressure (hypertension)** in many people. Also, you should not eat excessive amounts of animal fats. They may help deposit **cholesterol** in your arteries. This can block the arteries. If the heart's arteries are blocked, you will have a **heart attack**. If arteries in the brain are blocked, you will have a **stroke**.

You should eat well-balanced meals that keep your weight normal. An overweight body puts an extra load on the heart.

Exercise

Regular exercise such as running, jogging, and cross-country skiing helps control your weight (Fig. 25-23). It also strengthens your heart and arteries. Further, it helps relieve tension and helps you sleep. All of these factors build a healthier circulatory system.

Smoking

Smoking raises your blood pressure and increases your pulse rate. In other words, it makes your heart work harder. Yet, at the same time, smoking cuts down on the amount of oxygen delivered to your heart.

Smokers have three times the chance of a heart attack as non-smokers. Also, smoking plays a major role in strokes. And, it seems to promote hardening of the arteries.

Rest and Tension

It's easy to say "get lots of rest; relax." However it isn't always as easy for someone to do those things. Yet adequate rest and a relaxed outlook on life are important for a healthy circulatory system. Continued tension and lack of enough sleep can raise the blood pressure.

Here are some suggestions. Don't smoke. Avoid stimulants like coffee, particularly from mid-afternoon on. Exercise regularly. Eat well-balanced, regular meals. Set aside times in the day for relaxation. All these, combined, will help relieve tension and promote sleep.

Section Review

1. Outline the importance of diet for a healthy circulatory system.
2. Name three exercises that promote fitness of the circulatory system.
3. Describe the effects of smoking on the circulatory system.
4. Why is it important to have enough rest and relaxation?

Main Ideas

1. The circulatory system has four main parts: blood and lymph, tubes, capillaries, and the heart.
2. The body has two main blood circuits: the lung circuit and the systemic circuit.
3. The lymphatic system fights infection.
4. The heart is really two pumps in one.
5. Physically fit people have a low HPS.
6. There are three kinds of blood cells: red cells, white cells, and platelets.
7. There are four main blood types: A, B, AB, and O.
8. A healthy circulatory system depends mainly on your lifestyle.

Glossary

arteriole	are-TEER-ee-ole	a small artery
atrium	AY-tri-um	a chamber of the heart that receives blood
hypertension		high blood pressure
lymph	limf	the fluid in the lymph ducts
platelets		tiny blood cells that speed up clotting
spleen		an organ that stores red blood cells and destroys old ones
ventricle	VEN-tri-kel	a chamber of the heart that pumps blood
venule	VEN-ule	a small vein

Study Questions

A. True or False

Decide whether each of the following statements is true or false. If the sentence is false, rewrite it to make it true. (Do not write in this book.)

1. The blood pressure in veins is higher than in arteries.
2. Leaky valves can cause varicose veins.
3. The left ventricle pumps blood to the lungs.
4. The first heart sound ("lub") is caused by the closing of the valves between the atria and ventricles.

B. Completion

Complete each of the following sentences with a word or phrase that will make the sentence correct. (Do not write in this book.)
1. Lymph consists of plasma and ▓▓▓ .
2. Blood is carried throughout the body by the ▓▓▓ circuit.
3. ▓▓▓ molecules carry oxygen in the red blood cells.

C. Multiple Choice

Each of the following statements or questions is followed by four responses. Choose the correct response in each case. (Do not write in this book.)
1. Tiny arteries called
    a) venules   b) lymph ducts   c) arterioles   d) capillaries
2. Blood is pumped to the body by the
    a) right atrium   c) right ventricle
    b) left atrium   d) left ventricle
3. Red blood cells are made in the
    a) spleen   b) liver   c) blood   d) bone marrow
4. Suppose a person with Type B blood needs a blood transfusion. This person could be given blood that is
    a) Type B or Type O   c) Type B only
    b) Type B or Type AB   d) Type O only

D. Using Your Knowledge

1. Imagine some blood in the right atrium of your heart.
    a) Describe the path it follows from the right atrium to the left atrium.
    b) What happens to the blood along this path?
2. Imagine some blood in the left atrium of your heart.
    a) Describe the path it follows from the left atrium to the right atrium. Assume that this blood all went to the intestines.
    b) What happens to the blood along this path?
3. A person had a resting pulse rate of 76 beats/min. After this person exercised for six months, this rate became 56 beats/min. How many fewer beats does the heart make in a day?
4. The blood helps to keep the body temperature constant. How?

E. Investigations

1. Design and do an experiment to study the effects of coffee on the pulse rate.
2. Design and do an experiment to study the effects of a certain exercise on the pulse rate.
3. An important lifesaving method is called cardio-pulmonary resuscitation (CPR). Find out what this is and how it is carried out.
4. Find out how blood pressure is measured. What is meant by

systolic pressure and diastolic pressure? What is a good reading for a person of your age?
5. Some people need an artificial heart pacemaker. Find out what this is. Also, find out why some people need them.
6. Two serious blood diseases are leukemia and anemia. Select one of these. Research its causes, symptoms, effects, and treatment.

26 The Excretory and Reproductive Systems

THE EXCRETORY SYSTEM
26.1   Excretory Organs
26.2   Care of Your Urinary System

THE REPRODUCTIVE SYSTEM
26.3   The Male Reproductive System
26.4   The Female Reproductive System
26.5   Fertilization and Development
26.6   Care of Your Reproductive System

The Excretory System

Your body makes a large amount of metabolic waste each day. (Metabolic waste is formed during processes such as respiration and digestion.) If this waste builds up in your body, it can poison you.

Your circulatory system acts as a "waste collection service". It collects wastes and carries them to "dump sites". These "dump sites" are called **excretory organs**. **Excretion** is the process by which your body gets rid of metabolic waste. The first half of this chapter deals with the excretory organs and how they work.

26.1 Excretory Organs

The **excretory organs** are the lungs, skin, liver and kidneys.

The Lungs

Respiration produces two wastes. These are carbon dioxide and water. You learned in Chapter 24 how the **lungs** get rid of carbon dioxide. The lungs also help get rid of excess water. Breathe on a cold mirror. What do you see? How did this water get into your lungs?

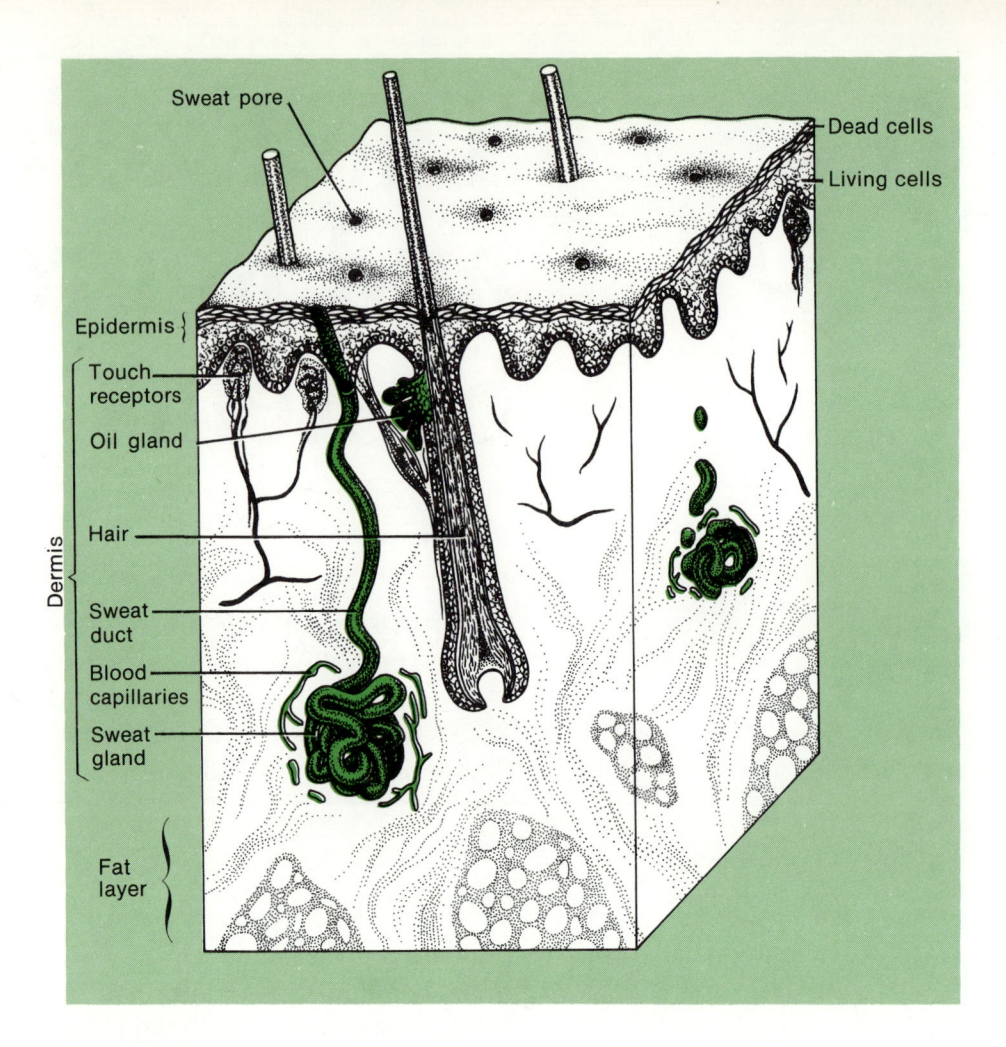

Fig. 26-1 Sweat glands in the skin help with excretion.

The Skin

Sweat glands in the skin help with excretion (Fig. 26-1). These glands are made of coiled tubes. They take water, salts, and urea from the blood. These substances are dumped out of the **sweat pores**. We call this liquid **sweat**.

Body odour is not caused by sweat itself. Instead, bacteria on the skin feed on the sweat. The wastes produced by the bacteria cause body odour.

As skin helps with excretion, it also helps cool the body. Evaporation of water from the skin uses up heat. This helps keep you cool.

The Liver

The **liver** has many important functions in your body. You learned in Chapter 23 that it makes bile. Bile helps in the digestion of fats. The liver also stores excess glucose. Also, it makes vitamin A and proteins. Further, it plays an important role in excretion.

During digestion, proteins are broken down into amino acids. Many of these amino acids are used by your body to make proteins. The ones that are not used are carried to the liver. It changes them into **urea**. The urea is put back into the blood. It is eventually taken from the blood by the kidneys and skin.

The Kidneys

The **kidneys** are the main excretory organs of the body (Fig. 26-2,A). The kidneys remove excess minerals and urea from the blood. They also control the water content of the blood.

The work of the kidneys is done by **nephrons** [NEF-rons]. Each kidney has about one million of these. The blood enters the nephrons. In the nephrons, wastes diffuse from the blood into **tubules**. These tubules take the wastes to the inside of the kidney. From there, the wastes go to the **ureter**. These wastes are called **urine**. Urine is about 96% water, 2% urea, and 2% wastes such as salt.

The kidneys are part of the **urinary system** (Fig. 26-2,B). The urinary system collects, stores, and eliminates urine. The kidneys form urine from substances in the blood. The urine travels down the ureters to the **bladder**. The bladder stores the urine for several hours. Then it is excreted through the **urethra** [you-REE-thra].

Section Review

1. What is excretion?
2. Describe the role of each of the following organs in excretion: lungs, skin, liver, kidneys.
3. Describe the urinary system.

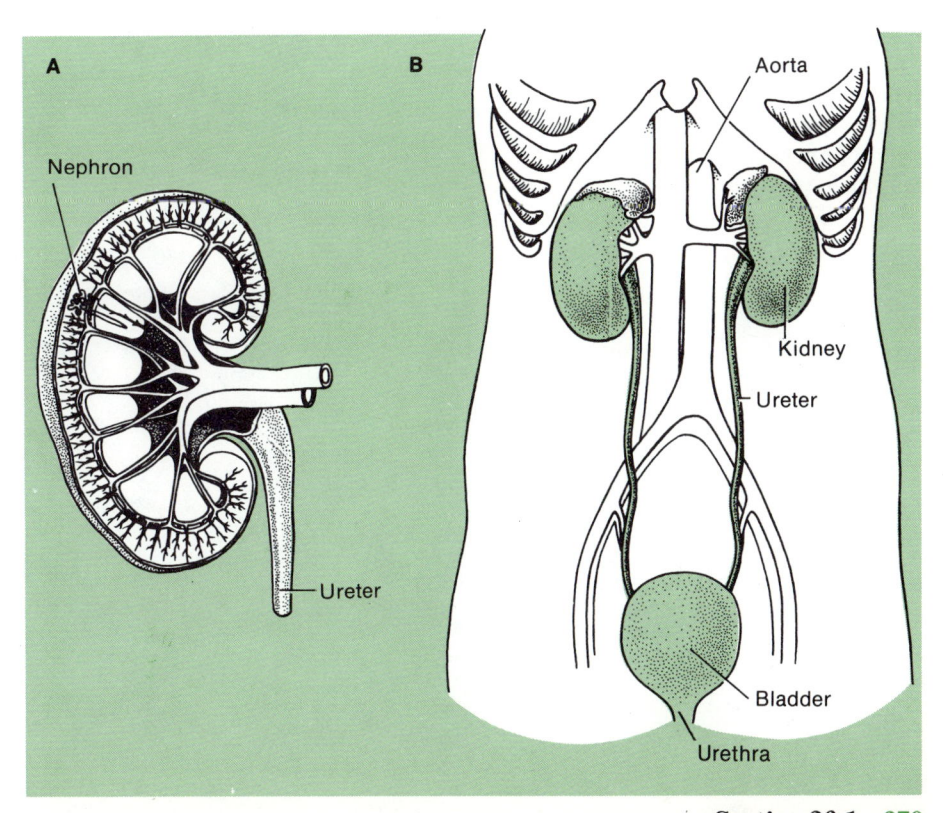

Fig. 26-2 The kidney (A) and urinary system (B)

26.2 Care of Your Urinary System

Your kidneys are busy! They recycle 95% of the water you drink. They filter wastes such as urea and salts from the blood. And they control the amount of glucose, salt, and other substances in the blood. But to do these jobs, your kidneys must be kept in good shape.

Excess salt can damage the kidneys. There is more than enough salt in cereals and other prepared foods. Don't add salt to any foods you are preparing.

Alcohol, when used regularly for a long period of time, can damage the kidneys. Heavy drinking bouts can do the same.

A diet that is too high in proteins puts a strain on the kidneys. And, too little water can cause parts of the kidneys to become clogged. Eat well-balanced meals and drink lots of water.

You may not think that smoking can affect your urinary system. But it can. Chemicals in the smoke get into your blood in the lungs. Some of these chemicals circulate to the urinary system. There they can cause cancer of the kidneys and bladder. Smokers have about 10 times the chance of getting cancer of the bladder as do non-smokers.

Kidney Machines

Sometimes a person's kidneys may stop working completely. They may have been damaged in an accident. They may have been

Fig. 26-3 An artificial kidney or kidney machine. Note the 2 tubes going to the patient's arm. One carries blood from an artery to the machine. The other carries purified blood from the machine to a vein.

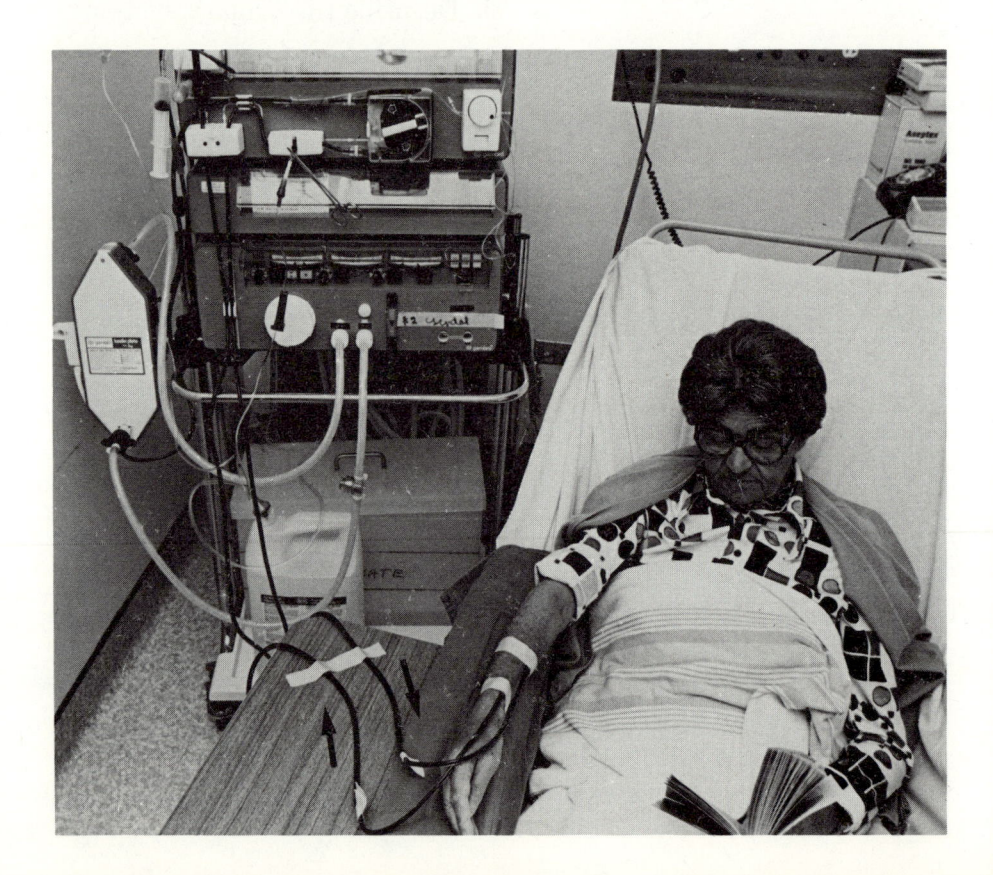

damaged by a disease. Or, they may have been damaged by a poor lifestyle. Regardless of the cause, a person cannot live for long without kidneys. In just 2 or 3 d the tissues become waterlogged. And poisonous wastes begin to affect vital organs.

Many hospitals now have **kidney machines**. The patient's blood is pumped through the machine. It filters out urea and other substances. Some patients need 2 or 3 treatments a week. Each treatment lasts for 4 to 6 h (Fig. 26-3).

Kidney machines use **dialysis tubing**. You used it in some activities in Chapter 6. Kidney machines are also called dialysis machines.

Kidney Transplants

The use of kidney machines in hospitals is inconvenient and expensive. Smaller units for home use have been developed. But they, too, are expensive and take a great deal of time out of one's life. A **kidney transplant** may solve these problems.

Figure 26-4 shows how a donated kidney is placed in a patient's body. The donated kidney must match closely the tissue type of the patient. Otherwise, the donated kidney will be rejected. Do you know where the donated kidneys come from?

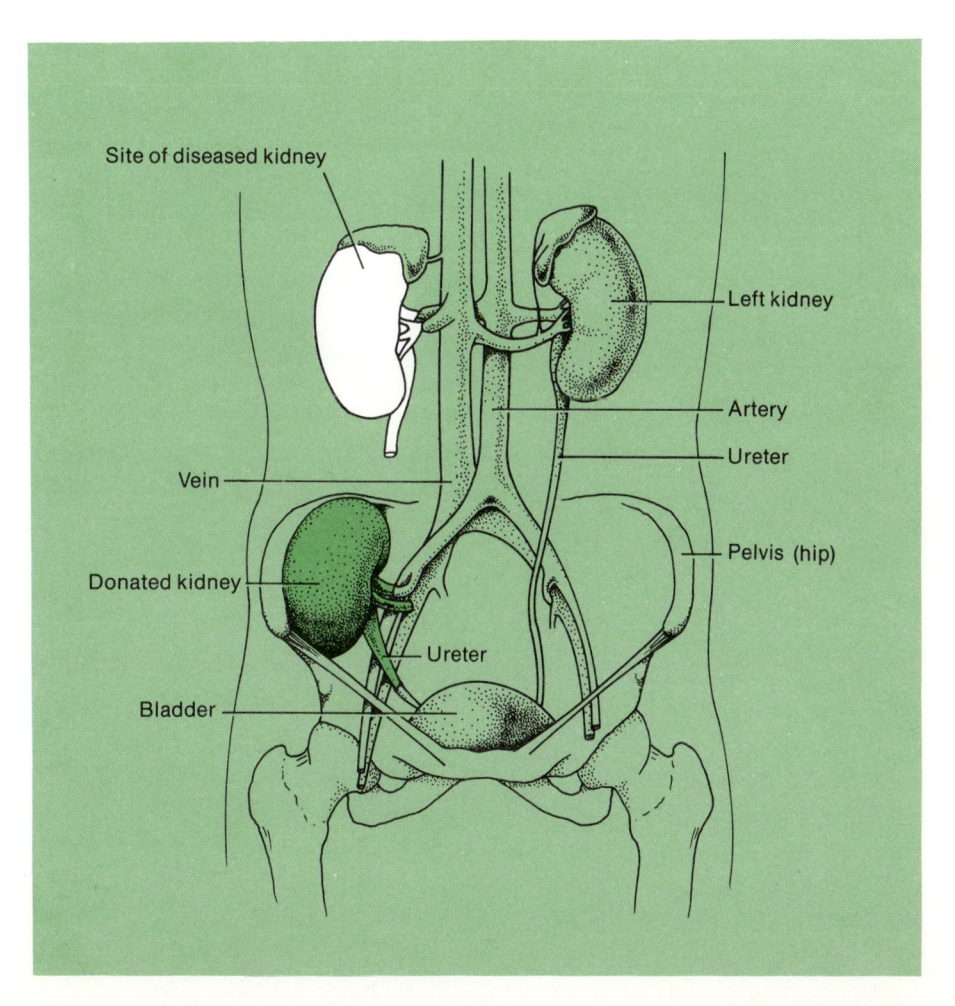

Fig. 26-4 A kidney transplant. Note how the donated kidney is "hooked up" into the circulatory system.

Section 26.2

Section Review

1. Describe 5 things that can damage kidneys.
2. What does a kidney machine do?
3. Why is a tissue match important in a kidney transplant?

The Reproductive System

So far in Unit 6 you have studied four systems: digestive, breathing, circulatory, and excretory. If one of them stops working, you would die without special medical help. This is not so, however, for the reproductive system. You don't need it to stay alive. But you need it to help create life.

Humans reproduce by **sexual reproduction**. This involves the union of a **male gamete (sperm)** with a **female gamete (egg)** (Fig. 26-5). The end-product is a new human.

26.3 The Male Reproductive System

The male reproductive system is shown in Figure 26-6. The main reproductive organs are the **testes** (singular: **testis**). They are also called **testicles**. They make the **male gametes**, or **sperm**. The testes are located outside the body cavity. They are in a sac of skin called the **scrotum**. The cooler temperature outside the body is needed for proper sperm production.

When the sperm are released from the testes, they travel down the **sperm ducts**. These ducts carry the sperm to the **urethra**. The urethra is in the **penis**. Note that the urethra also carries urine from the bladder during urination.

As the sperm passes along the sperm ducts and urethra, fluids are added to them by three glands. These are the **seminal vesicles** (sperm chambers), **prostate gland**, and **Cowper's gland**. The mixture of sperm and the fluids from these glands is called **semen** [SEE-men]. The semen is ejaculated from the penis through the urethra.

The testes make male sex hormones. These hormones give the typical characteristics of the mature male: thick facial and body hair, a deep voice, and muscular development.

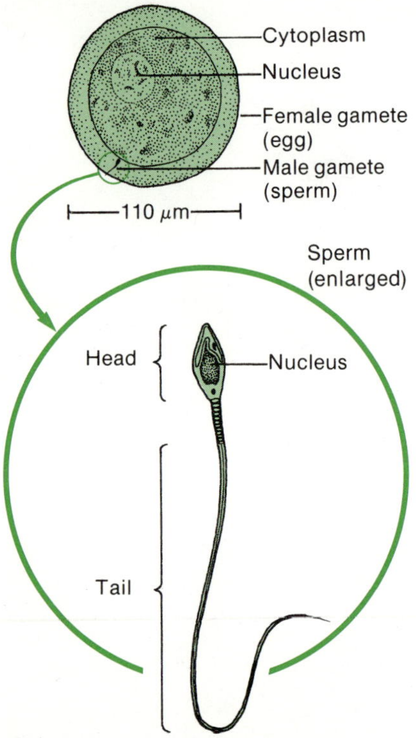

Fig. 26-5 The human gametes. The egg and sperm are about 1/200 as large as shown in the top drawing. An egg is barely visible to the unaided eye.

Section Review

1. Where are sperm made?
2. Describe the path followed by sperm during ejaculation.
3. What is semen?
4. How do male sex hormones affect the body?

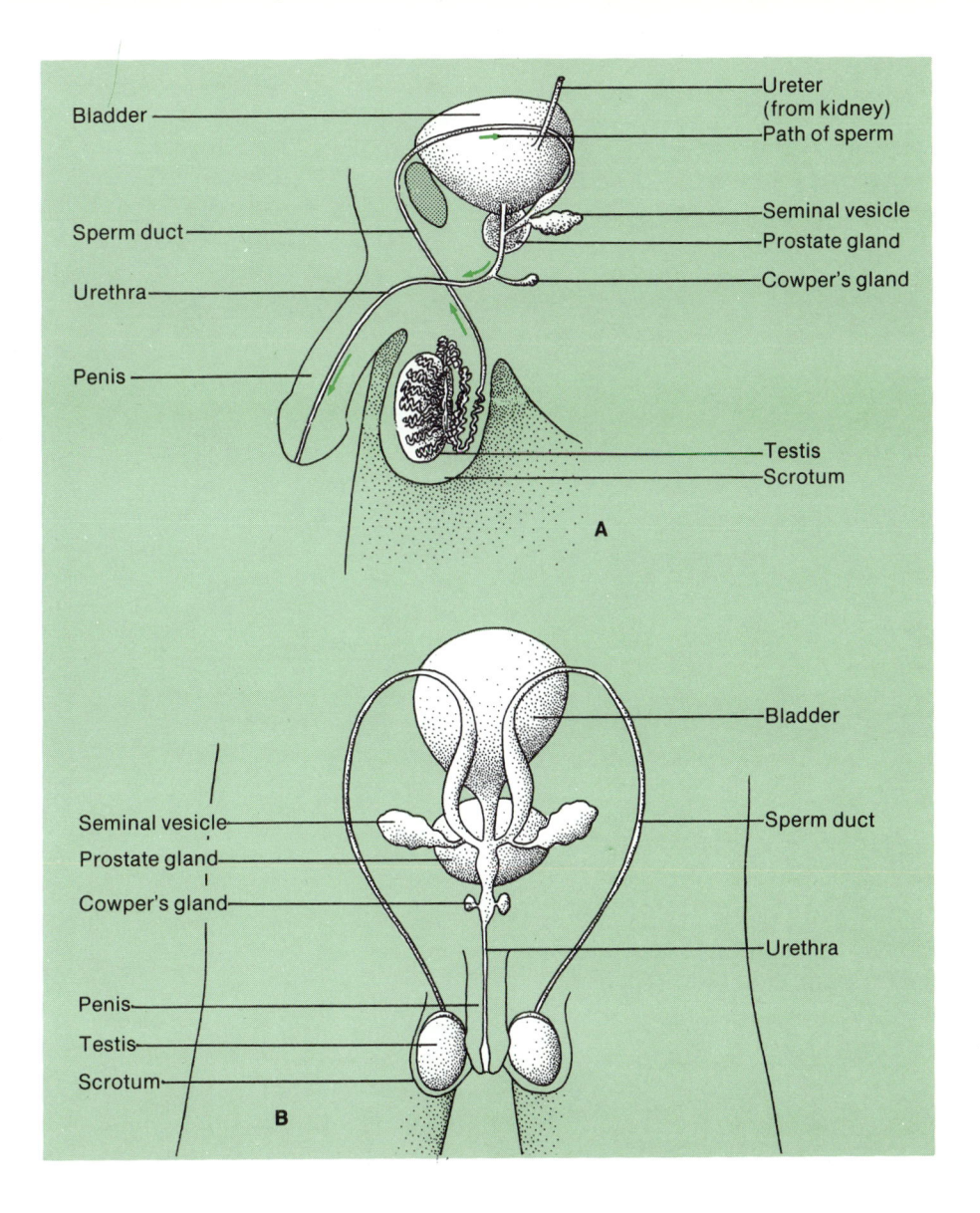

Fig. 26-6 The male reproductive system: side view (A); front view (B)

26.4 The Female Reproductive System

The female reproductive system is shown in Figure 26-7. The main reproductive organs are the **ovaries**. They make the **female gametes**, or **eggs**. Unlike the male testes, the ovaries are located within the body cavity. When a female is born, each ovary contains about 300 000 partly developed eggs.

The ovaries are not connected to the **oviducts (Fallopian tubes)**. However, the oviducts are lined with cilia (tiny hairs). When an egg is released by an ovary, the beating of the cilia draws the egg into an oviduct. The egg then moves down the oviduct to the **uterus** [YOO-ter-us].

If the egg is not fertilized, it passes through the neck of the uterus,

Section 26.4 383

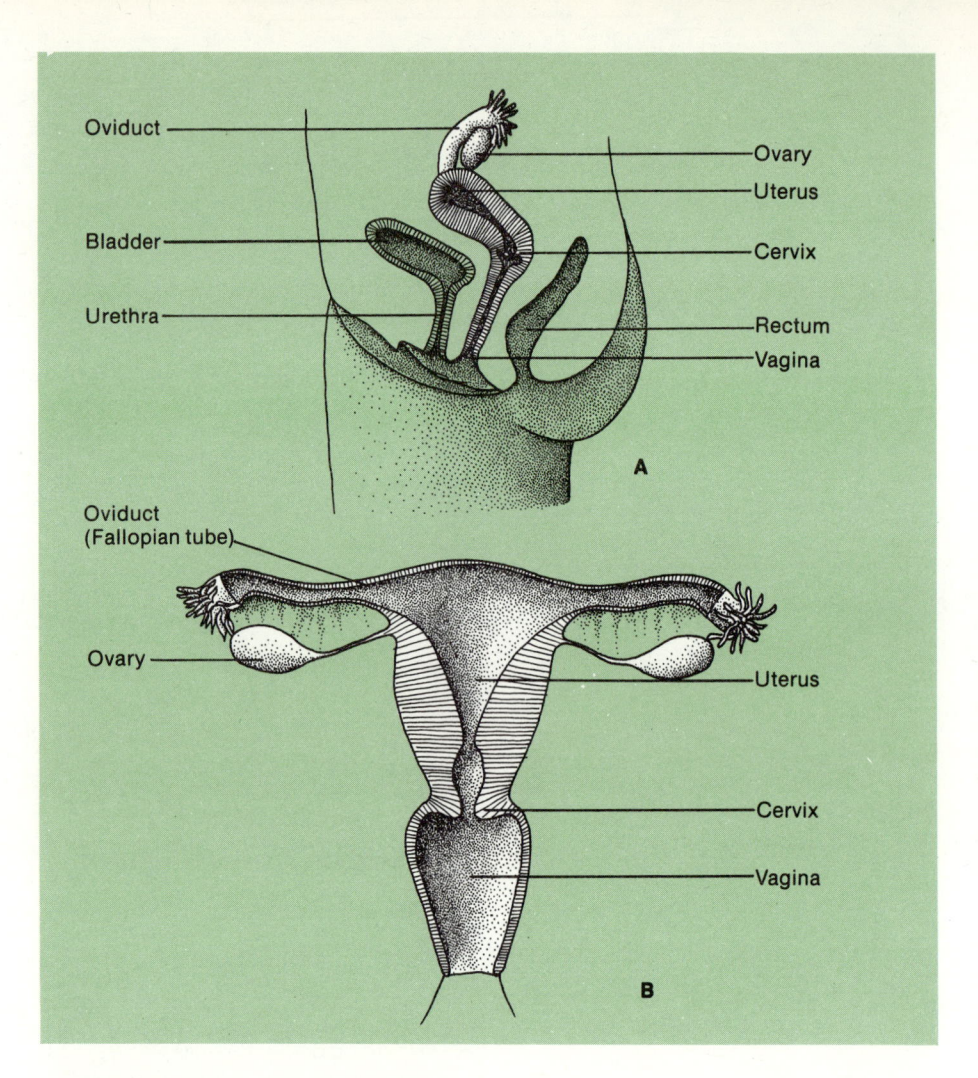

Fig. 26-7 The female reproductive system: side view (A); front view (B)

called the **cervix** [SIR-viks]. From there it enters the **vagina**. Then it is discharged from the body.

Note that, unlike the male, the urethra is separate from the reproductive system.

The ovaries make two hormones. They give the typical characteristics of the mature female: broad pelvis and developed breast. And they help prepare the uterus (or womb) for a baby.

The Menstrual Cycle

A mature female goes through a cycle called the **menstrual cycle** [MEN-strul]. This cycle averages 28 d. Follow Figure 26-8 as we discuss this cycle.

The Cycle Once every 28 d or so, an egg matures in an ovary. At the same time, the walls of the uterus begin to thicken. The uterus is getting ready to receive and nourish the egg. Next, the egg is released from the ovary. This is called **ovulation** [ov-you-LAY-shun]. The egg enters one of the oviducts. This tube moves the egg to the uterus. By this time, the uterus is swollen with blood.

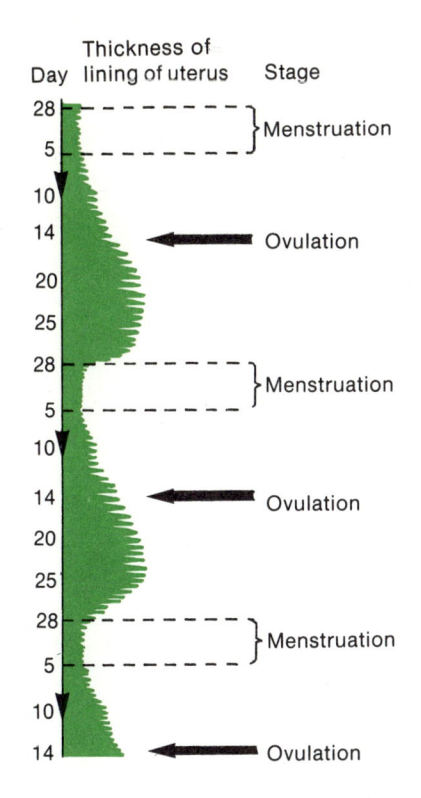

Fig. 26-8 The menstrual cycle. Not all women have a 28 d cycle. Also, the length of the cycle may change from time to time in an individual.

If the egg is fertilized by a sperm, development occurs as described in Section 26.5. If the egg is not fertilized, the menstrual cycle continues.

About 14 d after ovulation, the lining of the uterus breaks down. (The egg was not fertilized. Therefore, the lining is not needed.) This state is called **menstruation** [men-STRAY-shun]. The menstrual discharge consists of mucus, blood, and dead cells. It lasts for 4-5 d. It leaves the body through the vagina.

Shortly after menstruation, another egg matures in an ovary. And the cycle begins again.

When Can Pregnancy Occur? As Figure 26-8 shows, ovulation takes place about half way between two menstrual periods. It usually starts about 14-15 d from the *start* of menstruation. An egg lives for 1-2 d after ovulation. If live sperm are in the oviduct during this time, the egg is usually fertilized. That is, the female becomes **pregnant**.

At What Age Does Menstruation Occur? Some girls begin to menstruate as early as age 9. Some don't begin until they are as old as 21. Most, however, begin around age 13 to 14.

Menstruation usually stops at age 40 to 50. Once it does stop, the woman can no longer become pregnant.

Section Review

1. Where are eggs made?
2. Describe the path followed by an egg that does not get fertilized.
3. What is the function of each of the two female sex hormones?
4. Describe the menstrual cycle.
5. At what stage in the cycle can pregnancy occur?
6. In what age range does menstruation occur?

26.5 Fertilization and Development

Fertilization

For reproduction to occur, an egg and a sperm must unite. This process is called **fertilization**.

The **semen** from the male is ejaculated from the **penis** into the **vagina** of the female. The sperm move from the vagina up the **uterus** and into the **oviducts**. Fertilization usually occurs in an oviduct. One sperm enters an egg. This forms a **zygote**. Now a membrane forms around the zygote. This keeps other sperm out.

Development

The uterus thickens with many blood vessels. It is now ready to receive the zygote. It takes about 4-5 d for the zygote to move from the oviduct to the uterus. As it moves, the zygote grows by cell division (Fig. 26-9). By the time it reaches the uterus, it consists of about 100 cells. This group of cells becomes attached to the uterus. The group of cells is now called an **embryo** [EM-bree-oh].

A special tissue called the **placenta** [pla-SEN-ta] forms where the cells are attached to the uterus. In the placenta the capillaries of the mother's uterus are in close contact with the embryo. Oxygen and nutrients diffuse from the uterus into the embryo. And, carbon dioxide and other wastes diffuse from the embryo into the uterus. The mother then eliminates these from her body.

Development of the embryo continues for about 9 months (Fig. 26-10). In the last parts of development, the embryo is called a **fetus** [FEET-us].

Section Review

1. Make a summary of fertilization.
2. Describe the development of a new human from the zygote stage to the fetus stage.

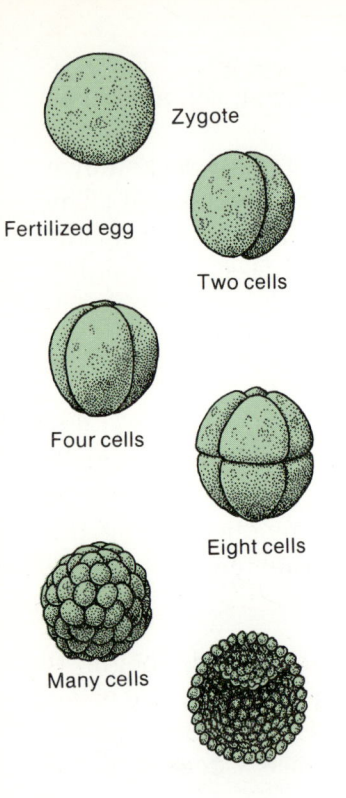

Fig. 26-9 The zygote grows by cell division (mitosis).

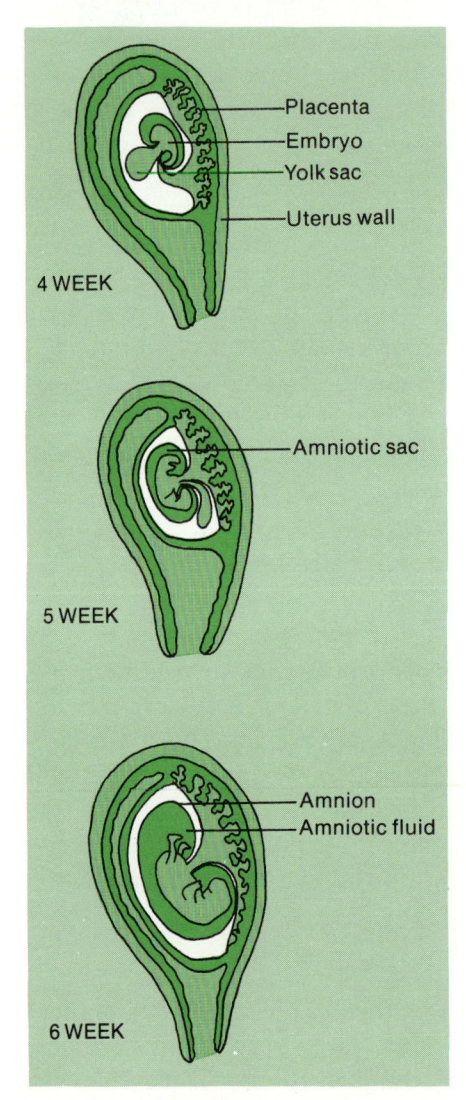

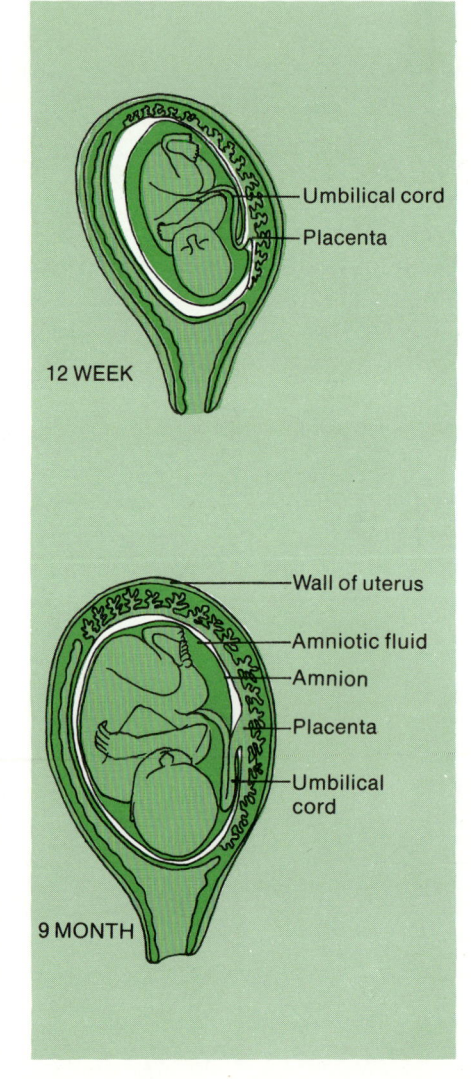

Fig. 26-10 Development of the embryo

386   Chapter 26

26.6 Care of Your Reproductive System

Venereal diseases [veh-NEER-ee-al] are the most serious diseases that affect the reproductive system. The three venereal diseases are **syphilis** [SIF-ill-is], **gonorrhea** [gone-or-EE-ah], and **genital herpes**. All three are usually spread by sexual contact with infected persons.

Syphilis

This disease goes through three stages. In the **first stage**, a hard sore appears on the genital organs. In the male, this sore is on the penis. Therefore it can be easily seen. In the female it is in the vagina, or in the moist tissue just outside the vagina. Therefore it cannot be easily seen. In fact, it may go unnoticed, since these sores are not usually painful. The sores usually appear 2-4 weeks after a person becomes infected.

The **second stage** begins 6 weeks to 6 months after the first stage. The disease has now invaded the tissues. Usually a rash appears over the skin. It is easily seen on the chest and abdomen. It even appears on the palms and soles. The rash is painless and is not itchy. The second stage lasts from a few weeks to many months. The infected person may have a mild sore throat, slight fever, and headache.

The disease now enters a **resting period**. The infected person seems healthy. In fact, the infected person may think the disease has gone away. Then the **third stage** begins. It can begin in a few weeks. Or, it may not begin for 25-30 years. In this stage the heart can be damaged. Also, the brain can be affected, causing blindness and insanity.

Treatment Syphilis responds well to penicillin. This is particularly so in the early stages. However, the best treatment is prevention. This disease often goes undetected until damage has been done.

Gonorrhea

This infection is the most common venereal disease. Like syphilis, it is usually spread by sexual contact.

In the male, symptoms first appear in the urethra (see Figure 26-2). A burning sensation results during urination. Also, a large amount of pus is formed for 2-3 months. The infection spreads through the reproductive system. And the infected person often becomes sterile.

In the female, symptoms first appear in the tissues outside the vagina. Then the disease spreads through the reproductive system. Here, too, sterility can result.

In both sexes, the disease may affect the bladder, rectum, mouth, joints, kidneys, bones, heart valves, brain, blood, and eyes.

Treatment This disease responds well to several antibiotics. However, different strains require different antibiotics. Once again, prevention is the best treatment. Health officials are concerned about the increasing number of cases of this serious disease.

Genital Herpes

This disease is a close relative of the viruses that cause cold sores and shingles. In fact, the sores look much like cold sores. They appear first on the genital organs. Genital herpes, like other venereal diseases, is spread mainly by sexual contact.

The virus can go into a resting stage for long periods of time. In fact, the infected person may think that it is gone. Then it becomes active when the infected person is overtired, sick, or under stress. You may have noticed that cold sores often appear at such times.

Treatment There is no cure. However, doctors can provide medication that relieves the symptoms.

Section Review

1. **a)** Describe the three stages of syphilis.
   b) What is the cure for syphilis?
2. **a)** State the symptoms of gonorrhea.
   b) What is the cure for gonorrhea?
3. What is genital herpes?

Main Ideas

1. Excretion is the process by which your body gets rid of metabolic wastes.
2. The lungs, skin, liver, and kidneys are excretory organs.
3. Excess salt, alcohol, proteins and smoking can damage the urinary system.
4. The testes of a male make sperm.
5. The ovaries of a female make eggs.
6. Fertilization is the union of an egg and a sperm.
7. The zygote (fertilized egg) develops into an embryo.
8. Three serious venereal diseases are syphilis, gonorrhea, and genital herpes.

Glossary

cervix	SIR-viks	the neck of the uterus
embryo	EM-bree-oh	the developing zygote

excretion		the elimination of metabolic wastes
menstruation	men-STRAY-shun	the discharge of a mixture of blood, mucus and dead cells
ovary		the organ that makes eggs
placenta	pla-SEN-ta	a special tissue between a fetus and its mother
semen	SEE-men	a mixture of sperm and fluids
testicle		the organ that makes sperm
uterus	YOO-ter-us	the womb of a female
zygote		a fertilized egg

Study Questions

A. True or False

Decide whether each of the following statements is true or false. If the sentence is false, rewrite it to make it true. (Do not write in this book.)

1. Sweat glands excrete urea.
2. Urine is about 96% water.
3. Male gametes are called testes.
4. Genital herpes can be cured with penicillin.

B. Completion

Complete each of the following sentences with a word or phrase that will make the sentence correct. (Do not write in this book.)

1. The process by which you get rid of metabolic wastes is called _____.
2. The liver changes amino acids into _____.
3. The filtering action of kidneys is done by _____.
4. Smoking can cause cancer of the _____ and _____.
5. Urine and sperm both travel down the _____.
6. The most common venereal disease is _____.

C. Multiple Choice

Each of the following statements or questions is followed by four responses. Choose the correct response in each case. (Do not write in this book.)

1. The kidneys are connected to the bladder by the
   a) ureters   **b)** urethra   **c)** oviducts   **d)** sperm ducts
2. Sperm are produced in the
   a) prostate gland   **b)** semen   **c)** testes   **d)** sperm ducts

3. Eggs are produced in the
   - **a)** oviducts
   - **b)** ovaries
   - **c)** vagina
   - **d)** uterus
4. The egg is usually fertilized in the
   - **a)** vagina
   - **b)** uterus
   - **c)** womb
   - **d)** oviduct

D. Using Your Knowledge

1. Why must wastes be removed from the body.
2. Why do unwashed feet often have an unpleasant odour?
3. Why do bacteria live so well on your feet and underarms?
4. Look closely at the sweat gland in Figure 26-1. Use your knowledge of diffusion to explain how this gland takes wastes from blood capillaries.
5. What would happen if male hormones were injected into a mature woman?

E. Investigations

1. Find out how a kidney machine takes wastes from a patient's blood.
2. Find out where donor kidneys come from for kidney transplants.
3. Find out how it is possible to have twins.

27 Structural Systems

THE SKELETAL SYSTEM
27.1 Activity: The Human Skeleton
27.2 Bones and Cartilage
27.3 Activity: The Parts of a Long Bone
27.4 Joints and Ligaments
27.5 Care of Your Skeletal System

THE MUSCULAR SYSTEM
27.6 Muscles and Tendons
27.7 Activity: How Muscles Work
27.8 Care of Your Muscular System

THE SKIN
27.9 The Structure and Functions of Skin
27.10 Activity: A Study of Skin
27.11 Care of Your Skin

Imagine yourself without bones. What would you look like? Now imagine yourself without muscles. What would you look like? Could you move? Finally, imagine yourself without skin. What would you look like?

You have three **structural systems**. They are the **skeletal system**, **muscular system**, and **skin**. Together they give you your basic shape. They also protect vital parts like the brain, lungs, and heart. And they let you move from place to place (Fig. 27-1).

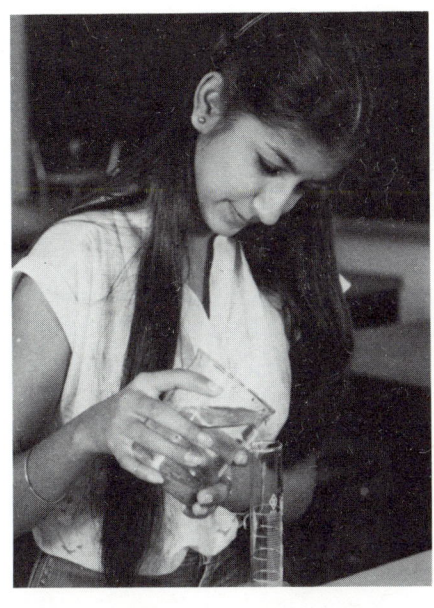

Fig. 27-1 Bones, muscles, and skin give you your shape and structure. Bones and muscles let you move.

The Skeletal System

Your skeletal system is made up of 206 bones. The largest bone is your thigh bone. The smallest is a tiny bone in your ear. Your skeleton has four functions:
1. It is the frame for your body.
2. It protects internal organs.
3. It works with muscles to let you move.
4. Some bones make red blood cells. In fact, they make over a thousand million cells a day!

Introduction 391

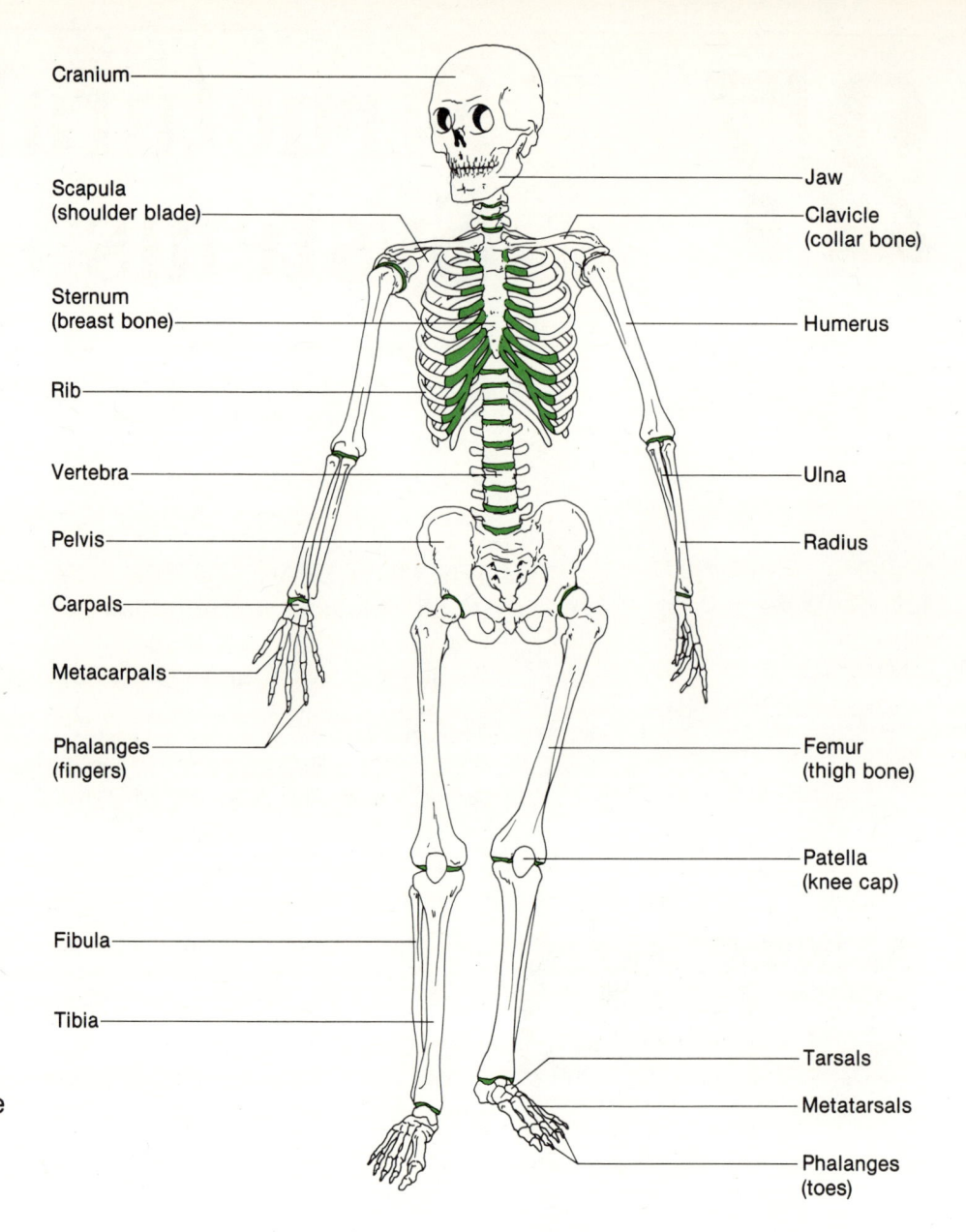

Fig. 27-2 The main bones of the human skeleton. The green shows main areas of cartilage.

Labels on figure: Cranium, Jaw, Scapula (shoulder blade), Clavicle (collar bone), Sternum (breast bone), Humerus, Rib, Vertebra, Ulna, Pelvis, Radius, Carpals, Metacarpals, Phalanges (fingers), Femur (thigh bone), Patella (knee cap), Fibula, Tibia, Tarsals, Metatarsals, Phalanges (toes)

27.1 ACTIVITY The Human Skeleton

In this activity you will use a model and/or Figure 27-2 to study the human skeleton. What are the main bones? How do they function? How do joints work?

Problem

What are the parts of the skeleton and how do they work?

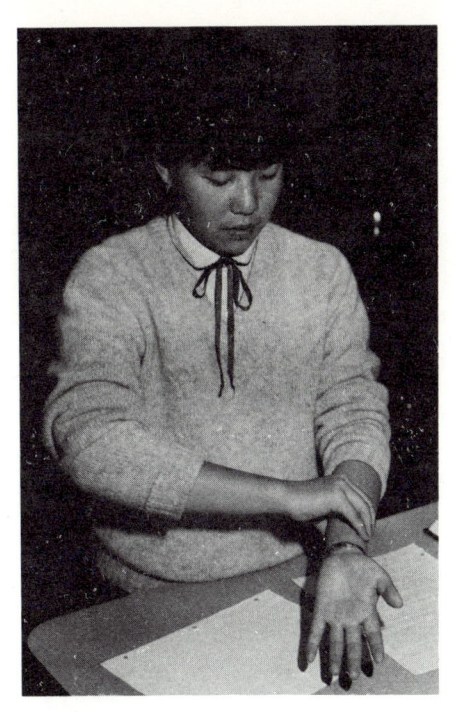

Fig. 27-3 Which bone moves, the radius or the ulna?

Materials

model of a human skeleton

Procedure

a. Find in the model and in your body each of the bones shown in Figure 27-2.
b. Hold your left hand in front of you as shown in Figure 27-3. Place your right hand on the left forearm. Find the radius and ulna.
c. Now turn your left hand back and forth (see arrow). Which bone moves the most, the radius or the ulna?
d. Study the shoulder joint of the model.
e. Now swing one of your arms around in a circle. Note how the shoulder joint works.
f. Find another joint that looks and works like the shoulder joint.
g. Study the knee and elbow joints. How do they differ from the shoulder joint?
h. Your ears and nose have cartilage in them. Bend your ears and nose. Describe cartilage.
i. Study the spine of the model. How does it move?

Discussion

1. The arm and leg have similar bones. Copy Table 27-1 into your notebook. Complete the table by matching the leg bones and the arm bones.

   Table 27-1 Comparing an Arm and a Leg

Arm	Leg
Humerus	
Radius	
Ulna	
Carpals	
Metacarpals	
Phalanges	

2. Why is one bone in your arm called the radius?
3. The shoulder and hip joints are called **ball and socket joints**. Why?
4. The knee and elbow joints are called **hinge joints**. Why?
5. What advantage does a ball and socket joint offer?
6. Why would you not want a ball and socket joint in your knees or elbows?
7. What is the main difference between cartilage and bone?

27.2 Bones and Cartilage

Your skeleton is made of two parts, **bones** and **cartilage** [KAR-tih-lij].

Bones

You may find it difficult to believe that a hard bone is alive. But it is. A bone is made of living **bone cells** in a non-living mixture of minerals and protein. The minerals are calcium and phosphorus compounds. They make the bones hard. The proteins make the bones less brittle.

Tiny **canals** run through bones (Fig. 27-4). They contain blood vessels that carry blood to and from the bone cells to keep them alive. The outside of bones is covered by an **outer living layer**. This layer repairs breaks in bones.

Some bones are hollow. Others contain **marrow**. It is a spongy solid made of living bone cells, blood vessels, and nerves. Marrow makes red blood cells, some white blood cells, and platelets. You may have seen marrow in the hollow centres of bones that came from the meat market.

Cartilage

Cartilage is softer than bones. It is also more flexible. A few months before you were born, most of your skeleton was cartilage. However, most of this cartilage changed to bone as you developed.

You still have some cartilage in your skeleton. You have some in your ears, nose, voice box, windpipe and chest. You also have cartilage over the ends of many bones. And, there is cartilage between the vertebrae of your backbone. These pieces of cartilage are called **discs**. In Figure 27-2 the main sites of cartilage are shown in green.

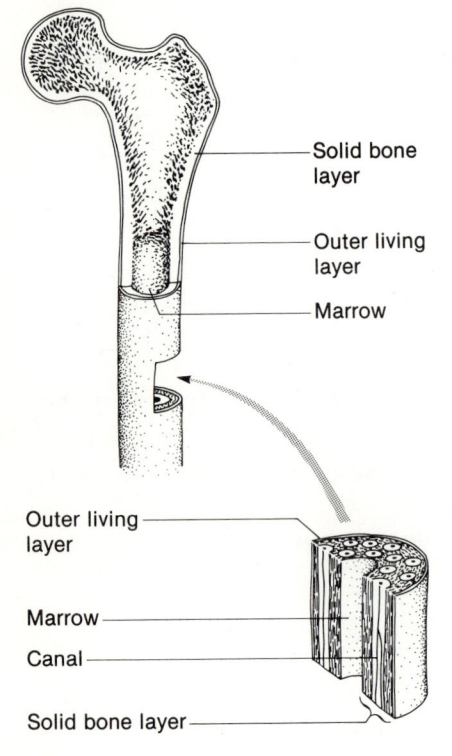

Fig. 27-4 Structure of a bone

Section Review

1. Describe the structure of a bone.
2. What is bone marrow?
3. Describe cartilage.
4. Name the main locations of cartilage.

27.3 ACTIVITY The Parts of a Bone

In this activity you will study a bone. Look for evidence that it was alive. See how many parts you can find.

Problem

Can you find the parts of a bone?

Materials

piece of beef bone (as in Figure 27-5)
microscope slide
cover slip
microscope
scalpel
dropper
dissecting needle
hand lens

Procedure

a. Find the following parts of the bone: outer layer, solid bony layer, canals, marrow.
b. Study each part with a hand lens.
c. Use the scalpel and dissecting needle to get a small piece of marrow.
d. Make a wet mount of the piece of marrow.
e. Study the mount under low then medium power. Look for blood cells.

Fig. 27-5 Can you find the parts of a bone?

Discussion

1. Describe the solid bony layer.
2. Describe bone marrow as you saw it.
3. How do blood cells get from the bone marrow to the blood?

27.4 Joints and Ligaments

A **joint** occurs where one bone meets another. Most joints allow movement. However, some, like those joining the bones in the cranium, do not move (Fig. 27-6).

Table 27-2 lists the main types of moveable joints and their functions.

Table 27-2   Moveable Joints

Type of joint	Examples	Type of motion
Ball and socket	Hip; shoulder	Rotates and moves side to side
Hinge	Knee; elbow	Up and down
Pivot	Between skull and spine	Turns
Gliding	Between vertebrae	Twists; bends

In order for a joint to operate, the bones must be held together. Special tissues called **ligaments** do this. A layer of **cartilage** covers the end of the bones that meet in joints. A special fluid lubricates the two bone surfaces. This fluid, together with the cartilage, keeps the ends of the bones from wearing out by rubbing on one another. This fluid is like motor oil in one respect. When it is cold, it becomes thick. Perhaps this is why cold weather produces "stiff joints".

Section 27.4

Fig. 27-6 Types of joints. How many kinds of moveable joints are there?

CRANIUM
Fixed joints
Hinge joint
ELBOW
Pivot joint
Ball and socket joint
Gliding joints
VERTEBRAE
HIP

Fig. 27-7 Injuries are common in contact sports. How can they be lessened?

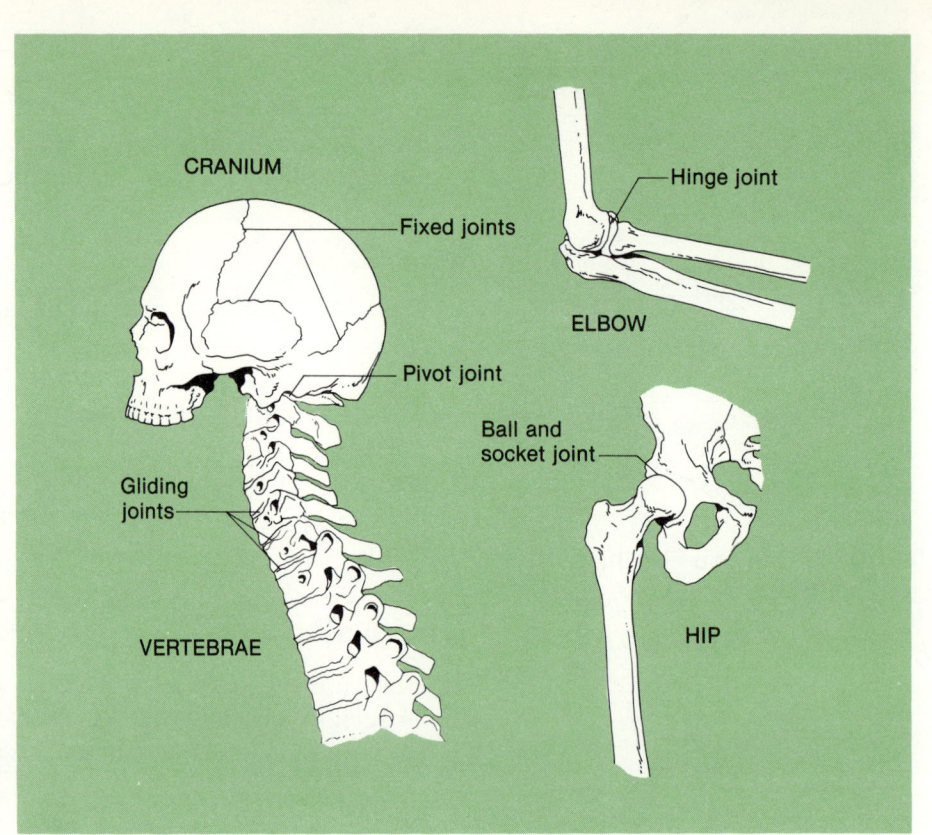

Section Review

1. Where does a joint occur?
2. Where in your body are there some joints that do not move?
3. Name four types of moveable joints. Find examples of each in your body. Experiment to find out the type of motion each type of joint allows.

27.5 Care of Your Skeletal System

Your skeletal system is the frame for your body. It protects vital organs like the brain. It works with muscles to let you move. And it makes blood cells. Clearly it is worth looking after!

Bones are made strong by calcium and phosphorus compounds. Therefore calcium and phosphorus should be included in your diet. Milk is the best source of these two elements.

Until recently most people believed that adults didn't need much milk. Now doctors recommend that even older adults drink 2 or 3 glasses of milk a day. It can prevent brittle bones and the humped back of old age.

The skeletal system is often damaged in sports activities (Fig. 27-7). Contact sports like hockey and football break bones. They also

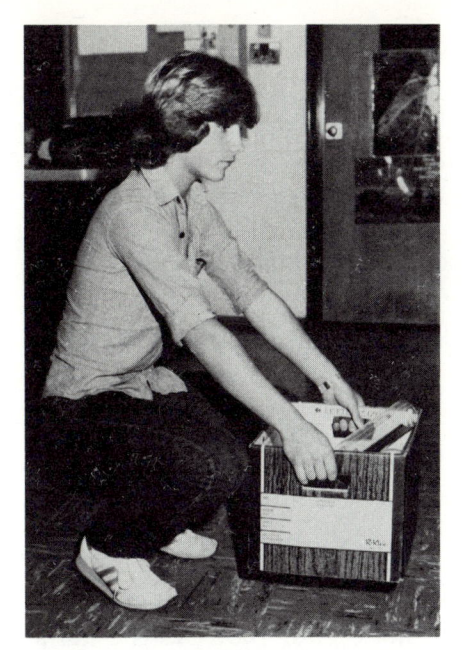

Fig. 27-8 Keep the back straight and let your legs do the lifting.

damage cartilage and tear the ligaments in joints. Yet these sports are fun and useful in many ways. To help avoid injury, do four things: Get in top physical shape. Wear the best protective equipment. Know how to play the game properly. *Always* warm up before entering the game.

Even joggers have skeletal injuries. Damaged cartilage in the knees and back is common. Torn ligaments also happen. Yet most jogging injuries can be avoided. Warm up. Wear proper shoes. Stay within your physical limits.

Skiing accidents break bones, damage cartilage, and tear ligaments. Yet these, too, can be largely avoided. The advice is the same: Ski within your limits. Get in top physical shape before hitting the hills. Warm up before that first run. And *never* ski hills beyond your ability.

Lifting heavy masses often damages cartilage in the back. One should always lift with the legs, not the back, as shown in Figure 27-8.

It is important to look after your skeletal system when you are young. Damaged joints, torn ligaments, and damaged cartilage seem to heal fast. And the pain goes away. But often problems return in middle age as **arthritis** [ar-THRI-tis]. This painful and crippling disease of the joints can often be prevented by avoiding injuries when you are young.

Section Review

1. Describe how bones can be made strong.
2. List some skeletal injuries that can result from contact sports.
3. How can most sports injuries be prevented?
4. What is arthritis? How is it often caused?

The Muscular System

27.6 Muscles and Tendons

Your body has over 700 **muscles**. Some are attached to bones in your skeleton. Some of these are attached directly to the bones. Others are attached to the bones by means of **tendons**. You can feel tendons in your wrists, elbows, knees, and ankles. Other muscles are in blood vessels, your heart, digestive tract, and other organs.

You use muscles every time a part of your body moves. Muscles help you run, shake your head, and blink. They also help food move through your digestive tract. And they move blood through your body.

How Muscles Work

Muscles can only pull. They cannot push. For this reason, muscles occur in pairs. When one member of a pair pulls, a limb moves in a certain way. When the other pulls, the limb moves back to its original position. Clearly, one member of a pair must be relaxed when the other is pulling.

Your arm has a pair of such muscles. These muscles move your forearm up and down. The muscles are the **biceps** and **triceps**. Figure 27-9 shows how they move your forearm.

Types of Muscles

You have three types of muscles. They are **skeletal, smooth,** and **cardiac**. All three work in the same way. On a signal from the brain, the muscle contracts. This causes it to pull on a limb or other body part.

Skeletal muscles are under voluntary control. You can decide when you want to raise your arm or move your foot. As the name suggests, these muscles are attached to your skeleton. Most of them are attached with tendons. Skeletal muscles move your body parts.

Smooth muscles are found in the walls of the digestive tract and blood vessels. These muscles are under involuntary control. They function normally without you deciding that they should.

Cardiac muscles form a major part of the heart. These muscles contract at regular intervals. This creates the beating of the heart. These muscles are under involuntary control.

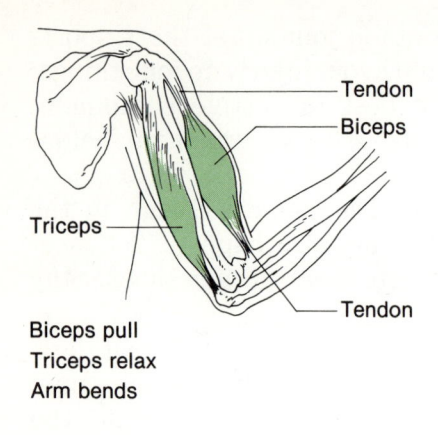

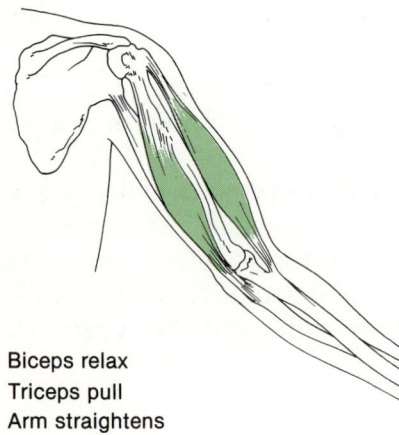

Fig. 27-9 How the arm muscles work

Biceps pull
Triceps relax
Arm bends

Biceps relax
Triceps pull
Arm straightens

Section Review

1. Explain how muscles work.
2. Name the three types of muscles. What does each type do? Are they voluntary or involuntary?

27.7 ACTIVITY How Muscles Work

Muscles occur in pairs. One member of the pair moves a body part one way. The other moves the body part back. In this activity you will find out how the biceps-triceps pair moves your arm.

Problem

How do the biceps and triceps work?

Materials

your body

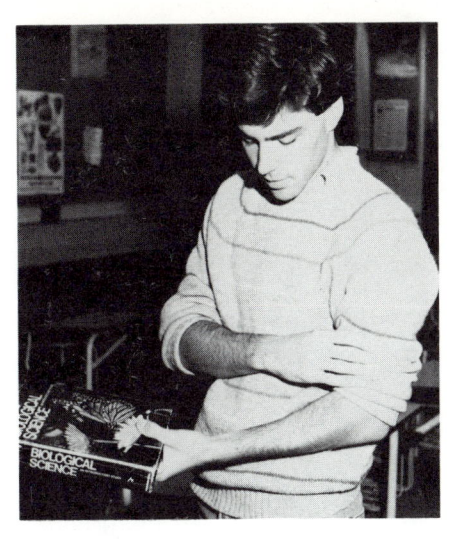

Fig. 27-10 How do your arm muscles work?

Procedure

a. Pick up a book in your left hand.
b. Put your right hand over your left biceps as shown in Figure 27-10.
c. Slowly raise the book by moving only your forearm. Note what happens to the biceps.
d. Now slowly lower the book by moving only your forearm. Note what happens to the biceps.
e. Repeat steps (c) and (d) with your right hand on your left triceps. Note what happens to the triceps.
f. Repeat step (e). But this time ask your partner to lift up on the book as you are trying to lower it. Note what happens to the triceps.

Discussion

1. What three things happen to the biceps when it contracts?
2. What happens to your arm when the biceps contracts?
3. What happens to your biceps when you lower your arm?
4. Were the triceps needed in Step (e) to lower your arm? Why? What is your proof?
5. Were the triceps needed in Step (f) to lower your arm? Why? What is your proof?
6. Find another pair of muscles in your body that works like the biceps-triceps pair.

27.8 Care of Your Muscles and Tendons

Fig. 27-11 Don't "fool around" with weights. Get advice from your gym teacher.

Regular exercise helps keep muscles in shape. But the exercise must be well-balanced. Running helps mainly the leg muscles. It does little for the upper body. Therefore, it should be combined with pushups, situps, and weight training (Fig. 27-11). *Always get professional advice on a program that is good for you.*

Heavy regular exercise makes muscles grow thicker. When they are thicker, they are stronger. Muscles do not grow by cell division. Instead, the cells simply get larger.

Have you heard of pulled muscles and tendonitis? These are injuries of the muscular system. They can generally be avoided with a proper warmup.

Section Review

1. What is meant by a well-balanced exercise program?
2. How do muscles grow?

The Skin

27.9 The Structure and Functions of Skin

You learned in Chapter 26 that skin helps with excretion. But it also has many other functions.

Structure

The outer layer of the skin is called the **epidermis** [ep-ee-DER-mis]. It has two parts (see Fig. 26-1). The outer part consists of dead cells. They are always being lost from your body. You may have seen them peeling off someone with a bad sunburn. The inner part of the epidermis consists of living cells. These replace the dead cells that are lost.

The living cells of the epidermis contain pigments like **melanin** [MEL-a-nin]. These give your skin its colour. People with a lot of melanin have brown or black skin. People with little melanin have white skin. Melanin absorbs the burning rays of the sun. This protects the next layer of the skin from sunburn.

The **dermis** contains nerves, blood vessels, hair cells, oil glands, and sweat glands. At the bottom of it is a **fat layer**. This layer helps insulate your body.

Functions

With the skeleton and muscles, skin gives your body its shape. Skin also protects inner parts of your body. It keeps diseases out. And it prevents those parts from drying out.

Skin is also a sense organ. Note the touch receptors in Figure 26-1. These nerve endings sense heat, cold, pain, and pressure. They give you your sense of touch.

You learned in Chapter 26 that skin is important in excretion. It helps your body get rid of excess water, salts, and urea. These are collected by the **sweat glands**. Then they are dumped out the **sweat pores** as sweat.

Skin is your body's air conditioning system. The skin contains many blood vessels. When the body is hot, these blood vessels swell up. This makes them move closer to the surface of the skin. Then heat can more easily radiate from the blood to the air. This cools your body. When the body is cool, the blood vessels shrink. As a result, they move deeper into the skin. Then your body does not lose as much heat.

On very hot days, the air may be warmer than your blood. Now your body cannot cool itself as we just described. Instead, it must cool itself by **sweating**. The sweat flows out onto the skin. There it evaporates. This evaporation cools the skin and, in the long run, the whole body.

Section Review

1. Describe the epidermis of the skin.
2. What is melanin?
3. Describe the dermis layer of the skin.
4. List and explain five functions of skin.

27.10 ACTIVITY A Study of Skin

Have you ever looked closely at your skin? Is it the same all over your body? What do freckles and moles look like when magnified? Can you see sweat pores? How does sweating cool your body?

Problem

What parts of skin can you see with a hand lens?

Materials

hand lens	water	printing pad
dropper	rubbing alcohol	

Procedure

a. Look at the back of your hand with the hand lens. Sketch what you see. Label as many parts as you can.
b. Look at the palm of your hand with the hand lens. Sketch what you see.
c. Study your fingertips with a hand lens.
d. Press the fingertips of one hand on the printing pad. Now press them on a piece of paper to get your fingerprints. Compare yours to those of 2 or 3 classmates.
e. Look at as many as you can of freckles, moles, scars, calluses, scrapes, cuts, and blisters. Record what you observe.
f. Put a few drops of water on the back of one hand. Fan that area vigorously. Record what you observe.
g. Repeat step (f) using a few drops of rubbing alcohol.

Discussion

1. List the differences you saw between the skin on the back of your hand and the palm. What are the reasons for these differences?
2. Of what value to you are the special patterns on your fingertips?
3. Why can fingerprints be used for identification?

4. How do cuts, scars, and blisters affect the epidermis?
5. What do steps (f) and (g) prove?

27.11 Care of Your Skin

Always keep your skin clean. Oils can clog sweat and hair pores. Then bacteria enter and cause infections. Soap and water are generally all you need for cleaning your skin.

Protect your skin from the sun. This is especially important if you have fair skin. A sunburn is always damaging to the skin. A dark tan may look good. But repeated tanning causes skin to age and wrinkle. You may look great today with your tan. But when you are 40 you may look 60 with your wrinkles! Besides, ultraviolet rays from the sun can cause **skin cancer**.

You can protect exposed skin from the sun with a lotion that contains a **sunscreen**. A common sunscreen is **p-aminobenzoic acid**, or **PABA**. Look for the **Sun Protection Factor** on the bottle (Fig. 27-12). The higher the factor, the better the protection. For example, a factor of 2 gives you twice your natural protection from sunburn. But a factor of 12 gives you twelve times your natural protection.

The skin is often attacked by many diseases such as **acne** and **eczema** [EK-see-ma]. Always get medical help for such diseases. Self-treatment can cause serious problems.

Fig. 27-12 Buy suntan lotions that tell you the Sun Protection Factor. Use a high factor if you are fair skinned. Also, use a high factor if you have no tan.

Section Review

1. What can sun do to your skin?
2. What is a sunscreen?
3. Explain the Sun Exposure Factor found on sunscreen lotions.

Main Ideas

1. There are three structural systems: the skeletal system, the muscular system, and the skin.
2. The skeletal system consists of bone and cartilage.
3. There are four types of moveable joints: ball and socket, hinge, pivot, and gliding.
4. Ligaments hold joints together.
5. Many muscles are attached to bones with tendons.
6. There are three types of muscles: skeletal, smooth, and cardiac.
7. Sports injuries can often be prevented with warmups, proper equipment, and proper training.
8. The skin protects the body, helps with excretion, and cools the body.
9. Excess sun can damage the skin.

Glossary

arthritis	ar-THRI-tis	a disease of the joints
cartilage	KAR-tih-lij	a flexible part of the skeleton
ligament	LIG-a-ment	a tissue that holds bones together
marrow		a spongy solid in the centre of many bones
melanin	MEL-a-nin	a dark pigment in the skin
tendon		a tissue that joins muscles to bones

Study Questions

A. True or False

Decide whether each of the following statements is true or false. If the sentence is false, rewrite it to make it true. (Do not write in this book.)
1. The shoulder is a hinge joint.
2. In your forearm, the radius rotates around the ulna.
3. Red blood cells are made in the bone marrow.
4. The bones in a joint are held together by ligaments.
5. Muscles can only pull; they cannot push.
6. The epidermis of the skin contains only dead cells.

B. Completion

Complete each of the following sentences with a word or phrase that will make the sentence correct. (Do not write in this book.)
1. The knee is a _____ joint.
2. The skeleton has two main parts, bones and _____.
3. The place where two bones meet is called a _____.
4. The arm muscles used in pushups are the _____.

C. Multiple Choice

Each of the following statements or questions is followed by four responses. Choose the correct response in each case. (Do not write in this book.)
1. The bones in a joint are separated from one another by
   a) cartilage   b) ligaments   c) muscles   d) tendons
2. Ball and socket joints are found in the
   a) shoulder and spine   c) elbow and knee
   b) shoulder and hip   d) neck and spine

3. A fair-skinned person is going to the beach for the first time this year. This person should use a tanning lotion with a Sun Protection Factor of
   a) 2   b) 4   c) 8   d) 14

D. Using Your Knowledge

1. Pregnant women should have lots of calcium and phosphorus in their diets. Why?
2. Hockey players often suffer knee injuries. Many, like Bobby Orr, must have the cartilage removed. Then they may need operations in later years to remove tiny pieces of bone. Why?
3. Survival experts tell us never to work up a sweat on an outing in cold weather. Why?

E. Investigations

1. Design an experiment to show how a pair of muscles operates your knee joint.
2. Try an experiment to find out which arm muscles are used in pushups and in chinning yourself.
3. Your nails and hair are part of the skin. Find out
   a) what they are made of;
   b) if they are dead or alive;
   c) how they grow.
4. Research the causes and treatment of acne.

28 Control Systems

THE NERVOUS SYSTEM
28.1 What Does the Nervous System Do?
28.2 Nerve Cells
28.3 Activity: Testing Your Reflex Action
28.4 The Central Nervous System
28.5 Activity: Testing Your Reaction Time
28.6 The Autonomic Nervous System

THE ENDOCRINE SYSTEM
28.7 Endocrine Glands and Hormones
28.8 Some Endocrine Glands
28.9 Activity: The Effects of Caffeine
28.10 Care of Your Control Systems

Your stomach growls and you say "I'm hungry." You smell your favourite food and your mouth waters. You touch a hot object and, without thinking, jump back. You get excited and your heart beats faster. You get nervous and you begin to sweat. Someone shouts and you turn your head.

Something must control how your body responds to such stimuli. And something must control things like your heartbeat. There are two "somethings". They are your **nervous system** and your **endocrine system** [EN-do-krin].

Fig. 28-1 What controls learning? What controls your reaction time? What controls your coordination?

The Nervous System

You are not born knowing how to play ball. You must learn how to catch, throw, and bat. You must practice until you become coordinated. And you must be able to react quickly.

What controls learning? What controls coordination? What controls reaction time (Fig. 28-1)? Let's find out.

28.1 What Does the Nervous System Do?

You learned in Chapter 2 that response to stimuli is a characteristic of living things. A **stimulus** is a change in an organism's environment that causes activity in the organism. Such activity is called a **response**. Thus we say that organisms respond to stimuli. For

Fig. 28-2 The nervous system has two main parts, the central nervous system and the outer nervous system. There are over 150 000 km of nerves in your body.

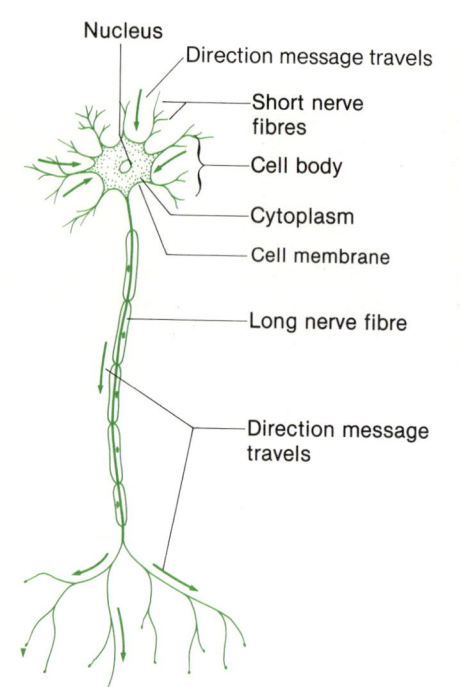

Fig. 28-3 One type of nerve cell, the motor nerve cell

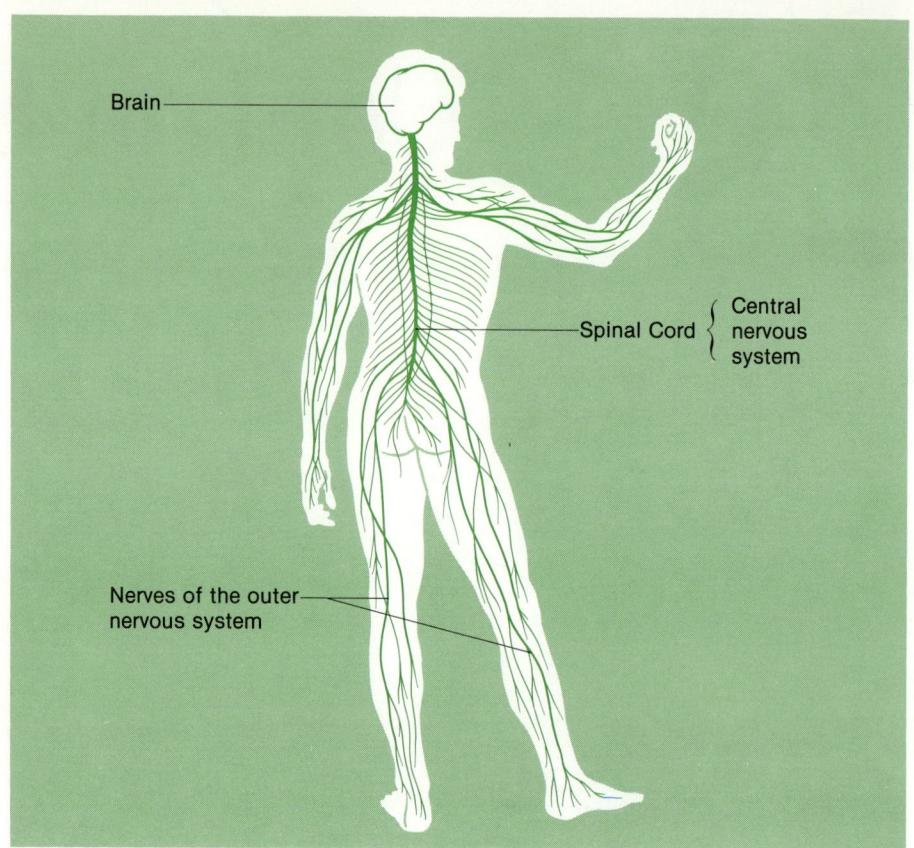

example, if you touch something hot (the stimulus) you respond by quickly pulling back. Such a response is called **behaviour**. In fact, everything you do is called your behaviour. Throwing a ball, laughing, running, reading, and watching television are examples of your behaviour.

Your behaviour is controlled by your **nervous system** (Fig. 28-2). It *receives* information from your environment. It *stores* some of that information. And it *sends* information to parts of your body. In other words, it controls how you respond to stimuli. That is, it controls your behaviour.

Section Review

1. What is a stimulus?
2. What is a response?
3. What is behaviour?
4. What does your nervous system do?

28.2 Nerve Cells

Your nervous system is made up of **nerve cells**. These are the longest cells in your body. In fact, the fibres from some of these cells run the

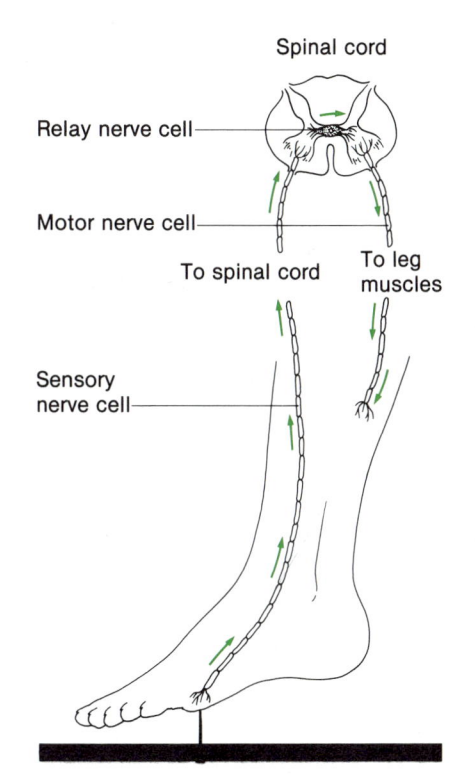

Fig. 28-4 The three types of nerve cells work together. Can you see how?

entire length of your leg! A **nerve** is not just one nerve cell. It is a bundle of fibres from many nerve cells.

There are three major types of nerve cells. Figure 28-3 shows a motor nerve cell. Sensory nerve cells and relay nerve cells look much like it.

Sensory nerve cells pick up messages from the skin or from sense organs like the tongue and eyes. Then they send these messages to the brain or spinal cord.

Motor nerve cells carry messages from the brain and spinal cord to muscles and glands.

Relay nerve cells connect the sensory nerve cells and motor nerve cells. They are in your brain and spinal cord.

Now, let's see how the three types of nerve cells work together. Suppose you accidentally step on a nail (Fig. 28-4). Sensory nerve cells pick up a message from the skin of your foot. This message travels up the sensory nerve cells to your spinal cord. Relay nerve cells in your spinal cord transfer, or relay, the message to motor nerve cells. The motor nerve cells send a message to your leg muscles. Then your muscles lift your foot from the nail. This often happens before you feel pain. It may even happen before damage is done. That's because messages can travel at speeds up to 160 m/s!

Section Review

1. Describe a nerve cell.
2. What is a nerve?
3. Name the three types of nerve cells. Then state what each type does.

28.3 ACTIVITY Testing Two Reflex Actions

Suppose you stepped on a nail. You would quickly pull your foot off the nail. You wouldn't think about it first. You wouldn't have to decide to lift your foot. It would just lift.

Such an automatic act is called a **reflex action**. In this activity you will study two reflex actions. But don't worry. We're not going to ask you to step on a nail!

Problem

What is a reflex action?

Materials

ball of paper
sheet of clear plastic

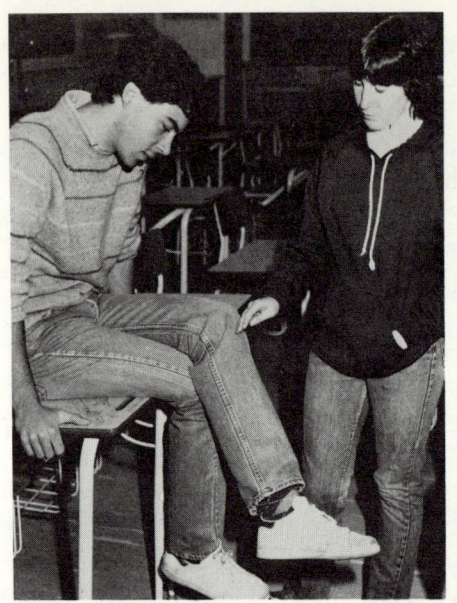

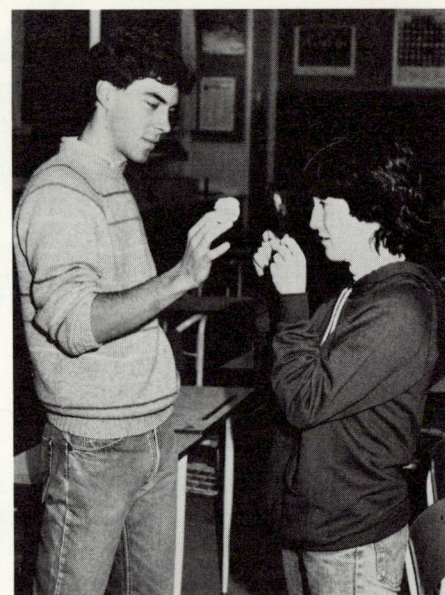

Fig. 28-5 Studying two reflex actions

Procedure

a. Cross your legs as shown in Figure 28-5. Make sure that the top leg swings freely.
b. Hit the tendon just below the kneecap of the top leg. Use the edge of your hand. Use a fast sharp blow. What does your leg do?
c. Repeat step (b) several times.
d. Hold the sheet of clear plastic about 10 cm in front of your face as shown in Figure 28-5.
e. Ask your partner to throw a ball of paper at the plastic. Watch the ball as it is thrown at you. What did your eyes do?
f. Repeat step (e) several times. Try to keep from blinking.

Discussion

1. In the knee experiment, what is the stimulus? What is the response?
2. In the ball of paper experiment, what is the stimulus? What is the response?
3. A **reflex action** is defined as a quick, unlearned, automatic response. Were your responses in these experiments reflex actions? Why do you say so?

28.4 The Central Nervous System

Your nervous system has two main parts. They are the **central nervous system** and the **outer nervous system** (see Figure 28-2). The central nervous system consists of the **brain** and **spinal cord**. The

408   Chapter 28

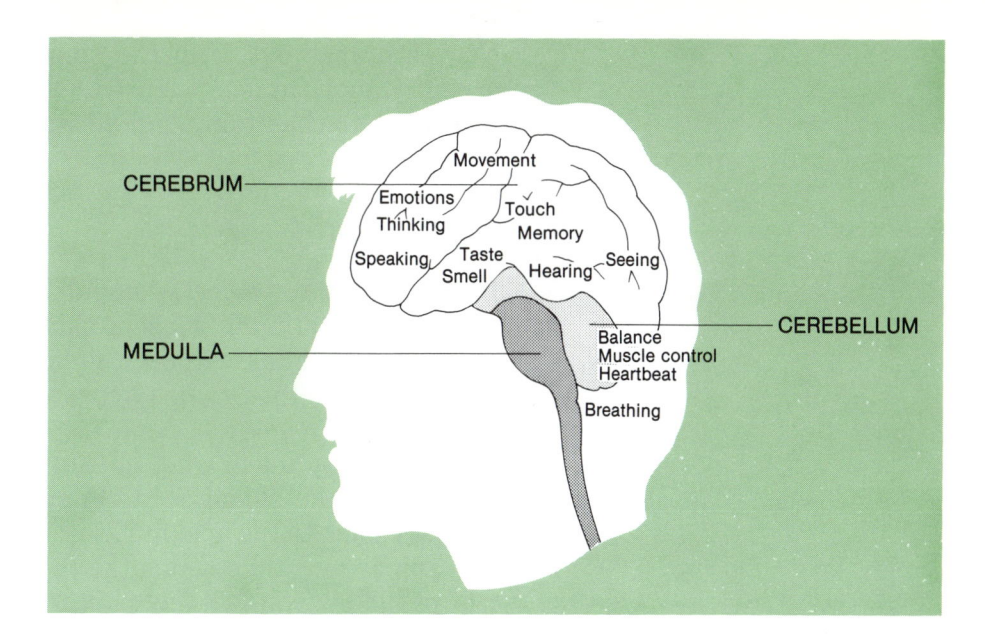

Fig. 28-6 The 3 main parts of the brain and their chief functions

outer nervous system consists of all the nerves outside the central nervous system. These nerves link the central nervous system with the rest of the body.

The Brain

Your **brain** is made of about ten thousand million (10 000 000 000) nerve cells. It consists of three main parts. These are the **cerebrum** [ser-EE-brum], **cerebellum** [ser-uh-BELL-um], and **medulla** [mi-DUL-a]. Together, these parts make our best computers look like toys.

The **cerebrum** controls thought, emotions, learning, and some voluntary movements. Also, it is your memory bank. And it controls the senses of seeing, hearing, smelling, tasting, and touching. As Figure 28-6 shows, it is the largest part of the brain. The superior intelligence of humans is due to the well developed cerebrum.

The **cerebellum** coordinates your muscles. Without it, you couldn't walk, run, swim, or even talk. The cerebellum also helps you keep your sense of balance.

The **medulla** is located at the top of the spinal cord. It controls the automatic functions of your body. Some of these are breathing, heartbeat, muscular action in the digestive tract, and blood pressure.

The Spinal Cord

Your backbone consists of many bones called **vertebrae**. You can feel these in the centre of your back. You can see what they look like in Figures 27-2 (page 392) and 27-6 (page 396). Your **spinal cord** runs down a channel in the vertebrae. The vertebrae protect the spinal cord.

The spinal cord runs from the brain most of the way down your backbone. It is the link between your brain and the rest of your body.

Section 28.4

Nerves branch out from the spinal cord to all parts of your body except your head (see Figure 28-2). Thus messages can travel from your body to the brain. And messages can travel from your brain to your body.

Reflex Actions

If you touch a hot object, you pull back without thinking. If someone scares you, you jump without thinking. If someone throws a ball at your face, you blink without thinking. These are all **reflex actions**. In a reflex action, the brain is bypassed to save time. Let's see how this works:

Suppose you touch a hot object. Nerve cells in your fingers send a message to the spinal cord. To save time, the spinal cord sends a message back to the arm muscles. It tells them to move your fingers. The spinal cord then sends the message on to your brain. Now you will yell. Now you will feel pain. And now you will look to see what has happened.

Section Review

1. What are the functions of the cerebrum? the cerebellum? the medulla?
2. How is the spinal cord protected?
3. What is the main function of the spinal cord?
4. Explain how a reflex action saves time.

28.5   ACTIVITY   Testing Your Reaction Time

A glass slips out of your hands. Can you catch it before it hits the floor? Or are you in trouble? Well, that depends on your **reaction time**.

Your eyes see the glass slip. A message is sent through nerve cells to the spinal cord. The spinal cord sends it to the brain. Then the brain sends a message to your arm muscles. And they reach out for the glass. If a reflex action occurs, the message will only go as far as the spinal cord. But in either case, whether you catch the glass or not depends on your reaction time. That's the time it takes you to react to the event. How good is your reaction time?

Problem

How good is your reaction time?

Materials

30 cm ruler

Procedure

a. Copy Table 28-1 into your notebook.
b. Work with a partner. Call one of you "Person A" and the other "Person B".
c. Person A: Hold your hand out with your thumb and first finger about 1 cm apart (Fig. 28-7).
d. Person B: Hold the ruler as shown in Figure 28-7. The zero end should be between Person A's thumb and finger.
e. Person A: Keep your eyes on the region of your thumb and finger.
f. Person B: Drop the ruler.
g. Person A: Try to catch the ruler by moving only your thumb and finger.
h. Person A: Record in your table the distance on the ruler at which you caught it.
i. Repeat this 9 times using the same hand.
j. Person A: Repeat this procedure using your other hand.
k. Change persons and do the whole experiment again.
l. Calculate the average distance for the 10 trials for each hand. Record your answer in your table.

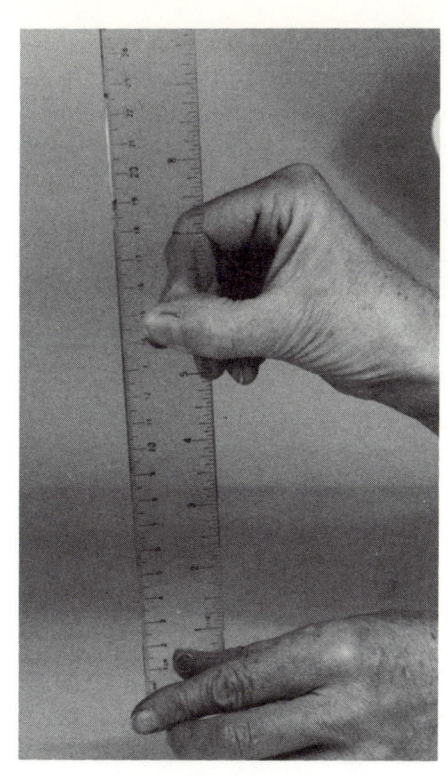

Fig. 28-7 How quickly can you catch the ruler?

Table 28-1 Reaction Time

Trial	Left hand (cm)	Right hand (cm)
1		
2		
3		
•		
•		
•		
10		

Discussion

1. Compare the reaction times of your left and right hands. Try to account for the differences.
2. How did your reaction time change with practice?
3. Compare your reaction time with that of your partner.
4. Do you think that this activity used a reflex action? Or, was the brain involved? Why do you say so?

28.6 The Autonomic Nervous System

Look back to Figure 28-2. Do you see two nerve cords, one on each side of the spinal cord? They are the main part of the **autonomic nervous system** [aw-toe-NOM-ik].

Section 28.6 411

The autonomic nervous system is part of the outside nervous system. That is, it is not part of the central nervous system. It has several jobs. But most of these do not involve the central nervous system. It controls such things as heartbeat and movement of muscles in the digestive system.

The autonomic nervous system works without you thinking about it. You don't need to think about your heartbeat. Nor do you have to decide when your stomach valve will open.

Section Review

1. What kinds of jobs does the autonomic nervous system do?
2. Why can't the central nervous system do these jobs?

The Endocrine System

The **endocrine system** works with the nervous system to control all body functions. These systems do much the same jobs. And they depend on one another to do those jobs.

28.7 Endocrine Glands and Hormones

Endocrine Glands

Your body has many glands. You have already met some glands. The salivary glands and pancreas are examples. These glands have ducts (tubes) that carry the chemicals from the glands to body organs.

The **endocrine glands** are **ductless**. They put chemicals directly into your bloodstream. Then the blood takes them throughout the body. Your endocrine system consists of many glands. Figure 28-8 shows some of them. These glands make hormones. Some glands make only one hormone. Others make several.

Hormones

Hormones [HOR-mohnz] are special chemicals that control certain body functions. They keep the body functions going at the right speed. They speed up activities, when necessary. And they slow down activities, when necessary. They keep the total body working together. Each hormone has a special job to do. Only a very small amount of a hormone is needed by the body.

Section Review

1. How do endocrine glands differ from other glands?
2. What are hormones?

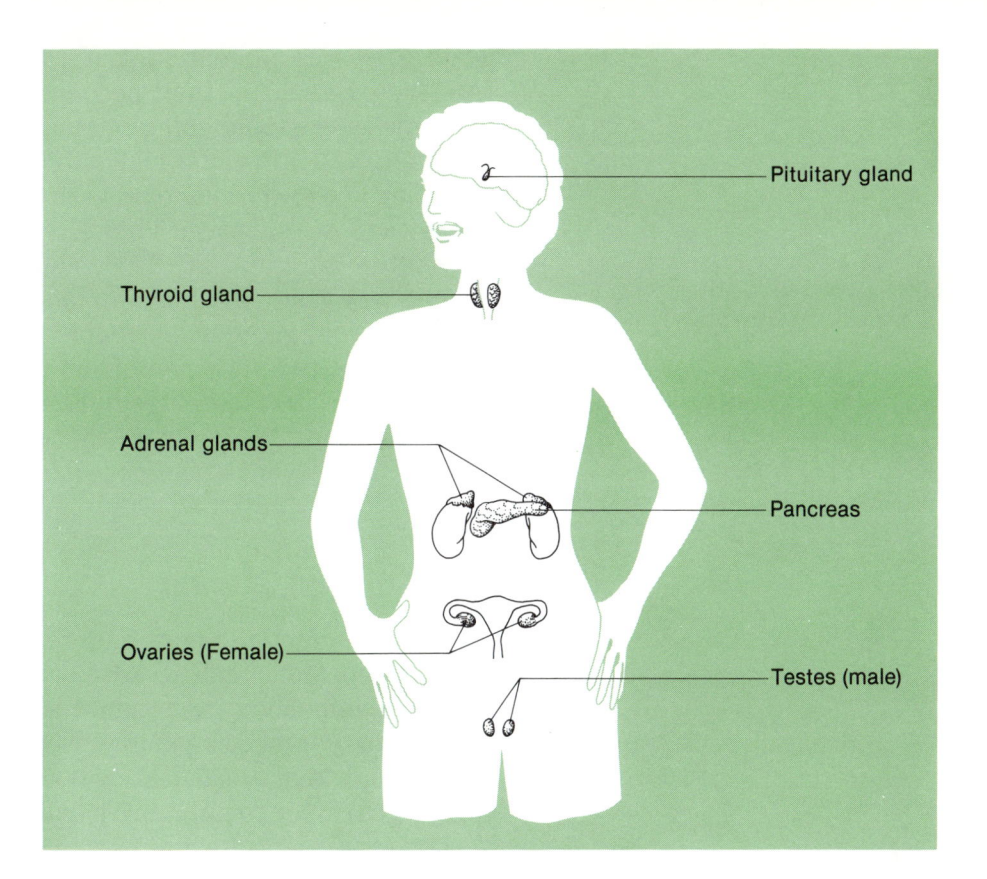

Fig. 28-8 Some of the endocrine glands. The glands are not connected. Therefore the endocrine system is not a true system.

28.8 Some Endocrine Glands

This section describes the endocrine glands shown in Figure 28-8. It also outlines their main functions.

The Pituitary Gland

The **pituitary gland** [pi-TOO-i-ter-ee] is a pea-sized gland at the base of the brain. It is often called the "master gland". That's because it makes hormones that control the other endocrine glands.

This gland also makes a **growth hormone**. This hormone controls your growth. If you have the right amount of it, you grow to normal size. If you have too little, you will become a dwarf. And, if you have too much, you will become a giant.

The Thyroid Gland

The **thyroid gland** [THI-roid] is in your neck. It is near your voice box. It makes a hormone that controls the rate of cellular processes such as respiration. If you have the right amount of it, all processes work at just the right speed.

Sometimes this gland makes too much hormone. People with this condition are said to have an **overactive thyroid**. They usually have

Section 28.8 413

large appetites and eat a lot. Yet they lose weight. Their heartbeat rate increases. And they become nervous and jittery. Their body processes are simply going too fast.

Sometimes this gland makes too little hormone. People with this condition are said to have an **underactive thyroid**. These people lack energy and are sluggish. They have poor appetites and eat little. Yet they often gain weight. That's because their body processes slow down. They just don't burn up food very quickly.

The thyroid gland needs **iodine** to make its hormone. Most people get all the iodine they need from their food. But in the Great Lakes region and a few other regions of Canada, food does not supply enough iodine. At one time thyroid disease was common in these regions. To prevent it, iodine is added to table salt. This salt is called **iodized salt**.

The Adrenal Glands

An **adrenal gland** [ah-DREE-nal] sits on the top of each kidney. They make a hormone called **adrenalin** [ah-DREN-al-in]. This hormone prepares your body for action. Large amounts are poured into your body when you become angry, frightened, or excited. This hormone makes your heart beat faster. It raises your blood pressure. And it causes your liver to put more sugar into your blood. This sugar gives you extra strength for fighting (or running away).

The Sex Glands

Females have **ovaries**. Males have **testes**. These are the **sex glands**.

The ovaries make female sex hormones. They control the functioning of the sex organs. They also control menstruation. And they give the typical characteristics of the mature female: broad pelvis and developed breasts.

The testes make male sex hormones. They control the functioning of the sex organs. They also give the typical characteristics of the mature male: thick facial and body hair, a deep voice, and muscular development.

Glands in the Pancreas

The pancreas itself is not an endocrine gland. But it has some tiny endocrine glands in it. These glands make a hormone called **insulin**. This hormone controls the rate at which your body uses sugar. If your body makes too little insulin, it cannot use the sugar in the blood. The result is excess sugar in your blood. This condition is called **diabetes**.

Section Review

1. Why is the pituitary gland called the master gland?
2. How does the growth hormone affect growth?

3. What are the symptoms of an overactive thyroid? an underactive thyroid?
4. Why do we need iodine in our diet?
5. What does adrenalin do in our bodies?
6. State the functions of the male and female sex hormones.
7. What is the function of insulin?

28.9 ACTIVITY The Effects of Caffeine

Most people know that coffee and tea contain **caffeine**. But did you know that chocolate and most cola drinks do too? How does this drug affect your body (Fig. 28-9)?

Problem

How does caffeine affect your body?

Materials

cup of strong coffee
blood pressure apparatus (optional)

Procedure

a. Sit quietly for 2 or 3 min.
b. Find and record your resting pulse rate (see Section 25.6, page 364).
c. Drink a cup of strong coffee.
d. Continue to sit quietly for about 0.5 h. Find your pulse rate every 5 min. Record your results.
e. If blood pressure apparatus is available, your teacher may measure the blood pressure of 2 or 3 students before and after they drink coffee.

Discussion

1. What was your resting pulse rate?
2. Describe how the coffee (caffeine) affected your pulse rate.
3. Were some people affected more than others? Why might this be so?
4. What part of your nervous system was affected by the caffeine?
5. What hormone was dumped into the blood in greater amounts because of the caffeine?
6. Why can too much coffee keep people awake at night?

Fig. 28-9 Coffee: a relaxing drink?

Section 28.9   415

28.10 Care of Your Control Systems

Damage to the Central Nervous System

Each year thousands of Canadians are crippled by injuries to the brain and spinal cord. Many of these are caused by car accidents. And most of them can be prevented. You should *always* wear a seatbelt. Even in a low speed collision, your brain can be damaged if your head strikes the windshield or dashboard.

Most car accidents are caused by drinking drivers, speeding drivers, and aggressive drivers. How would you deal with these problem drivers?

Car accidents often break the spinal cord. Sometimes the break occurs high in the neck. In this case the person often dies. If, however, the person lives, the legs and arms will likely be paralyzed. This person is called a **quadriplegic** [kwod-ri-PLEE-jik]. If the break occurs further down, the legs and/or trunk will be paralyzed. This person is called a **paraplegic** [par-ah-PLEE-jik].

Many people damage their brains and spinal cords in sports activities. Again, most of these can be prevented. Diving into shallow water has crippled thousands (Fig. 28-10). Always check the depth before you dive.

Wear a helmet in contact sports like hockey. And always wear a helmet when you are on a motorbike. In fact, the wearing of a helmet is a good idea for horseback riders and bicycle riders as well.

Fig. 28-10 Never dive into unknown waters. Check the depth first.

Medical Checkups

Have regular checkups. Your doctor will tell you how often you need one. During a checkup a doctor will make sure all your endocrine glands are working properly.

Suppose your pancreas is not making enough insulin. Your doctor will discover this by testing for sugar in your urine. If the diabetes is mild, diet and pills can control it. If the diabetes is more serious, injections of insulin may be needed. In either case, early detection is important. Otherwise, organs in your body can be damaged.

Most problems of the endocrine system can be controlled with medication.

Drugs

Some drugs excite you. They are called **stimulants**. Others depress you. They are called **depressants**. But all drugs can harm you. That's because they act on your control systems. They upset the normal balance of body processes. Table 28-2 lists some common stimulants and depressants. You can read more about these in Chapter 37.

You may find this rule hard to follow. But, if you want a healthy body and a long life, try it: *Avoid the use of drugs unless they are prescribed by a doctor.*

Table 28-2 Some Drugs

Stimulants (speed up body activities)		Depressants (slow down body activities)	
caffeine	nicotine	alcohol	barbiturates
cocaine	amphetamines	morphine	codeine
marijuana	LSD	opium	heroin
angel dust		marijuana	LSD
		angel dust	

Section Review

1. What is a quadriplegic? a paraplegic?
2. List three ways you can prevent injuries of the central nervous system.
3. How is diabetes treated?
4. What is a stimulant? a depressant?

Main Ideas

1. There are two control systems: the nervous system and the endocrine system.

2. Your behaviour is controlled by your nervous system.
3. A reflex action is a quick, unlearned, automatic response.
4. Your nervous system has two main parts: the central nervous system and the outer nervous system.
5. The autonomic nervous system works without you thinking about it.
6. Endocrine glands make hormones.
7. Hormones control certain body functions.
8. A person can be paralyzed if the brain or spinal cord is damaged.
9. Drugs affect your control systems.

Glossary

cerebellum	ser-uh-BELL-um	the part of the brain that controls muscles
cerebrum	ser-EE-brum	the part of the brain that controls thought, emotions, learning, memory, and the senses
endocrine gland	EN-do-krin	a gland that makes hormones
hormone	HOR-mohn	a chemical that controls certain body processes
medulla	mi-DUL-a	the part of the brain that controls automatic functions
paraplegic	par-ah-PLEE-jik	a person with the legs and/or trunk paralyzed
quadriplegic	kwod-ri-PLEE-jik	a person with arms and legs paralyzed

Study Questions

A. True or False

Decide whether each of the following statements is true or false. If the sentence is false, rewrite it to make it true. (Do not write in this book.)

1. Activity by an organism is called a stimulus.
2. Motor nerve cells carry messages from the central nervous system to parts of the body.
3. You can speed up your learned reflex actions by practicing.
4. The medulla of the brain controls thought.
5. Your heartbeat is controlled by the autonomic nervous system.

B. Completion

Complete each of the following sentences with a word or phrase that will make the sentence correct. (Do not write in this book.)
1. The two control systems are the ▓▓▓▓ system and the ▓▓▓▓ system.
2. An organism's response to a stimulus is called its ▓▓▓▓ .
3. Learning is controlled by the ▓▓▓▓ of the brain.
4. Endocrine glands make ▓▓▓▓ .

C. Multiple Choice

Each of the following statements or questions is followed by four responses. Choose the correct response in each case. (Do not write in this book.)
1. Your muscles are coordinated by the
   a) medulla **b)** cerebrum **c)** cerebellum **d)** spinal cord
2. The spinal cord is part of the
   a) central nervous system **c)** outside nervous system
   b) autonomic nervous system **d)** endocrine system
3. A flash of light in your eyes is called a
   a) response **b)** behaviour **c)** reflex action **d)** stimulus
4. The "master gland" is the
   a) adrenal gland **b)** pancreas **c)** pituitary gland **d)** thyroid gland
5. Marijuana is
   a) a stimulant only **c)** both a stimulant and depressant
   b) a depressant only **d)** neither a stimulant or a depressant

D. Using Your Knowledge

1. Humans have a more developed cerebrum than other animals. What evidence do we have of this?
2. Suppose a large (or even a medium-sized) bear rushed into the room right now. What changes would your autonomic nervous system cause in your body? Why?
3. A diver hit the bottom of a pool and broke his spinal cord high in the neck. How will this affect his body?

E. Investigations

1. Select any hormone present in the human body. Prepare a report of about 200 words on its source and functions.
2. Select one of the drugs in Table 28-2. Prepare a report of about 150 words on its effects on human behaviour.

Unit 7: Continuity

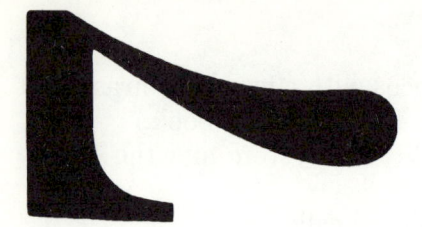

No two individuals of the same species are exactly alike. Yet, for each species, the offspring look like their parents in most respects. Apparently the parents pass on to their offspring some kind of plan. What is the nature of this plan? How do offspring inherit characteristics from their parents?

In this unit you will find out about the plan. You will also find how the field of genetics affects our lives. And you will investigate a related topic, evolution.

CHAPTER 29
Basic Genetics

CHAPTER 30
Genes, Chromosomes, and DNA

CHAPTER 31
Importance of Genetics

CHAPTER 32
Evolution

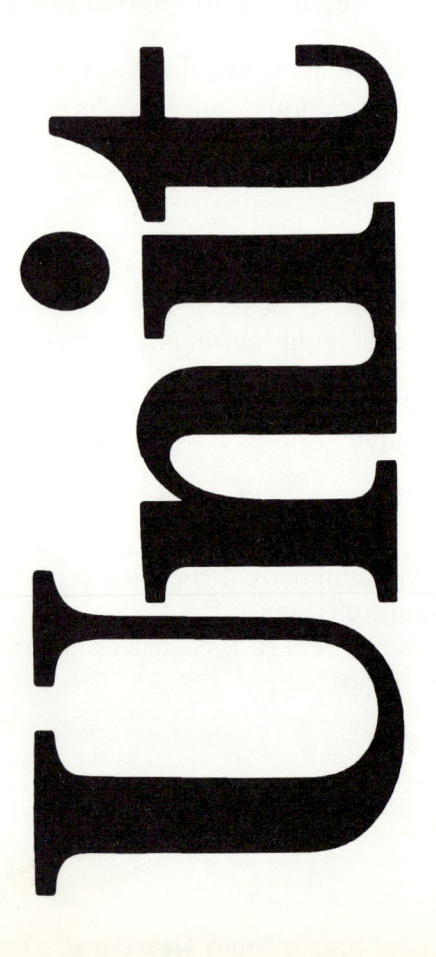

Fig. 29-0 Every organism inherits certain characteristics from its parents. What characteristics has this colt inherited from its mother?

29 Basic Genetics

29.1 Mendel's Experiments
29.2 Activity: Genetics of Corn
29.3 Mendel's Law of Dominance
29.4 Mendel's Law of Segregation
29.5 Understanding Mendel's Laws Using Symbols
29.6 The Punnett Square
29.7 Activity: PTC Tasting
29.8 Blended Inheritance

Have you ever wondered why you "have your father's nose"? Or, why you look a lot like your sister or brother? All of us have noticed that families tend to look alike (Fig. 29-1). Children seem to get some **traits**, or characteristics, from their parents. The passing on of traits from generation to generation is called **heredity** or **inheritance**. The study of heredity is called **genetics**. In this chapter you will begin a study of genetics. You will learn about the first experiments in genetics.

29.1 Mendel's Experiments

Gregor Mendel was a monk who lived in Austria in the middle of the 19th century. He was a teacher who taught math and science. As a hobby, Mendel did experiments with small animals and plants. His

Fig. 29-1 It is easy to tell that these two children are sister and brother. Why do families tend to look alike?

most important experiments were done with peas from his garden. Mendel looked at how traits were inherited in his pea plants. Let's look at the results of some of his experiments.

Mendel's Pea Plants

Why did Mendel choose pea plants for his study of inheritance? First, he noticed that not all pea plants looked the same. They had different characteristics, or **traits**. Some were tall, while others were short. Some had wrinkled seeds, while others had smooth seeds. Mendel noticed seven pairs of traits that he could study. He chose plants with opposite forms of a trait as parents. For example, he crossed tall pea plants with short pea plants. This crossing of two different types of the same species is called **hybridization**. The offspring are called **hybrids** (Fig. 29-2).

Pea plants are good for hybridization experiments. Why? Pea plants are easy to grow. In nature, they are self-pollinating. This means that the pollen from the anther of a pea flower is transferred to the pistil of the *same* flower. Why was self-pollination important for Mendel's experiment? Mendel cut off the stamens of some pea flowers before they could self-pollinate. Then, he could choose pollen from different pea plants to fertilize these flowers (Fig. 29-3).

Mendel's Results

In one experiment, Mendel took pollen from some tall pea plants. He fertilized the flowers of short pea plants with this pollen. Mendel had crossed two types of pea plant. They were different in one trait only.

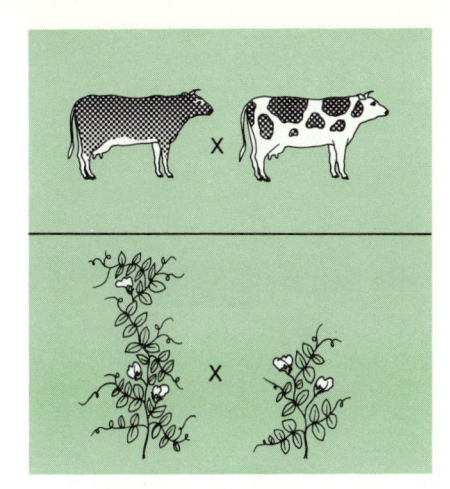

Fig. 29-2 These are two examples of hybrid crosses. The cattle are members of the same species. But they differ in colour. The pea plants are also of the same species. But they differ in height. The offspring of these crosses will be hybrids.

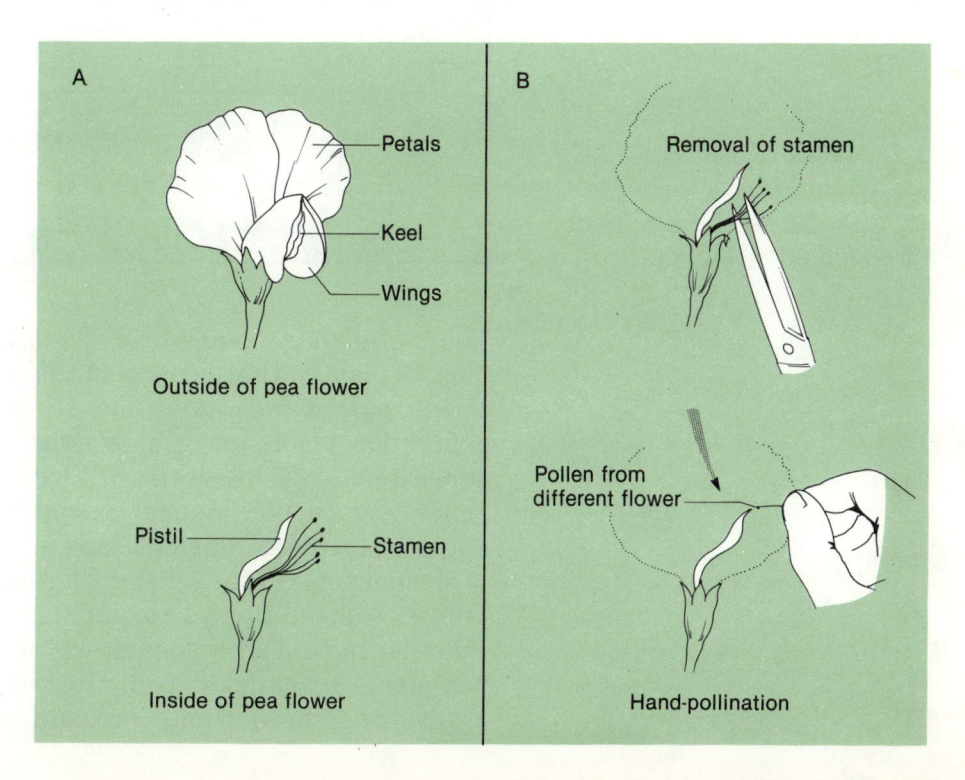

Fig. 29-3 Pea plants were good for Mendel's hybridization experiments. The stamen and pistils are protected from the wind by the wings and keel (A). Mendel cut off the stamens of some pea flowers before self-pollination. He then put pollen from another plant on the pistil (B).

Section 29.1

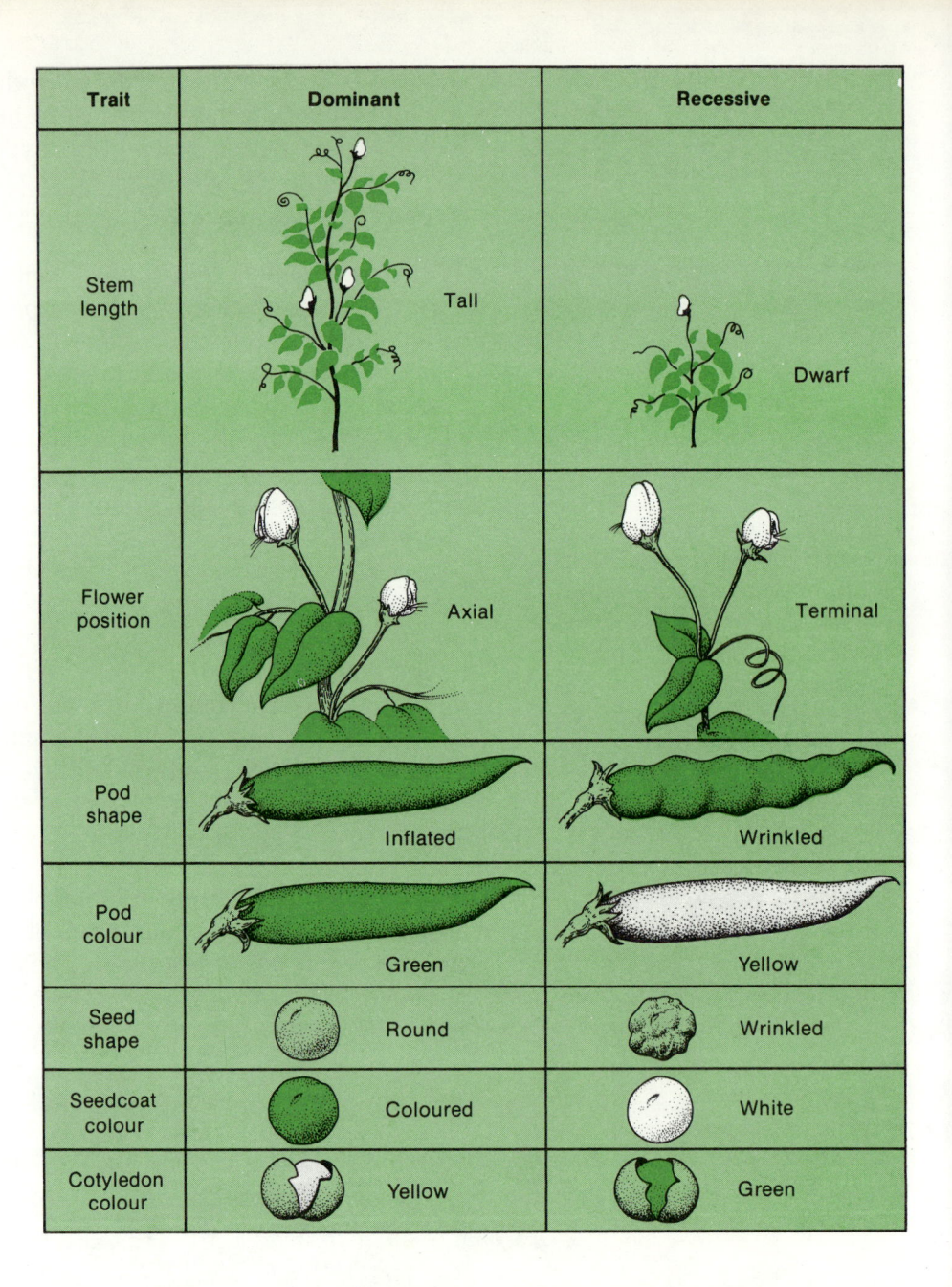

Trait	Dominant	Recessive
Stem length	Tall	Dwarf
Flower position	Axial	Terminal
Pod shape	Inflated	Wrinkled
Pod colour	Green	Yellow
Seed shape	Round	Wrinkled
Seedcoat colour	Coloured	White
Cotyledon colour	Yellow	Green

Fig. 29-4 Some of the pairs of traits studied by Mendel

This is called a **monohybrid cross**. Let's see what happened when Mendel planted the seeds from this monohybrid cross.

One parent was tall. The other parent was short. The first generation of offspring of a hybrid cross is called the **first filial generation** or **F_1 generation**. What height would you expect the F_1 generation to be? Mendel thought that the offspring pea plants might be medium in height. Was Mendel's prediction right?

Mendel found that *all* the F_1 generation were tall. The trait for tallness seemed to be stronger than the trait for shortness. Mendel said that the tall trait dominated the short trait. He called tallness in pea plants a **dominant trait**. He called shortness a **recessive trait**. Figure 29-4 shows some dominant traits that Mendel found in pea

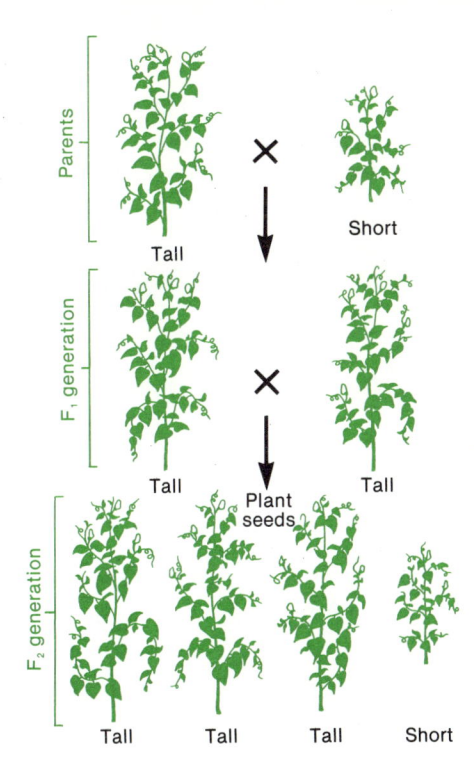

Fig. 29-5 Mendel's experiment with tall and short pea plants. The F_1 generation were all tall plants. When the F_1 hybrids self-pollinate, ¾ of the offspring are tall and ¼ are short.

plants. Why weren't the pea plants a mixture of their parent's heights? What happened to the short trait? What does dominant trait really mean?

Further Experiments

Mendel did some more experiments to try to find the answers to these questions. He let the F_1 generation pea plants self-pollinate. As you know, all the F_1 plants were tall. Mendel planted the seeds from the F_1 cross. He found that three-quarters of the plants that grew from these seeds were tall plants. But, one quarter of the offspring were short plants (Fig. 29-5). This generation of plants is called the **second filial** or **F_2 generation**.

Green pods is also a dominant trait in pea plants. What do you think Mendel found when he crossed a plant with green pods and a plant with yellow pods? He found that all the F_1 generation had green pods. When he let the F_1 generation self-pollinate, he found that three-quarters of the F_2 generation had green pods. The rest of the F_2 generation had yellow pods. This is the same pattern that Mendel found when he crossed the plants with different heights. He found the same pattern for all seven pairs of traits shown in Figure 29-4. The dominant trait is the stronger trait. It shows up more often in the F_2 generation. The opposite, or recessive, trait shows up less often in the F_2 generation.

You have read about the results of Mendel's experiments. Mendel drew some important conclusions from these results. You will learn about Mendel's rules in Section 29.3.

Section Review

1. What is genetics?
2. Why are pea plants good for hybridization experiments?
3. What is a monohybrid cross?
4. What is a dominant trait? Name two dominant traits in pea plants.

29.2 ACTIVITY Genetics of Corn

Corn plants have pairs of opposite traits much the same as Mendel's pea plants. Corn seedlings can be green or albino (colourless). Does this pair of traits follow the same pattern of inheritance that Mendel found for his pea plants? This activity will help you answer this question. Note: You will not be able to answer the Discussion questions until you have read the rest of this chapter.

Problem

How are green and albino traits inherited in corn?

Fig. 29-6 Plant the corn seeds about 5 cm deep and 5 cm apart.

Materials

potting soil
enough containers to hold 8 seeds, planted 4-5 cm deep and 4-5 cm apart
corn seed that will grow to be the F_2 generation of a cross between green and albino corn

Procedure

a. Fill the containers to a depth of 7-8 cm with potting soil.
b. Plant 8 corn seeds of the F_2 generation about 5 cm deep and 5 cm apart. Label the containers (Fig. 29-6).
c. Water the soil when it seems dry.
d. Look at the containers after 7, 10, and 14 d. Write down the number of green and albino corn seedlings.
e. Pool the class results.
f. If you do not have corn seeds to plant, you can use the class results in Table 29-1 to answer the discussion questions.

Table 29-1 Some Class Results

Colour of seedling	Number of F_2 seedlings
Green	185
Albino	55

Discussion

1. What trait do you think is dominant in corn: green leaf colour or albino leaf colour?
2. Do your results show that the inheritance of leaf colour in corn is the same as height in Mendel's pea plants? Explain your answer.
3. Draw a Punnett Square of a cross between a green corn plant (GG) and an albino corn plant (gg). Will the F_1 offspring be green or albino?
4. Draw a Punnett Square of the F_1 cross from Question 3 (Gg x Gg). How many of the offspring are green? How many are albino? How do the results of the Punnett Square compare with the class results?

29.3 Mendel's Law of Dominance

As you know, Mendel found some interesting patterns in the results of his pea experiments. He found what he called dominant and recessive traits. What happens to the recessive trait in the F_1 generation? And why does it reappear in the F_2 generation?

Genes

Mendel concluded that the parent pea plants must have many "bits" or units of information. These bits of information are passed on to their offspring. Mendel called these units of information **factors**. Today, we call them **genes**.

A tall plant has the gene for tallness. This gene will be passed on to its offspring. A short plant has the gene for shortness. This gene will be passed on to its offspring.

Mendel knew that inheritance in pea plants was not this simple. He knew that the tall F_1 pea plants could have short offspring. If tall pea plants only have the gene for tallness, how can they have short offspring? The short F_2 plants get the gene for shortness from their parents. So, the tall F_1 parents must have the gene for shortness. But they must also have a tall gene because they are tall. Therefore Mendel concluded that each pea plant must have two genes for each trait. In pea plants, the gene for height has two forms (tall and short).

Law of Dominance

The tall F_1 plants are hybrids for the gene for height. They have the dominant gene for tallness. They also have the recessive gene for shortness. When the short gene is present at the same time as the tall gene, it is not expressed. The plant will be tall even though the short gene is present. Any gene that expresses itself when the other partner gene does not is called a **dominant gene**. The dominant gene masks the presence of its partner gene. Mendel called this idea the **Law of Dominance**.

Section Review

1. What is a gene?
2. Why is the tall gene in pea plants called a dominant gene?

29.4 Mendel's Law of Segregation

How do parents pass on genes to their offspring? Genes are passed on to the offspring through the reproductive cells or gametes. In the last section you learned that pea plants have two genes for each trait. We call these partner genes. This is true for other organisms as well. If an offspring is to end up with two genes for each trait, each parent must pass on only one of its genes. When the gametes are being formed in the parent, the partner genes separate into different cells. A pea plant gamete may have the gene for yellow pods. Or, it may have the gene for green pods. But it will not have both genes. In this way, when gametes join to form a zygote, the offspring will have a pair of genes for pod colour.

Mendel used these observations to write the **Law of Segregation** [SEG-ruh-gay-shun]. This law tells us that any organism has a pair of genes for each trait. The partner genes are separated, or **segregated**, when gametes are formed. This law applies to humans as well.

Section Review

1. How are genes passed from parents to offspring?
2. What does the Law of Segregation tell us?

29.5 Understanding Mendel's Laws Using Symbols

Mendel needed an easy way to write down his ideas about genes. He used letters of the alphabet as symbols for genes. Mendel used a capital letter to stand for a dominant trait. And he used a small letter to stand for a recessive trait. Therefore the gene for tallness in pea plants is T. And the gene for shortness is t.

Let's use these symbols to look more closely at Mendel's results. In Section 29.1, you learned that Mendel crossed a tall pea plant and a short pea plant. You know all pea plants have two genes for each trait. A pure tall pea plant can have only tall offspring. Therefore it must have two genes for tallness. We can write this as TT. These symbols show what genes for height are present in the pea plant. They represent the **genotype** [JEN-o-type] of the plant.

A pure short pea plant can have only short offspring. Therefore it must have two genes for shortness. The genotype of a short pea plant can be written as tt.

How can we use these symbols to find out the genotype of the F_1 generation? The Law of Segregation tells us that all of the gametes of the tall parent must have the gene T. Also, all of the gametes of the short parent must have the gene t. Figure 29-7,A shows how we can write this cross using symbols.

As you can see, all the offspring have the genotype Tt. They are hybrids for the height gene. Will these plants be tall or short? We know that the T gene is dominant. The Law of Dominance tells us that the offspring will be tall. The appearance of the plant is called its **phenotype** [FEE-no-type]. The F_1 offspring have a tall phenotype.

Can two pea plants with the same phenotype have different genotypes? The tall parent has the genotype TT. The tall F_1 offspring has the genotype Tt. Both plants are tall. They both have the same phenotype. The phenotype only tells us the appearance of an organism. It does not tell us the exact genotype.

We can also use gene symbols to help us to understand what happens when the F_1 plants self-pollinate. We know that the genotype of the F_1 can be written as Tt. Let's call one F_1 parent "Plant A" and the other "Plant B". The F_1 cross can be written as:

$$\text{Tt (Plant A)} \quad \times \quad \text{Tt (Plant B)}$$

The gametes of the parent plants will have only one gene for height.

428 Chapter 29

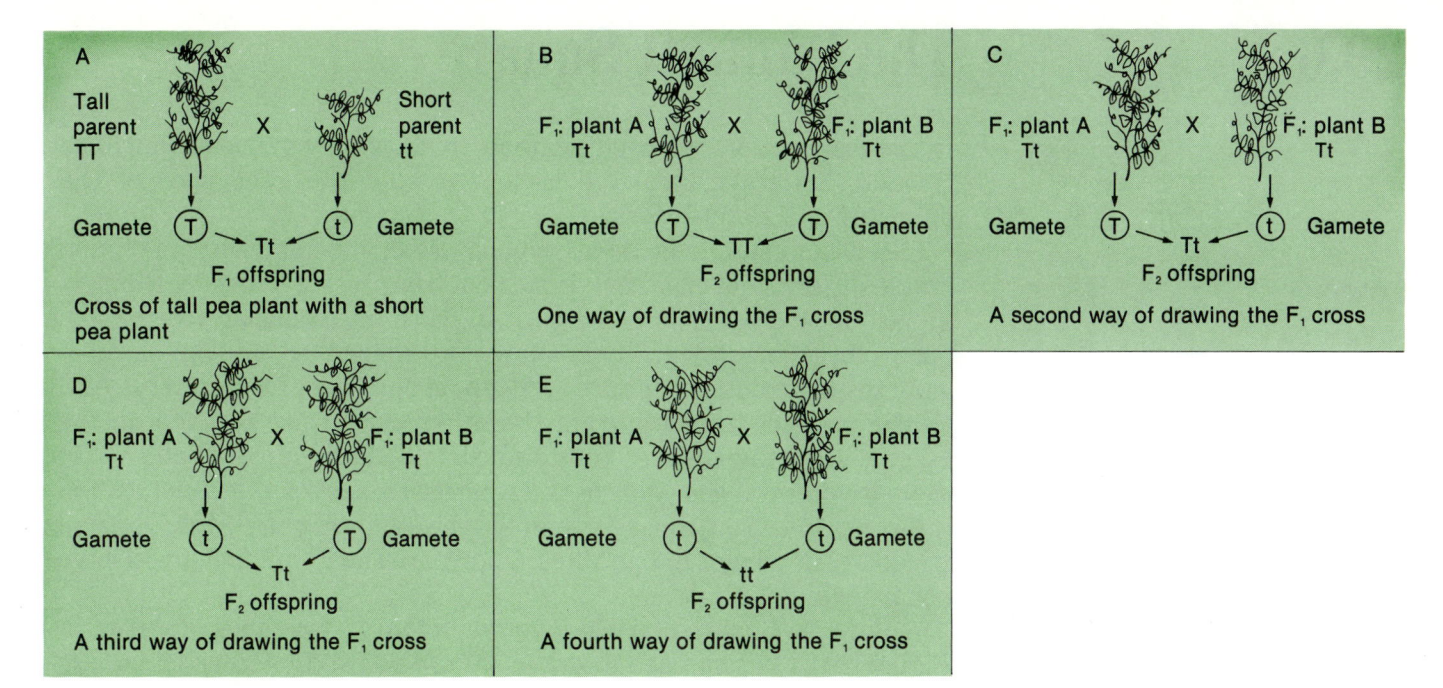

Fig. 29-7 Symbols can be used to help us understand Mendel's pea plant experiments.

Figure 29-7,B shows what happens if the gametes of both parents get the T gene. The genotype of the offspring will be TT. What phenotype will this plant have?

Suppose the gamete of Plant A contains the gene for tallness, T. But the gamete of Plant B has the t gene (Fig. 29-7,C). The F_2 seedling will have the genotype Tt. We know that the T gene is dominant. Therefore the offspring will have a tall phenotype.

There are two more ways to draw this cross. In one case, the offspring has the genotype Tt and is tall (Fig. 29-7,D). But if the gametes of both parents have the t gene, the offspring will be tt (Fig. 29-7,E). This plant will be short.

Using symbols, we have shown that there are four ways that the F_1 parents can hybridize to produce the F_2 generation. The genotypes of the F_2 offspring may be TT, Tt, or tt. The Tt genotype will occur twice as often as the TT or tt genotype (Fig. 29-7, C & D). What will the phenotypes be? Both the TT and Tt offspring will be tall. Only the tt offspring will be short. Of the crosses that we drew, 3 out of 4 result in tall offspring.

This is exactly what Mendel found. He found that three-quarters of the F_2 plants in his crosses were tall while one-quarter were short. We have just explained Mendel's results using symbols for the two genes for height.

Section Review

1. What is the difference between genotype and phenotype?
2. Two pea plants have the same phenotype for height. Must they have the same genotype? Explain your answer.

Section 29.5

29.6   The Punnett Square

In the last section, we used symbols to show how gametes join. The Punnett Square makes it easier to predict the genotypes of the offspring in a simple cross.

Like pea plants, fruit flies are good organisms for studying simple crosses. Fruit flies have many pairs of contrasting traits. A Punnett Square for a cross between a fruit fly with pure normal wings and a pure short-winged fruit fly is shown in Figure 29-8. All the possible gametes from one parent are written on one side of the square. And all the possible gametes from the other parent are written on the other side of the square. You take each of these gametes and write them together in the grids of the square. This gives the genotypes of the offspring. You can see that all the offspring are Ww. Normal wings are dominant in fruit flies. Therefore the F_1 offspring will have normal wings.

A Punnett Square of a cross between two F_1 fruit flies is shown in Figure 29-9. You can see that the square shows that three-quarters of the F_2 generation will have normal wings and one-quarter will have short wings. The Punnett Square can be used to predict the results of simple crosses. It can also be used to solve harder genetic problems.

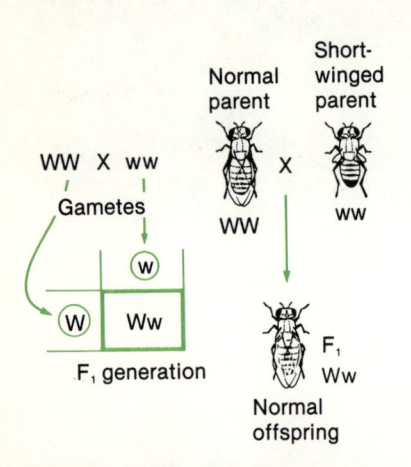

Fig. 29-8 Punnett Square of a cross between a fruit fly with normal wings and a fruit fly with short wings. W stands for the dominant gene for normal wings. w stands for the recessive gene for short wings.

Section Review

1. a) What is a Punnett Square?
   b) How can it be used to predict the results of simple crosses?
2. Why are fruit flies good organisms for studying simple crosses?

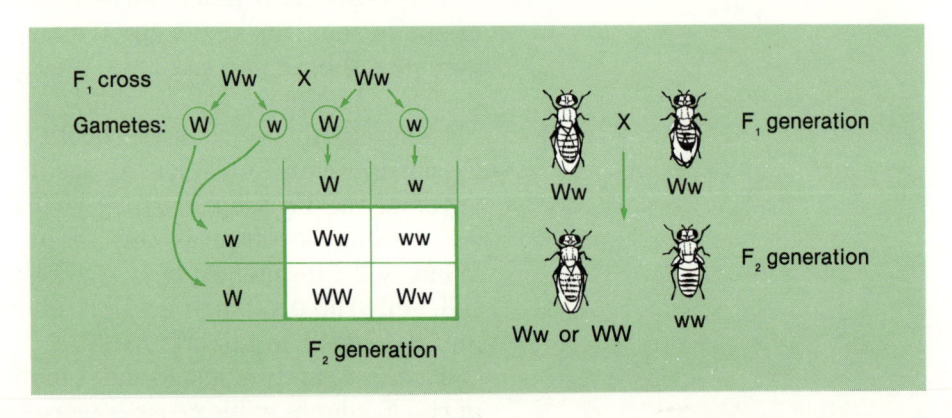

Fig. 29-9 Punnett square of the cross between two F_1 offspring shown in Fig. 29-8.

29.7   ACTIVITY   PTC Tasting

You have read that fruit flies and pea plants have many pairs of contrasting traits. The same is true for humans. For example, some

430   Chapter 29

people can taste a substance called PTC. These people are called "tasters". Other people cannot taste PTC. They are called "non-tasters". PTC tasting is an inherited trait. The taster gene (T) is dominant to the non-taster gene (t). In this activity, you will find out if you are a taster. You will also test your family.

Problem

Are you and your family tasters or non-tasters?

Materials

PTC test paper

Procedure

a. Get a piece of PTC test paper for yourself. Also get a strip for each person in your family that you can test. (e.g. mother, father, brothers, sisters, grandparents).
b. Place the strip on your tongue, towards the back. Do not chew the strip. You should be able to tell right away if you are a taster. Throw away the strip.
c. Have your family do step (b).
d. Write down the results for each person tested.
e. Compare your results with those of your classmates.

Discussion

1. The taster gene (T) is a dominant gene. Are any of your family non-tasters? What must their genotype be?
2. Tasters can have two different genotypes. What are they? (Hint: Remember the tall F_1 pea plants.)
3. Can you tell from the test of your parents what the genotypes of your family should be? You might use a Punnett Square to answer this question.

29.8 Blended Inheritance

When Mendel crossed a tall and a short pea plant, he did not find medium-sized offspring. The two genes for height were not blended together in the offspring. As you know, Mendel wrote the Law of Dominance based on these observations.

Is the Law of Dominance always true? Botanists have found plants which show exceptions to the Law of Dominance. Figure 29-10 shows a cross between a red four o'clock flower and a white four o'clock flower. Botanists thought that all of the F_1 generation would be red. Instead, they found all pink flowers! The red and white genes act together to produce a pink colour in the offspring. This is called **blended inheritance**.

Is Mendel's Law of Segregation true for four o'clock flowers? As you can see in Figure 29-10, when you mate two pink F_1 flowers, the two genes for colour segregate. Therefore Mendel's Law of Segregation is true even in plants showing blended inheritance.

Section Review

1. Did scientists think the red or the white gene in four o'clock flowers was dominant? Were they right?
2. What is blended inheritance?
3. How do we know that Mendel's Law of Segregation is still true in four o'clock flowers?

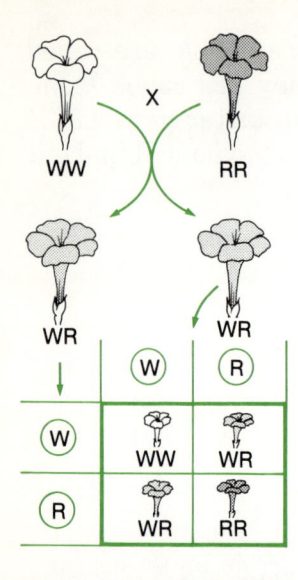

W = gene for white flowers
R = gene for red flowers

Fig. 29-10 When four o'clock flowers are crossed, the traits for flower-colour are blended. In the F_2 generation, the traits segregate.

Main Ideas

1. Heredity is the passing on of traits from generation to generation. Genetics is the study of heredity.
2. Parents pass on traits, or genes, to their offspring.
3. Adult organisms have two copies of all genes. They pass on one copy of each gene to their offspring through the gametes.
4. When a gene expresses itself while its partner does not, it is said to be a dominant gene.
5. Mendel's Law of Segregation tells us that partner genes are separated when gametes are formed.
6. The genotype of an organism tells us what genes it has.
7. The phenotype of an organism tells us about its appearance.
8. Blended inheritance is an exception to the Law of Dominance.

Glossary

genes	jeenz	units of inheritance
genetics	je-NET-iks	the study of heredity
genotype	JEN-o-type	the genes of an organism
hybrid	HI-brid	an organism that is not pure for a given trait
phenotype	FEE-no-type	the appearance of an organism due to its genes
segregate	SEG-ruh-gayt	separate

432 Chapter 29

Study Questions

A. True or False

Decide whether each of the following statements is true or false. If the sentence is false, rewrite it to make it true. (Do not write in this book.)

1. Hybrids are the offspring of a cross between members of two different species.
2. When Mendel crossed a pure tall pea plant (TT) and a pure short pea plant (tt), he found all the offspring were tall.
3. The phenotype of an organism tells us exactly what genes it carries.
4. Mendel's Law of Dominance is always true.
5. A cross between two pea plants that are only different in one trait is called a monohybrid cross.

B. Completion

Complete each of the following sentences with a word or phrase that will make the sentence correct. (Do not write in this book.)

1. Gregor Mendel studied the inheritance of different traits in _____.
2. A gene which is expressed when its partner gene is not is called a _____ gene.
3. Genes are passed from parent to offspring through the _____.
4. A _____ _____ makes it easier to predict the offspring of a simple cross.
5. Flower colour in four o'clock flowers is an example of _____ inheritance.

C. Multiple Choice

Each of the following statements or questions is followed by four responses. Choose the correct response in each case. (Do not write in this book.)

1. Mendel explained heredity in pea plants by assuming
    a) all traits are dominant
    b) each parent has two genes for each trait
    c) offspring have more genes than their parents
    d) all traits are recessive
2. The offspring of a cross between a pea plant with smooth pods (SS) and a pea plant with wrinkled pods (ss) are represented by
    a) SS   b) ss   c) Ss   d) SS or ss
3. If the F_1 generation from the cross in question 2 are allowed to self-pollinate, the offspring (F_2) will
    a) all have wrinkled seeds
    b) all have smooth seeds

c) be three-quarters wrinkled and one-quarter smooth
   d) be one-quarter wrinkled and three-quarters smooth
4. The Law of Dominance tells us that
   a) recessive genes cannot be passed on to the offspring
   b) dominant genes can mask the presence of recessive genes
   c) recessive genes are never expressed
   d) hybrids do not have dominant genes
5. In a cross between two pink four o'clock flowers, the offspring will be
   a) all pink
   b) half red and half white
   c) red, white, or pink
   d) all red or all white

D. Using Your Knowledge

1. Suppose white fur is a recessive trait in guinea pigs. What will be the result of crossing a pure brown-haired guinea pig and a pure white-haired guinea pig?
2. Some students in a biology class mated two fruit flies with red eyes. They looked at the eye colour of 1176 offspring. They found that 392 of the offspring had brown eyes. The other 784 had red eyes. What is likely the genotype of the parents? (Use symbols to answer this question.)
3. Tongue-rolling is a dominant trait in humans. Will the children of two people who cannot roll their tongues be "tongue-rollers"? Why or why not?
4. Draw a Punnett Square to show the offspring of a cross between a hybrid plant with coloured seeds and a pea plant with white seeds. What are the possible genotypes of the offspring? Why is it important that you know that the one parent is a hybrid for the seed colour genes?

E. Investigations

1. Examine Figure 29-11. This shows a cross between two guinea pigs. Design an experiment to explain the results of this cross.
2. Look at a seed catalogue. Pick out some flowers or seeds that you think are hybrids. Explain your choice.
3. You know that PTC tasting is a dominant trait in humans. Find out some other dominant or recessive traits in humans. Can you predict the genotype of your family members for some of these traits?

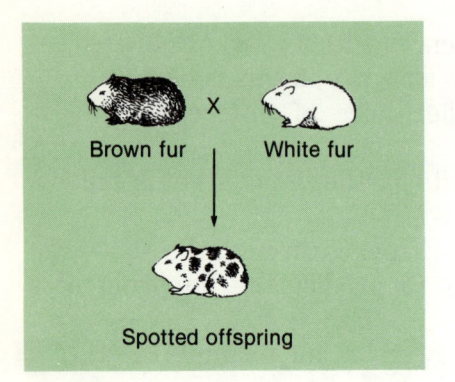

Fig. 29-11 Can you explain the results of this cross?

30 Genes, Chromosomes, and DNA

30.1 Chromosomes and Meiosis
30.2 Meiosis and Mendel's Laws
30.3 Linkage
30.4 Activity: Law of Independent Assortment
30.5 Sex and Chromosomes
30.6 Sex Linkage
30.7 Activity: Sex Linkage in Humans: Colour Blindness
30.8 Mutations
30.9 DNA: The Genetic Material

In the last chapter, you learned about Mendel's early experiments in genetics. Mendel described *what* happened in his pea plant crosses. But, biologists of Mendel's time did not know *why* he got those results. Where are Mendel's "factors" or genes located in the cell? Today, biologists understand that Mendel's factors are carried on the **chromosomes**. In this chapter, you will find out how this discovery was made. You will also learn something about how genes work.

30.1 Chromosomes and Meiosis

A Review of Mitosis

You studied a process called mitosis in Section 8.2 (page 93). Here is a list of the main ideas from that section.
- The nucleus of a cell contains chromosomes.
- Chromosomes are the genetic material of the cell.
- During mitosis the nucleus forms an *exact copy* of itself.
- Mitosis ensures that the daughter cells are *exactly like* the parent cells. Each daughter cell has the same number and kind of chromosomes as the parent cell.
- Every body cell in an organism has the same number of chromosomes. For example, all normal human cells have 46 chromosomes. All onion cells have 16. All crayfish cells have 200. And all fruit fly cells have 8.

The Need for a New Type of Division: Meiosis

You began life when a **sperm cell** from your father united with an **egg cell** from your mother. A cell called a **zygote** was formed. Suppose the sperm cell and egg cell each had 46 chromosomes like all human body cells. Then the zygote would have 92 chromosomes. You developed from the zygote by mitosis. Therefore all your body cells would have 92 chromosomes. But this is not so. If it were, you would not be a human.

What, then, happens? When the gametes were being formed, their chromosome number is cut in half. The sperm cell now has 23 chromosomes. And the egg cell also has 23 chromosomes. Then, when they unite, they form a zygote with 46 chromosomes. This zygote develops into a normal human with 46 chromosomes in all the body cells.

The process that reduces the chromosome number is called **meiosis** [mi-OH-sis]. Let's see how this process works.

Stage 1 of Meiosis

Figure 30-1 shows the first stage of meiosis. Follow the diagrams carefully from top to bottom. Read the explanations beside the diagrams as you do so. Note: The chromosomes in each body cell normally occur in pairs. Such a pair is called a **homologous pair** [ho-MOL-o-gus]. Thus the 46 chromosomes in a human cell make up 23 homologous pairs. One chromosome of each pair comes from the father. The other comes from the mother. Each chromosome of a pair carries genetic information for the same traits as its partner. The genetic information is at specific points called **genes**.

The two chromosomes of a pair carry genes for the same traits at the same positions. However, the two chromosomes are not genetically identical. For example, both chromosomes may carry information on hair colour. But one chromosome may carry genes for brown hair. And the other may carry genes for blond hair. The offspring's hair colour is determined by the combination of these two pieces of genetic information.

Stage 2 of Meiosis

Figure 30-2 shows the second stage of meiosis. Follow the diagrams carefully from top to bottom. Read the explanations beside the diagrams as you do so.

In this stage, mitosis occurs except for one thing. The chromosomes do not make copies of themselves. (They did this back in Stage 1.)

As you can see, the gametes have only half the number of chromosomes as the parent cell had. Note, too, that each gamete has only one chromosome from each homologous pair. In other words, the homologous chromosomes have separated, or **segregated**, during meiosis.

Homologous pair

Homologous pair

The body cell that produces gametes has homologous pairs of chromosomes.

The homologous chromosomes pair up.

Chromatid

Each chromosome makes a copy of itself. Each part is called a chromatid.

The chromosome pairs line up at the centre of the cell.

The chromosome pairs separate.

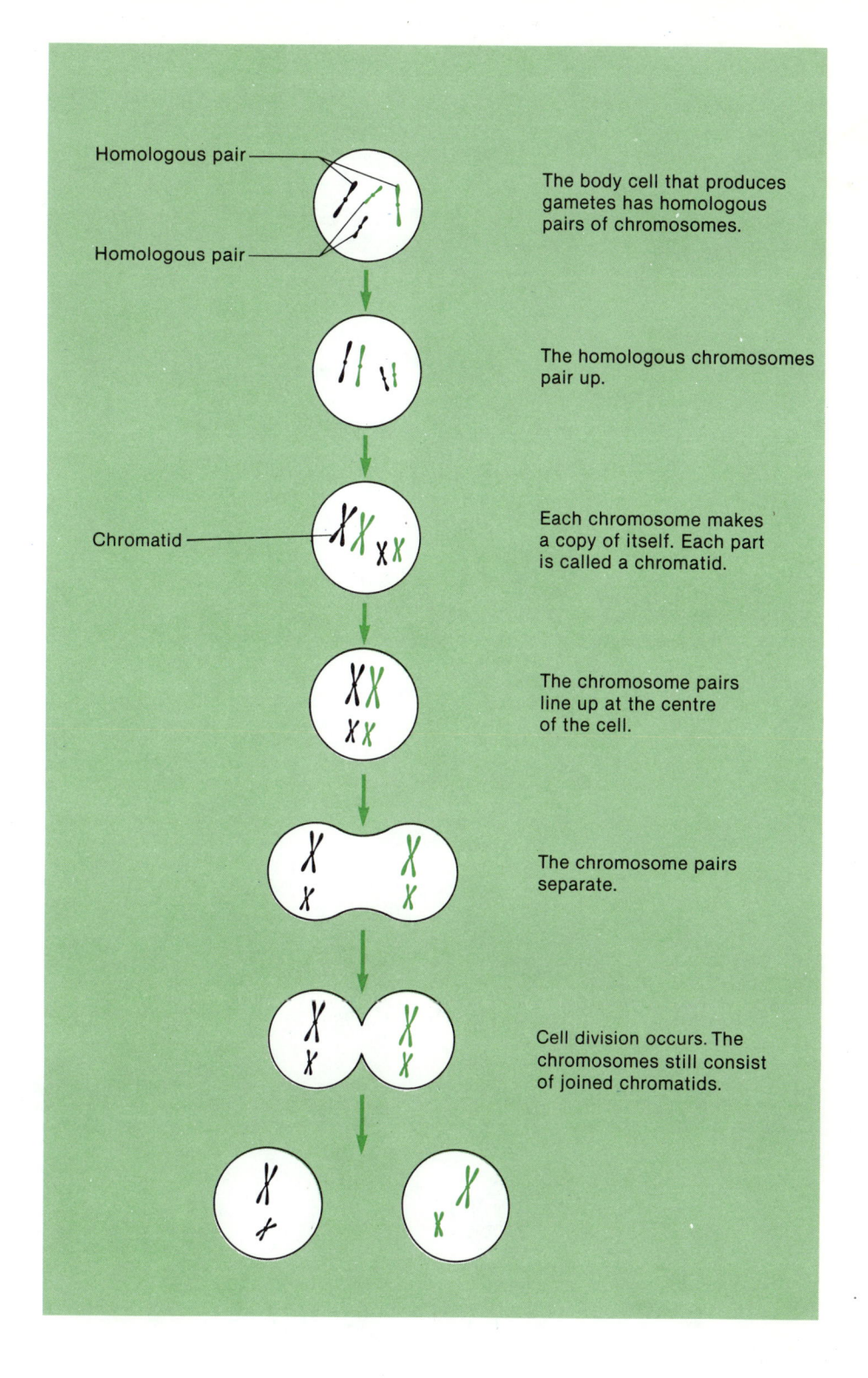

Cell division occurs. The chromosomes still consist of joined chromatids.

Fig. 30-1 Stage 1 of meiosis for an organism with 4 chromosomes in its body cells

Section 30.1 437

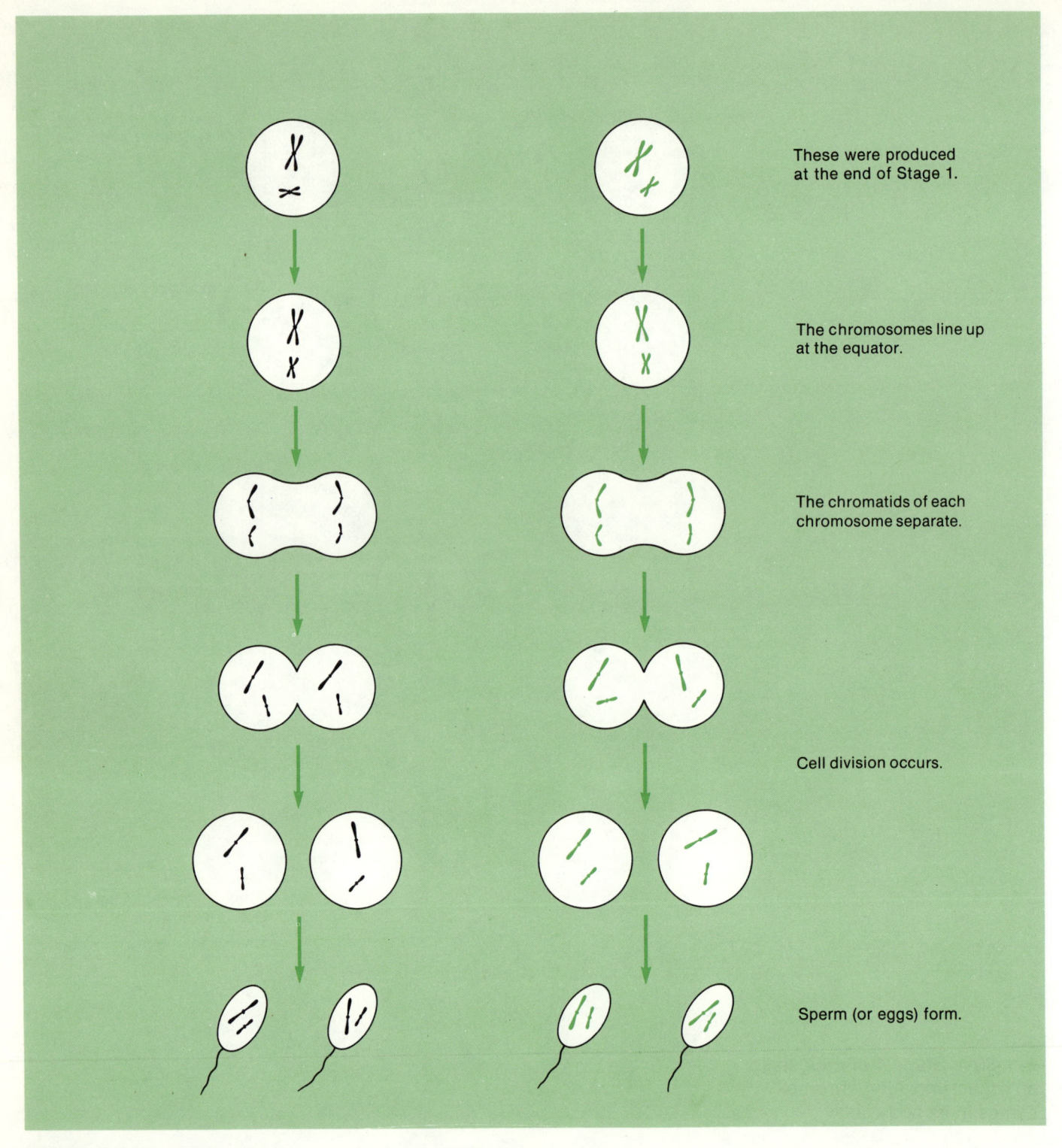

Fig. 30-2 Stage 2 of meiosis

These were produced at the end of Stage 1.

The chromosomes line up at the equator.

The chromatids of each chromosome separate.

Cell division occurs.

Sperm (or eggs) form.

Section Review

1. What happens during mitosis?
2. Why is meiosis needed?

3. What is a homologous pair?
4. What is a chromatid?
5. What does meiosis accomplish?

30.2 Meiosis and Mendel's Laws

Early in this century, a student named Walter S. Sutton studied Mendel's results. He compared Mendel's results with what scientists knew about meiosis. Go back and read Section 29.4 again (Mendel's Law of Segregation). Do you see any similarities between Mendel's laws and what you have just learned about meiosis?

Sutton noticed that chromosomes occur in pairs in cells. Mendel's "factors" also occur in pairs. Chromosomes separate or segregate during meiosis. Mendel found that pairs of "factors" segregate when gametes are formed.

Other scientists found Sutton's results interesting. Thomas H. Morgan did some important experiments using fruit flies. These experiments helped to prove that genes are found on the chromosomes. You will learn more about Morgan's experiments later in this chapter.

Genes and Chromosomes

You know that all the body cells in an organism have the same number of chromosomes. For example, all human cells have 46 chromosomes. And all fruit fly cells have 8 chromosomes. You also know that chromosomes are found in pairs. Genes are also found in pairs. A parent carries two genes for a trait.

Study Figure 30-3. There are 14 pairs of chromosomes in all the cells of the garden pea plant. Figure 30-3 shows the pair of chromosomes that has the gene for height. One member of this pair

Fig. 30-3 Height in pea plants is controlled by two genes on a pair of chromosomes. These partner chromosomes segregate at meiosis into separate gametes.

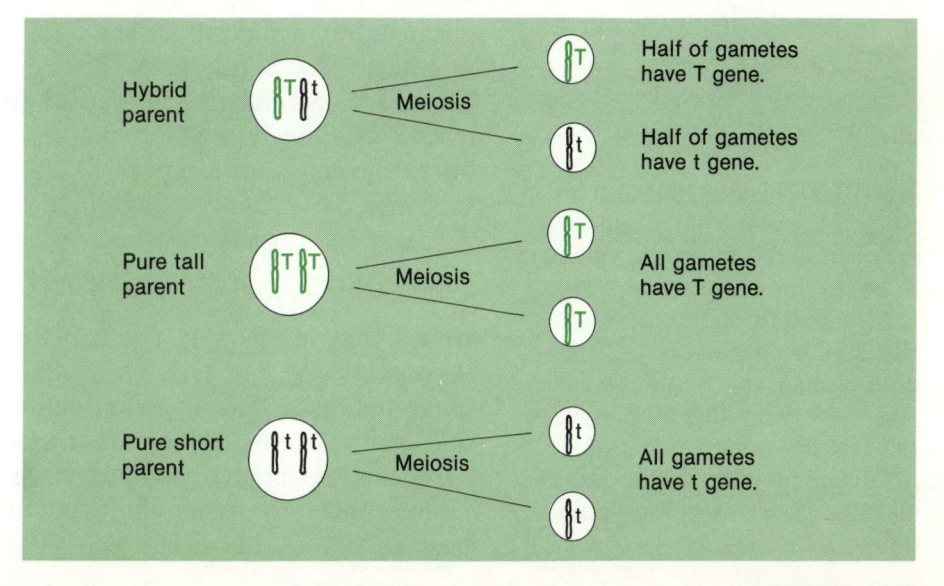

has the gene for tallness (T). The partner chromosome has the gene for shortness (t). During meiosis, these partner chromosomes segregate (separate). The chromosome carrying the T gene goes to one of the gametes. And the t gene ends up in another gamete. In fact, of all the gametes made by this pea plant, half will have the T gene. The other half of the gametes will have the t gene. What type of gametes will be produced by a pure tall pea plant (TT)? Or a pure short pea plant (tt)?

Gametes only have one chromosome from each pair of chromosomes in the parent. How many chromosomes does a pea plant gamete have? It must have half the number of chromosomes of the parent, or seven. What happens when two gametes meet to form a zygote? The zygote will have the same number of chromosomes as the parent (Fig. 30-4). Meiosis is important for keeping the number of chromosomes and genes constant from generation to generation.

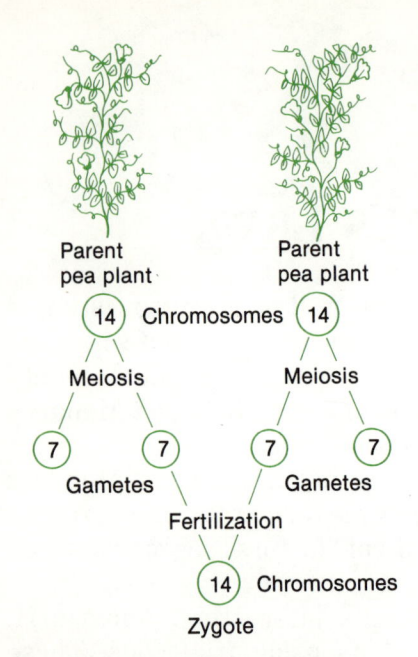

Fig. 30-4 Meiosis is important for keeping the number of chromosomes constant from generation to generation.

Section Review

1. What important observations did Walter S. Sutton make?
2. How many chromosomes do human gametes have?
3. What type of gametes will be produced by a hybrid tall pea plant (Tt)? How many gametes of each type will be produced?
4. Why is meiosis important?

30.3 Linkage

Fig. 30-5 This is a diagram of chromosome II from a human cell. Some of the genes known to be on this chromosome are shown as letters and numbers beside the chromosome. This kind of diagram is called a linkage map. From this map, we can tell that the gene for insulin (INS) will be inherited together with the gene for hemoglobin (Hb).

What Is Linkage?

When Mendel studied pea plants, he looked at the inheritance of more than one trait in the same plant. He always found that the traits were *not* inherited together. Instead, he found that traits were inherited independently. By chance, Mendel missed an important observation. Think about the number of genes that there must be in humans. Now think about the number of pairs of chromosomes. There are certainly more than 23 genes in humans! In fact, there are over 3000 *known* genes. Scientists know that there are many genes on a chromosome (Fig. 30-5). These genes tend to be inherited together. This idea is called linkage.

Law of Independent Assortment

Genes that are not on the same chromosome tend to be inherited separately. At meiosis, they are said to be assorted (separated) independently. This observation has been called the **Law of Independent Assortment**. Independent assortment of chromosomes at meiosis is one important reason why organisms of the same species do not all look the same. You will learn more about what this means in the next section of this chapter.

440 Chapter 30

Section Review

1. What does linkage mean?
2. What is the Law of Independent Assortment? Why is it important?

30.4   ACTIVITY   Law of Independent Assortment

You have learned that genes on different chromosomes tend to be inherited separately. In this activity, you will use homemade chromosomes to help you understand independent assortment. You know that pea plants can be tall or short. And the pods of pea plants may be green or yellow. The genes for these two traits are on different chromosomes in the pea plant gamete. Using your homemade chromosomes, you will follow these pea plant genes through meiosis. By comparing your data with the class results, you should be able to draw some important conclusions about independent assortment.

Problem

Can you demonstrate the Law of Independent Assortment?

Materials

green and black wool   sticky labels
paper clips   small box

Procedure

a. Make your pea plant chromosomes as shown in Figure 30-6. Use the sticky labels to attach the genes to the chromosomes. Use symbols to represent the 4 genes:
T = tall, t = short, G = green pods, g = yellow pods.
b. Place all 4 chromosomes into the box on your desk.
c. Put 4 pieces of paper on your desk. Label them gametes 1, 2, 3, and 4. You are responsible for two of these gametes. Your partner is in charge of the other two gametes.
d. Section 30.1 (Meiosis) should help you with the rest of this activity. Reach into the box and remove one chromosome. Reach in again and draw out a second chromosome. If it is the same colour, put it back and draw again. No peeking! When you have two chromosomes of different colours, your partner can remove the other two chromosomes from the box. You have just taken your chromosomes through Stage 1 of meiosis. The partner chromosomes have separated from each other.
e. Remove the paper clip from your black chromosome. Place one strand of wool, or chromatid, in each gamete. Do the same for your green chromosome. Now, you should both have made two gametes, each carrying one height gene and one pod colour gene. This was like Stage 2 of meiosis. Why?

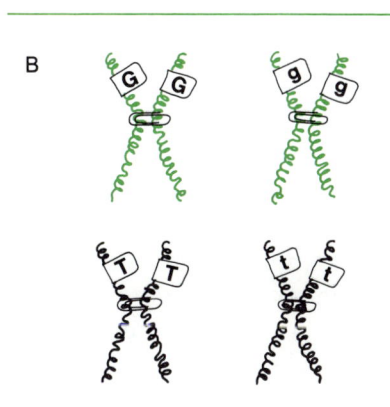

Fig. 30-6 How to make your chromosomes. (A) Cut 4 strands of black wool. The black chromosomes carry the gene for height. Put the T or tallness gene on 2 strands of wool, or chromatids. Put the t or shortness gene on the other 2 black chromatids. The green chromosomes carry the gene for pod colour. Attach sticky labels as shown. (B) Bring the two matching chromatids together with paper clips as shown. You have made two pairs of homologous chromosomes!

f. Write down the genotype of all 4 of your gametes.
g. Your teacher will help you pool the class results.

Discussion

1. Does everyone in the class end up with the same four gametes? What are the possible combinations of gametes?
2. The Law of Independent Assortment works in the first stage of meiosis. How does this experiment show this?
3. What offspring would you expect from a cross of a tall, yellow-podded pea plant and a short, green-podded pea plant? This is called a **dihybrid cross**. Your teacher will show you how to use a Punnett Square to answer this question.

30.5 Sex and Chromosomes

Will it be a boy or a girl? People are always trying to guess the sex of an unborn child (Fig. 30-7). How is sex determined? The answer to this question was first discovered by doing experiments with fruit flies.

Morgan's Experiments

You read earlier that Thomas S. Morgan did some important experiments with fruit flies. These helped to prove that genes are located on the chromosomes. Morgan looked at the chromosomes from some cells of the fruit fly. He noticed that all the cells had four pairs of chromosomes. Three of the pairs of chromosomes looked exactly the same in cells from both male and female fruit flies. But the fourth pair of chromosomes is not the same. Figure 30-8 shows that in the cells of the male fruit fly, one member of the fourth pair is bent like a hook. This chromosome is called the **Y chromosome**. The straight partner is called the **X chromosome**. The female fruit fly has two X chromosomes in all her body cells. The male fruit fly has one X

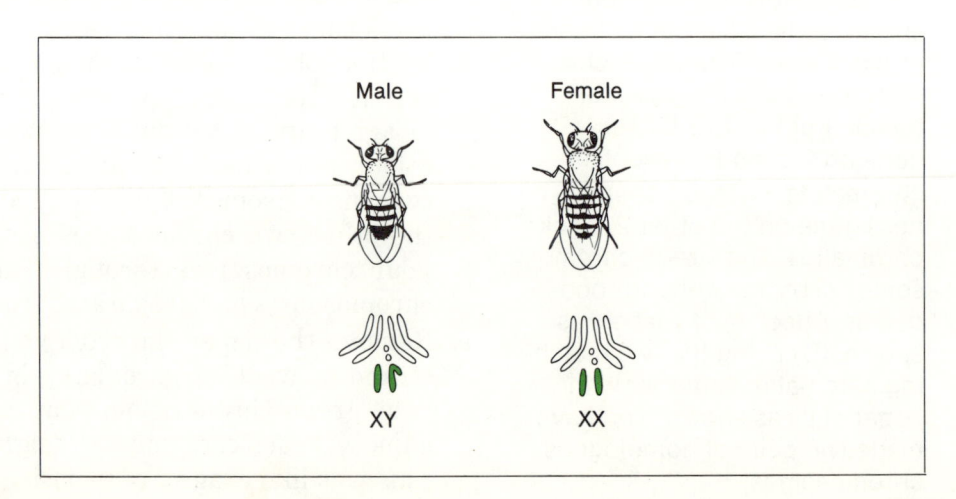

Fig. 30-7 People often wonder what the sex of an unborn baby will be. How do the chromosomes determine sex?

Fig. 30-8 The chromosomes of the fruit fly. The male fruit fly has one X and one Y chromosome. The female fruit fly has 2 X chromosomes.

and one Y chromosome in all his body cells. The X and Y chromosomes are called the **sex chromosomes**. The other three matching pairs of chromosomes are called the **autosomes**.

How Do the Sex Chromosomes Determine Sex?

After meiosis, all gametes of the fruit fly will have received one of each pair of autosomes. Each gamete also receives a sex chromosome. Study Figure 30-9,A. What sex chromosome will *all* the gametes (eggs) of the female fruit fly receive? All the eggs will contain an X chromosome. But all gametes (sperm cells) of a male fruit fly do not receive the same sex chromosome. Half of the gametes will receive an X chromosome and half will get a Y chromosome.

Now, what happens when two gametes meet to form a zygote at fertilization? Figure 30-9,B shows that, when an X-carrying sperm fertilizes an egg, the zygote will be XX. What is the sex of the zygote? It is female. If a sperm carrying a Y chromosome unites with an egg, the zygote will be XY, or male. The chances of the zygote being male or female are equal. Half of all newborn fruit flies should be male and half should be female.

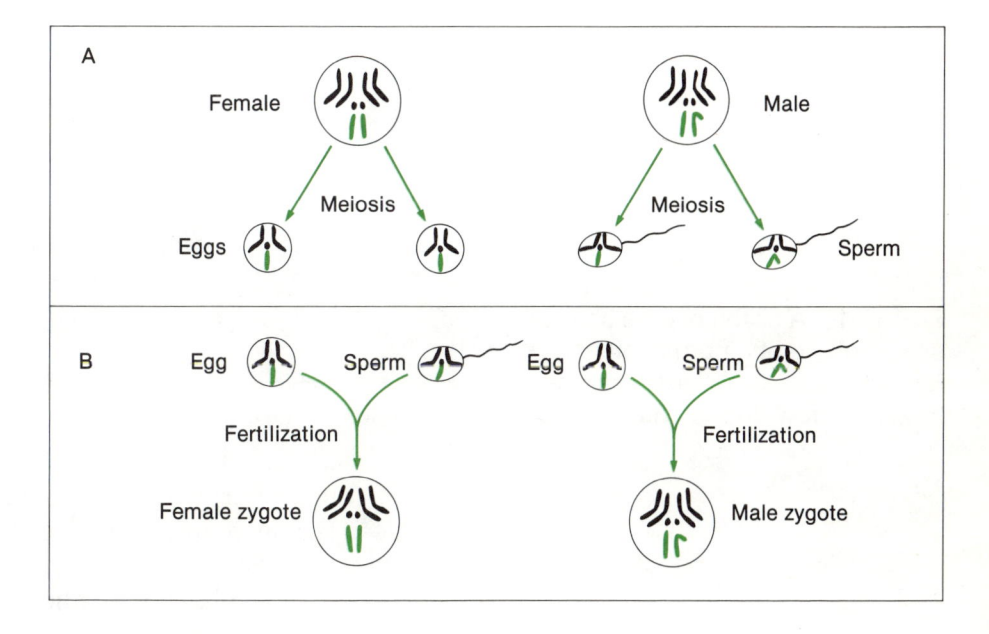

Fig. 30-9 How does a fruit fly become male or female? At meiosis, chromosomes segregate into the gametes. The male gametes do not all have the same chromosomes as the female gametes. How might they differ (A)? We expect half of all newborn fruit flies to be female and half to be male (B). Why?

Sex in Humans

You have just learned how sex is determined in fruit flies. Does this also apply to humans? As you know, human cells have 23 pairs of chromosomes. One of these pairs determines the sex of a human baby. The other 22 pairs of chromosomes are autosomes. Figure 30-10 shows a chart of the chromosomes of a woman and a man. The chromosomes are arranged in pairs. This kind of chart is called a **karyotype** [KAR-ee-o-type]. Can you see the difference between the

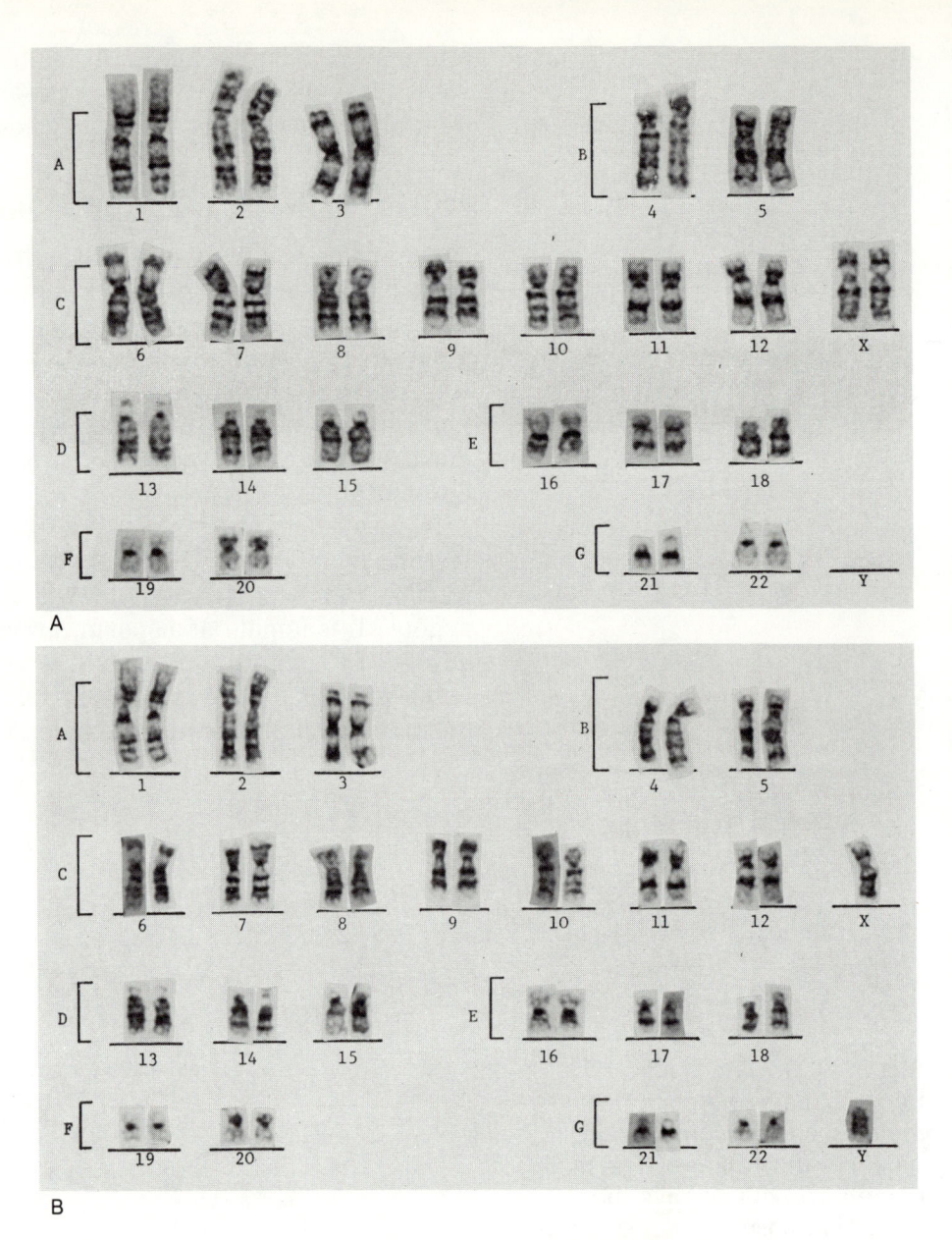

Fig. 30-10 A chromosome chart or karyotype of a normal human female (A); a normal male karyotype (B). How do they differ?

male and female karyotype? The female has two X chromosomes. The male has one X and one Y chromosome. Half of all the male sperm will carry a Y chromosome. Half will carry an X chromosome (Fig. 30-11). This explains why about half of all babies born are male while about half are female. Can you explain why?

Section Review

1. What important difference did Morgan notice when he examined the chromosomes from male and female fruit flies?
2. a) What sex chromosome do fruit fly eggs carry?
   b) What sex chromosome do fruit fly sperm carry?
3. How does the number of boys born compare with the number of girls?

Fig. 30-11 Sex in humans is determined by a pair of sex chromosomes. Two X chromosomes produce a female child. An X and a Y chromosome make a male.

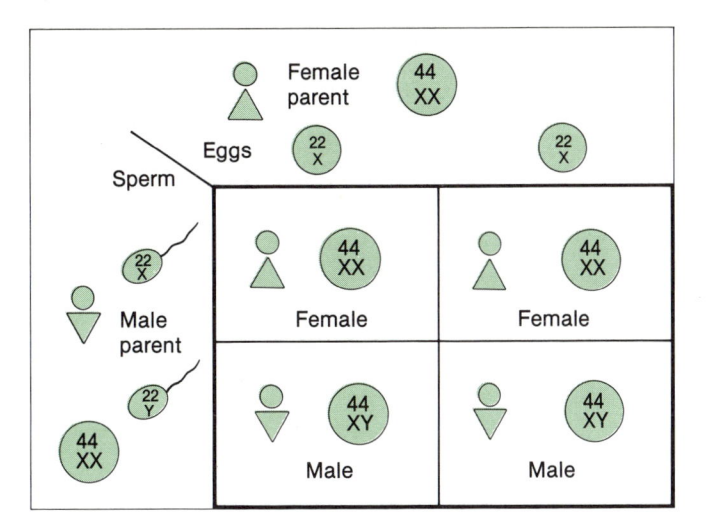

30.6 Sex Linkage

Morgan discovered the sex chromosomes in fruit flies. Do you think that these sex chromosomes only carry the genes for sex? Using fruit flies, Morgan showed that the X chromosome carries other genes as well.

Normally fruit flies have red eyes. But Morgan discovered some flies with white eyes. He mated red-eyed and white-eyed flies. The results showed that red-eyes was a dominant trait while white eyes was recessive. But, Morgan made a very interesting observation. Only male fruit flies had white eyes. Morgan showed that this was because the gene for white eyes is on the X chromosome (Fig. 30-12). A male fly has white eyes if the single X chromosome in his cells has the white-eye gene. His Y chromosome does not have an eye colour gene to mask the effect of the white-eye gene. But, females have two

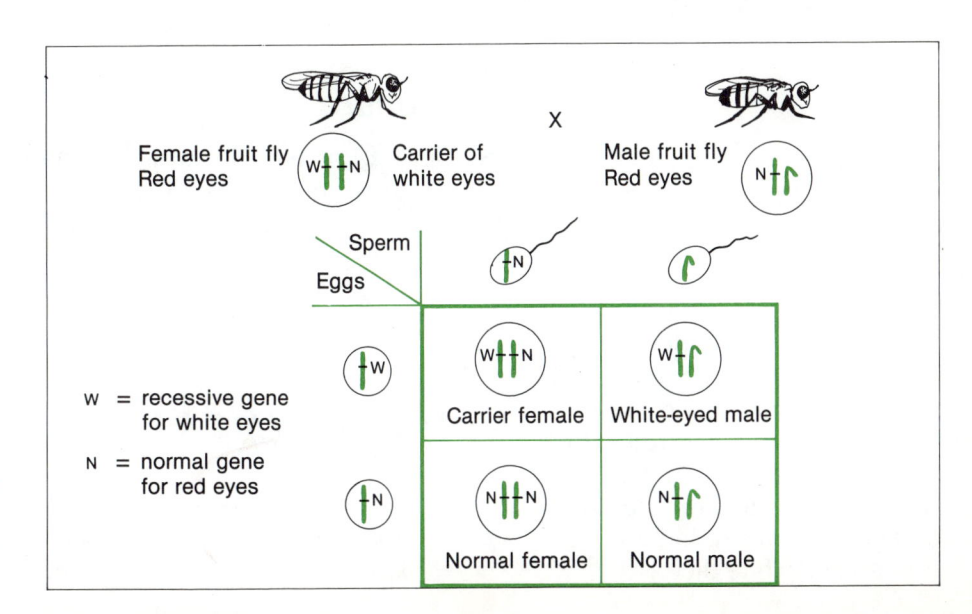

Fig. 30-12 Morgan discovered that the gene for white eyes was on the X chromosome. He analyzed offspring from crosses like the one shown here.

Section 30.6   445

X chromosomes. They *can* have a normal red-eye gene to dominate the white-eye gene. These females are said to be **carriers** of the trait for white eyes. Because the gene for white eyes is on the X chromosome, it is said to be **sex-linked**. The inheritance of white eyes in fruit flies is an example of **sex linkage**.

Section Review

1. What did Morgan discover about the inheritance of white eyes in fruit flies?
2. If a gene is sex-linked, what chromosome is it on?
3. Half of the male offspring of a female fruit fly that carries the white-eye gene will have white eyes. Is this true? (*Hint:* Study Figure 30-12.)

30.7 Sex Linkage in Humans: Colour Blindness

There are many examples of genes that are carried on the X chromosome in humans. One well known trait is red-green colour blindness. People with this type of colour blindness cannot distinguish between red and green. In this activity you will use dice to see how colour blindness is inherited in humans.

Problem

How is colour blindness inherited in humans?

Materials

red dice   white dice   pen and paper

Procedure

a. Copy Figure 30-13 into your notebook. (Don't copy the examples under "Child 1" and "Child 2".)
b. Choose a lab partner. One member of the team will roll the dice. The other member will record the results.

Fig. 30-13 We have put in an example for Child 1 and Child 2. Child 1 is a colourblind male. (An even number was rolled on both the red and white dice.) Child 2 is a normal female. (An odd number was rolled on both the red and white dice.)

Child 1	Child 2	Child 3	Child 4	Child 5	8
CB XY	N XX N				

CB = colourblind gene   X = X chromosome
N = normal gene   Y = Y chromosome

c. You are going to use the dice to find out the genotype of 8 children of an imaginary couple. The mother is a carrier of red-green colour blindness. Red-green colour blindness is a recessive gene. This means that the mother has normal vision. However, she can pass the colour-blind gene on to her offspring. The father is a male with normal vision.

d. Roll the red dice. If the dice shows an *even* number, the father passes on the Y chromosome to the offspring. If the dice shows an *odd* number, the father passes on the normal X chromosome to his child. Record the result under "Child 1' in your table.

e. To find out what chromosome the mother passes on to the first child, roll the white dice. If the dice shows an *even* number, the mother passes on the gene for colour blindness. If the dice shows an *odd* number, she passes on the normal X chromosome. Record this result in your table.

f. Repeat steps (d) and (e) until you have the genotype for 8 children.

g. Your teacher will help you pool the class results.

Discussion

1. How many boys and how many girls did your couple have? Is this what you would expect? What was the ratio of boys to girls in the class results?
2. Are there any colour-blind girls among the children? Why or why not?
3. How many colour-blind males are there among all the children in the class results? How many normal boys? Is this what you would expect? Why or why not?

30.8 Mutations

Where did the gene for colour blindness come from? And how did fruit flies get white eyes? You have learned that genes are passed on from generation to generation with no change. But, there are exceptions to this rule. Sudden changes in genes are known to take place. These changes are called **mutations**. The white-eyed fruit fly was the result of a mutation in the gene for eye colour. Also, colour blindness is likely the result of a mutation.

Many geneticists think that mutations take place all the time. Such mutations occur naturally. For example, a short-tailed cat may appear in a barnyard full of long-tailed cats. Or, a white deer may occur in a population of brown deer (Fig. 30-14).

Mutations can also be made to happen by exposing cells to x-rays or certain chemicals (Fig. 30-15). Most mutations are harmful to the organism. However, some mutations are helpful. You will learn more about mutations when you read about natural selection in Chapter 32.

Fig. 30-14 One of these deer carries mutant genes. How does this deer differ from most other members of its species?

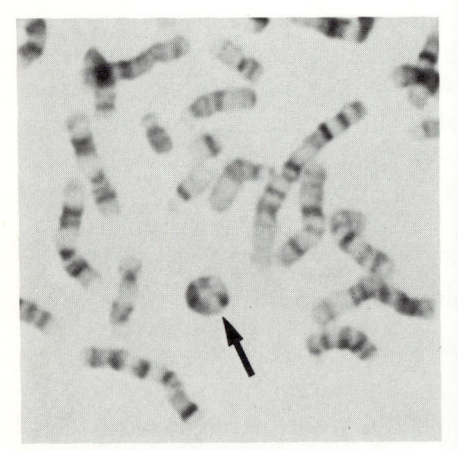

Fig. 30-15 This chromosome is abnormal. The change may have been caused by exposure to X-rays.

Section Review

1. What is a mutation?
2. How do mutations occur?

30.9 DNA: The Genetic Material

You know that genes are carried on the chromosomes. You also know that changes, or mutations, can occur in these genes. What are chromosomes made of? What kind of change causes a mutation?

Scientists know that chromosomes contain large molecules of **DNA**. DNA is a complicated molecule that is shaped something like a spiral staircase. This shape is called a **double helix**. Figure 30-16 shows that the double helix is made up of two backbones like the banisters on the staircase. These backbones are made of chemicals called phosphates and sugars. The most important part of the DNA molecule is the "steps". The steps are made up of pairs of chemicals called **bases**. There are 4 different bases in DNA. These 4 bases form a sort of 4-letter alphabet that spells out the genotype of an organism.

The arrangement of the bases on the DNA, together with another similar chemical called **RNA**, results in certain proteins and enzymes being made in the cells. For example, the arrangement of the bases at the gene for eye colour may result in a protein being made that gives a child blue eyes. Changes in the DNA alphabet result in changes in the organism. These changes are what we call a mutation.

Section Review

1. What is DNA? Describe what you know about the structure of DNA.
2. What kind of change can cause a mutation?

Main Ideas

1. Genes, or Mendel's "factors", are carried on partner chromosomes.
2. During meiosis, partner chromosomes segregate into different gametes.
3. Gametes have only one chromosome from each pair of chromosomes in the parent.
4. There are many genes on any one chromosome. Genes that are on the same chromosome tend to be inherited together.
5. Sex is determined by a pair of sex chromosomes (X and Y).
6. Genes located on the X chromosome are sex-linked.
7. Changes in genes are called mutations.
8. Chromosomes are made of large molecules of DNA.

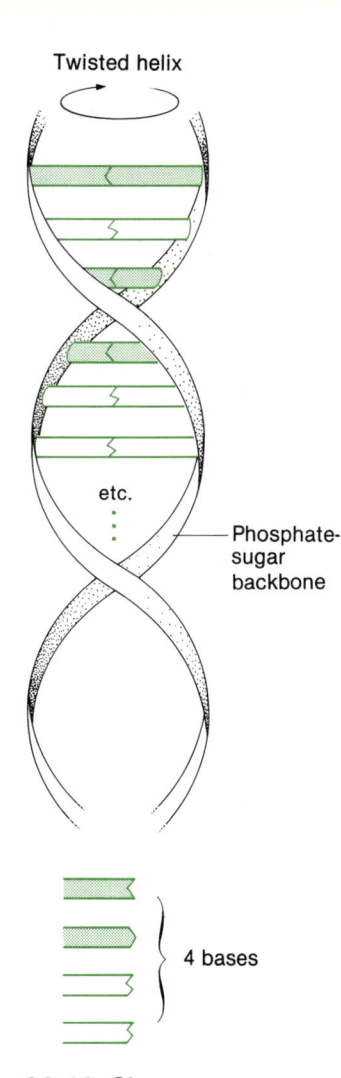

Fig. 30-16 Chromosomes are made of DNA, a molecule shaped something like a spiral staircase. Notice that the 4 bases pair up with each other to form the "steps".

Glossary

autosome	AW-toh-sohm	any chromosome except for the sex chromosomes
karyotype	KAR-ee-o-type	an ordered chart of the chromosomes
linkage		genes on the same chromosome are inherited together
meiosis	mi-OH-sis	the process of nuclear division that reduces the chromosome number by half
mutation	myoo-TAY-shun	a change in a gene

Study Questions

A. True or False

Decide whether each of the following statements is true or false. If the sentence is false, rewrite it to make it true. (Do not write in this book.)
1. Human gametes have 46 chromosomes.
2. The Law of Independent Assortment says that genes on different chromosomes are inherited separately.
3. Each chromosome carries only one gene.
4. All the gametes of a male fruit fly have a Y chromosome.
5. Colour blindness is a sex-linked trait in humans.

B. Completion

Complete each of the following sentences with a word or phrase that will make the sentence correct. (Do not write in this book.)
1. Chromosomes separate or _____ during meiosis.
2. Fruit flies have _____ pairs of autosomes.
3. A red-eyed female fruit fly that has one copy of the gene for white eyes is a _____ of the white-eye gene.
4. Red-green colour blindness is a _____ trait in humans.
5. DNA is made of _____ , sugars and 4 different _____ .

C. Multiple Choice

Each of the following statements or questions is followed by four responses. Choose the correct response in each case. (Do not write in this book.)
1. Suppose that the gene for green pods is on the same chromosome as the gene for tallness in pea plants. If this is so, then
    a) pod colour and height would tend to be inherited together
    b) pod colour and height would tend to be inherited separately
    c) green pods would be dominant to tallness
    d) tallness would be dominant to green pods
2. In a sample of sperm from a human male
    a) all the sperm will have an X chromosome
    b) all the sperm will have two X chromosomes
    c) half the sperm will have an X chromosome and half will have a Y chromosome
    d) all the sperm will have a Y chromosome
3. A cross between a normal red-eyed male fruit fly and a female carrier of the white-eye gene will
    a) produce only red-eyed sons and white-eyed daughters
    b) produce only white-eyed sons and red-eyed daughters
    c) produce half normal and half white-eyed sons
    d) produce all white-eyed offspring

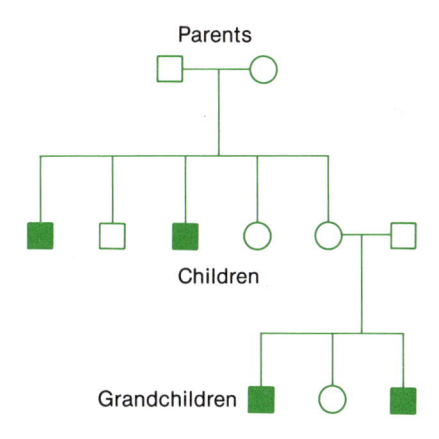

Fig. 30-17 A pedigree showing the inheritance of hemophilia in humans.

5. Mutations are
   a) usually harmful to the organism
   b) always caused by chemicals or x-rays
   c) a special kind of DNA
   d) dominant traits that occur naturally

D. Using Your Knowledge

1. Explain the difference between the two words in each pair:
   a) meiosis — mitosis
   b) chromosome — DNA
   c) gamete — zygote
   d) linkage — sex linkage
2. Examine the pedigree in Figure 30-17. The squares stand for males and the circles for females. A coloured symbol means that that person has the blood-clotting disease hemophilia. What can you tell about the inheritance of hemophilia from this pedigree?
3. Explain blended inheritance using your knowledge of how DNA works.

E. Investigations

1. Write a report on some chemicals that are known to cause mutations. Are we exposed to any of these chemicals?
2. Find a pedigree of the Royal British family in a genetics book. Queen Victoria was thought to be a carrier of hemophilia. How would you deduce which other members of the royal family are carriers of this disease?
3. Find out how many boys and girls are born each year at several hospitals in your neighbourhood. Is the ratio of boys to girls what you expected? Explain.

31 Importance of Genetics

31.1 Genetic Disorders in Humans
31.2 Multiple-factor Inheritance: Effect of Many Genes
31.3 Activity: Inheritance of Height
31.4 Genetics and Environment
31.5 Activity: Investigating the Inheritance of Some Human Traits
31.6 Plant Breeding
31.7 Animal Breeding
31.8 Trends in Genetics

You have learned about Mendel's laws. And you have read about genes and chromosomes. Why bother learning genetics? You may be wondering: "Is genetics important to me?"

Scientists are finding out that genetics plays a very important role in our lives (Fig. 31-1). In this chapter, you will learn about some of the many ways that genetics affects you and other people. You will also read about the very important role that the study of genetics will play in the future.

31.1 Genetic Disorders in Humans

Fig. 31-1 How can the study of genetics help you and other people?

You know about some traits in humans that are inherited according to Mendel's laws. For example, the ability to taste the chemical PTC is inherited as a simple dominant trait. You also learned about red-green colour blindness. This is a sex-linked trait in humans.

Colour blindness is not a very serious condition. However, there is a very long list of genetic diseases in humans that can be very serious. There are two basic types of genetic disorders:

1. Diseases resulting from a mutation in a gene. These genes are usually recessive and inherited according to Mendel's laws.
2. Diseases resulting from having the wrong number of chromosomes.

What are some examples of these two types of genetic disorders? What can geneticists do to help people who have these diseases?

Sickle Cell Anemia

Normally, our red blood cells are a round, smooth shape. They flow easily through the veins and arteries. **Sickle cell anemia** is a serious

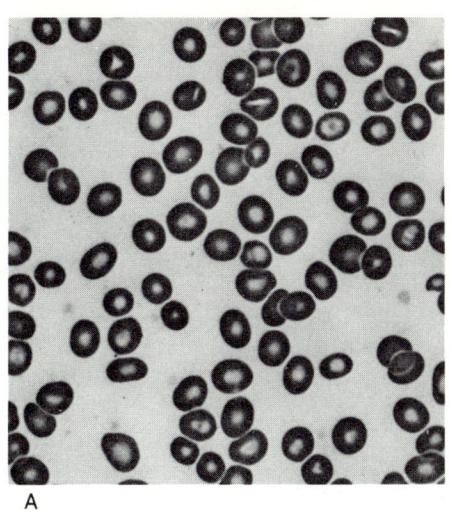

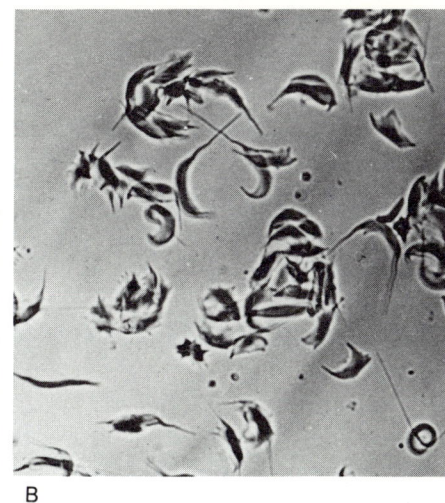

Fig. 31-2 Normal red blood cells are smooth and round (A). Cells from a person with sickle-cell anemia are a crescent-shape (B).

blood disease. The red blood cells are a sickle or crescent shape (Fig. 31-2). The abnormal sickle cells often get clogged in the arteries and veins. This means that the red blood cells cannot do their job of carrying oxygen to the tissues. This results in damage to some body organs, pain, and sometimes death.

What causes the red blood cells to have a sickle shape? Geneticists have found out that the sickle shape is due to a change in the red blood cell protein, **hemoglobin**. This change is caused by a mutation in the gene for hemoglobin. The sickle cell anemia gene is recessive. If people have two copies of this recessive gene, they have sickle cell anemia. People with only one copy of the sickle cell gene are normal, but can pass the gene on to their offspring. These people are **carriers** of sickle cell anemia (Fig. 31-3).

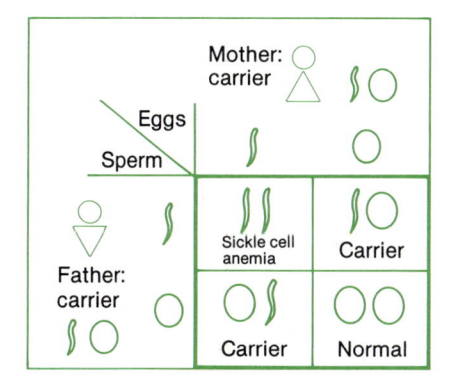

Sickle cell anemia is more common in people of African ancestry. About one out of every ten black Americans carries the gene. A lot fewer people actually have the disease.

Geneticists have been learning about sickle cell anemia for a long time. Figure 31-4 shows how a person's blood can be tested to find out if he/she has the sickle cell gene. If people with sickle cell anemia are given proper treatment, many years can be added to their lives.

More Recessive Genetic Disorders

Another disease that geneticists have studied a lot is **Tay-Sachs disease**. This serious disease is common only among Jewish children whose ancestors came from Eastern Europe. Tay-Sachs disease is caused by a mutation in the gene for a special protein. This protein plays an important role in the nervous system of humans. Children with Tay-Sachs disease are very ill. They usually die before they are five years old. You will read about how geneticists can help families that have children with Tay-Sachs disease later in this chapter.

There are many diseases caused by recessive genes that geneticists do not understand as well as sickle cell anemia or Tay-Sachs disease. For example, **cystic fibrosis** is a disease that affects about

Fig. 31-3 The gene for sickle-cell anemia is recessive. A Punnett Square shows what you might expect if two carriers of the sickle cell gene marry and have children. What percentage of the children would you expect to have sickle cell anemia?

Section 31.1 453

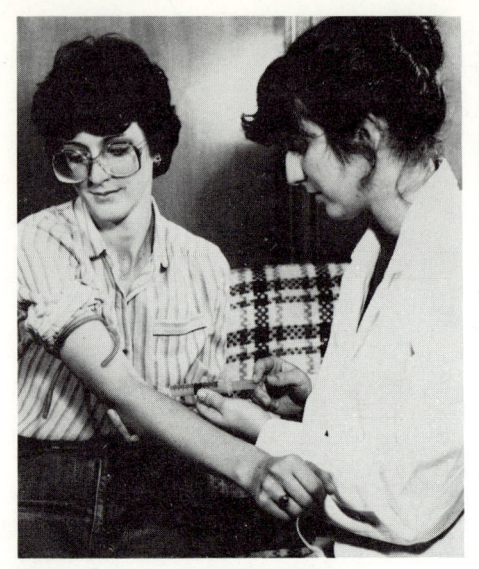

A 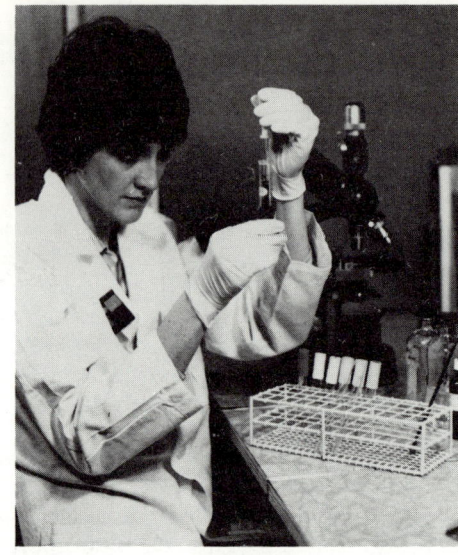   B 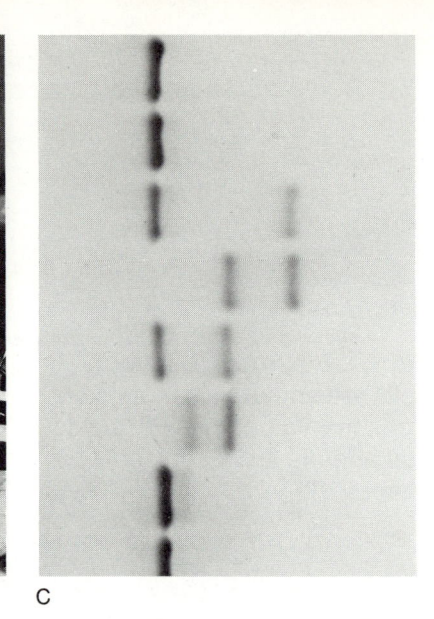   C

Fig. 31-4 Geneticists have developed tests for the sickle cell gene. A blood sample is taken from the patient (A). The hemoglobin is isolated from the blood (B). The hemoglobin is tested. The abnormal hemoglobin makes a different pattern than normal hemoglobin (C).

one in every 1600 children born in North America. People with this disease have problems with their lungs and digestive system. Research has helped to find ways to treat cystic fibrosis (Fig. 31-5). Scientists are working hard to understand the genetic basis of this disease.

Down's Syndrome

A person with Down's Syndrome has 47 instead of 46 chromosomes in every body cell. Usually a baby with more than the normal number of chromosomes will die before birth. But when there are 3 number 21 chromosomes, the baby survives with a special condition.

No one knows the exact causes of Down's Syndrome. But when the mother is over 35 years old, the chances of having a Down's Syndrome child seem to increase. The extra chromosome may come from the father or the mother. During meiosis, the chromosomes fail to separate properly. Geneticists can help couples predict the chances of conceiving a child with this condition (see Section 31-8).

People with Down's Syndrome share certain physical traits such as small bodies and a characteristic facial appearance. These people are mentally retarded. However, the seriousness of the condition varies with the individual. Early stimulation programs help many Down's Syndrome children to reach their full potential and to participate happily in society.

Fig. 31-5 People with cystic fibrosis must inhale antibiotics through a special respirator to make breathing easier. This also prevents infection.

Section Review

1. What are the two basic types of genetic disorders?
2. What causes sickle cell anemia?
3. What is Tay-Sachs disease?
4. What is different about the karyotype of a person with Down's syndrome?

Fig. 31-6 A karyotype of a child with Down's syndrome

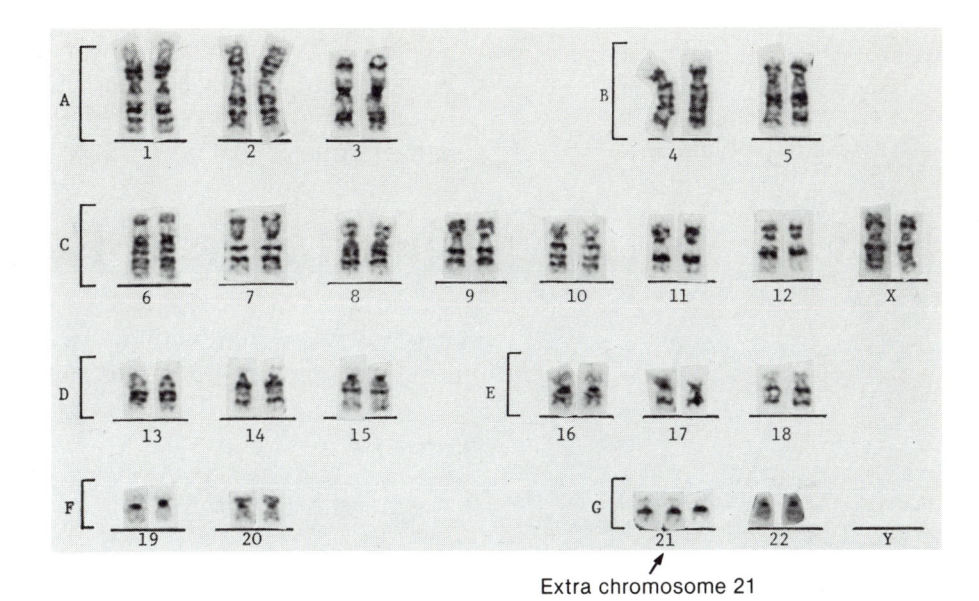

Extra chromosome 21

31.2 Multiple-Factor Inheritance: Effect of Many Genes

In the last section you learned about some genetic disorders that are caused by the inheritance of a single abnormal gene. There are some normal traits in humans that are inherited in a simple fashion as well (Fig. 31-7). However, very few human traits are controlled by a single pair of genes. Most human inheritance is much more complicated than Mendel would have predicted!

Look around the classroom. Is everyone tall or short? Or are there a whole range of heights? Does everyone have either brown or blue eyes? Or are there all different shades of eye colour? Most human traits, including height and eye or skin colour, are controlled by several genes or factors. The inheritance of height is an example of **multiple-factor** or **multiple-gene inheritance** (Fig. 31-8).

Multiple-factor inheritance is hard to study. However, geneticists do understand that it is different combinations of a series of genes that results in a range of eye colour from blue to brown.

Fig. 31-7 Some traits in humans are controlled by a single pair of genes. What traits do you have?

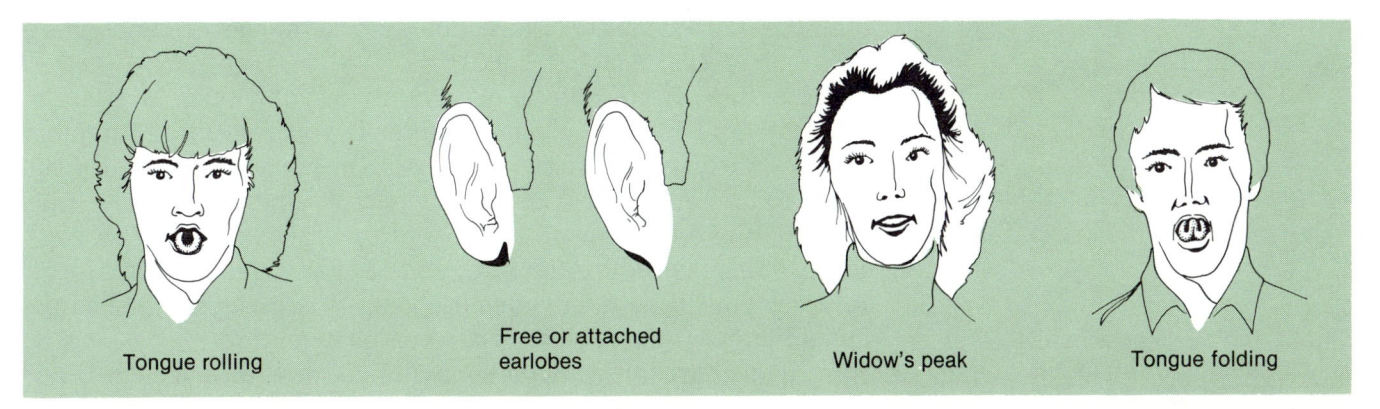

Tongue rolling

Free or attached earlobes

Widow's peak

Tongue folding

Section 31.2 455

Section Review

1. What is multiple-factor inheritance?
2. Name some human traits that are controlled by more than one pair of genes.

31.3 ACTIVITY Inheritance of Height

In this activity, you will do two things that should help you to understand how several genes can produce a whole range of heights.

Problem

How is height inherited?

Materials

measuring tape
graph paper
dice
pen and paper

Procedure A

a. Choose a lab partner. Using a measuring tape, measure your partner's height. Write your partner's height on a sheet of paper and give it to your teacher.
b. Your teacher will make a chart with the number of students of different heights sorted out into columns. Use your graph paper to graph these results. An example is shown in Figure 31-9.

Procedure B

a. Let's pretend that 4 genes control height in humans (A, B, C, and D). Let's also assume that everyone grows to a basic height of **150 cm** no matter what height genes they have.
b. You are going to figure out the height genotype of an imaginary person. You know that all genes come in pairs. In this case, one member of each pair of height genes does *not* change the basic height. Let's call these genes A^o, B^o, C^o, and D^o. But, the other member of each pair of genes adds 2 cm to the basic height. These genes are called A^2, B^2, C^2, and D^2.
c. Roll the dice. If you roll an even number, your imaginary person inherits A^o from one of his parents. If you roll an odd number, he/she inherits A^2. Write your result in your notebook. Do this again to find out what A gene your person inherits from the other parent. Write this down.
d. Repeat this procedure for B, C, and D height genes. Don't forget to write your results in your notebook. When you're finished you should have a genotype for your person's height.
e. Figure out how tall your person will be. For the Genes A^2, B^2, C^2,

Fig. 31-8 Height is an example of multiple-factor inheritance. This means that people can have a whole range of heights.

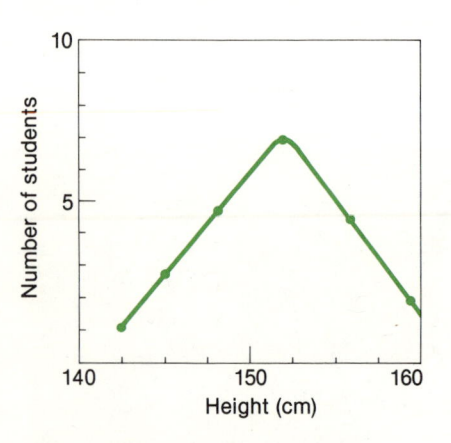

Fig. 31-9 Your class results may look something like this.

and D^2 add 2 cm to the basic height of 150 cm. For $A°$, $B°$, $C°$, and $D°$, add nothing to this height. Here are some examples:

$A°A°B°B^2C°C^2D^2D^2E°E°$ = 158 cm
$A^2A^2B°B^2C°C°D°D°E°E°$ = 156 cm
$A^2A^2B^2B^2C^2C^2D^2D°D^2E°E^2$ = 166 cm

f. Your teacher will help you to pool the class results.
g. Draw a graph as you did in Procedure A.

Discussion

1. What is the most common height in your class? Is there a range of heights among your classmates?
2. Is there a similarity between the two graphs you drew in Procedure A and Procedure B?
3. Do you think height is controlled by many genes? How might the results of this experiment help you answer this question?

31.4 Genetics and Environment

You know that height is controlled by several genes. But are genes the only factors controlling a person's height? Suppose a person has the genes that will allow him/her to grow tall. Do you think he/she will grow tall if he/she does not receive a well-balanced diet?

Geneticists know that our environment plays a very important role in determining the way we are (Fig. 31-10). Studies have shown that people of your generation are, on the average, taller than people of your great-great grandparents' time. This change is probably due to a better, well-balanced diet. The food you eat is an important part of your environment.

Bright people often tend to have bright children. Is this because they pass genes for intelligence on to their children? Or is it because

Fig. 31-10 People can be very different depending on the environment in which they live.

Section 31.4

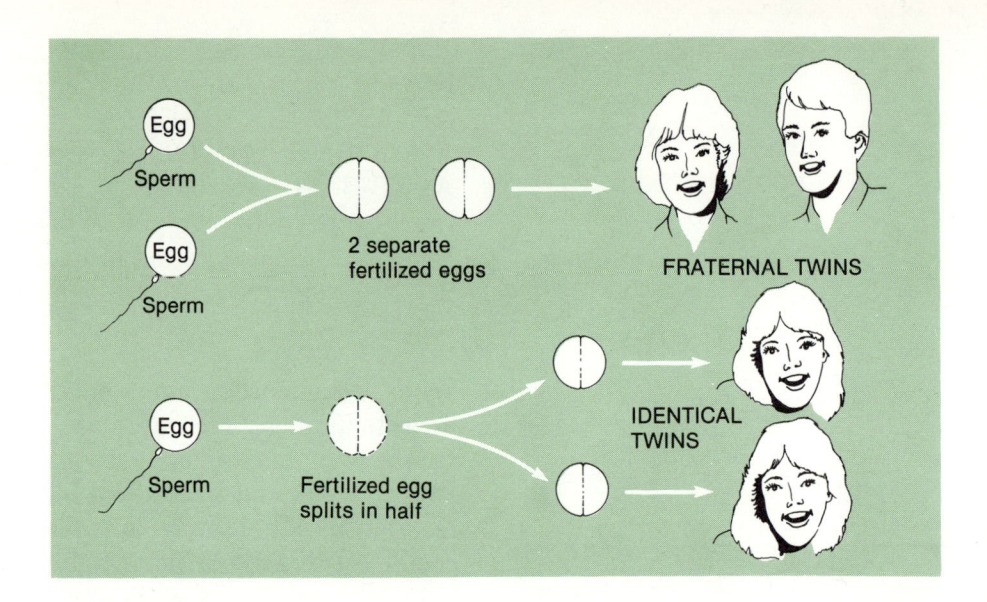

Fig. 31-11 Identical twins develop from a single fertilized egg. Fraternal twins come from two separate fertilized eggs.

bright parents tend to provide a better learning environment for their children? Or is it a combination of both? This is a very hard question to answer. It is difficult to separate the effects of genetics and the environment.

Studying Genetics Using Twins

One way to study the effects of genetics and the environment is to look at inheritance in twins. How do twins arise?

Usually a woman produces only one egg at a time. But sometimes two eggs will be produced. If both these eggs are fertilized, twins will be born. Twins coming from separate eggs are really like siblings who are born at the same time. They are called **fraternal twins**. They need not look more like each other than you do like your brother or sister. They can even be of different sexes.

Twins can also arise from a single egg. After fertilization, the egg splits into two halves as shown in Figure 31-11. These two halves then develop into embryos resulting in **identical twins**. They have *exactly* the same genes.

Identical twins do not have different genes. They differ only in their environments. Geneticists can use this fact to help separate the effects of heredity and environment. Twin studies have helped to show that intelligence is determined by *both* genetics and environment.

Section Review

1. Are genes the only factors controlling height in humans? What else plays an important role in determining many human traits?
2. What is the difference between fraternal twins and identical twins?
3. Why are twin studies useful for studying genetics and the environment?

31.5 ACTIVITY Investigating the Inheritance of Some Human Traits

By now, you know that human heredity is not simple! For this activity, you will do a survey of some traits in your classmates. Also, you will try to answer some questions about these traits. Are the traits determined mostly by genetic factors? What kinds of genetic factors? Does environment play a role in determining the trait?

Problem

How are human traits inherited?

Materials

pen and paper

Procedure

a. Copy Table 31-1 into your notebook.
Note: You will need more space in your table to record all your results.

Table 31-1 Inheritance of Some Human Traits

Class-mate	Eye colour	Hair colour	Hair texture (curly or straight)	Ear lobes (free or attached)	Teeth (crooked or straight)	Face shape (round or square)	Other
1							
Family							
2							
Family							
3							
Family							
4							
Family							
5							
Family							
Myself							
Family							

b. Interview at least five of your classmates. Ask them questions that will allow you to fill in the blanks of the table. Under "family", you should include information about parents, brothers, and sisters. Make sure you include yourself and your family in your survey.

c. Make up your own trait for the last column in the table.

Discussion

1. Is there any trait in the table that seems to be caused by a dominant gene? a recessive gene? What evidence do you have for your answers?
2. Are there any traits which might be examples of multiple-factor inheritance? Why?
3. What traits may be influenced by the environment? (*Hint:* Can a trip to the hairdresser change the texture or colour of your hair? What other factors change the colour of hair?)

Fig. 31-12 How are these plants and animals useful to humans?

31.6 Plant Breeding

The study of genetics is not only useful in helping to understand human inheritance. Our knowledge of genetics can also be applied to plants and animals. Why is this important?

Many plants and animals are very useful in our lives (Fig. 31-12). We use them for food, clothing, and medicines. For the next two sections of this chapter, you will learn how the science of genetics helps to produce plants and animals best suited to human needs.

Selective Breeding

You know that Mendel chose certain pea plants as parents for his crosses. In other words, he *selected* his parent pea plants. **Selective breeding** is the choosing of a specific individual plant or animal as a parent. People have been breeding plants and animals for a very long time. Organisms with useful traits are selected and bred to produce offspring. A **breed** is a kind of plant or animal that is developed through selection (Fig. 31-13). Our knowledge of the laws of heredity has helped us to understand how to make breeding more efficient.

New Types of Plants

Mutations occur in plants as they do in other organisms. Sometimes a mutation will result in a plant with desirable traits. This plant can then be selectively bred to develop a new type of plant (Fig. 31-14). Delicious apples, seedless grapes, and pink grapefruit are all examples of plant mutations.

Have you ever noticed extra large blueberries or strawberries at the grocery store? These berries have a whole extra set of chromosomes. This condition is known as **polyploidy**. There are several examples of polyploid plants, many of which produce larger fruit. Why is this useful?

Through selection and breeding, geneticists are working to develop more and better food for the growing world population. For example, plants with larger heads of grain have been produced. Plants that are resistant to diseases have been selected. Scientists have also been searching for new types of plants that can grow in poor soil. This kind of research has been called the "green revolution". It is very important that scientists continue to work hard on these kinds of problems.

Section Review

1. What is selective breeding?
2. Define polyploidy. Give some examples of polyploid plants that are useful to humans.
3. What special kinds of plants are geneticists trying to develop? Why?

Fig. 31-13 The sheepdog is a breed of dog. The sheepdog is often used to herd sheep and other farm animals. What traits do you think were selected in the development of this breed?

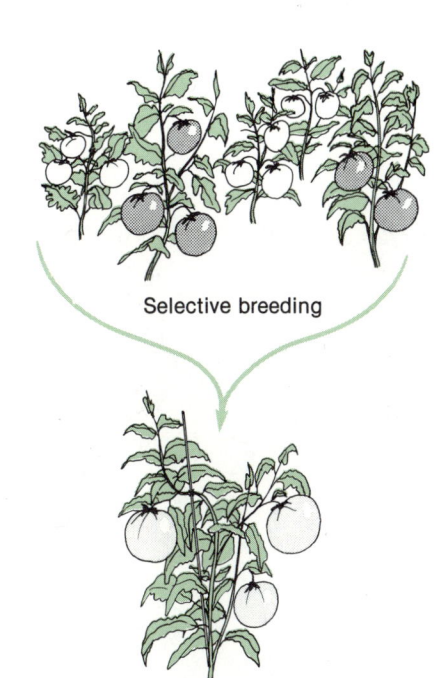

Fig. 31-14 The selective breeding of plants with desirable traits results in the development of new and better types of plants.

31.7 Animal Breeding

Look at the two cattle in Figure 31-15. Which animal would you rather get close to? Hornless cattle are easier for farmers to deal with. Where do you think hornless cattle came from?

The same principles of selective breeding that you learned about for plants can also be applied to animals. The absence of horns in cattle is one example of a helpful mutation in a domestic animal.

There are many different breeds of domestic animals. They all have certain traits that make them useful. For example, Shorthorn cattle are known for their excellent beef. They live well in many parts of Canada. However, they live poorly in hot, dry climates such as the south-western United States. The **Brahman** is an Indian breed that lives very well in hot climates. Brahman bulls were bred to Shorthorn cows. The mating of two different breeds is known as **crossbreeding**. The crossbred offspring of the Brahman and Shorthorn are known as the **Santa Gertrudis** breed. These cattle are good beef producers and can withstand hot dry climates.

Can you think of another common example of a hybrid animal? The mule is a strong agile animal that is the result of a cross between a horse and a donkey (Fig. 31-16). Why is this cross different from the mating of different breeds of cattle? The horse and the donkey are two different species. Their offspring, the mule, is sterile.

There are many other examples of selective breeding of domestic animals. Sheep with short legs have been selected to prevent them from jumping fences. Horses with long legs have been bred as racehorses. Our knowledge of genetics aids in the development of useful breeds of animals.

Fig. 31-15 Breeds of hornless cattle have been developed by selection. Do you think this was a useful thing to do?

Fig. 31-16 The mule is a hybrid between a horse and a donkey. How are these animals useful to humans?

Section Review

1. What is the Santa Gertrudis breed of cattle? How was it developed?
2. Give three examples of selective breeding of domestic animals.

31.8 Trends in Genetics

Can we use our knowledge of genetics to improve ourselves as well as plants and animals? The idea of selective breeding to produce "superior" people is not a new one. But who can decide what traits in humans are "superior"? Are blue eyes better than brown eyes? Today, geneticists are aware that our understanding of heredity in humans must not be abused. There are many ways this knowledge can be very useful to people.

Genetic Counselling

Earlier you read about some genes that cause genetic diseases. A genetic counsellor is a person who is trained to help families who think they may have a genetic problem (Fig. 31-17). Suppose a couple visit a genetic counsellor because there has been Tay-Sachs disease in their families. They are concerned that their unborn child may have Tay-Sachs disease. How can the genetic counsellor help these people?

The counsellor can check the family histories of the couple to see if any relatives have had children with Tay-Sachs disease. The counsellor can then predict the chances of the couple having a child with this genetic disease. Blood samples may be taken from the couple. Carriers of Tay-Sachs disease can be detected by a simple blood test.

If the couple were carriers of the gene for Tay-Sachs disease and the woman was pregnant, the counsellor might arrange for the mother to have **amniocentesis** [am-ni-oh-sen-TEE-sis]. This procedure

Fig. 31-17 A genetic counsellor helps people understand genetic problems.

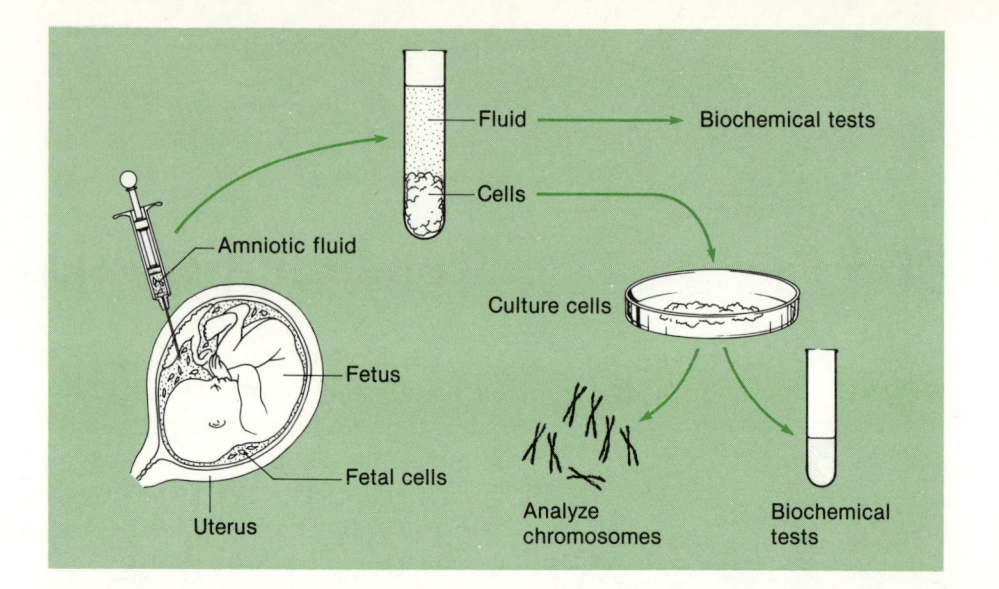

Fig. 31-18 Amniocentesis involves taking a sample of amniotic fluid from a pregnant woman. This fluid contains fetal cells that can be grown in culture and analyzed.

is used to find out about a baby before it is born. Figure 31-18 shows how cells of the fetus are obtained and examined. A biochemist could tell the genetic counsellor whether or not the fetus had Tay-Sachs disease.

A lot of other diseases can be detected by amniocentesis or by other simple tests. PKU is a rare, inherited disease. People with PKU cannot digest a certain amino acid. If they are not helped very early in life, they will become severely retarded. There is a simple, cheap test for PKU. It is given routinely to babies at birth in several places. If PKU is present, the diet of the baby is changed. The child will develop normally. This kind of testing is called **genetic screening**. Counsellors can use the results of these tests to help people.

Genetic Engineering

Recently, scientists who work with DNA have made some very exciting discoveries. They have been able to take DNA from cells and isolate certain genes! They can then use these genes to change the DNA in a living cell. This kind of experiment is called **genetic engineering** or **recombinant DNA** research.

How do scientists isolate genes? Genetic engineers do a lot of work with simple organisms like bacteria and yeast. These organisms contain small rings of DNA. Figure 31-19 shows that these rings are useful for isolating, or **cloning**, genes. Many genes from different organisms have now been cloned.

How useful are these experiments? Scientists have isolated a strain of bacterium that is able to digest oil. This bacteriun may be useful in cleaning up pollution from oil spills. The gene for human **insulin** has been cloned in bacteria. The bacteria then produce insulin. Insulin is needed by **diabetics** (people who cannot metabolize sugar). These insulin-making bacteria may be used to produce large amounts of insulin to help diabetics.

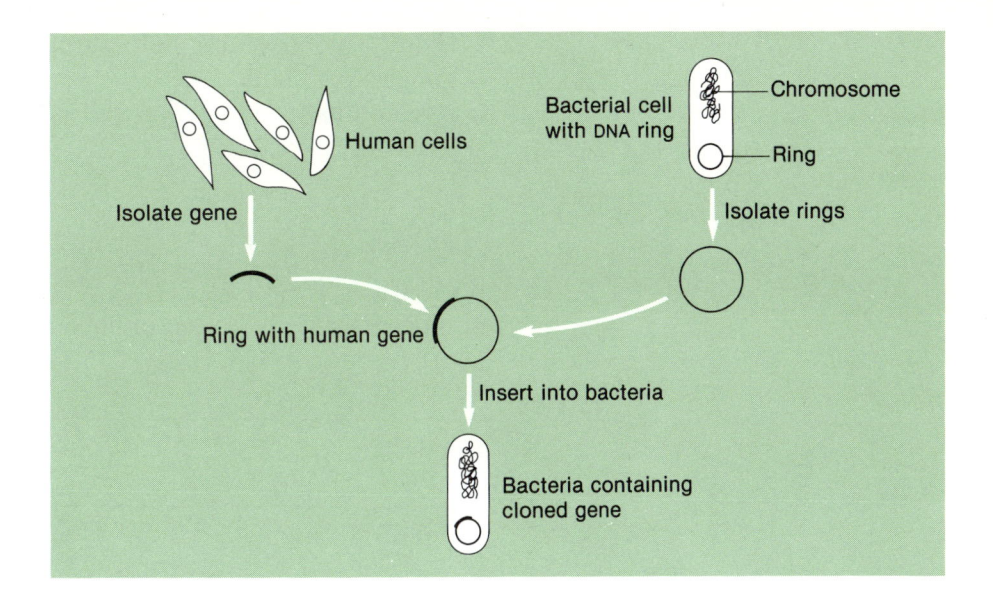

Fig. 31-19 This diagrams shows how a human gene can be cloned in a bacterial cell.

In the future, geneticists hope to be able to take cloned genes and change defective human cells. In this way, some genetic diseases may be cured. However, scientists have a lot of work to do before they reach this goal.

Some people do not like the idea of recombinant DNA research. Perhaps a bacterium carrying a harmful gene could be used in a bad way. Governments in Canada and the United States have written guidelines for scientists working in this field. Labs are carefully licensed and supervised. Recombinant DNA research is very important. We must be careful that the results of this exciting research are not abused.

Section Review

1. What is genetic counselling?
2. Describe the procedure of amniocentesis.
3. What is genetic screening? Give an example.
4. How can recombinant DNA research be helpful to people?

Main Ideas

1. Human genetic disorders may result from a mutation in a gene.
2. Some genetic disease is caused by extra chromosomes.
3. Many traits in humans are determined by several genes.
4. The environment plays an important role in determining the way we are.
5. The study of twins has helped geneticists understand the roles of heredity and the environment.
6. Selective breeding has helped produce plants and animals best suited to human needs.

Main Ideas   465

7. Genetic counsellors help families who think they may have a genetic problem.
8. Recombinant DNA research is leading to many new advances in genetics.

Glossary

amniocentesis	am-ni-oh-sen-TEE-sis	a procedure used to find out about a baby before it is born
breed		plants or animals developed through selection
cloning		a procedure for isolating genes
crossbreeding		the mating of two different breeds
genetic engineering		changing the DNA in a living cell

Study Questions

A. True or False

Decide whether each of the following statements is true or false. If the sentence is false, rewrite it to make it true. (Do not write in this book.)
1. Sickle cell anemia is a disease that results from having the wrong number of chromosomes.
2. Tay-Sachs disease is most common among people of African ancestry.
3. The inheritance of height is an example of multiple-factor inheritance.
4. Twins are **always** genetically identical.
5. The **Santa Gertrudis** breed of cattle is the result of crossbreeding.

B. Completion

Complete each of the following sentences with a word or phrase that will make the sentence correct. (Do not write in this book.)
1. Large strawberries with an extra set of chromosomes are _____ .
2. _____ is used to find out about a baby before it is born.
3. Heredity and the _____ both play important roles in determining the way we are.
4. The mule is a _____ between a horse and a _____ .
5. The gene for human insulin has been _____ in bacteria.

C. Multiple Choice

Each of the following statements or questions is followed by four responses. Choose the correct response in each case. (Do not write in this book.)

1. Sickle cell anemia is the result of
   a) having an African ancestry
   b) a mutation in the gene for hemoglobin
   c) lack of oxygen
   d) abnormal veins and arteries
2. Eye colour is determined by
   a) several genes acting together
   b) one pair of genes
   c) a dominant gene
   d) the environment
3. Identical twins
   a) have exactly the same environment
   b) may be of different sexes
   c) are no more alike than siblings
   d) are identical genetically

D. Using Your Knowledge

1. Duchenne muscular dystrophy is a recessive sex-linked disease in humans. If a woman is a carrier of this disease, will any of her daughters have muscular dystrophy? How about her sons? Why?
2. People with Klinefelter's syndrome have an extra X chromosome. Could this disease be detected by amniocentesis? Why or why not?
3. Do you think foot size is a multiple-factor trait? Why or why not?
4. Why is heredity in humans harder to study than it is in plants and animals?
5. How might the environment cause people to be taller today than in past generations?

E. Investigations

1. Can you think of some cases in which height is *not* affected by many genes? *Hint:* Go to the library and read about achondroplasia.
2. Find out about sperm banks for domestic animals. What is artificial insemination? Prepare a report on this special type of selective breeding.
3. Write to or visit a hospital in your area. Do they have an amniocentesis program? Try to collect some pamphlets on amniocentesis and genetic counselling. What are the real problems with these programs?
4. Arrange a class debate on recombinant DNA research. Discuss the usefulness and the problems of this branch of genetics.
5. Research Canada's role in breeding rust-resistant wheat.

32 Evolution

32.1 The Diversity of Life: Evolution
32.2 The Darwin-Wallace Theory
32.3 Activity: Understanding Natural Selection
32.4 More Factors in Evolution
32.5 Activity: Protective Colouration

Fig. 32-1 Fossils of entire plants and animals or parts of organisms have been studied by scientists. Fossils tell a tale of past life on earth.

Think for a moment about the number of different species of organisms on earth. Over 1 500 000 species have been named, and countless thousands remain to be discovered. Scientists know that each of these species fits into a certain environment. As a result, three questions come to mind.

1. Why are there so many species of organisms?
2. Why does each species fit into a certain environment?
3. Is there a relationship between the number of species (the diversity of life) and the environment?

You already know that the environment is an important factor in determining the nature of organisms. In this chapter you will learn why many scientists think that *changes* in the environment are important in explaining the diversity of life.

This chapter deals with the **Theory of Evolution**. Like any theory, this theory is a set of ideas put together (by reasoning from known facts) to explain something. It is important to remember that the theory itself is not a fact. That is, it is not beyond question. In fact, the theory of evolution has undergone change and will probably change more as new discoveries are made.

32.1 The Diversity of Life: Evolution

What is Evolution?

Life has existed on earth for a very long time. Scientists think that the earth is about 5 000 000 000 years old and that life on earth began about 3 000 000 000 years ago. What evidence do scientists use to make these estimates? Evidence of past life can often be found in rocks in the form of fossils (Fig. 32-1). By studying fossils, scientists have found out that many species have died out or become **extinct** throughout the history of the earth (Fig. 32-2). But they also know that new species have appeared! What causes these changes?

Scientists know that the environment has changed during the history of the earth. How do you suppose a large change in the environment might affect an organism? A cactus has many features or adaptations that allow it to thrive in a desert (Fig. 32-3). Would these adaptations help the cactus survive in a warm, moist climate?

Today, many scientists believe that species of organisms survive *because they change.* These changes take place over many thousands of years. Changes in species of organisms throughout time is called **evolution**. Several scientists have proposed theories to explain how and why species of organisms change or evolve. Let's see what one of these says.

Lamarck's Theory: Does It Explain Evolution?

One of the first scientists to think about evolution was the French biologist, Jean Lamarck. He proposed his theory in the early 1800s. Lamarck thought that the environment was an important factor in evolution. As you will see, this agrees with modern theories of evolution. However, Lamarck's ideas about how this occurred have been proven wrong. His reasoning, however, is interesting. Let's look at it.

Lamarck used an example to explain his theory (Fig. 32-4). He suggested that giraffes used to have short necks. They ate the leaves on the lower parts of trees. Soon this food became scarce. The

Fig. 32-2 Almost 200 000 000 years ago, many species of dinosaurs roamed the earth. Dinosaurs are now extinct. What caused this change?

Fig. 32-3 What characteristics do these organisms have that allow them to survive in their environments?

Section 32.1

Fig. 32-4 Lamarck developed a theory to explain how traits like the long neck of a giraffe might evolve. How is this theory incorrect?

giraffes began stretching their necks to reach the leaves on the higher parts of the trees. As a result, their necks became longer. The long neck trait was then passed on to the next generation. Over many years, the modern long-necked giraffe evolved.

Think about what you know about heredity. A giraffe may well stretch its neck to reach leaves. But will the genes that determine neck length be changed? The giraffe would still pass on genes for a short neck to its offspring. Do you think the man in Figure 32-5 will have children with muscular bodies?

Scientists have done experiments to show that Lamarck's theory is not correct. The tails were cut off long-tailed male and female mice for many generations. Lamarck's theory suggests that the offspring should have short tails because the tails are not being used. None of the offspring, however, had short tails. If this experiment were conducted for thousands of years, would the results be different? We are not sure.

Lamarck thought that organisms change *in order to survive* in their environment. You have just learned, however, that this is not necessarily true. What, then, is the explanation?

Section Review

1. What important message do scientists see in the fossil record?
2. Define evolution.
3. **a)** Describe Lamarck's theory.
   b) Describe an experiment that shows that his theory is not correct.

Fig. 32-5 This man has lifted weights to change the shape of his body. Do you think he will pass these traits on to his children?

32.2 The Darwin-Wallace Theory

In 1859, both Charles Darwin and Alfred Wallace published a theory to explain the diversity of life. Darwin thought and wrote more about the theory than Wallace did. Therefore he is usually given the most credit for these important ideas.

Darwin was a naturalist who sailed around the world studying plants and animals. He was fascinated by the diversity of life and collected many specimens to study in his laboratory. Out of his studies he made a very important observation: There are often similarities between the many diverse species of plants and animals. For example, Figure 32-6 shows how three different species of animals have similar bone structure. On the basis of such observations, Darwin proposed that today's living species are descendants of species that lived on earth in the past. That is, today's organisms have changed or evolved from earlier species. This idea is called the **Theory of Organic Evolution**.

Darwin then gave thought to this question: Why have so many species evolved?

Natural Selection

Throughout this book you have seen how organisms seem to be suited to their environments. Darwin proposed that organisms that are well adapted to their environments will survive and produce offspring. The traits in the organisms that allow them to survive are passed on from generation to generation. Organisms that are not well adapted to their surroundings will generally not live long enough to reproduce. Therefore undesirable traits will not be passed on to the next generation. Darwin called this process **natural selection**.

What evidence did Darwin see in nature that natural selection was taking place? First, Darwin noticed that most organisms produce

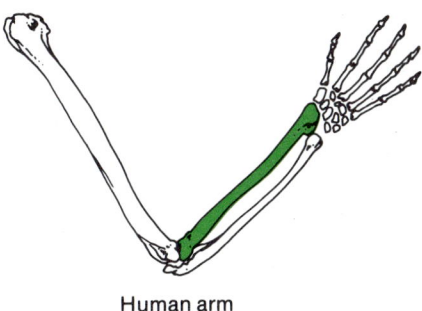

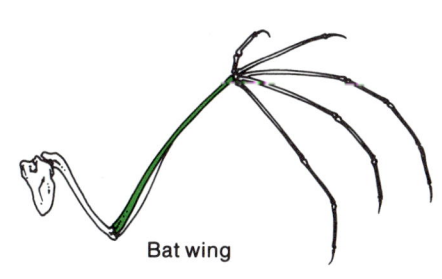

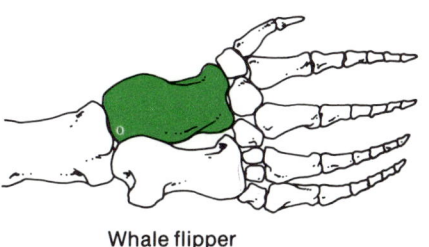

Fig. 32-6 These are three forelimbs from very different organisms. The bone structure shows the same basic pattern.

Section 32.2   471

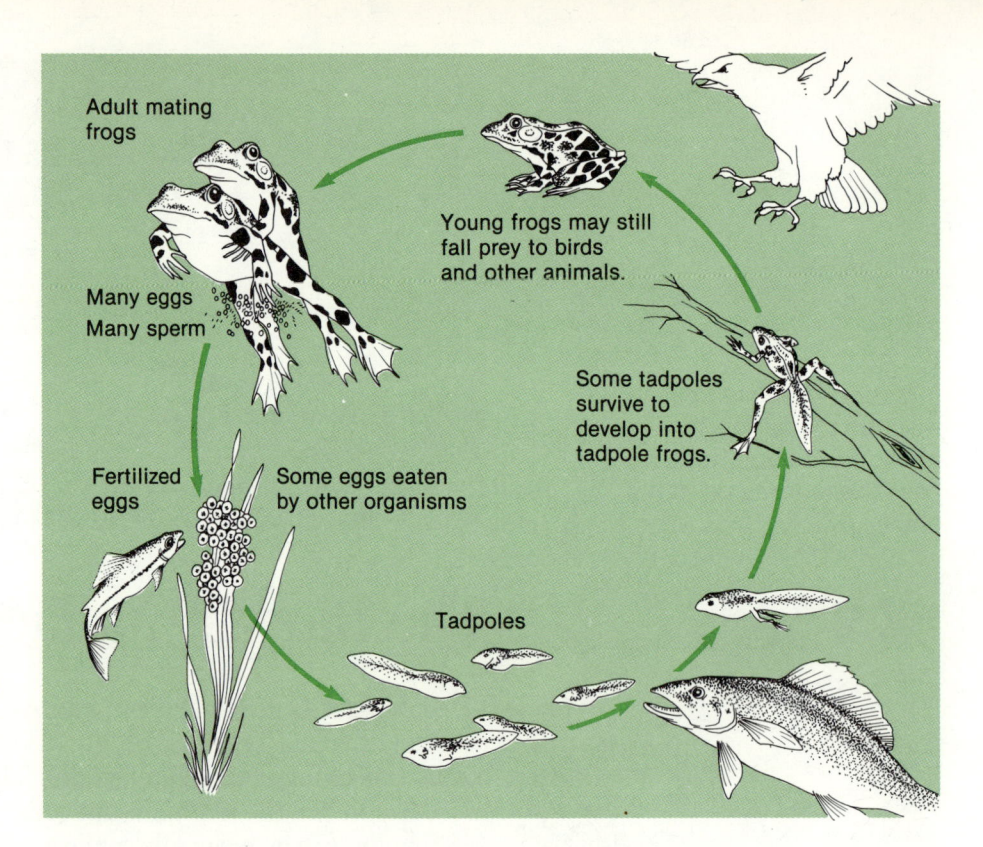

Fig. 32-7 Not all of the offspring produced by frogs survive to become adults. What features might help tadpoles and young frogs survive?

many offspring. But not all of these offspring survive. A female frog produces hundreds of eggs. Of those that are fertilized, only a few develop into zygotes. Also, only a few of the zygotes become tadpoles. Further, only the best adapted or most "fit" tadpoles survive to become adults (Fig. 32-7).

Darwin also noticed that not all members of a species are the same. That is, there is **variation** among numbers of a species. Sometimes this variation gives a particular organism an advantage over another organism. Suppose, for example, that some deer in a herd have longer legs than other members of the herd. These deer will be able to escape predators better than deer that have shorter legs. Therefore, the long-legged deer will live longer and, as a result, produce more offspring than other deer. In the next generation, more deer will have longer legs.

Lamarck also noticed that organisms varied. How did Darwin's and Lamarck's observations on variability differ? Lamarck thought that organisms developed changes *in order to adapt to the environment*. However, Darwin noticed that organisms vary *whether they need to or not*. Whether this variation is helpful or harmful to the species depends on the environment.

Where did variation within species come from? How does natural selection explain the development of new species? Darwin and Wallace did not answer these questions. They were unable to do so since scientists of the time did not understand genetics. However, in Section 32.4, you will find out how the Darwin-Wallace theory has changed due to our understanding of genetics.

Section Review

1. How does the Darwin-Wallace theory differ from Lamarck's theory?
2. What is natural selection?
3. Why is variation an important factor in evolution?

32.3 ACTIVITY Understanding Natural Selection

In this activity, you will use red and green chips to help you understand how natural selection may work. The chips represent a population of organisms of the same species. However, there is variation among members of this species. The red chips represent organisms that are better adapted to survive in the environment. They produce more offspring each generation than the organisms represented by the green chips. You will see that your "chip population" will change over several generations of this selection.

Problem

How might natural selection work?

Materials

about 100 red chips   paper bag
about 100 green chips   pen and paper

Procedure

a. Copy Table 32-1 into your notebook.

Table 32-1 Natural Selection

Generation	Red chips	Green chips	Total chips
0	50	50	100
1			
2			
3			
4			

b. Place 50 red chips and 50 green chips in the bag.
c. Shake the bag to mix the chips. Draw out a sample of 50 chips. These chips represent organisms that will reproduce. Set aside the rest of the chips.
d. Count the number of green and red chips. Record these numbers in your table.

Section 32.3   473

e. The red chips are better adapted to survive than the green chips. Suppose that they will produce twice as many offspring as the green chips. Put the *same number* of green chips back in the bag as you drew out of the bag. Put back twice the number of red chips. For example, if you drew out 22 green chips and 28 red chips, put back 22 green chips and 56 red chips.
f. Shake the bag to mix the chips. Draw out 50 chips. You are now sampling the second generation of chips. Repeat steps (d) and (e) for this generation of chips as well.
g. Repeat the sampling procedure for a third and fourth generation.

Discussion

1. Describe the trend in the number of green versus red chips from generation to generation.
2. Why does this experiment demonstrate how natural selection may work?
3. Suppose the red organisms continued to have this kind of advantage for thousands of generations. What do you think would eventually happen in this population?

32.4 More Factors in Evolution

Since Darwin's time, scientists have learned a great deal about heredity. They now understand how organisms might acquire traits that would help them to survive in a new environment. Organisms may vary because of **gene mutations** [myoo-TAY-shuns]. In fact, it is now thought that gene mutations are the main cause of changes that lead to evolution.

Mutations

Can you think of a mutation that may be helpful to an organism? Suppose a certain species of bird is well adapted to its environment. Most of the members of this species have small beaks. However, a few birds have a mutation that makes their beaks large and sharp. This mutation is neither harmful nor helpful since these birds eat small soft berries. But what will happen if the environment changes? Suppose the bushes that produce the soft berries die out and that the only food left in the new environment is hard nutty fruits. Now a large sharp beak will be an advantage in this new environment (Fig. 32-8). As a result, the birds with the beak mutation will probably be able to get more food than the other birds. They will survive to produce more offspring. The mutant gene will be passed on to the next generation. Over many generations, more variation may occur. A new species of bird may evolve.

Fig. 32-8 How might this bird's beak help it to survive?

Fig. 32-9 It is thought that the ancestors of the camel may have migrated from North America to other parts of the world. Populations survived in the mountains of South America. These populations developed into the llama (A). Other populations survived in the deserts of Africa (B). How do these animals differ? How have they adapted to their environment?

Migration

What other factors may allow changes in populations of animals to occur? **Migration** [my-GRAY-shun] is the movement of organisms from one environment to another. How might natural selection and migration interact to cause changes to occur?

By migrating to a new area, an organism is changing its environment. New traits caused by mutations may be an advantage in the new environment (Fig. 32-9). Offspring with these mutations are more likely to survive. After several generations, all members of the new population may have different traits than the old population. They are no longer the same as the original population. This is one way new species may evolve.

Section Review

1. Describe how mutations and the environment might interact to result in the evolution of new species.
2. What is migration? How might it aid in the evolution of new species?

32.5 ACTIVITY Protective Colouration

Many characteristics help protect organisms from their enemies (Fig. 32-10). For example, animals that can blend in with their environment are not easily seen by predators. As a result, they have a better chance of surviving to reproduce and pass on this desirable trait to the next generation. This kind of characteristic is called

Section 32.5 475

protective colouration. Could such characteristics be adaptations which have allowed the survival of the species? How can we be sure?

In this activity, you will see how blending in with the environment can protect an organism from predators. That is, you will see how those organisms with the genes for the protective colour will survive to pass those genes on to the next generation.

Problem

Does protective colouration protect organisms from predators?

Materials

brown yarn (at least 1 m)
green yarn (at least 1 m)
tray of earth
stop watch
scissors

Procedure

a. Choose a partner.
b. Make a bunch of green and brown "worms" by cutting the yarn into pieces of 5 cm long. Make at least 20 of each colour of worm.
c. One member of the pair will be the predator. This person should leave the classroom. The other member of the team will now place the worms in the earth environment. Mix them up well with the earth but do not bury them completely.
d. Call the predators back into the classroom.
e. Time the predator for 15 s as he/she picks worms from the earth environment. Work as fast as you can!
f. At the end of 15 s, count the number of green and brown worms collected. Give your result to your teacher.
g. Pool the class results.

Discussion

1. Is green or brown a protective colour in the earth environment? Why? What colour of worm will survive best to produce offspring?
2. In 1850 in Manchester, England, most peppered moths were light in colour with dark spots. Some of the moths were mostly dark with white spots. Today, most of the moths are dark with white spots. Very few light coloured moths survive. How might you explain this difference? Think about how human activity has changed the environment in cities over the years.
3. Discuss how protective colouration is an example of natural selection.

Fig. 32-10 These organisms have characteristics that protect them from their enemies. What are these adaptations?

Main Ideas

1. Organisms have many characteristics that allow them to survive in certain environments.
2. Lamarck thought organisms changed in order to survive in an environment. Experiments have shown that this idea is not correct.
3. The Theory of Organic Evolution suggests that today's species have evolved or changed over many generations.
4. Organisms that are well adapted to their environment survive to pass on desirable traits to the next generation. This process is called natural selection.
5. Gene mutations are likely the main cause of the variability that leads to evolution.
6. Migration and natural selection may act together to result in the formation of new species.
7. Protective colouration is a special kind of characteristic that helps protect organisms from predators.

Glossary

evolution		changes in species of organisms throughout time
migration	my-GRAY-shun	the movement of organisms from one environment to another
mutation	myoo-TAY-shun	a sudden change in genes
variations		differences among members of a species

Study Questions

A. True or False

Decide whether each of the following statements is true or false. If the sentence is false, rewrite it to make it true. (Do not write in this book.)

1. The Theory of Evolution is a fact.
2. Evolution is the changes in species of organisms throughout time.
3. A giraffe probably got its long neck by stretching up into trees to reach food.
4. Darwin proposed that today's living species are descendents of species that lived on earth in the past.
5. All members of a species are exactly the same.

B. Completion

Complete each of the following sentences with a word or phrase that will make the sentence correct. (Do not write in this book.)

1. A ▓▓▓▓▓ is a set of ideas put together (by reasoning from known facts) to try to explain something.
2. Evidence of past life can often be found in rocks in the form of ▓▓▓▓▓.
3. The main cause of changes that lead to evolution is thought to be ▓▓▓▓▓.
4. The movement of organisms from one environment to another is called ▓▓▓▓▓.

C. Multiple Choice

Each of the following statements or questions is followed by four responses. Choose the correct response in each case. (Do not write in this book.)

1. Organisms that are well adapted to their environments are more likely to survive than organisms that are not well adapted to their environment. This process is called
   a) evolution **b)** natural selection **c)** gene mutation **d)** variation
2. An Arctic bird called a ptarmigan is brown in the summer and white in the winter. This is an example of
   a) protective colouration   **c)** evolution
   b) natural selection   **d)** variation
3. Rabbits lose a great deal of body heat through their ears. Desert rabbits have long ears and Arctic rabbits have short ears. This difference may be due to
   a) variation   **c)** the Theory of Evolution
   b) migration   **d)** natural selection

D. Using Your Knowledge

1. What are the differences between natural selection and the artificial selection or breeding you read about in Chapter 31? What similarities, if any, are there between these two processes?
2. When Darwin travelled to the Galapagos Islands, he noticed 13 different species of finches. Each species had a different kind of beak for eating particular foods. How would Darwin explain this variation? How might Lamarck explain this diversity?
3. Many organisms have organs that are no longer thought to be useful. For example, the appendix in humans has no known function. Are such organs evidence for evolution? Explain your answer.
4. Monarch butterflies are very distasteful to birds. The Viceroy butterfly looks much like the Monarch butterfly. This is known as mimicry [MIM-i-kree]. How might this mimicry help the Viceroy survive? Do you think mimicry is a result of natural selection? Why?

E. Investigations

1. Go to the nearest museum of natural history. Take along a notebook and pencil. Find out how fossils are formed. Make sketches of two or three extinct animals. Why do you think they became extinct?
2. Take a walk through a wooded area or field near your home. Make notes on organisms that have special characteristics to help them survive in this environment. Makes notes on any examples of protective colouration that you find.
3. Hugh DeVries was a Dutch botanist who proposed a theory to explain the diversity of organisms. Find out about DeVries' theory. How does it relate to the theories you have learned about?
4. Find out why many of our common antibiotics are less effective against bacterial diseases now than they were 20 years ago.
5. Many people, including some scientists, believe that life on earth originated through an act of creation. They believe that a group of species was created. Then these species evolved to form the wide variety of species present today. According to this viewpoint, skunks may have always had scent and turtles may have always had hard shells. Find out more about this viewpoint. Then write a paper of about 200 words in which you analyze this viewpoint.

Unit 8

Ecology

Every living thing on earth interacts with other living things. It also interacts with its non-living environment — air, water, and soil, for example. Even humans interact with their living and non-living environments. Can you think of ways you depend on other living things? How do they depend on you? How do you depend on your non-living environment? And how do you affect it?

Ecology deals with the interrelationships between organisms and their environments. This unit is about ecology. In the first two chapters you will study the basic ideas of ecology. Then, in the next two chapters, you will see how those ideas work in natural ecosystems.

CHAPTER 33
How Living Things Interact

CHAPTER 34
Matter and Energy in Ecosystems

CHAPTER 35
Aquatic Ecosystems

CHAPTER 36
Terrestrial Ecosystems

Fig. 33-0 The deer interact with both their living and non-living environments. How do the deer depend on plants? How do plants depend on the deer? How do the deer depend on the water? Do the deer affect the water in any way? What animals interact with the deer?

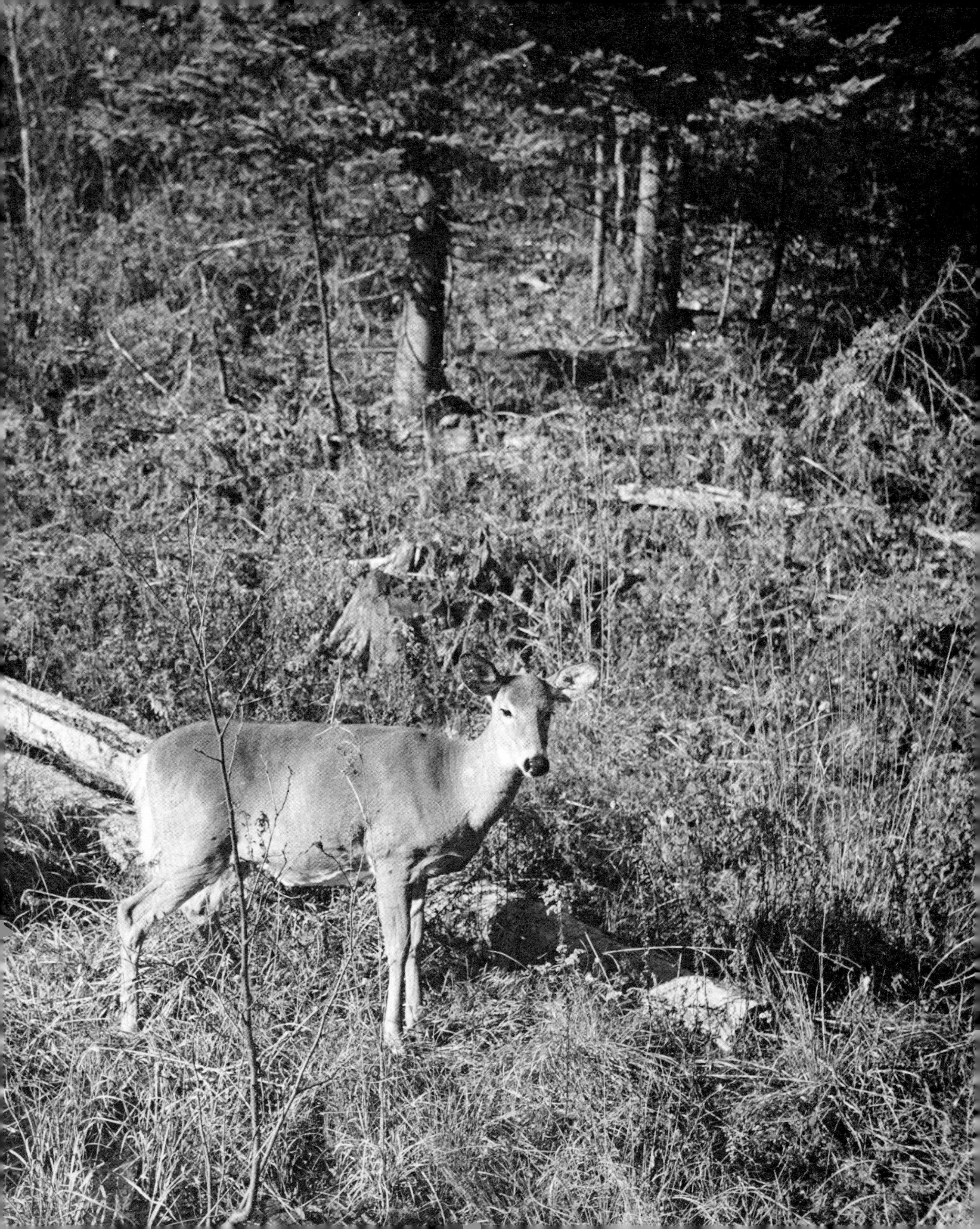

33 How Living Things Interact

33.1 What Is Ecology?
33.2 Activity: Making a Model Ecosystem
33.3 Habitat and Niche
33.4 Feeding Levels
33.5 Food Chains and Food Webs
33.6 Activity: Feeding Levels in a Meadow or Lawn
33.7 The Pyramids of Life
33.8 More Feeding Relationships

Environment... ecology... habitat... ecosystem... food chain... We often hear these words today. That's because people are becoming concerned about their environment. They care about clean air and clean water. They care about our forests and wildlife. But caring isn't enough. To protect our environment, we also need to understand how it works. And that's what this chapter is all about.

33.1 What Is Ecology?

A Definition of Ecology

Imagine that you are the person who photographed Figure 33-1. Everything around you is your **environment**. Part of your environment is **living**. The dog, cattle, trees, and grass are examples. And, part of your environment is **non-living**. Soil, air, water, wind, and light are examples.

Can you think of ways that you interact with the other animals? That is, how do you depend on them? And how do they depend on you? How do you interact with the trees? the grass? And how do you interact with the non-living factors?

No organism lives completely on its own. It depends on other organisms and they depend on it. It also depends on the non-living environment in which it lives. **Ecology** is the study of the relationships among organisms and between organisms and their environments. You will meet many interesting ecological relationships in this chapter and the next three.

Fig. 33-1 What organisms make up the living environment? What factors make up the non-living environment?

From Individual to Biosphere

We can study how organisms interact at four levels: population, community, biome, and biosphere (Fig. 33-2). Let's see what these terms mean.

482 Chapter 33

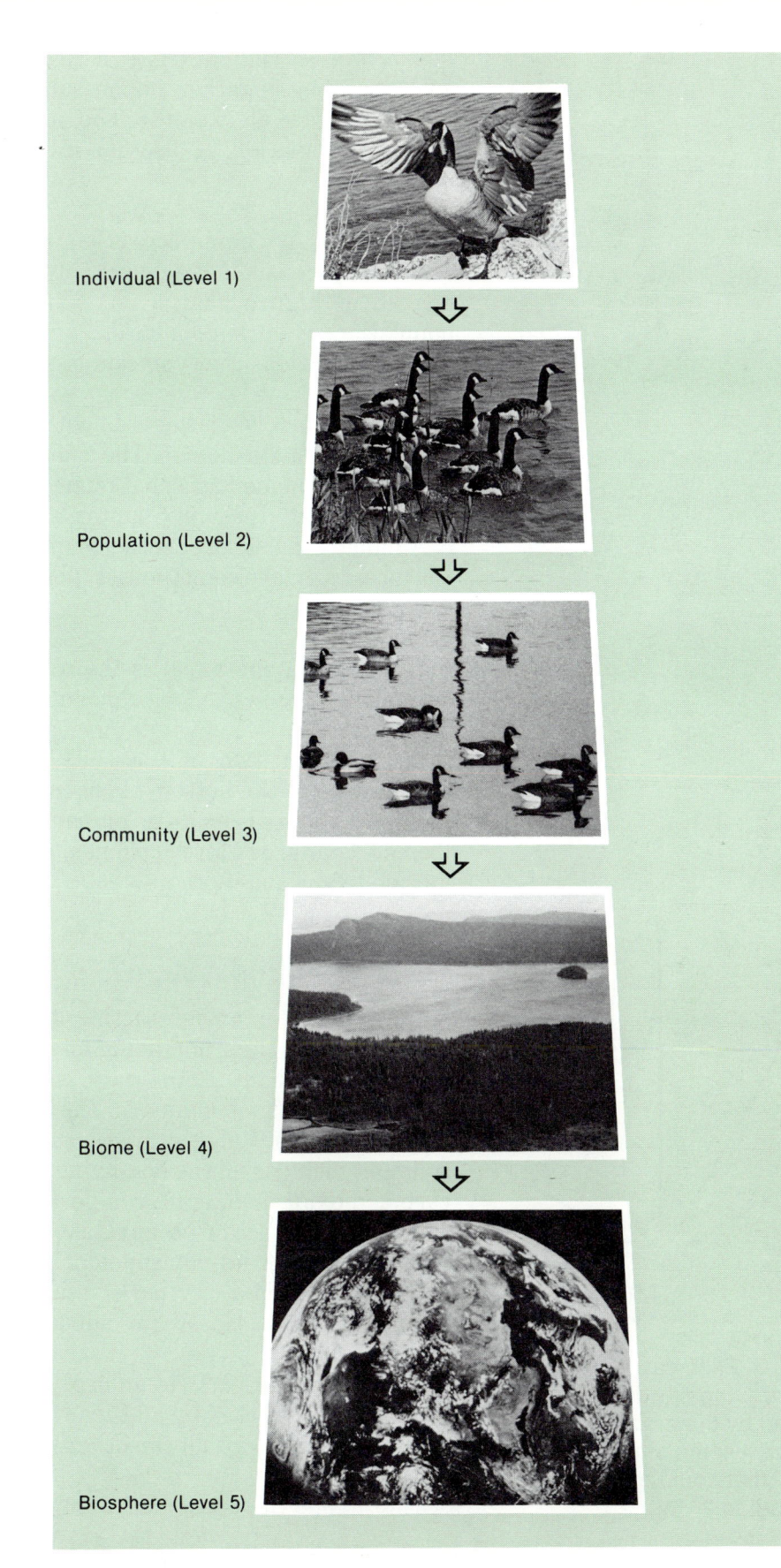

Fig. 33-2 Individuals make up a population; populations make up a community; communities make up a biome; and biomes make up the biosphere.

Individual (Level 1)

Population (Level 2)

Community (Level 3)

Biome (Level 4)

Biosphere (Level 5)

Section 33.1 483

Population A **population** is a group of individuals of the same species, living together in the same area. The geese in level 2 of Figure 33-2 are an example. The pond with the geese in it also has a bullfrog population. It also has a water lily population and a perch population.

Community A **community** is *all* the living things in an area. A community consists of several populations. The pond community in level 3 of Figure 33-2 is an example. It is made up of a goose population, a duck population, a water lily population, and many other populations that you cannot see.

Biome A **biome** is a large area with a characteristic climate. Canada has only five main biomes. The coniferous forest in the north is one of these. Each biome has a characteristic set of plants and animals. You will investigate these in Chapter 35.

A biome consists of several communities. For example, the coniferous forest has lake communities, pond communities, bog communities, and many others.

Biosphere The **biosphere** is the region on earth in which life exists. Organisms live in the lower parts of the atmosphere. They also live in almost all bodies of water on earth. They live on the soil and in the first metre or two of the soil. This thin layer from the lower atmosphere to the bottom of oceans makes up the biosphere.

The biosphere is made up of many biomes. Among them are desert biomes, coniferous forest biomes, grassland biomes, tundra biomes, marine biomes, and others.

What Is an Ecosystem?

An **ecosystem** is an interacting system that consists of groups of organisms and their non-living environment. Thus an ecosystem has two parts, a **living community** and a **non-living environment**. The living community includes all the living things in the ecosystem — animals, plants, fungi, protists, and monerans. The non-living environment includes all the non-living factors in the ecosystem — water, carbon dioxide, soil, light, temperature, wind, moisture, and others.

An ecosystem can be any size. Any community of living things interacting with its environment is an ecosystem. A woodlot is an ecosystem. So is a city park, a lake, a pond, and the whole Arctic tundra. Even a classroom aquarium or a handful of soil is an ecosystem.

Remember, all parts of an ecosystem are interrelated (Fig. 33-3). Each part is affected by all the other parts. Therefore, if one part is changed in any way, all the other parts will be changed also.

An Example Suppose, for example, that a lumber company com-

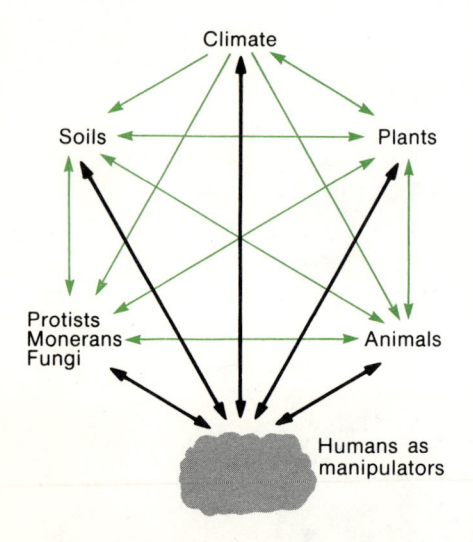

Fig. 33-3 Each part of an ecosystem, living or non-living, is affected by the other parts. Can you name some other non-living factors that could be placed in this diagram?

pletely cleared a forested area of its trees. What effects would this have on the forest ecosystem? The trees would no longer add humus to the soil, since they would no longer drop leaves onto the ground. Snails, slugs, earthworms, bacteria and fungi thrive on leaf litter. Therefore, they would decrease in numbers. Some species might vanish completely. Animals that feed on these organisms would also be affected. Soil erosion might occur, since the leaf canopy would no longer be present to absorb the energy of a heavy rainfall. If the soil was washed away, many plant species would disappear. The animals that eat these plants would move away or die of starvation. Plants such as ferns that require shade and moisture, would die.

On the other hand, many species of plants that require bright sunlight might then be able to grow in the area. Grasses, goldenrod, and other sun-tolerant plants might gradually become established. Shrub and tree species that could not grow in the shade of the forest might appear. New insect populations might be established. New bird populations might appear. Mammal species, such as white-tailed deer, might increase in numbers. But the original ecosystem would be gone, perhaps forever.

The chain of events that occurs when one factor in an ecosystem is changed is long and involved. However, these events are certain to occur. Try now to imagine further changes that would occur in a forest that has been cleared. Do you think that neighbouring ecosystems would be affected? Would a nearby stream, pond, or meadow change in any way as a result of lumbering activities in the forest?

Section Review

1. What makes up an organism's environment?
2. What is ecology?
3. Explain the terms population, community, biome, and biosphere.
4. What is an ecosystem?
5. Name 5 ecosystems.

33.2 ACTIVITY Making a Model Ecosystem

In this activity you will build a **model ecosystem**. Then you will study how it works. The basic principles you learn here also apply to natural ecosystems such as lakes and forests. How many of these principles can you discover?

Problem

What are the basic parts of an ecosystem? How does an ecosystem work?

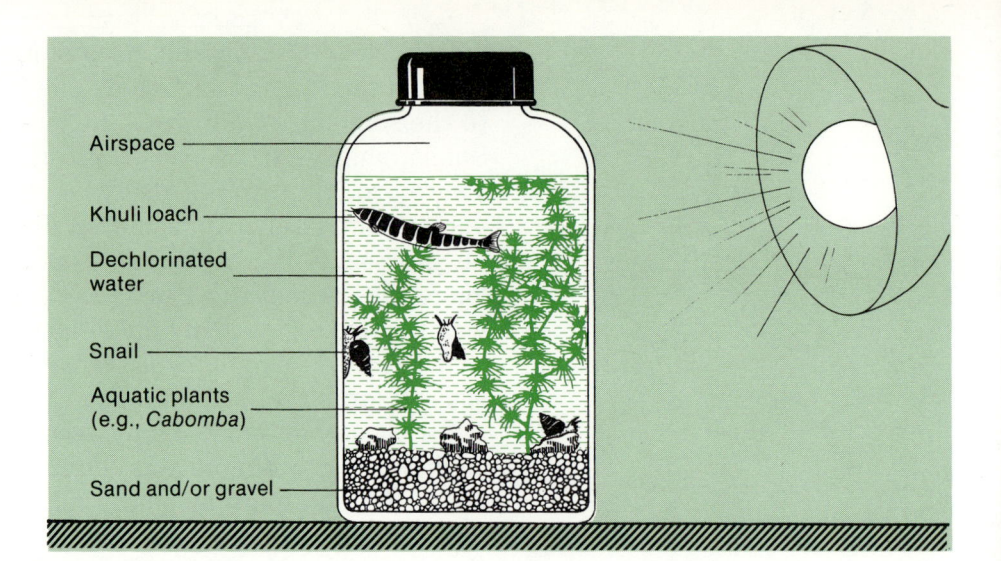

Fig. 33-4 A classroom ecosystem. Where do the plants, fish, and snails get their nutrients? Why is the light needed?

Materials

large bottle or jar with top (at least 3-4 L)
table lamp with 60 W bulb
khuli loach or another plant-eating fish
strands of an aquatic plant (3 or 4)
pond snails (8-10)
clean gravel and/or sand

Procedure

a. Place sand or gravel to a depth of 2-3 cm in the jar.
b. Fill the jar with water. If you use tap water, let the jar stand with the top removed for 48 h. This lets the chlorine leave the water.
c. Add a few strands of an aquatic plant such as *Cabomba*.
d. Add 8-10 pond snails to the water.
e. Place a khuli loach in the water.
f. Put the top on the jar and seal it tightly.
g. Place a table lamp with a 60 W bulb in it close to the jar as shown in Figure 33-4.
h. Place the setup in a location away from windows and other places where light and temperature conditions change greatly during the day.
i. Leave the lamp on 24 h per day. Or, place it on a timer that provides at least 16 h of light per day. Do not depend on your memory to turn the light off and on!
j. A healthy ecosystem will have a pale green colour in the water. This colour is caused by algae, which are food for the fish and snails. As the days pass, move the light closer if the water does not develop a green colour. Move it further away if the green becomes intense.
k. Observe your ecosystem closely from time to time for several months. Make careful notes of any changes that occur.

Discussion

Many days may pass before you can answer all these questions.
1. Your ecosystem is called a *closed* ecosystem. How does this differ from natural ecosystems?
2. Why do you think we used a closed ecosystem for this activity?
3. How do the plants and algae get carbon dioxide for photosynthesis?
4. How do the organisms get oxygen for respiration?
5. How do the plants get the nutrients they need?
6. How do the animals get the nutrients they need?
7. Why is the light required?
8. What do you think will happen if a fish or snail dies?

33.3 Habitat and Niche

Habitat

The **habitat** of an organism is the place in which it lives. An ecosystem, such as a woodlot, has many habitats. For example, the habitat of an earthworm is the rich woodlot soil. The habitat of a land snail is the moist leaf litter. The habitat of a porcupine is a hollow tree. The habitat of a bluejay is the branches of the trees.

Habitats may overlap. For example, the porcupine may seek out a meal of bark in the branches that are also the habitat of the bluejay. However, since these animals do not eat the same food, no problems result from the overlap of their habitats.

Niche

The **niche** of an organism is its total role in the community. For example, the niche of a deer is to feed on grass and other plants, to

Fig. 33-5 Can you describe the habitat and niche of each of these animals?

become food for wolves, to provide blood for blackflies and mosquitoes, to fertilize the soil with nutrients, and so on. The niche of a frog in a pond is to feed on insects, to become food for snakes and other animals, and many other things.

Comparing Habitat and Niche

Many people confuse habitat and niche. This may help you to remember the difference: Think of the habitat as the "address" of the organism. And think of the niche as its "occupation", or "job" (Fig. 33-5).

If two species have the same habitat and similar niches, they will **compete** with one another. For example, mule deer and elk often live in the same mountain valley. That is, they have the same habitat. Both species eat grass and other plants. Both are preyed upon by wolves and are attacked by many of the same parasites. That is, they have similar niches. Clearly they will compete for available space and food in the valley.

Section Review

1. What is the difference between habitat and niche?
2. What happens if two species have similar niches in the same habitat?
3. Describe the habitat of the snails in your model ecosystem.
4. What is the niche of the snails in your model ecosystem?

33.4 Feeding Levels

All ecosystems have feeding levels called producers, consumers, and decomposers. Let's see what these are.

Producers

All living things need energy. Plants, many monerans, and many protists contain chlorophyll. Therefore they can carry out photosynthesis. That is, they store some of the sun's energy in starch molecules. They make, or *produce*, their own food. Therefore they are called **producers**.

Producers are an important food group in ecosystems. Why is this so? Can you name some producers in a woodlot ecosystem? in a pond ecosystem?

Consumers

All ecosystems have **consumers**. These are animals that feed on plants or on other animals. Animals that feed on plants are called

Fig. 33-6 This ground squirrel eats grass, seeds, and a variety of plants. It also eats grasshoppers. What kind of consumer is it: herbivore, carnivore, or omnivore?

herbivores [HUR-bi-vohrs]. Deer, rabbits, mice, cattle, and rabbits are herbivores. Animals that feed on other animals are called **carnivores** [CAR-ni-vohrs]. Wolves, polar bears, eagles, and mountain lions are carnivores.

Some animals are both herbivores and carnivores. For example, a fox eats berries as well as mice (Fig. 33-6). Such animals are called **omnivores** [OM-ni-vohrs]. Are you a herbivore, carnivore, or omnivore?

Carnivores that feed on live animals are called **predators**. The animals that are eaten are called **prey**. The marsh hawk in Figure 33-7 preys on mice. It is the predator. Mice are its prey.

Animals that feed on dead organisms are called **scavengers**. Snails and crayfish are scavengers in ponds. They eat dead plants and animals. Crows, magpies, and vultures are also scavengers. You may have seen one of these birds feeding on an animal carcass on the roadside. Some animals may be predators sometimes and scavengers at other times. A wolf is an example.

Decomposers

Ecosystems also contain a feeding level called **decomposers**. These organisms are mainly bacteria and fungi such as yeasts and moulds. They feed on non-living organic matter such as dead plants and animals and animal wastes.

All organisms eventually die. And all animals produce wastes. If decomposers were not present, dead organisms and wastes would soon smother the earth. But decomposers perform an even more important function. As they break down or feed on the organic matter, they return valuable nutrients to the ecosystem. These nutrients can now be used again to help producers grow.

Fig. 33-7 A predator-prey relationship. This marsh hawk is looking for mice or large insects like grasshoppers.

Section 33.4 489

Section Review

1. What are producers?
2. What is a consumer?
3. How do herbivores, carnivores, and omnivores differ?
4. What are predators and prey?
5. What is a scavenger?
6. **a)** What are decomposers?
   b) Give two reasons why decomposers are important.
7. **a)** What are the producers in your model ecosystem?
   b) Name the consumers in your model ecosystem. What do they eat?
   c) How do you know your model ecosystem contains decomposers?

33.5 Food Chains and Food Webs

Organisms in an ecosystem may be linked together in feeding relationships called **food chains**. For example, clover is food for the groundhog. The groundhog, in turn, is food for the fox. This simple food chain in shown in Figure 33-8.

Many food chains follow the pattern shown in Figure 33-8. That is, they begin with producers and move through herbivores to carnivores. Some food chains, however, begin with dead plants or animals. An example is a crayfish feeding on a dead fish in a pond. In a sense, even this food chain began with a producer. The fish, when it was alive, was in a food chain that began with a producer.

Fig. 33-8 Many food chains begin with producers. Some begin with dead plants or animals.

PRODUCER
Clover

HERBIVORE
Groundhog

CARNIVORE
Coyote

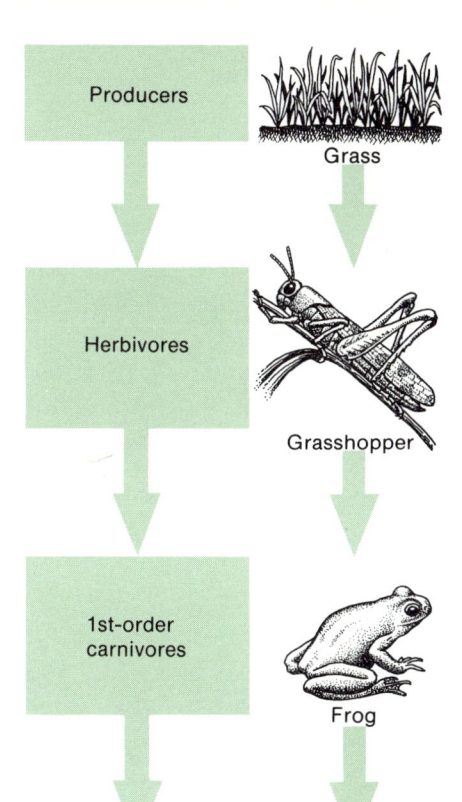

Most animals are in more than one food chain. In other words, they have more than one source of food. For example, a rabbit could be part of the following food chains:

Grass → Rabbit → Fox
Clover → Rabbit → Fox
Clover → Rabbit → Wolf
Grass → Rabbit → Owl
Grain → Rabbit → Human

Some food chains can be quite long. That's because they have several "orders" of carnivores. Look at Figure 33-9. What is meant by **first-order carnivore, second-order carnivore,** and **top carnivore**? Food chains can be connected to form a **food web** (Fig. 33-10).

Section Review

1. What is a food chain?
2. What is the general pattern for food chains that begin with producers?
3. What is meant by these terms: first-order carnivore, second-order carnivore, top carnivore?
4. What is a food web?

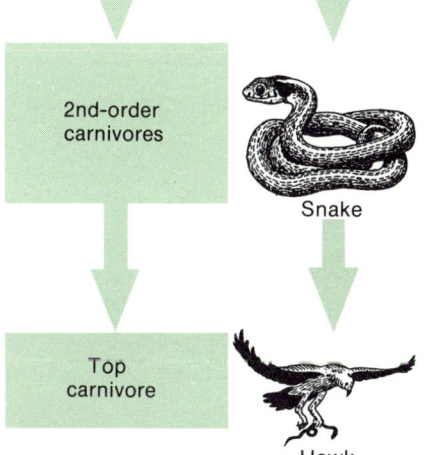

Fig. 33-9 The general pattern for a food chain

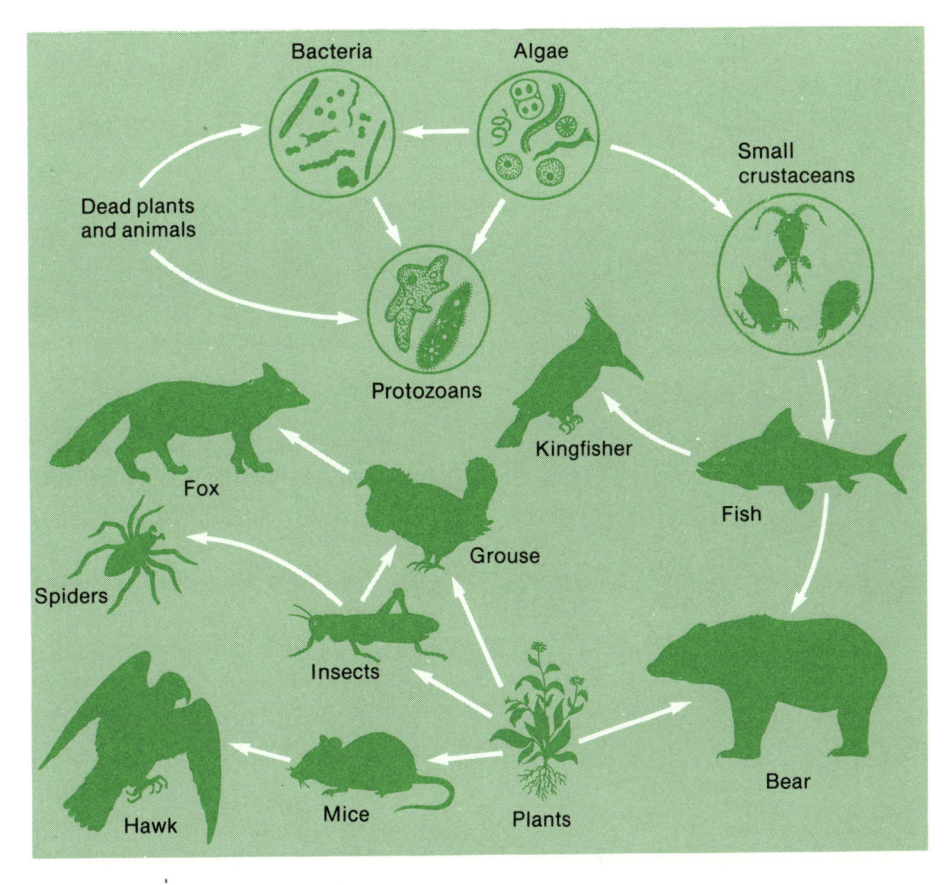

Fig. 33-10 A food web. Follow the arrows to find out who eats who.

Section 33.5   491

33.6 ACTIVITY Feeding Levels in a Meadow or Lawn

There are three main feeding levels: producers, consumers, and decomposers. Consumers, in turn, can be divided into many smaller levels. Herbivores, first-order carnivores, second-order carnivores, and scavengers are examples.

In this activity you will visit a meadow or lawn. A meadow (unmown area) is best since it will have more life in it. At the site you will search for organisms. How many can you find? What feeding levels are they in?

Problem

What organisms and feeding levels are present in a lawn or meadow?

Materials

jars with tops (5)
hand lens
white tray
trowel
sweep net

Procedure

a. Copy Table 33-1 into your notebook. Make it a full page long.

Table 33-1 Study of a Meadow or Lawn

Species (name or sketch)	Feeding level and sub-level

b. Look at the vegetation closely. How many kinds of plants are present? If you know their names, put them in your table. If you don't, make a sketch of them. What feeding level do the plants belong to? Put your answers in the table.

c. Use the trowel to collect some topsoil and plant litter. Put this in the tray. Spread it out. Then search for animals (Fig. 33-11). Watch them with the hand lens. You may wish to put them in jars. Then they are more easily examined. Put their names (or sketches) in the table. What feeding level and sub-levels are the animals in? Put your answers in the table.

Note: You won't get all the sub-levels. But if you watch the animals closely, you will get some of them.

d. Sweep the vegetation with the sweep net as shown in Figure 33-12. Transfer the animals to jars. Watch them with the hand

Fig. 33-11 Sorting through soil and plant litter for animals

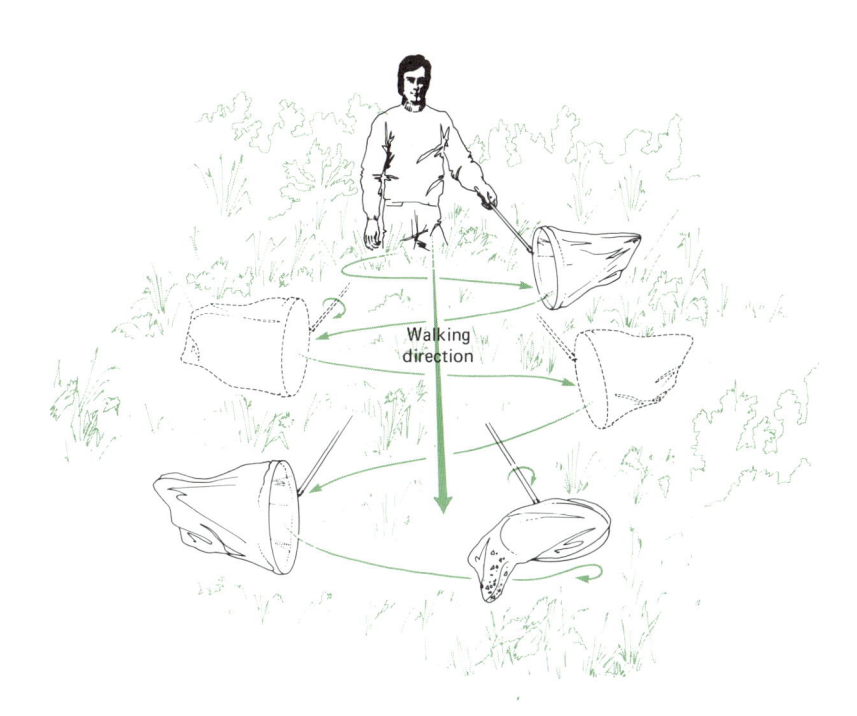

Fig. 33-12 Move the net from side to side as you walk along. Note how the animals are trapped when you are done.

lens. Put their names (or sketches) in the table. What feeding level and sub-levels are the animals in? Put your answers in the table.

e. Are there any decomposers present? What is your evidence?
f. Release all animals when your study is complete.

Discussion

1. List as many food chains as you can for this ecosystem.
2. What would happen if all the producers died?
3. What might happen if all the herbivores died?
4. What larger animals may be involved in the food chains of this ecosystem?

33.7 The Pyramids of Life

Pyramid of Numbers

Consider this simple food chain:

$$\text{Wheat} \longrightarrow \text{Mouse} \longrightarrow \text{Owl}$$

A mouse that eats only wheat must eat a few hundred kernels a day to stay alive. However, the owl needs only five or so mice a day. Many food chains show a similar numerical relationship. That is, many producers feed fewer herbivores which, in turn, feed still fewer carnivores. Such relationships are often represented by a **pyramid of numbers** (Fig. 33-13).

Not all pyramids of numbers have such a regular shape. For example, you may have seen tent caterpillars feeding on an apple

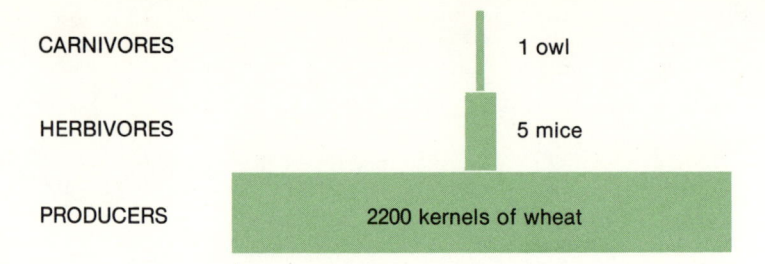

Fig. 33-13 A pyramid of numbers

CARNIVORES	1 owl
HERBIVORES	5 mice
PRODUCERS	2200 kernels of wheat

tree (Fig. 33-14). One apple tree feeds thousands of caterpillars. They, in turn, feed just a few thrashers (a bird). And, several hundred lice may feed on the thrashers. This pryamid of numbers is shown in Figure 33-15. Can you draw the pyramids for these food chains?

Dead fish → Moulds

Grass → Rabbits → Wolves → Fleas

Pyramid of Biomass

A pyramid of numbers has one fault. It treats all organisms as though they were the same. It ignores differences in size. Yet, to a hungry fox, size is important. One rabbit makes a better meal than one mouse!

To get around this fault, ecologists use a **pyramid of biomass**. Each level in the pyramid shows the **biomass** (total mass of the organisms) at that level (Fig. 33-16). Here is an example that uses a food chain of which you may be a part:

Grain → Chicken → Human

Chickens eat grain, and humans eat chickens. About 3 g of grain are needed to form 1 g of chicken. Yet that 1 g of chicken will form less than 0.1 g of human.

Fig. 33-14 This "tent" is home for hundreds of tent caterpillars. These animals eat the leaves of the tree.

Section Review

1. What is a pyramid of numbers?
2. Why do some pyramids of numbers not have a pyramid shape?
3. What is a pyramid of biomass?

PARASITES	1500 lice
CARNIVORES	5 thrashers
HERBIVORES	4000 tent caterpillars
PRODUCERS	1 apple tree

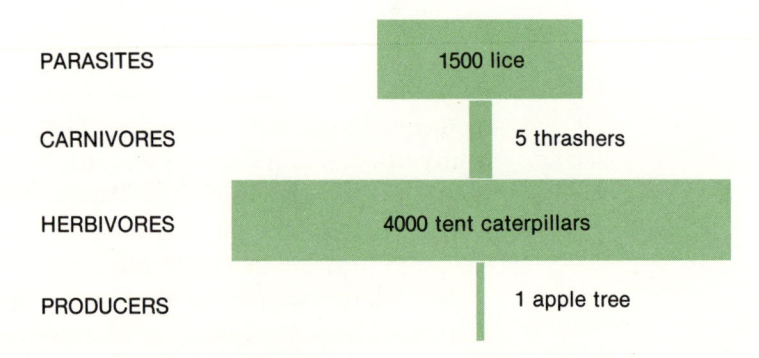

Fig. 33-15 Another pyramid of numbers. Parasites are organisms that feed on another organism without killing it.

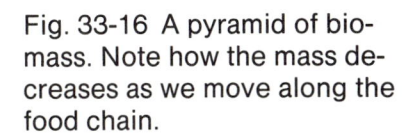

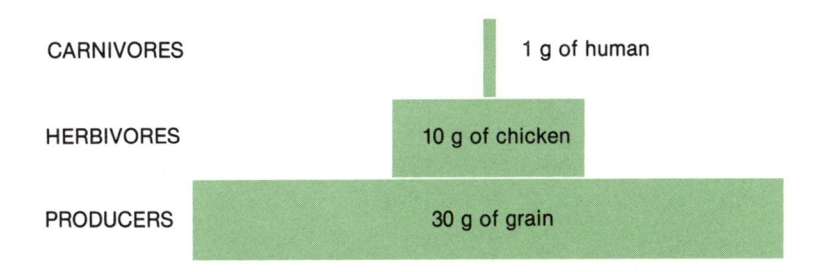

Fig. 33-16 A pyramid of biomass. Note how the mass decreases as we move along the food chain.

33.8 More Feeding Relationships

There are three feeding relationships in which organisms help provide food for other organisms without being killed and eaten themselves. These are parasitism, mutualism, and commensalism. In all three, two different species live together in close association. Let's see how they differ.

Parasitism

Parasitism [PAIR-uh-si-tiz-em] is a relationship between two organisms in which one organism benefits and the other suffers harm. The organism that benefits is called the **parasite**. The organism that is harmed is called the **host**. The mosquito is a parasite. When you provide it with a blood meal, you are the host.

Fleas, lice, ticks, and blackflies are common parasites. Humans, other mammals, and birds are their common hosts. Tapeworms are parasites that live in the human intestine. Many plant diseases are also parasites. Rusts, smuts, and mildews are examples.

Most parasites do not kill their hosts. If they did, they would lose their food supply. Besides, it isn't nice to kill a host who just gave you a free meal!

Mutualism

Mutualism [MEW-choo-uh-liz-em] is a relationship between two organisms in which both organisms benefit. For example, the relationship in a lichen [LIKE-en] is mutualism. A lichen consists of an alga and a fungus, growing together in a close relationship (Fig. 33-17). The alga contains chlorophyll. Therefore it can make food. It shares this food with the fungus. The fungus, in turn, provides the alga with water, minerals, and support.

Another example of mutualism occurs on the roots of legumes (peas, beans, alfalfa, clover, soybeans). This example was mentioned in Section 10.1, page 118. A bacterium called *Rhizobium* lives in lumps on the roots of legumes (see Figure 10-2, page 118). These bacteria change atmospheric nitrogen into nitrates. The legume plants absorb this nitrate and use it to make proteins. In return, the legumes provide the *Rhizobium* with food, water, and a home.

Fig. 33-17 This lichen looks like one organism. However, it is really two, a fungus and an alga living in a relationship called mutualism.

Section 33.8

Commensalism

Commensalism [kuh-MEN-suh-liz-em] is a relationship in which one organism benefits and the other neither benefits nor suffers harm. An example occurs in the forests of northern Canada (Fig. 33-18). A lichen called "Old man's beard" uses the spruce tree as a home. It gets water and some minerals from the bark. But it usually does not harm the tree. Nor does the tree get any benefit from the lichen. Sometimes the lichens get very dense. Then they do harm the trees. What would you now call the relationship?

Section Review

1. What is parasitism? Use an example to illustrate your answer.
2. What is mutualism? Use an example to illustrate your answer.
3. What is commensalism? Use an example to illustrate your answer.

Fig. 33-18 Commensalism. The spruce tree is not harmed. Nor does it get any benefit. The lichen, Old man's beard, receives several benefits.

Main Ideas

1. Ecology is the study of the relationships among organisms and between organisms and their environments.
2. An ecosystem is an interacting system that consists of groups of organisms and their non-living environment.
3. The habitat of an organism is the place in which it lives.
4. The niche of an organism is its total role in the community.
5. Ecosystems have three main feeding levels: producers, consumers, and decomposers.
6. Organisms may be linked together through food chains.
7. Parasitism, mutualism, and commensalism are close relationships between two organisms.

Glossary

biome		a large area with a characteristic climate
biosphere		the region on earth in which life exists
community		all the living things in an area
ecosystem	EE-koh-sis-tem	an interacting system consisting of living and non-living parts
population		a group of individuals of the same species
scavenger		an organism that feeds on dead organisms

Study Questions

A. True or False

Decide whether each of the following statements is true or false. If the sentence is false, rewrite it to make it true. (Do not write in this book.)
1. Ecology deals only with relationships between organisms.
2. A population is made up of several communities.
3. The habitat of an animal is the place where it lives.
4. Carnivores are animals that eat both plants and animals.
5. A parasite usually kills its host.

B. Completion

Complete each of the following sentences with a word or phrase that will make the sentence correct. (Do not write in this book.)
1. A biome is a large area with a characteristic _____.
2. Grass is called a _____ because it makes its own food.
3. Animals that prey on live animals are called _____.
4. A food web is made of many food _____.

C. Multiple Choice

Each of the following statements or questions is followed by four responses. Choose the correct response in each case. (Do not write in this book.)
1. A groundhog lives in a burrow and roams through nearby fields seeking food. This is a description of the groundhog's
   a) niche   b) habitat   c) community   d) ecosystem
2. A skunk eats living grubs and worms. It also eats fruits of some plants. The skunk is best described as an
   a) omnivore   b) herbivore   c) carnivore   d) scavenger
3. Consider this food chain:
   algae → small animals → minnows → trout → humans
   The trout in this food chain are
   a) second-order carnivores   c) top carnivores
   b) first-order carnivores   d) third-order carnivores

D. Using Your Knowledge

1. What is the difference between a community and an ecosystem?
2. Name five populations you could find in a forest.
3. Identify each of the following as a population, community, or biome:
   a) the porcupines in a woodlot;
   b) the deer in a meadow;
   c) all the living things in a forest;

d) the trout in a lake;
   e) the prairies of western Canada.
4. Complete the following food chains:
   a) in a meadow: grass ⟶ crickets ⟶
   b) in a garden: seedlings ⟶ earthworms ⟶
   c) on a mountain: grass ⟶ mountain goat ⟶
   d) in a pond: algae ⟶ aquatic insects ⟶
5. Explain why decomposers are important to all ecosystems.
6. Draw pyramids of numbers for these food chains:
   dead fish ⟶ moulds
   grass ⟶ rabbits ⟶ wolves ⟶ fleas
7. Select five foods that you ate during the last three meals. Draw the food chains that include those foods and you.

E. Investigations

1. Visit an ecosystem such as a woodlot, meadow, pond or lake. Write a report of about 200 words on its feeding levels and food chains.
2. Write a short paper (about 100 words) on a parasite that causes a human disease.
3. Find what you think is an ecological problem in your area. Write a paper of about 500 words on its causes, effects, and possible solutions.

34 Matter and Energy in Ecosystems

34.1 Non-Living Factors in Ecosystems
34.2 Activity: Effect of Temperature on Germination and Seedling Growth Rate
34.3 Energy Flow in Ecosystems
34.4 Nutrient Cycles in Ecosystems
34.5 Activity: Upsetting the Balance in an Ecosystem

In Chapter 33 you learned how living things interact with one another. You also learned that living things interact with their non-living environment. In this chapter you will learn more about the non-living environment and how it affects living things.

34.1 Non-Living Factors in Ecosystems

The fox in Figure 34-1 was photographed in Gros Morne National Park in Newfoundland. Think about this animal's environment. What non-living factors affect the fox? in what ways?

Fig. 34-1 This fox is in a clearing in the forest. What non-living factors affect the fox?

Section 34.1 499

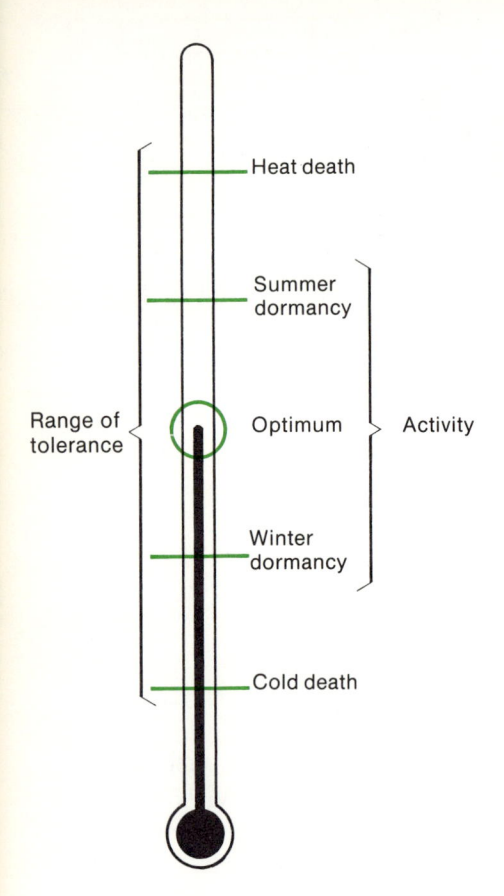

Fig. 34-2 Range of tolerance to temperature. Note that the animal goes dormant (inactive) if the environment gets too hot or too cold. Why is this?

Terrestrial Ecosystems

The fox lives in a **terrestrial** (land) **ecosystem**. Such ecosystems are affected by five main factors. They are temperature, moisture, light, wind, and soil conditions.

Each organism has a **range of tolerance** for each of these factors (Fig. 34-2). This range depends on the factor and on the organism. When the range is exceeded, in either direction, the animal suffers. Within each range of tolerance there is a point at which the organism lives best. This is called the **optimum**. However, conditions are seldom at the optimum. Therefore organisms with the broadest range of tolerance generally survive best.

In Chapter 36 you will learn about the interdependence of living and non-living factors in terrestrial ecosystems.

Aquatic Ecosystems

Many non-living factors also affect **aquatic** (water) **ecosystems**. For example, life in a stream may be affected by temperature, speed of the water, nature of the bottom, light, oxygen, and many other factors (Fig. 34-3). A fast cool stream with lots of oxygen will likely support trout. But a sluggish, warm, polluted stream will not.

Life in a pond or lake may be affected by temperature, light, depth, nature of the bottom, oxygen, and many other factors. And life on the seashore may be affected by wave action, the salt content of the water, the tides, as well as many other factors.

In Chapter 35 you will learn about the interdependence of living and non-living factors in aquatic ecosystems.

Section Review

1. What five non-living factors affect life in terrestrial ecosystems?
2. What is meant by each of these terms: range of tolerance; optimum?

Fig. 34-3 What kinds of organisms can live in this fast water?

3. What non-living factors affect life in a stream? in a pond or lake? on the seashore?

34.2 ACTIVITY Effect of Temperature on Germination and Seedling Growth Rate

When a seed is planted in the proper environment, it **germinates**. The tiny plant that forms is called a **seedling**. Both the germination of the seed and the growth of the seedling are affected by many non-living factors. Temperature is one of these.

In this activity you will study the effects of temperature on the germination and growth of several types of plants. Gardeners plant certain seeds in the early spring while the soil is still cool. They plant others in the late spring when the soil is much warmer. See if you can find out why they do this.

Problem

How does temperature affect germination and seedling growth rate?

Materials

trays about 10 cm x 30 cm (4)
potting soil
thermometer
seeds of radish, spinach, tomato, pea, bean, squash, corn (any 5 or 6 will do)

marking pen
markers (20-25)
refrigerator
incubator
water

Procedure

a. Fill the 4 trays with soil.
b. Label the trays "freezer", "refrigerator", "room", and "incubator".
c. Plant 8 radish seeds in 2 rows at one end of each tray (Fig. 34-4). Use the planting depth indicated on the seed package. Mark the rows.
d. Now plant 8 spinach seeds in 2 rows next to the radish in each tray. Mark the rows.
e. Continue this process with all the types of seeds provided.
f. Water the trays until the soil is damp but not soggy.
g. Put the "freezer" tray in the freezer and the "refrigerator" tray in the refrigerator. Put the "room" tray in a dark cupboard at room temperature. Put the "incubator" tray in an incubator at 35°C-40°C.
h. Copy Table 34-1 into your notebook.

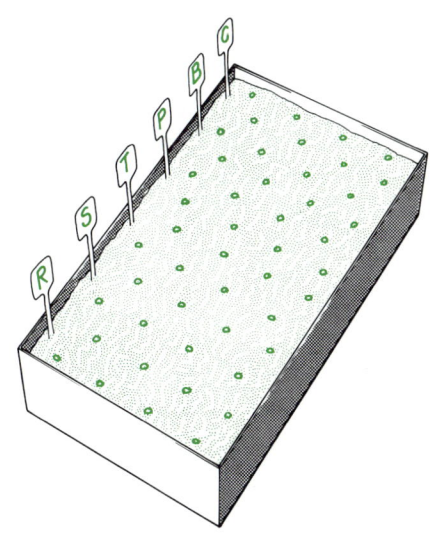

Fig. 34-4 Plant the seeds 4 in a row. Put the rows 2-3 cm apart.

i. Observe the trays every day. Water them, when necessary. Record in your table the day when seedlings first appear. Also, record the day when they are 1 cm tall (on the average).

Table 34-1 Temperature and Growth Rate

Species	Day plants first appear				Day plants are 1 cm tall			
	Freezer	*Refrigerator*	*Room*	*Incubator*	*Freezer*	*Refrigerator*	*Room*	*Incubator*
Radish								
Spinach								
Tomato								
Pea								
Bean								
Squash								

Discussion

1. Why was the "room" tray placed in the dark?
2. Did all seeds and seedlings respond to temperature in the same way? Why? (Hint: Use the information on range of tolerance and optimum temperature given in Section 34.1.)
3. Which seeds would you plant in the early spring? Why?
4. Which seeds would you plant in the late spring? Why?

Fig. 34-5 Have you ever taken a few moments to note the activity in an ecosystem?

34.3 Energy Flow in Ecosystems

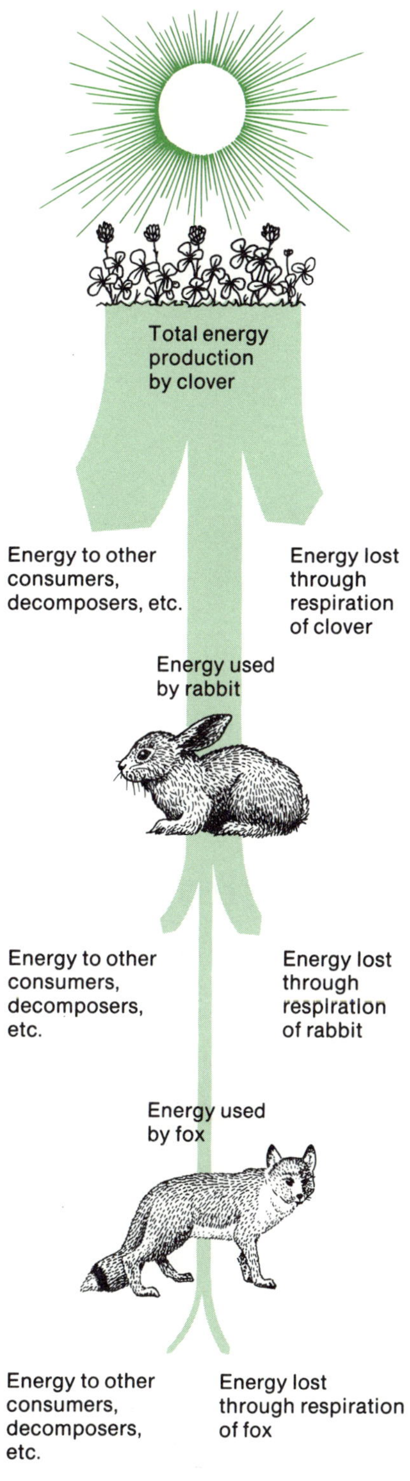

Spend a few quiet moments some day near the woods, beside a stream, in a meadow, or in a park (Fig. 34-5). You'll likely be surprised by the constant activity around you. Birds are constantly searching for food. Squirrels dart everywhere. Insects hop, crawl, and fly in all directions. Activity is the essence of life. And, in order to have activity, energy is needed. Where do organisms get their energy? What happens to energy in ecosystems?

Energy from the Sun to Carnivores

All the energy used by living things comes, in the first place, from the sun. Producers store some of the sun's energy in the foods they make by photosynthesis. They use some of this food for their own life processes. The rest is stored. Herbivores get their energy by eating producers. And carnivores get their energy by eating herbivores or other carnivores. In a sense, then, the sun's energy is passed along food chains to carnivores.

Energy Flow Is One-Way

The passing of energy along a food chain is not very efficient. A great deal is lost at each feeding level. As an example, consider this food chain:

$$\text{Clover} \longrightarrow \text{Rabbit} \longrightarrow \text{Fox}$$

Follow Figure 34-6 as you read on. The clover, through photosynthesis, stores some of the sun's energy in foods. Much of this energy is lost as heat through respiration by the clover. However, when the rabbit eats the clover, the rabbit gets some of the stored energy.

Like the clover, the rabbit loses much of the energy it took in through life activities such as respiration. When the fox eats the rabbit, the fox gets some of the energy stored in the rabbit. It does not get all this energy, however. For instance, bones cannot be digested by the fox.

The fox also loses much of the energy it took in through life activities such as respiration. Parasites and decomposers also use some of the fox's energy. In the end, little energy remains to be passed on to higher feeding levels.

As Figure 34-6 shows, **energy is gradually lost along a food chain** This energy leaves the food chain as heat. It cannot be recaptured by any organisms in the food chain. It is lost forever to that ecosystem. Thus energy flow is one-way along a food chain. **For an ecosystem to keep operating, energy must always enter it from the sun.**

Fig. 34-6 Energy flow along a food chain

Section 34.3 503

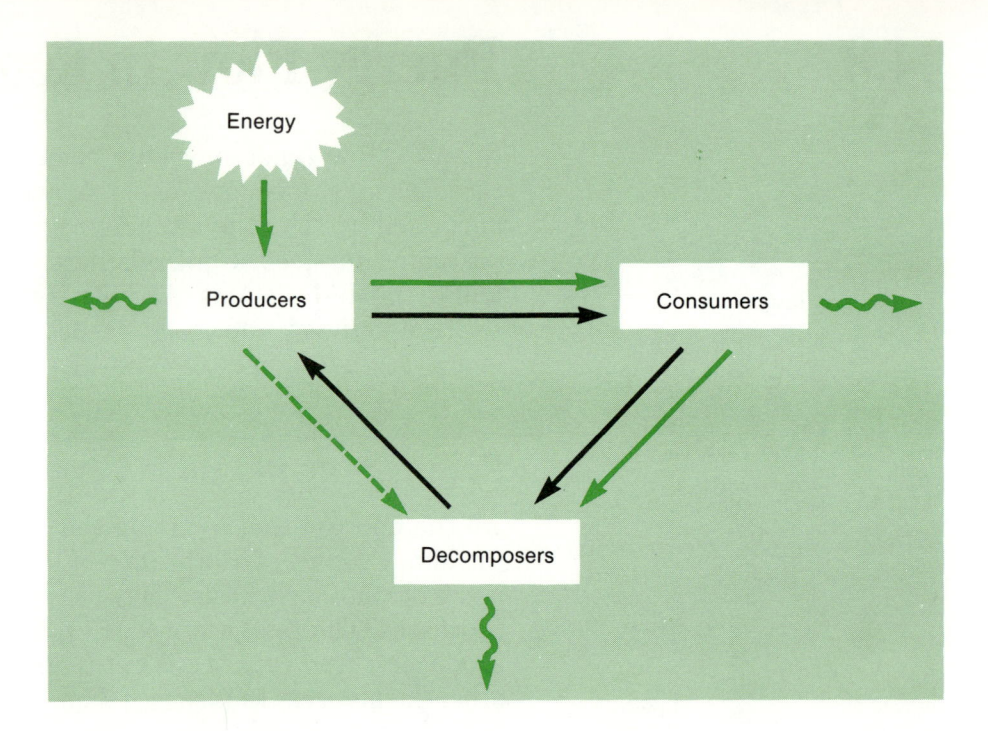

Fig. 34-7 Energy flow and nutrient cycles in an ecosystem. The green arrows show the direction of energy flow. The black arrows show the path of nutrient flow.

Section Review

1. Describe the path of energy from the sun to carnivores.
2. Describe how energy is lost along a food chain.
3. What is meant by "energy flow is one-way"?
4. Describe the flow of energy through your model ecosystem.

34.4 Nutrient Cycles in Ecosystems

An ecosystem needs more than energy in order to function. It needs over 20 different elements. The main ones are carbon, hydrogen, oxygen, nitrogen, phosphorus, potassium, and sulfur. Ecologists call these elements **nutrients** [NOO-tree-ents].

Figure 34-7 shows that energy is lost along a food chain. However, nutrients flow through the food chain back to the start. As you can see, producers pass the nutrients on to consumers. Then decomposers break down animal wastes and dead organisms. This releases nutrients so producers can use them again. In this way, **nutrients are recycled through an ecosystem**. The path each nutrient follows is called a **nutrient cycle**. Let's see how these cycles operate by looking at three basic cycles, those of water, carbon, and nitrogen.

The Water Cycle

The hydrogen and oxygen atoms in water are nutrients that organisms need. These nutrients are recycled through ecosystems as follows:

504 Chapter 34

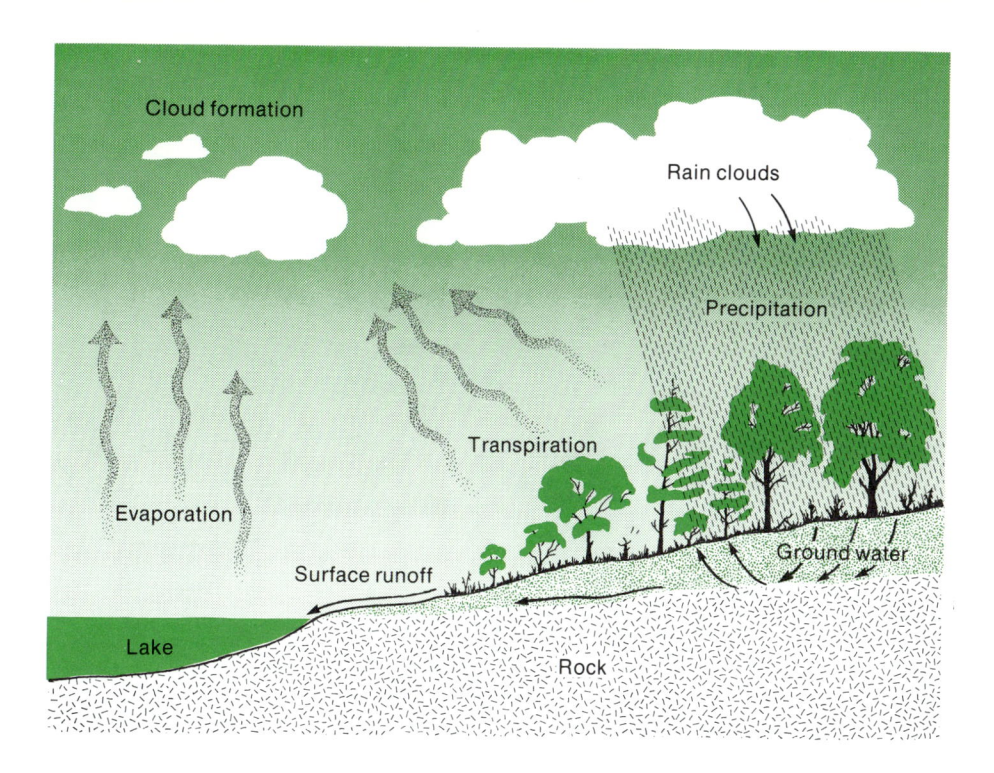

Fig. 34-8 The water cycle

Water vapour enters the atmosphere through transpiration from vegetation. It also enters the atmosphere by evaporation from bodies of water and the soil (Fig. 34-8). In the cool upper atmosphere this vapour condenses, forming clouds. In time, enough water collects in the clouds to cause precipitation. When this occurs, some of the water falling on the ground is absorbed. Some of it runs along the surface of the ground to a stream, pond, or other body of water.

Some of the water in the soil is absorbed through the roots of green plants. Thus animals can obtain water by eating green plants. Of course, they can also obtain it directly by drinking it from a body of water. When plants and animals die, they decompose. During this process, the water present in their tissues is released into the environment. Some of the soil water seeps to ground water level. Then it returns to a body of water.

The Carbon Cycle

Carbon is another nutrient that all organisms need. In fact, it is the basic building block of all living things. Like water, carbon moves through an ecosystem in a cycle (Fig. 34-9). Here is how it works:

Carbon is present in the atmosphere as carbon dioxide. Producers (plants and algae) use it to make food. Now the carbon is in the producers. Herbivores eat the plants, and carnivores eat the herbivores. Now the carbon is in animals. Both plants and animals respire. This returns carbon dioxide to the atmosphere. Decomposers break down dead plants and animals as well as animal waste. This, too, returns carbon dioxide to the atmosphere.

Some organic matter does not decompose easily. Instead, it builds

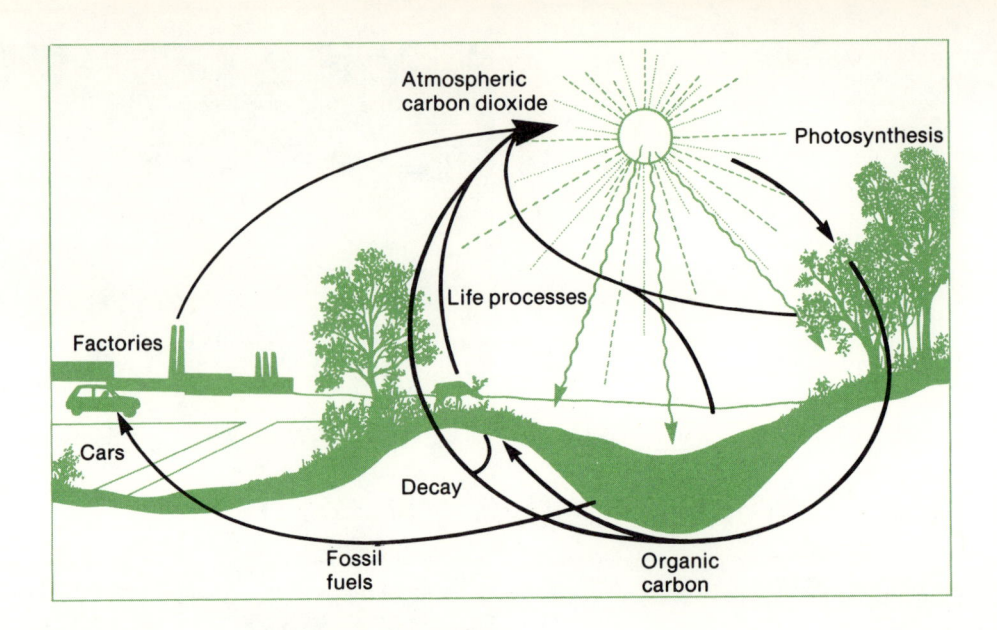

Fig. 34-9 The carbon cycle

up in the earth's crust. Oil and coal were formed from the build-up of plant matter millions of years ago.

At one time, the carbon cycle was almost a perfect cycle. That is, carbon was returned to the atmosphere as quickly as it was removed. Lately, however, the increased burning of fossil fuels has added carbon to the atmosphere faster than producers can remove it.

The Nitrogen Cycle

Nitrogen is another important nutrient. All living things need nitrogen to make proteins. Let's see how this nutrient is recycled in ecosystems (Fig. 34-10).

Almost 78% of the atmosphere is nitrogen (N_2). However, neither plants nor animals can use this form of nitrogen directly. The nitrogen must be in the form of a nitrate (NO_3). Then plant roots can absorb it. Lightning forms some nitrate. It causes oxygen and nitrogen in the atmosphere to join. *Rhizobium* bacteria can do the same thing. You may recall that these bacteria live on the roots of legumes (Section 33.8, page 495). Many bacteria and blue-green algae also form nitrates.

Plants use the nitrates that they absorb to make plant proteins. Animals get the nitrogen that they need to make proteins by eating plants or other animals.

When plants and animals die, bacteria change their nitrogen content to ammonia. The nitrogen in the urine and fecal matter of animals is also changed to ammonia by bacteria. Ammonia, in turn, is converted to nitrites and then to nitrates by bacteria. This completes the main part of the cycle. Bacteria convert some nitrites and nitrates to nitrogen (N_2) to complete the total cycle. The nitrogen cycle need not and often does not involve this last step.

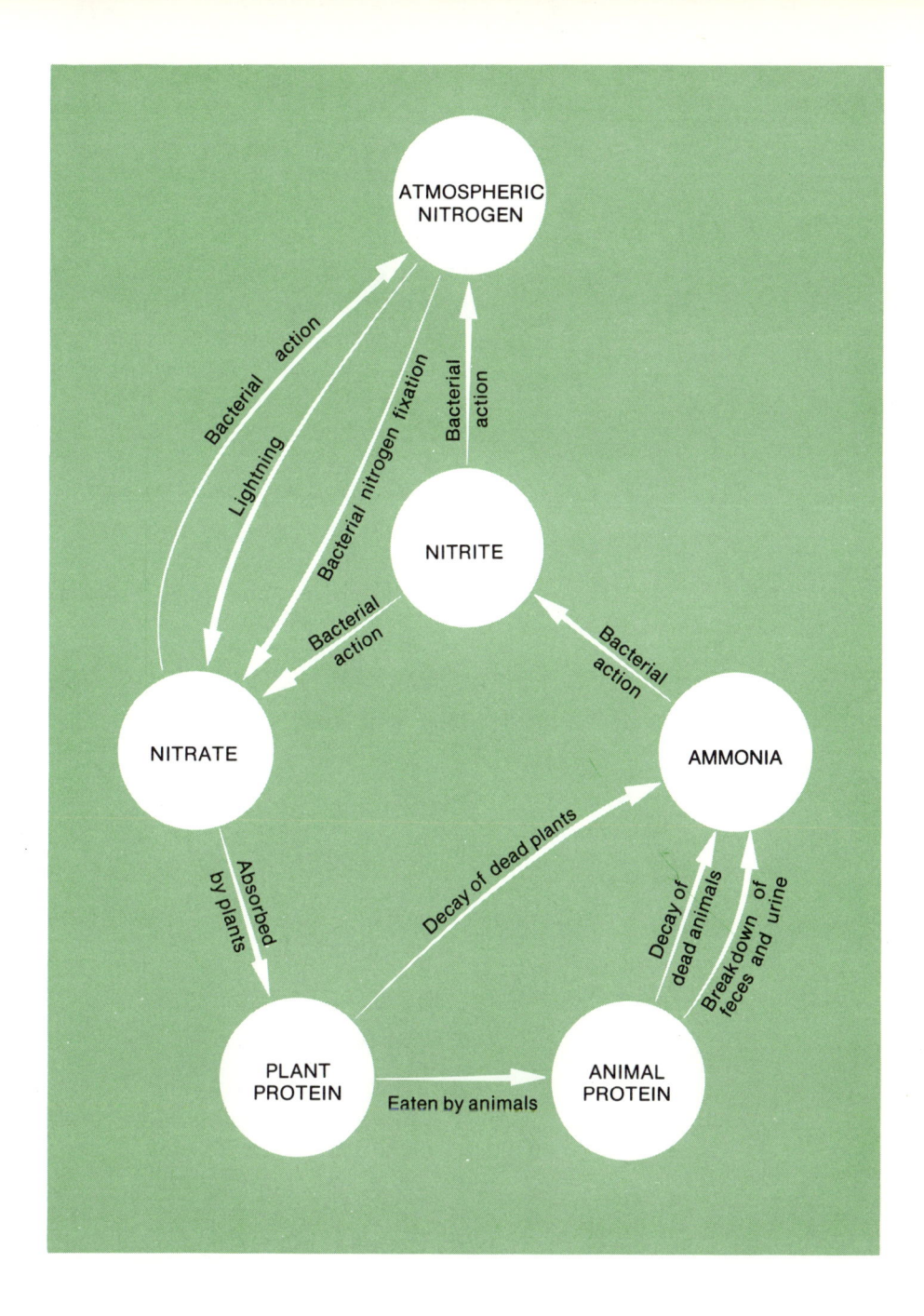

Fig. 34-10 The nitrogen cycle

Section Review

1. Name 7 main nutrients.
2. What is a nutrient cycle?
3. Describe the water cycle.
4. Describe the carbon cycle.
5. Describe the nitrogen cycle.
6. Explain how your model ecosystem can demonstrate nutrient cycles.

Section 34.4

34.5 ACTIVITY Upsetting the Balance in an Ecosystem

Nutrients pass through cycles in ecosystems. They pass from producers through consumers. Then they are returned to producers by decomposers. As a result, a **balance** usually exists in an ecosystem. That is, some of each nutrient is in the producers. Some of it is in the consumers. And some of it is in the free state, ready to be absorbed by producers.

What do you suppose would happen to an ecosystem if we added extra nutrients in the free state? How might this affect the cycles? the producers? the consumers?

In this activity you will make a simple aquatic ecosystem. Then you will add some lawn fertilizer to it. This fertilizer contains three main nutrients: nitrogen, phosphorus, and potassium.

Problem

How will lawn fertilizer affect the balance in an ecosystem?

Materials

wide-mouthed jars, with a capacity of at least 1 L (2)
pond water (2 L)

strands of *Cabomba* or other aquatic plant, each about 10-20 cm long (6)
pond snails (6)

Procedure

a. Fill both jars with pond water.
b. Add half the *Cabomba* to each jar.
c. Add 3 pond snails to each jar.
d. Label one jar "Control". Label the other jar "Experimental".
e. Add a *very small* pinch of lawn fertilizer to the experimental jar.
f. Place the jars, side by side, in a bright location.
g. Observe the jars each day for 2-3 weeks. Make notes on changes in the appearance of the *Cabomba*, and the snails.

Discussion

1. What are the producers in your ecosystems?
2. What are the consumers in your ecosystems?
3. a) What important invisible organisms are in your ecosystem?
   b) What is their role?
4. What is the purpose of the control?
5. Why are the jars placed side by side?

6. a) Describe the changes the fertilizer caused in your ecosystem.
   b) Explain these changes.
7. Explain why sewage can cause plant and algal growth in lakes.

Main Ideas

1. The living and non-living parts of an ecosystem interact.
2. Energy flow in ecosystems is one-way. It is not recycled.
3. Energy is lost along food chains. Thus energy must always enter ecosystems from the sun.
4. Nutrients are recycled in ecosystems.

Glossary

aquatic	ah-KWOT-ik	having to do with water
climax		the end of succession
nutrient	NOO-tree-ent	an element needed by organisms

Study Questions

A. True or False

Decide whether each of the following statements is true or false. If the sentence is false, rewrite it to make it true. (Do not write in this book.)
1. Life in a stream can be affected by the speed of the water.
2. Energy is recycled in ecosystems.
3. Nutrients are recycled in ecosystems.
4. Decomposers return carbon dioxide to the atmosphere.

B. Completion

Complete each of the following sentences with a word or phrase that will make the sentence correct. (Do not write in this book.)
1. The tiny plant that forms after a seed germinates is called a _____.
2. The elements that are needed in ecosystems are called _____.
3. Bacteria on the roots of legumes can form _____ from nitrogen and oxygen.

C. Multiple Choice

Each of the following statements or questions is followed by four responses. Choose the correct response in each case. (Do not write in this book.)

1. The temperature at which an organism lives best is called the
   a) optimum temperature
   b) range of tolerance
   c) summer dormancy
   d) winter dormancy
2. Herbivores get their energy
   a) directly from the sun
   b) from plants
   c) from carnivores
   d) from decomposers
3. A homeowner left the grass clippings on the lawn when it was mowed. She discovered that the lawn needed less fertilizer than it did when she collected the clippings. This is because
   a) energy flow is one way
   b) energy must enter the lawn ecosystem from the sun
   c) the clippings help retain water
   d) nutrients are recycled
4. The nutrient that is the basic building block of all living things is
   a) hydrogen
   b) oxygen
   c) nitrogen
   d) carbon

D. Using Your Knowledge

1. Many organisms live in the tidal zone of the rocky shore along the oceans. Make a list of the non-living factors that might affect these organisms.
2. An oxygen cycle exists in nature. Ecologists often consider it along with the carbon cycle. They call the result the **carbon-oxygen cycle**. Why does it make sense to do this?
3. It has been suggested that, to make the best use of food on this crowded planet, we should all become herbivores. In other words, we would no longer eat cattle, pigs, and fowl. Instead, we would eat the plants that these animals would normally have eaten. What do you think of this idea? Why?

E. Investigations

1. Find out what composting is. Build a simple compost heap (see Activity 10.2, page 119). Then compost some grass clippings in it. Now prepare a brief paper that explains how composting saves fertilizer and enriches the soil.
2. Plan and try an experiment to show the effects of light on germination and seedling growth.
3. Plan and try an experiment to show the effects of moisture on germination and seedling growth.

35 Aquatic Ecosystems

35.1 Freshwater Ecosystems
35.2 Oxygen, Carbon Dioxide, and pH
35.3 Other Chemical Factors
35.4 Activity: Measuring the pH of Water Samples
35.5 Activity: Analyzing an Unknown Water Sample
35.6 Activity: Effects of Organic Matter on Water Quality
35.7 Organisms as Indicators of Water Quality
35.8 Activity: Adaptations of Organisms

Oceans...lakes...ponds...bogs...rivers...marshes...estuaries...Aquatic ecosystems seem to be everywhere! But this is not surprising. Over 70% of the earth's surface is covered by water. Most of this, of course, is made up of the oceans. In fact, 97% of the world's water is salt water. Only 3% is fresh water. Of that 3%, 98% is frozen in the ice caps of Antarctica and Greenland. But the 2% that is not frozen makes a wide variety of freshwater ecosystems. This chapter looks at the ecology of **freshwater ecosystems**.

35.1 Freshwater Ecosystems

Freshwater ecosystems can be grouped into two categories, **standing waters** and **flowing waters**. A pond is an example of standing water. A river is an example of flowing water. In some cases it is difficult to classify a body of water as standing or flowing. For example, some rivers flow very slowly near their mouths. Thus they appear to be standing waters. Also, most ponds have inlets and outlets. As a result, they have a flow of water through them.

Types of Standing Waters

There are just two main types of standing waters that are wholly aquatic. These are **ponds** and **lakes**. The rest — **marshes, swamps, and bogs** — are partly terrestrial. In fact, they are often called **wetlands**. Table 35-1 describes the main types of standing waters, including wetlands.

All types occur right across Canada. Even the prairies are dotted with numerous ponds and small lakes. These have very high concentrations of dissolved nutrients. They are often called **sloughs** [sloos].

Fig. 35-1 Several springs feed this brook. The brook runs to a stream. The stream runs to a river. And the river runs to a lake.

Table 35-1 Types of Standing Waters

Name	Description
Pond	• Shallow • Light can reach bottom in most places • Considerable vegetation, mostly submerged
Lake	• Deeper than a pond • Light cannot reach bottom in many places • No vegetation in deeper areas
Marsh	• Very shallow • No open expanses of water • Contains "islands" of soggy land • Dominated by cattails, bulrushes, reeds, and grasses
Swamp	• Like a marsh, except "islands" of land have trees on them
Bog	• Waterlogged spongy area dominated by mosses • Contains acidic water and decaying vegetation
Slough	• Small lake or pond • Nutrient-rich • In low areas of prairies

Flowing Waters

You likely call flowing waters by such names as brook, creek, stream, and river. In general, brooks and creeks are the smallest, streams larger, and rivers are largest. Creeks and brooks are often fed by springs (Fig. 35-1). They may also drain ponds and small lakes. However, sooner or later, they join together to form streams. Streams, in turn, join together to form rivers. The rivers carry the water to a lake or, in some cases, to the sea.

Source of Oxygen The water of brooks, creeks, and streams usually rushes over rocks (Fig. 35-2). Therefore air circulates through it (aeration). Plants and algae are not an important source of oxygen in fast waters. In fact, most plants and algae are swept away by the current. However, a few producers do live in such waters. Clinging to the rocks are blue-green algae, green algae, diatoms, and mosses. These organisms photosynthesize. As a result, they add some oxygen to the water. However, the amount added is very small compared to that which enters by aeration.

Source of Food Producers may support some food chains in the fast water. For example, herbivores such as snails may graze on algae and mosses. The snails, in turn, may be eaten by carnivores such as fish. However, only a small part of the total food supply of a fast water ecosystem comes from these producers. Most of it enters the water from the surrounding land. Leaves, grass, twigs and other organic matter fall into the water or are washed in by a rain.

Fig. 35-2 Fast water gets aerated.

Fig. 35-3 Rivers are usually more nutrient-rich than other flowing waters. Therefore they support a large number and variety of consumers.

Scavengers feed on this organic matter. They, in turn, are eaten by carnivores. Also, insects and other small terrestrial animals often fall into the water. These are eaten by fish and other predators.

Rivers are generally wider, slower, warmer, and more nutrient-rich than streams, brooks, and creeks (Fig. 35-3). Some plants and algae often live in slower rivers. Also, decaying organic matter builds up on the bottoms of such rivers. In fact, a wide slow river has many of the properties of a pond.

Adaptations The organisms that live in flowing waters show unique adaptations to their habitats. For example, the stonefly nymphs that live under the rocks in fast streams are streamlined in shape. Also, they have hooks on their feet. In contrast, the sludge worms that live in the bottom ooze of slow rivers show no adaptations for holding on to the bottom material. However, they do show adaptations to the low oxygen concentrations of their habitat (see Section 35.8, page 524).

Section Review

1. What is the difference between standing and flowing waters?
2. **a)** How does a lake differ from a pond?
   b) How does a swamp differ from a marsh?
3. What is a slough?
4. What is a bog?
5. How do flowing waters get most of their oxygen?
6. Where do fast water ecosystems get most of their total food supply?

35.2 Oxygen, Carbon Dioxide, and pH

Life in aquatic ecosystems is affected by many non-living factors. Three of them are discussed here.

Oxygen

Most aquatic organisms need oxygen. They use oxygen that is dissolved in the water. But some kinds of pollution use up this oxygen. For example, suppose sewage is put into a small body of water. Bacteria feed on this sewage. As they feed, they use oxygen. They must respire. Therefore they take oxygen from the water. Then fish and other animals don't have enough oxygen. So they die.

How Much Oxygen Is Needed? Most trout need at least 10 µg/g of oxygen. (That's ten micrograms of oxygen in one gram of water.) However, chub need only 7 µg/g. And carp can live with only 1-2 µg/g of oxygen. In general, water should contain at least 5 µg/g of oxygen. If it doesn't, many kinds of animals will die (Fig. 35-4).

The amount of oxygen an animal needs depends on many factors. One factor is activity. An animal needs more oxygen when it is active than when it is resting. Another factor is the temperature. As the water temperature goes up, most animals need more oxygen. For example, a resting trout with a mass of 1 kg needs 55 mg of oxygen/h at 5°C. But, at 25°C it needs 285 mg/h.

Solubility of Oxygen in Water The amount of oxygen in water depends on the temperature. Table 35-2 shows this dependence. Note that water holds less oxygen at high temperatures.

Table 35-2 Solubility of Oxygen in Water (when air is the only source)

Temperature of Water (°C)	Solubility (µg/g)
0	14.6
10	11.3
20	9.1
30	7.5

Sources of Oxygen In streams and rivers the oxygen comes mainly from the air. As the water splashes over rocks, it picks up oxygen from the air.

Lakes and ponds get some of their oxygen from the air. It enters the water by diffusion. But most of the oxygen comes from plants and algae. They make oxygen by photosynthesis.

Fig. 35-4 You can measure the oxygen in water using a test kit. This person found 3 µg/g of oxygen in the water. Is this water good for most organisms?

Fig. 35-5 This water has an algal bloom. When the algae die, the oxygen level of the water may drop. Then fish may die.

Thermal Pollution Power generating plants and some industries put warm water into lakes and rivers. This is called **thermal pollution** [THER-mal].

Table 35-2 tells us that the warm water will not hold as much oxygen. Yet we know that animals need extra oxygen in warm water. As a result, the warm water kills some species of fish and other animals.

Sewage Pollution and Algal Blooms Much of **sewage** [SU-aj] was once part of living organisms. Sewage is a serious form of pollution. Sewage tends to remove oxygen from the water. Bacteria feed on the sewage and respire as they do so. Often they will reduce the oxygen level in the water to zero.

Sewage also adds nutrients to the water. **Nutrients** [NU-tree-ents] are elements that living things need for growth. Nitrogen [NI-tro-jen] and phosphorus [FOS-for-us] are examples. Extra nutrients make algae grow very quickly. Large amounts of algae build up in the lake. This is called an **algal bloom** [AL-gal]. When these algae are alive, they are good for the water. They produce oxygen by photosynthesis and are eaten by other organisms. Often, however, algae crowd themselves out of food and space and then die. Bacteria then feed on the dead algae. This uses up oxygen and kills other organisms (Fig. 35-5).

Carbon Dioxide

Most of the carbon dioxide in water comes from organisms. Most water has organisms in it. As they respire, they put carbon dioxide into the water.

Water near the surface usually has less than 10 $\mu g/g$ of carbon dioxide. But water near the bottom often has much more. Bacteria feed on organic matter in the bottom ooze. As they respire, they add carbon dioxide to the water at the bottom.

Polluted water often has over 25 $\mu g/g$ of carbon dioxide. This harms most living things. In fact, 50-60 $\mu g/g$ kills many species of animals.

pH

The pH of water is a measure of its acidity. Values for pH run from 0-14 (Fig. 35-6). A pH of 7 is neutral. A pH less than 7 is acidic. And a pH greater than 7 is basic. The lower the pH of water, the more acidic the water is. That is, water gets less acidic as its pH goes from 0 to 14. And it gets more basic at the same time.

A pH between 6.7 and 8.6 supports a good balance of fish types. Few fish species can live for long at a pH below 5.0 or above 9.0.

Acid spills by industries lower the pH. This can kill organisms. Also, smelters and coal-burning power plants put sulfur dioxide into the air. It reacts with water vapour in the air to form sulfuric acid.

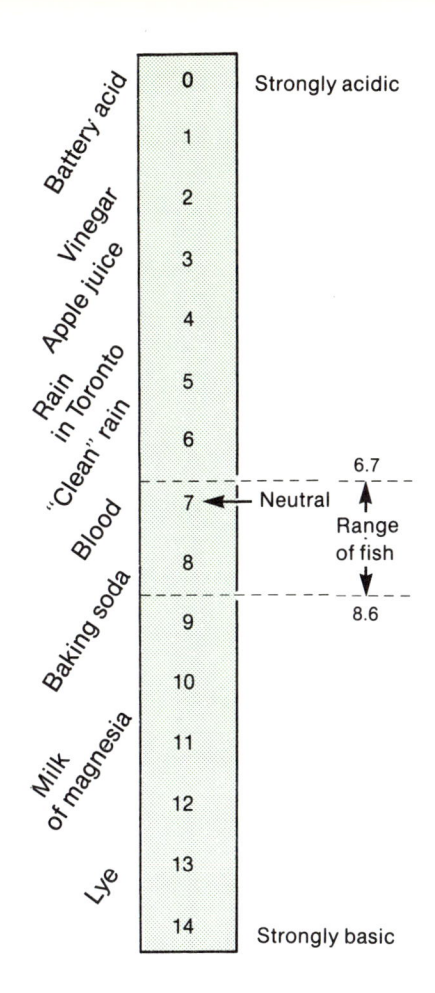

Fig. 35-6 The pH scale for acidity

When it rains or snows this acid gets into lakes and rivers. It is called **acid rain**. Its pH is usually below 4.0. It too can kill organisms.

Sewage usually raises the pH of water. Household sewage has many bases in it. Also, the decay of sewage by bacteria produces bases. Thus sewage can kill organisms by raising the pH.

Section Review

1. What is the least amount of oxygen water should contain?
2. How does temperature affect the solubility of oxygen in water?
3. a) Where does a stream get its oxygen?
   b) Where does a lake get its oxygen?
4. What is thermal pollution? How does it affect animals?
5. What is sewage pollution? How does it affect organisms in the water?
6. What is an algal bloom?
7. a) How does carbon dioxide get into water?
   b) What level of carbon dioxide harms organisms?
8. a) What is pH?
   b) Explain the pH scale.
   c) What pH range is good for fish?
9. What is acid rain? Why will it harm organisms?
10. How does sewage affect pH?

35.3 Other Chemical Factors

Four more factors that affect life in aquatic ecosystems are hardness, nitrogen, phosphorus, and chloride.

Hardness

Hardness in water is caused mainly by calcium and magnesium. Water in areas with sedimentary bedrock such as limestone and dolomite is usually quite hard. These rocks contain calcium and magnesium. They dissolve in the water as it runs over them.

Table 35-3 shows a scale for the degree of hardness. All living things need calcium and magnesium. Therefore hard water generally has more life in it than soft water.

Table 35-3 Scale of Hardness

Degree of hardness	Total hardness (µg/g)
Soft	0 — 60
Moderately hard	61 — 120
Hard	121 — 180
Very hard	Over 180

Water with a total hardness of about 250 µg/g is best for drinking. People who drink soft water most of the time are more likely to get diseases of the heart and circulatory system.

Nitrogen

Nitrogen is an element that all organisms need. Therefore it is called a nutrient. It makes organisms grow (Fig. 35-7). But sometimes it causes too much growth and algal blooms result. In such cases, nitrogen is called a **pollutant**.

Nitrogen (N) occurs in two main forms in water. One form is **ammonia** (NH_3); the other is **nitrate** (NO_3).

When dead plants and animals decay, ammonia is formed. It is also formed when fecal matter and urea break down. Thus ammonia in water suggests sewage may be present. In urban areas the sewage is likely human. In rural areas it may come from beef feedlots.

Many fertilizers also contain ammonia. Therefore runoff from fields and lawns pollutes water with ammonia.

Ammonia does little harm directly. But it converts to nitrate. Then the nitrate causes plants and algae to grow (see Nitrogen Cycle, page 507). If the total nitrogen is over 0.30 µg/g, an algal bloom may occur in a lake.

Fig. 35-7 A 4-9-15 fertilizer is 4% nitrogen, 9% phosphorus, and 15% potassium. All three nutrients are needed for plant growth.

Phosphorus

Like nitrogen, **phosphorus** is a nutrient. Phosphorus (P) usually occurs in water as a **phosphate** (PO_4). Its sources are much the same as those for nitrogen: sewage, animal wastes, and decaying plants and animals. Fertilizers usually contain phosphate. But runoff from fields and lawns contains little phosphate. The phosphate particles stick to soil particles. Therefore, unless **soil erosion** occurs, phosphates do not wash from fields and lawns.

Phosphate causes plants and algae to grow. If the total phosphorus is just 0.015 µg/g, an algal bloom may occur in a lake.

Chloride

The most common **chloride** in water is sodium chloride, or common salt. Sea water contains about 27 000 µg/g of chloride. Even fresh water contains some chloride. Clean rivers and lakes often contain 20-50 µg/g. But rivers polluted with sewage or street runoff can contain many times that amount. Values of 500-1500 µg/g are not uncommon.

Marine organisms are adapted to high chloride levels. Freshwater organisms are not. Therefore high chloride levels in fresh water can kill many forms of life.

Section 35.3

Section Review

1. **a)** What causes hardness in water?
   b) Why does hard water often contain abundant life?
   c) Why is very hard water best for drinking purposes?
2. In what forms does nitrogen occur in water?
3. What are the sources of nitrogen in water?
4. How much nitrogen is needed for an algal bloom?
5. In what form does phosphorus occur in water?
6. How much phosphorus is needed for an algal bloom?
7. Compare the amount of chloride in salt water with that in fresh water.

35.4 ACTIVITY Measuring the pH of Water Samples

You learned in Section 35.2 that pH is an important measure of water quality. In this activity you will measure the pH of several water samples. Then you will decide how the pH of each sample might affect animal life in the water.

Problem

How does the pH of the water vary with the nature of the water sample?

Materials

test tube
pH paper
water samples prepared or collected by students and teacher

Procedure

a. Copy Table 35-4 into your notebook.
b. Wash the test tube well.
c. Put 1-2 mL of distilled water (from the bottle labelled "Sample 1") in the test tube.
d. Find the pH of Sample 1 using the pH paper (Fig. 35-8). Record your results.
e. Repeat steps (b) to (d) using tap water (Sample 2).
f. Repeat steps (b) to (d) for a weak acid (Sample 3) and a weak base (Sample 4).
g. Repeat steps (b) to (d) for other samples provided. Your teacher may ask you to provide some samples. It will be interesting to test water from a local river or lake, from a swimming pool, and from an aquarium. You could also test water containing deter-

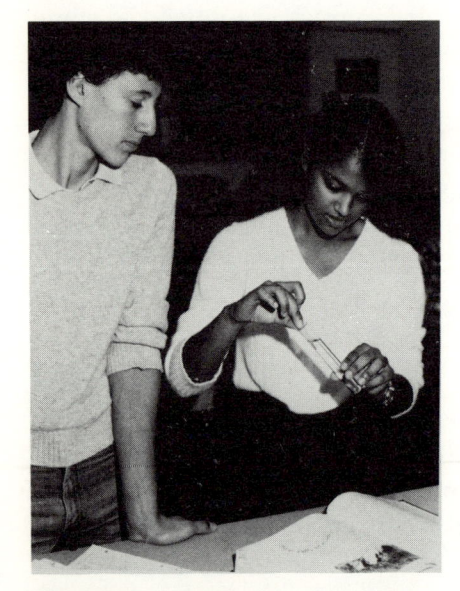

Fig. 35-8 Finding the pH using pH paper. Match the colours as quickly as you can. The colour can change in the air.

gents, water containing decaying organic matter like fish food, dish water, rain water, and melted snow.

h. Read the part on pH in Section 35.2 again. Now complete the last column in your table.

Table 35-4 The pH of Water Samples

Sample number	Nature of sample	pH of sample	Effect on living things
1	Distilled water		
2	Tap water		
3	Carbonic acid (a weak acid)		
4	Limewater (a weak base)		
5			

Discussion

1. What is the pH of distilled water? Why?
2. Did tap water have a different pH than distilled water? Why?
3. Which has the higher pH, carbonic acid or limewater? Why?
4. Try to give a reason for the pH of each sample from number 5 on.

35.5 ACTIVITY Analyzing an Unknown Water Sample

The aquarium for this activity contains water with certain chemical properties. It also has certain forms of life in it. As you know, the chemical factors (non-living) interact with the organisms. That is, they affect one another. Can you find out how?

Problem

How do chemical and biological factors interact in the aquarium?

Materials

selection of Hach water testing kits (oxygen, carbon dioxide, pH, hardness, ammonia, nitrate, phosphate, chloride)

an aquarium with enriched water, organisms, and a light thermometer ($-10°C$ to $110°C$)

Fig. 35-9 Don't stir up any sediment as you get your sample.

Procedure

a. Copy Table 35-5 into your notebook.
b. Get one of the water testing kits from your teacher. Carry out the test on the water in the aquarium (Fig. 35-9). Enter your data in the table that your teacher has prepared on the chalkboard.
c. Return the kit, obtain another, and repeat step (b).
d. Continue tests until the teacher announces that time is up. Your teacher will see that each test is performed at least 3 times so that accuracy can be checked.
e. Take the temperature of the water.
f. Record all the data in your table.
g. Make careful notes on other non-living properties of the aquarium ecosystem. These should include light intensity, clarity of the water, presence or absence of aeration, and the nature of any debris on the bottom.
h. Make careful notes on the biological properties of the aquarium ecosystem. These should include a description of the types and abundance of organisms present (fish, snails, plants, algae and small animals).

Table 35-5   Water Analysis

Factor	Trial 1	Trial 2	Trial 3	Average
Oxygen				
Carbon dioxide				
pH				
Hardness				
Ammonia				
Nitrate				
Phosphate				
Chloride				

Discussion

1. Account for the results of each test. For example, why was the oxygen concentration 8 µg/g instead of some other value?
2. Explain the effects of each result on living organisms. For example, if you obtained a pH of 9, what effect will that pH have on living organisms? Can the aquarium support a wide range of fish species?
3. Make an overall judgment on the quality of the water in the aquarium. Will it support a wide range of species or organisms? Is it polluted?

35.6 ACTIVITY Effects of Organic Matter on Water Quality

Organic matter is matter that contains carbon. Matter that is or was once part of a living organism is organic. Therefore, it contains carbon. It also contains nitrogen, phosphorus, and other elements. Many of these are nutrients. What do you think will happen if this kind of organic matter breaks down in water?

Problem

How does organic matter affect water quality?

Materials

small aquaria or pails of at least 5 L size (2)
several sprigs of an aquatic plant
100 W lamps (2)
organic matter (for example, fish food)
Hach water testing kits
thermometer ($-10°C$ to $110°C$)

Procedure

a. Fill both containers to within 2 cm of the top with water.
b. Add about 10 sprigs of aquatic plant to each container.
c. Place the containers side by side. Then mount a 100 W lamp over each one (Fig. 35-10).
d. Dump a can of fish food (or other organic matter) into one container.
e. Wait at least 2 or 3 d. Then test the water in each container. If possible, test for oxygen, carbon dioxide, pH, ammonia, nitrate,

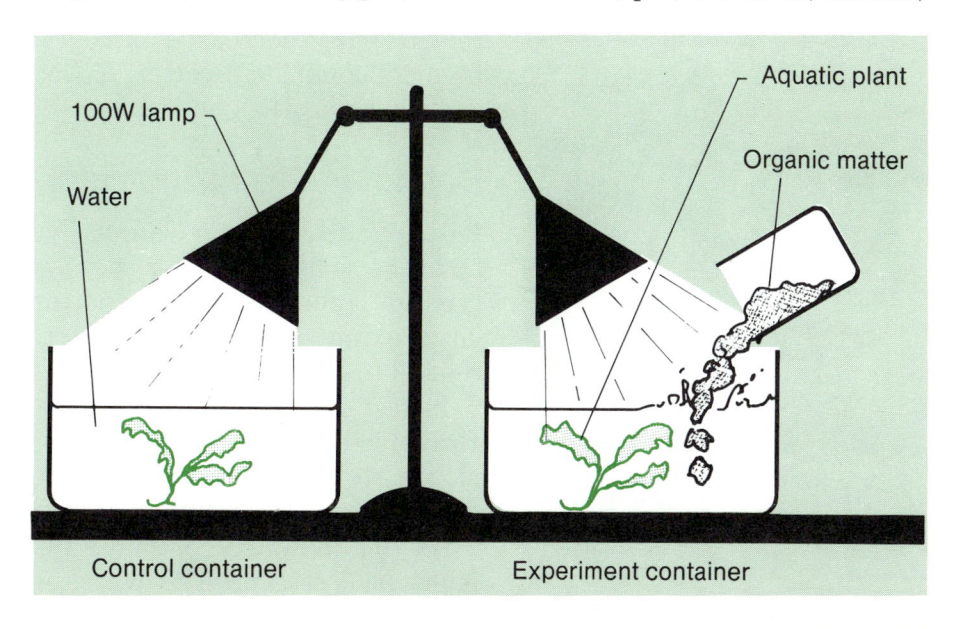

Fig. 35-10 How will organic matter affect water quality?

Section 35.6

and phosphate. Record your results in a table. Also record the water temperature.

f. Note any changes in the appearance of the water and plants.
g. If time permits, repeat the tests two weeks later. Also, note any changes in the appearance of the water and plants.

Discussion

1. No organic matter was added to one container. Why?
2. Explain any differences between the tests for the two containers.
3. Explain any differences in appearance between the two containers.
4. What difference did two weeks make? Why?

35.7 Organisms as Indicators of Water Quality

Index Species

A pollutant can kill all living things in an ecosystem. More commonly, though, the pollutant kills only certain species. Then the few species that are not killed increase in numbers. The species present in the highest numbers are called the **index species** of water quality. Biologists use many groups of organisms as index species. Among these are fish and bottom fauna (animals that live on, in, and near the bottom of the body of water).

Fish as Indicators of Water Quality

The dominant fish in unpolluted lakes are those that need clean, cool water. Among these are lake trout, whitefish, walleye, lake herring, and char (Fig. 35-11). As nutrients build up, most of these species remain, but in decreased numbers. Perch, black bass, pike, and smelt become dominant species. After further nutrient buildup, carp, sunfish, and catfish dominate.

Bottom Fauna as Indicators of Water Quality

Bottom fauna are animals that live on or near the bottom. They may be in the sediment, under rocks, and on submerged vegetation. The types of fauna found on the bottom of a stream or river depend to a large extent on the nature of the bed. This, in turn, is directly related to the speed of the water (see Table 35-6).

In fast water, all but the larger rocks are carried downstream. Gravel is deposited in slower regions. And sand settles out in still slower regions. Where there is little or no current, silt and mud are deposited on the bottom.

As you know, oxygen plays an important role in determining the fauna in a stream. The water in a fast stream is usually highly

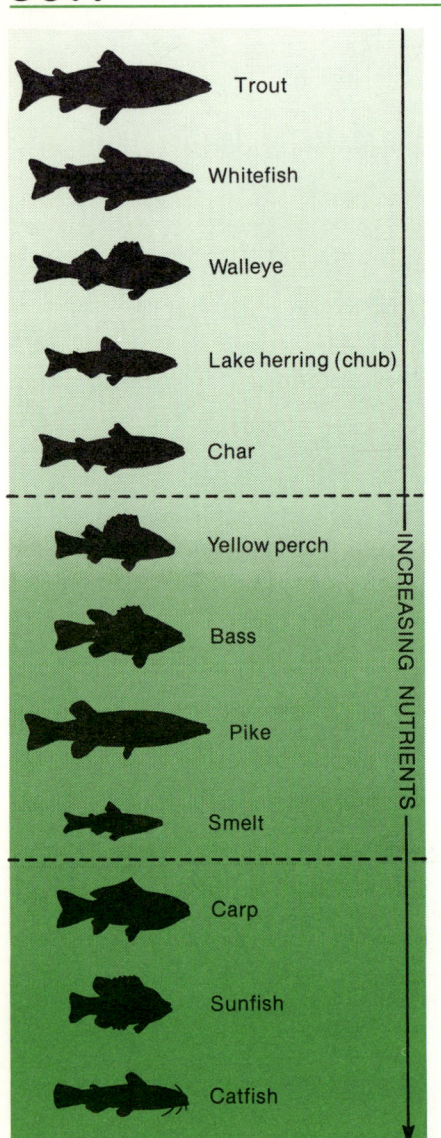

Fig. 35-11 Fish as indicators of water quality

(INCREASING NUTRIENTS)

Trout
Whitefish
Walleye
Lake herring (chub)
Char
Yellow perch
Bass
Pike
Smelt
Carp
Sunfish
Catfish

Table 35-6 Speed and Nature of the Bed

Speed	Nature of the bed
Over 100 cm/s	Rock
60-100 cm/s	Rocks and heavy gravel
30-60 cm/s	Light gravel
20-30 cm/s	Sand
10-20 cm/s	Silt
Under 10 cm/s	Mud

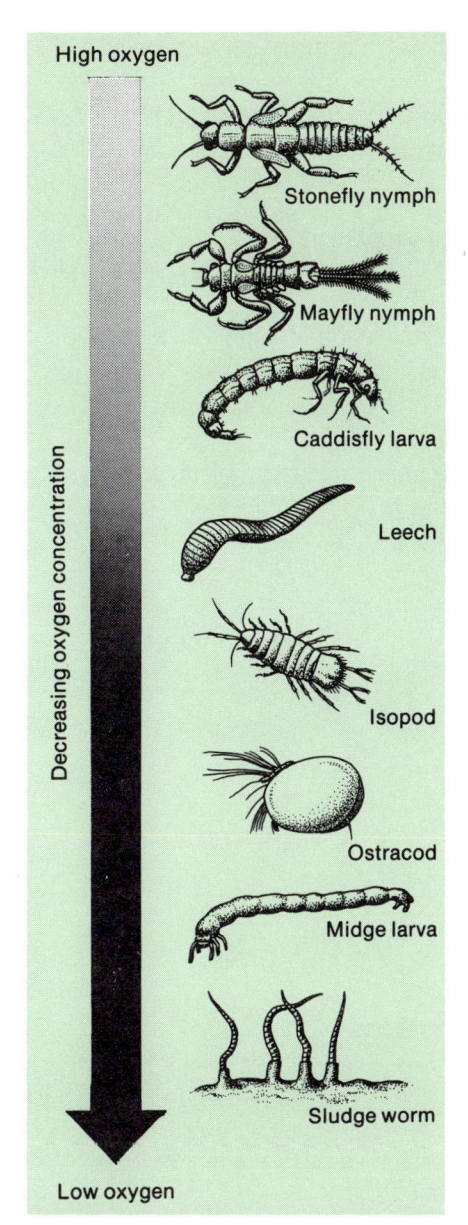

Fig. 35-12 Bottom fauna and their oxygen needs

oxygenated. Thus the bottom fauna in a fast area of a stream are those that require high oxygen conditions. They must also have adaptations that enable them to cling to the bottom. Otherwise, they would be swept downstream. This area of a stream has insects such as stoneflies, mayflies, and caddisflies (Fig. 35-12). All three are generally found under rocks and gravel. Sometimes a stream is polluted to the point where its oxygen concentration drops greatly. Then these animals will not be found in the stream.

At the bottom of Figure 35-12 are organisms that live in mud-silt bottoms. There the oxygen concentration is usually very low. The sludge worm is a segmented worm like the earthworm. It feeds on the muddy bottom just like an earthworm feeds on the soil in a garden. It builds a tube above a burrow extending down into the bottom ooze. While it feeds in the stream bottom, its tail sticks out of the tube. The tail sways back and forth. This aids the exchange of carbon dioxide and oxygen in the water. Blood vessels close to the body surface also help the sludge worm to get even the smallest traces of oxygen from the water.

The midge larva is another common bottom burrower. It too can tolerate low oxygen conditions. This larva occupies a tube that it makes from sludge material. Some species are red in colour. They are called bloodworms. However, they are not worms. They are insect larvae. You have likely seen the adult midge fly. It is an annoying little creature that looks like a small mosquito without piercing mouthparts. Clouds of midge flies often gather around your head on a humid, windless day.

Leeches, isopods, and ostracods generally live in muddy areas with slightly higher oxygen levels than those in which sludge worms and midge larva live. The fauna that live in mud-silt bottoms of streams can also be found on the bottoms of many lakes and ponds. Why is this so?

Section Review

1. What is an index species?
2. Name four fish species that require clean cool water.

Section 35.7   523

3. Explain how the speed of the water determines the nature of the bed in a stream.
4. How are sludge worms adapted to low oxygen conditions?
5. How are stoneflies adapted to fast water?

35.8 ACTIVITY Adaptations of Organisms

The stonefly lives in fast water. As Figure 35-13 shows, it is well adapted to that habitat. Its body is streamlined. Its legs are muscular. And it has claws on its feet. Without these adaptations the stonefly would be swept away by the current.

All organisms show adaptations to their environment. Can you find them on some aquatic invertebrates?

Problem
How are aquatic invertebrates adapted to their environments?

Materials
several species of aquatic invertebrates in petri dishes or beakers (some alive and some preserved)
hand lens
probe
field guides

Fig. 35-13 How is this stonefly adapted to fast water?

Procedure
a. Copy Table 35-7 into your notebook.
b. Get a specimen from the front bench.
c. Write its name in your table.
d. Write a description of its habitat in your table. You can find this information in one of the field guides.
e. Study the animal closely using the hand lens. Move the animal about, when necessary, with the probe. Do not damage it. Record its adaptations in your table.

Table 35-7 **Adaptations**

Organism	Habitat	Adaptations

Discussion
A completed table is your writeup for this activity.

524 Chapter 35

Main Ideas

1. There are two groups of freshwater ecosystems: standing waters and flowing waters.
2. Standing waters include ponds, lakes, marshes, swamps, bogs, and sloughs.
3. Flowing waters include brooks, creeks, streams, and rivers.
4. All organisms show adaptations to their environments.
5. Chemical tests and organisms are both indicators of water quality.

Glossary

algal bloom	AL-gal	the build-up of large amounts of algae
index species		the species present in the highest numbers

Study Questions

A. True or False

Decide whether each of the following statements is true or false. If the sentence is false, rewrite it to make it true. (Do not write in this book.)

1. A lake is a freshwater ecosystem.
2. A swamp has trees in it.
3. A stream gets most of its oxygen from producers.
4. Fish need more oxygen as the water becomes warmer.
5. Soft drinking water is better for your health than hard drinking water.
6. Stoneflies need cleaner water than midge larvae.

B. Completion

Complete each of the following sentences with a word or phrase that will make the sentence correct. (Do not write in this book.)

1. Wetlands include bogs, marshes, and _____.
2. Warm water holds _____ oxygen than cool water.
3. Extra nutrients in a lake can cause an _____ _____.
4. Chloride levels in rivers can be raised by _____ and _____ _____.

C. Multiple Choice

Each of the following statements or questions is followed by four responses. Choose the correct response in each case. (Do not write in this book.)

1. A lake has 2 µg/g of oxygen and 45 µg/g of carbon dioxide. This lake could support
   a) trout   b) mayflies   c) chub   d) little animal life
2. An industry put warm water into a stream. The water temperature changed from 10°C to 20°C. The oxygen level would likely
   a) increase by about 2 µg/g
   b) drop by about 2 µg/g
   c) drop by over 5 µg/g
   d) remain unchanged
3. A river contains 6 µg/g of ammonia, 4 µg/g of phosphate, and 900 µg/g of chloride. This river is most likely polluted with
   a) street runoff   b) fertilizer runoff   c) sewage   d) animal wastes
4. Two animals that can live in mud-bottoms of rivers are
   a) caddisflies and mayflies
   b) sludge worms and midge larvae
   c) sludge worms and mayflies
   d) caddisflies and stoneflies

D. Using Your Knowledge

1. Four decades ago mayflies were very abundant in towns along Lake Erie. In fact, shovels were used to remove them from streets. This problem no longer exists. Why?
2. Sludge worms are usually more common at the mouth of a river than at the headwaters. Why?
3. What would happen to a pond ecosystem if all the decomposers died? Why?
4. Imagine a pond with considerable vegetation on a sunny day. Table 35-8 shows data obtained where the water enters and leaves the pond. Account for the changes in each factor.

Table 35-8 Chemical Factors in a Pond

Factor	Water entering pond	Water leaving pond
Oxygen	7 µg/g	13 µg/g
Carbon dioxide	20 µg/g	5 µg/g
Nitrate	1.2 µg/g	0.1 µg/g
Phosphate	1.0 µg/g	0.5 µg/g

E. Investigations

1. Federal laws limit phosphates in laundry detergents to less than 5%. This has helped reduce pollution of lakes. However, dishwasher detergents are allowed to have up to 45% phosphate. The main reason is that the phosphate makes the dishes dry spotless. Discuss this matter with 2 or 3 people who have dishwashers. Then write a report with the title "Are Spotless Dishes More Important Than Clean Lakes?"

2. Select one pond invertebrate. Research its life cycle, feeding habits, and adaptations to its environment. Write a report of 300-500 words on your findings.
3. Do question 2 for a stream invertebrate.
4. Find out what branch of your provincial government is reponsible for water quality standards. Get a copy of those standards from the government. What levels of nitrates, phosphates, chlorides, sulfates, and mercury are recommended for drinking water?

36 Terrestrial Ecosystems

36.1 Five Factors that Affect Organisms
36.2 Activity: Nutrients, pH, and Plants
36.3 Activity: Comparing Two Terrestrial Sites
36.4 What Are Biomes?
36.5 The Tundra Biome
36.6 The Coniferous Forest Biome
36.7 The Temperate Deciduous Forest Biome
36.8 The Grasslands Biome
36.9 Mountain Biomes

Terrestrial ecosystems are those ecosystems that are based on land. They may be as small as a handful of soil or a fallen log. Or they may be as large as the Sahara Desert or the Arctic tundra.

You will begin this chapter by investigating small terrestrial ecosystems near your school. Then you will apply what you learned to the study of **biomes**, the largest terrestrial ecosystems.

36.1 Five Factors that Affect Organisms

Five main environmental factors affect organisms in terrestrial ecosystems. These are **temperature**, **moisture**, **wind**, **light**, and **soil conditions** (Fig. 36-1). You may recall that each organism has a range of tolerance for each of these factors (Chapter 34, page 500). Organisms with the broadest range of tolerance generally survive best. That's because they are best adapted to their environment.

The following are a few examples of such adaptations. You will see

Fig. 36-1 How is this plant affected by the five non-living factors? How has it adapted to them?

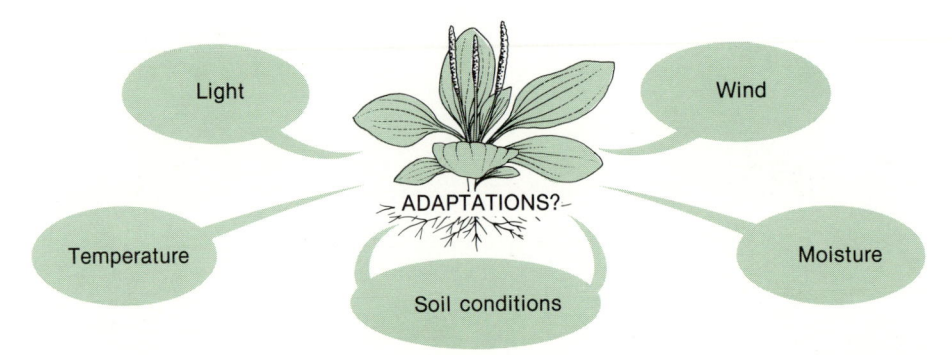

Fig. 36-2 Temperatures often reach 35°C–40°C where this cactus grows. Such high temperatures could kill the cactus if it was not adapted to them.

Fig. 36-3 How is the grasshopper adapted to high temperatures?

Fig. 36-4 Some adaptations of plants to low precipitation conditions

others in Activities 36.2 and 36.3. Also, you will read about adaptations in all the other sections of this chapter.

Temperature

Some plants have adapted to low temperatures. Spinach, for example, germinates and grows well only in early spring or late fall. Summer plantings are usually unsuccessful. Other plants have adapted to high temperatures. For example, cacti in our hot prairie regions have a thick outer coating (Fig. 36-2). This helps prevent water loss due to the heat of the day.

The grasshopper, too, has a thick outer coating, or exoskeleton (Fig. 36-3). This helps this dweller of hot meadows conserve moisture. Other animals lack structural adaptations like the grasshopper's. Instead they have behavioural adaptations. For example, groundhogs and ground squirrels seek shelter in burrows on hot days. And dogs seek shade and damp ground on hot days. Further, they pant to cool their bodies.

Moisture

There are two forms of moisture to which organisms must adapt in order to survive. They are **precipitation** and **relative humidity**.

Rain, snow, and other forms of precipitation provide plants and animals with much of their water. Therefore plants in areas of low precipitation must have adaptations for taking in and conserving the little moisture available. Figure 36-4 shows three of these adaptations. The adaptations of cacti and grasshoppers to temperature also let them live in low precipitation areas.

Some animals, like the spittlebug, have unique adaptations to low relative humidity (moisture in the air). This insect surrounds itself with juices from the plant on which it lives. In this way it can live in dry meadows without suffering from water loss.

Many have extensive fibrous roots for picking up water.

Some plants have deep tap roots for picking up water.

Many plants have narrow waxy leaves to reduce water loss.

Section 36.1 529

Wind

Wind affects the growth of plants. Strong winds can change their shape (Fig. 36-5). They can also damage or uproot plants. Also, they can dry them out.

Most plants that grow in windy areas, however, are adapted to the wind. They may, for example, be short and spreading instead of tall (Fig. 36-6). Or they may have extensive roots. Adaptations to lessen drying out include narrow waxy leaves. Most evergreens have these.

Fig. 36-5 This pine, like a weather vane, indicates the direction of the prevailing winds.

Fig. 36-6 This juniper (an evergreen shrub) is growing on a windy sand dune by a large lake. Why is it not uprooted by the wind?

Light

Plants must adapt to three aspects of light. These are the **intensity** (brightness), **duration** (how long the light shines), and the **colour** of the light. Adaptations to intensity are easily seen. For example, meadow plants usually have narrow leaves, whereas forest plants have broad leaves. The light is weak in the forest. Therefore broad leaves are needed to capture enough light to keep the plants alive (Fig. 36-7). Perhaps you have noticed, too, that sun-loving plants like geraniums and sunflowers turn their leaves towards the sun.

Soil Conditions

Nutrients and **pH** are just two of the many factors to which organisms must adapt. Legumes, like the black locust tree, can grow in nitrogen-poor soil. Legumes have nitrogen-fixing bacteria on their roots. These provide the legumes with needed nitrogen.

Sphagnum moss has adapted to acidic (low pH) soil. In contrast, oak trees and earthworms would die in such soil. They are not adapted to low pH conditions.

Fig. 36-7 This plant, like other forest plants, has broad leaves.

Section Review

1. Name 5 environmental factors to which terrestrial organisms must adapt.
2. Describe how one plant and one animal have adapted to high temperatures.
3. How are plants adapted to low precipitation?
4. Describe 3 adaptations of plants to windy conditions.
5. How have some plants adapted to low light conditions?
6. How are legumes adapted to low nitrogen conditions?

36.2 ACTIVITY Nutrients, pH, and Plants

Some plants need high nitrogen levels in the soil. Others need little nitrogen. Some plants need low pH soil (acidic). Others need high pH soil (basic). Can you find examples of such relationships?

Problem

How do soil nutrients and soil pH affect plants?

Materials

soil testing kits for pH, nitrogen, phosphorus, and potassium
copy of the *LaMotte Soil Handbook*
distilled water
collecting jars (4 or 5)
plant identification guide
spoon

Procedure

a. Copy Table 36-1 into your notebook.
b. Find a place where one particular plant is growing better than the others (Fig. 36-8).
c. Identify the plant. Put its name in your table.
d. Collect a spoonful or two of soil from the site.
e. Repeat steps (b) to (d) for 2 or 3 more sites.
f. Back in the classroom, test the soil samples for pH, nitrogen, phosphorus, and potassium. Put your values in your table.

Fig. 36-8 This area is dominated by bird's-foot trefoil, a legume. What soil conditions might be present?

Table 36-1 Nutrients, pH, and Plants

Plant	pH	Nitrogen	Phosphorus	Potassium

Discussion

1. Try to explain why each plant dominated its area. The *LaMotte Soil Handbook* will give you the information you need.
2. This activity shows that soil conditions affect plants. But the principles of ecology say that the plants also affect the soil. What evidence did you see of this fact?

36.3 ACTIVITY Comparing Two Terrestrial Sites

Everything in nature is interdependent. Change the environment and you change the organisms present. Change the organisms and you change the environment. See if you can prove this by comparing two sites.

Problem

Can you compare the living and non-living factors in two different terrestrial sites?

Some Suggested Sites

a) Area dominated by long grass versus area dominated by cut grass
b) Wooded area versus meadow (grassy area)
c) Coniferous tree area versus deciduous tree area
d) Low wet area versus sunny area of the schoolyard
e) Area with organically rich soil versus area with sandy soil
f) Sheltered area versus windy area
g) South-facing slope of a hill versus north-facing slope of a hill

Materials

The following is an ideal list. If your school does not have a certain piece of equipment, you can find substitutes. For example, a camera light meter can take the place of the light meter. Or, you can simply rate the light intensity on a relative scale such as: very light, bright, moderate, dim, very dim.

light meter
sling psychrometer (for humidity)
air thermometer
soil thermometer
soil test kit for pH, nitrogen,
    phosphorus, and potassium
soil sampler (or small shovel)
trowel
wind meter
soil sieves

tape measure (10 m)
sweep net
white tray
hand lens

Procedure

a. Copy Tables 36-2, 3, and 4 into your notebook. Make each table a full page in size. Record all data in these tables as soon as you get the data.
b. Select two sites for the comparison. They must differ in at least one obvious way. The sites need not be large.
c. Stake out study plots of identical size and shape at the two sites. An area of 100 m² for each study plot is more than adequate.

Table 36-2 Comparing Two Terrestrial Sites

Factor	Site 1			Site 2		
Light intensity						
Relative humidity						
Wind speed						
Air temperature	0.5 m	1.0 m	2.0 m	0.5 m	1.0 m	2.0 m
Soil temperature	2 cm	7 cm	15 cm	2 cm	7 cm	15 cm
pH						
Nitrogen						
Phosphorus						
Potassium						
Sieve results						
Soil profile						

Table 36-3 Animals and Their Adaptations

Animal	Relative abundance	Adaptations

Table 36-4 Plants and Their Adaptations

Plant	Relative abundance	Adaptations

Fig. 36-9 A close look at the northern boreal forest

d. Use the light meter to measure the average light intensity at each site.
e. Use the sling psychrometer to measure the relative humidity at each site.
f. Use the wind meter to measure the average wind speed at 3 levels at each site. Take readings 0.5 m, 1.0 m, and 2.0 m above the ground.
g. Measure the air temperature at each site. If the sun is shining, shade the bulb of the thermometer from the direct rays of the sun.
h. Use the soil thermometer to measure the soil temperature at depths of 2 cm, 7 cm, and 15 cm. Do this at both sites.
i. Determine the soil pH at each site. If possible, measure the concentrations of nitrogen, phosphorus, and potassium at each site.
j. Use the trowel to collect an "average" sample of soil from each site. Put the sample in the soil sieve. Separate it by shaking the sieve. Note the fraction of the total sample that ends up on each level of the sieve.
k. Use the soil sampler or shovel to determine the nature of the soil profile at each site. How deep is the litter layer? How deep is the topsoil?
l. Sweep each study plot thoroughly with the sweep net. Note the numbers of different species of organisms that are collected. Note, also, the relative abundance of each species. Is it abundant, frequent, occasional, or rare? Then note any adaptations to their environment that these organisms show.
m. Put a few scoops of soil in the white tray. Sort through it, looking for organisms. Again, note the numbers, relative abundance, and adaptations.
n. Examine the vegetation at each site. Note the numbers of different species, their relative abundance, and their adaptations.

Discussion

1. Write a paper that supports the ecological principle that everything in nature is interdependent. Use the findings of this field study. Pay particular attention to the differences in the living and non-living factors at the two sites.

36.4 What Are Biomes?

In Activities 36.2 and 36.3 you saw that living and non-living factors interact in terrestrial ecosystems. For example, plants affected the soil. And, soil affected the plants. This interdependence occurs in all ecosystems, even very large ones called biomes.

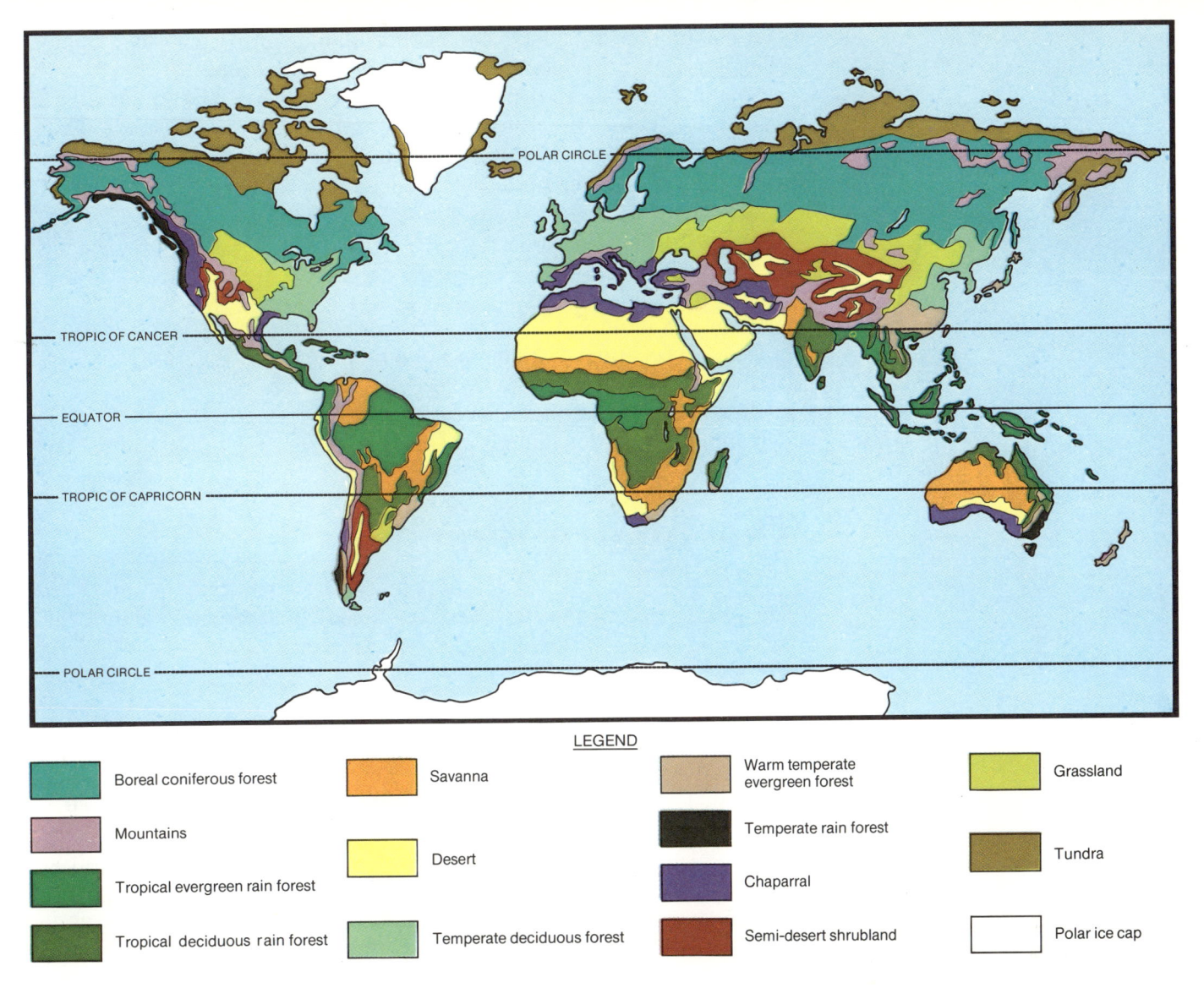

Fig. 36-10 The major biomes of the world

Legend:
- Boreal coniferous forest
- Mountains
- Tropical evergreen rain forest
- Tropical deciduous rain forest
- Savanna
- Desert
- Temperate deciduous forest
- Warm temperate evergreen forest
- Temperate rain forest
- Chaparral
- Semi-desert shrubland
- Grassland
- Tundra
- Polar ice cap

A Definition

A **biome** is a large region of the earth with a characteristic climate and climax community. The northern boreal forest is an example (Fig. 36-9). This biome extends across Canada and occurs in all provinces. It has a characteristic **climate** — cold and wet. This climate is determined by the yearly patterns in **temperature** and **precipitation**. This climate, in turn, creates a certain climax community. This community is most easily recognized by its dominant vegetation. For example, the northern boreal forest is dominated by conifers like black spruce.

Number of Biomes

There are only 10 or 15 major biomes on earth (Fig. 36-10). The actual number varies from ecologist to ecologist. There are zones

Section 36.4

between biomes in which one biome blends into the other. Some ecologists call these biomes. Others call them **ecotones**.

Canada has just five major biomes. These are the **tundra**, the **coniferous forest** (northern boreal and oceanic), the **temperate deciduous forest**, the **grasslands**, and the **mountains**. These are described in the remainder of this chapter. Can you find them in Figure 36-10?

Section Review

1. What is a biome?
2. What determines the climate of a biome?
3. How is a biome most easily recognized?
4. How many biomes are there on earth?
5. Name the 5 major biomes of Canada.

36.5 The Tundra Biome

The **Arctic tundra** is a vast treeless plain. It stretches beyond the northern forests to the edge of the arctic ice cap.

Non-Living Factors

The tundra climate is cold, windy, and dry. Even the wet summer months have only about 2.5 cm of precipitation. In the winter there is little snowfall. But the snow is constantly blown around. Therefore one gets the impression that snowfall is heavy. The tundra is often called a frozen desert.

This area has 24 h of daylight in midsummer. And, in midwinter, it has 24 h of darkness. Winter lasts 9 months. The spring thaw and Arctic summer are crowded into 3 months. In fact, the growing season is only about 60 d.

The soil thaws only to a depth of a few centimetres to half a metre. Below this lies the **permafrost**, soil that never thaws. This frozen layer is 600 m deep in spots. It prevents proper drainage of spring meltwater. As a result, vast marshy areas form on the land. Such areas are called **muskeg**. The freezing and thawing of soil has formed an interesting pattern on the tundra surface (Fig. 36-11).

Living Factors

Temperature limits the variety of organisms in the tundra. As a result, food chains and food webs are simple. And adaptations are easy to spot.

Vegetation Lichens, mosses, grasses, and herbs dominate the tundra (Fig. 36-12). A few stunted woody shrubs such as birches and

Fig. 36-11 This aerial photograph shows the quilt-like pattern of the tundra surface.

Fig. 36-12 The tundra in the summer. Note that no trees are present. A few shrubs appear in low areas.

willows grow in lower areas. Even 100-year old shrubs are less than 1 m tall. Their southern relatives are large trees.

Most tundra plants are **perennials**. They must grow for several seasons in order to store enough energy to flower. Also, most Arctic plants reproduce by **vegetative propagation**. Seeds have little chance of germinating in the tundra soil.

Animals Most tundra animals are white in the winter. Among these are the Arctic hare, grey wolf, lemming, Arctic fox, and ptarmigan. These animals change to darker colours during summer. This provides better camouflage (Fig. 36-13).

Arctic animals are protected from the cold by an insulating layer of fat. They also have air pockets trapped within long, dense fur or feathers (Fig. 36-14). These provide further insulation. Many animals beat the winter cold by migrating. For example, the Barren Ground caribou (Fig. 36-15) move southward to the tree line. And geese fly south to the ocean marshes.

Any tundra visitor will tell you that the most common animals are black flies, mosquitoes, and deer flies. However, the most important animal may be the lemming. This small rodent is a herbivore. It forms a key link in many Arctic food chains and webs. It is eaten by the Arctic fox, weasels, bears, wolves, and birds of prey such as hawks and the snowy owl.

Section 36.5

Fig. 36-13 The summer and winter plumage of a ptarmigan

Section Review

1. Where is the tundra?
2. **a)** List the 5 non-living factors that affect all terrestrial ecosystems (see Section 36.1, page 526).
   b) Describe each of these factors for the tundra.
3. Name the dominant types of vegetation in the tundra.
4. Describe the camouflage of Arctic animals.
5. How are Arctic animals protected against the cold?
6. Why is the lemming so important in the tundra?

Fig. 36-14 The musk ox has insulating wool beneath the hairy robe. The Inuit name for the musk ox is oomingmak, meaning "the bearded one". (Reproduction of art by Glen Loates courtesy of Setaol Incorporated.)

Fig. 36-15 Caribou means "shoveller". These animals use their hoofs to dig for food through the snow.

36.6 The Coniferous Forest Biome

Moving south from the tundra one meets clumps of dwarf trees. These are scattered in sheltered nooks. Finally one reaches a distinct **tree line**. This is the edge of the **northern boreal forest**. This vast **coniferous** (cone-bearing) **forest** stretches across Canada. Parts of it occur in all provinces, the Yukon, and the Territories (Fig. 36-16).

Another coniferous forest occurs in Canada. It is along the Pacific coast south of Alaska. Therefore it is called the **oceanic coniferous forest**. As you will see, this forest is quite different from the northern boreal forest.

Non-Living Factors

Northern Boreal Forest Average monthly temperatures are higher than the tundra. The growing season varies from 60 to 150 d. Summer days are shorter but warmer than they are in the tundra. Most important, the ground thaws completely. The winters are less severe and shorter. Snowfall is heavier. But total precipitation is still low.

During the last ice age, glaciers gouged depressions in the land. These filled with water, forming the countless lakes, swamps, and bogs of the northern woods.

Decomposers work slowly in the cold wet soil. Thus **peat** (partly decayed organic matter) is common throughout the woods.

Oceanic Coniferous Forest The climate and mountains along the Pacific ocean combine to produce this forest in British Columbia. It

Section 36.6 539

Fig. 36-16 The northern boreal forest. The great northern woods abound with lakes, swamps, and bogs.

differs greatly from the northern boreal forest. Average monthly temperatures range from 2°C to 18°C. The ground is frost-free for 120 to 300 d. And precipitation can exceed 600 cm/a! The warmth, rain, and high humidity nourish giant coniferous trees. Among them are Douglas fir, sitka spruce, and western red cedar (Fig. 36-17).

Living Factors (Northern Boreal Forest)

Vegetation **Coniferous trees**, or **conifers**, dominate the northern boreal forest. These cone-bearing trees include black spruce, white spruce, jack pine, white pine, and tamarack. All but the tamarack are **evergreen**. In other words, they keep their needles during the winter.

Conifers are well-adapted to the poor soil, low temperatures, and limited rainfall. Their leaves are reduced to needles. These needles have a waxy outer skin that reduces water loss. The needles can also withstand freezing.

Fig. 36-17 Giant red cedars and Douglas firs dominate these woods near Vancouver, B.C. (*left*)

Fig. 36-18 Few plants can survive in the dense shade of this black spruce forest. (*right*)

The evergreens form a dense canopy all year round. Thus little light reaches the forest floor. As a result, the common plants on the forest floor are ferns and mosses. Lichens are also common (Fig. 36-18). And fungi such as mushrooms serve as decomposers among the fallen needles.

Animals Boreal animals must survive the long, cold winter. The ground is frozen and the snow presents some problems. Foxes, wolves, and moose are common. They have thick winter fur. The snowshoe hare is also well-adapted. It turns white and has large tufts of fur on its feet that serve as snowshoes. The wolverine preys on the hare. It has spreading toes that let it run swiftly over the snow (Fig. 36-19).

Moose, too, are adapted for dealing with deep snow. They wade through it on stilt-like legs. Or, if it gets too deep, several moose get together and trample it down. Then they can reach tree shoots, brush, and twigs. Those same long legs serve the moose well in the summer. Moose wade in lakes and marshes where they browse on aquatic plants (Fig. 36-20).

As in the tundra, the most common animals are insects. Black flies, mosquitoes, and deer flies attack anything with blood. Moose stand neck-deep in water to escape them. Other insects attack trees. Outbreaks of spruce budworm and larch sawfly have wiped out vast areas of trees in many parts of Canada. Parts of scenic Cape Breton have been almost denuded of trees.

Small birds called warblers are common in the boreal forest. They feed largely on the abundant insects. Seed-eating birds are also common. They have specially adapted beaks for getting seeds from the cones (Fig. 36-21). Blue grouse and spruce grouse feed directly

Fig. 36-19 The wolverine is a fast predator of the boreal forest.

Section 36.6

Fig. 36-20 The moose is the symbol of the boreal forest. In fact, this biome is often called the "moose-spruce biome".

Fig. 36-21 The crossbill has a strong curved bill for cutting through the cone scales. Its tongue then reaches the seeds inside.

on the needles of the conifers. These needles contain little nutrient. Therefore these birds must eat constantly during the winter.

Section Review

1. **a)** Name the two types of coniferous forest in Canada.
   b) Where do these occur?
2. Describe the environment of the northern boreal forest.
3. Describe the environment of the oceanic coniferous forest.
4. **a)** What are conifers?
   b) What are evergreens?
5. Name three animals of the boreal forest and describe their adaptations.

36.7 The Temperate Deciduous Forest Biome

The **temperate deciduous forest** occurs on the south-east edge of the boreal forest. Only the most southern parts of Ontario and Québec are in this biome. However, as Figure 36-10 shows, most of the south-eastern United States occurs in it. This biome is recognized by its temperate (moderate) climate and deciduous trees. **Deciduous** trees drop their leaves in the winter.

Fig. 36-22 The trillium is just one of the spring flowers that cover the forest floor.

Fig. 36-23 The temperate deciduous forest. The trees in this particular area are mainly sugar maple and beech.

Fig. 36-24 The woodpecker has four toes per foot. Two toes point backwards. Why?

Non-Living Factors

Average annual precipitation in this biome is 75 to 125 cm. This precipitation falls fairly evenly through four distinct seasons. The climate is moderate. And the growing season is as long as six months.

The winters are short. But they are cold enough to greatly reduce growth and photosynthesis. Trees in this biome lose their leaves in the winter to conserve water. These leaves decay rapidly on the forest floor. The resulting soil is rich with humus.

Living Factors

Vegetation Because the soil is rich, much of this biome has been cleared for farming. Little of the original forest remains. Where it does, it is dominated by sugar maple, beech, and oak (Fig. 36-23). The southernmost areas have hickory, sycamore and other Carolinian species. The rich soil of the forest floor supports a wide variety of ferns, mosses, and wild flowers (Fig. 36-22).

Animals Tree dwellers abound in the deciduous forest. Squirrels, chipmunks, tree frogs, and woodpeckers are examples. They find shelter and food among the trees. All such animals are well-adapted to a life in trees. The woodpecker, for example, has opposing toes and a tail that can be used as a prop (Fig. 36-24).

Unlike conifers, deciduous trees are a rich source of food for animals. The buds and twigs store a great deal of food. Deer feed on the leaves, buds, fruit, and seeds of trees and shrubs. They feed along the edge instead of deep in the forest (Fig. 36-25). Rabbits, mice, and other rodents eat bark and small plants.

Carnivores include owls, hawks, weasels, and large mammals like bobcats. Some mammals, like racoons, skunks, and the red fox are omnivores (Fig. 36-26). They eat many types of food, both plants and

Section 36.7

Fig. 36-25 The ecotone, or edge, between a deciduous forest and a meadow. A wide variety of animals live here.

Fig. 36-26 The red fox is a common omnivore of the deciduous forest.

animals. The red fox, for example, feeds on mice, large insects, fish, eggs, berries, and even grass.

The rich soil supports a host of organisms. Just one square kilometre of soil litter can have over 120 species of invertebrates — spiders, insects, millipedes, centipedes, earthworms, and many others.

Section Review

1. What does "temperate deciduous forest" mean?
2. Describe the non-living factors that are characteristic of the temperate deciduous forest.
3. What tree species dominate this forest?
4. How is a woodpecker adapted for sitting on the trunks of trees?
5. What do deer eat?
6. Give an example of an omnivore and its food.

36.8 The Grasslands Biome

As Figure 36-10 shows, the **grasslands**, or **prairies**, is a small part of Canada. It is much more extensive in the United States. Nonetheless, it grows an important part of the world's food supply.

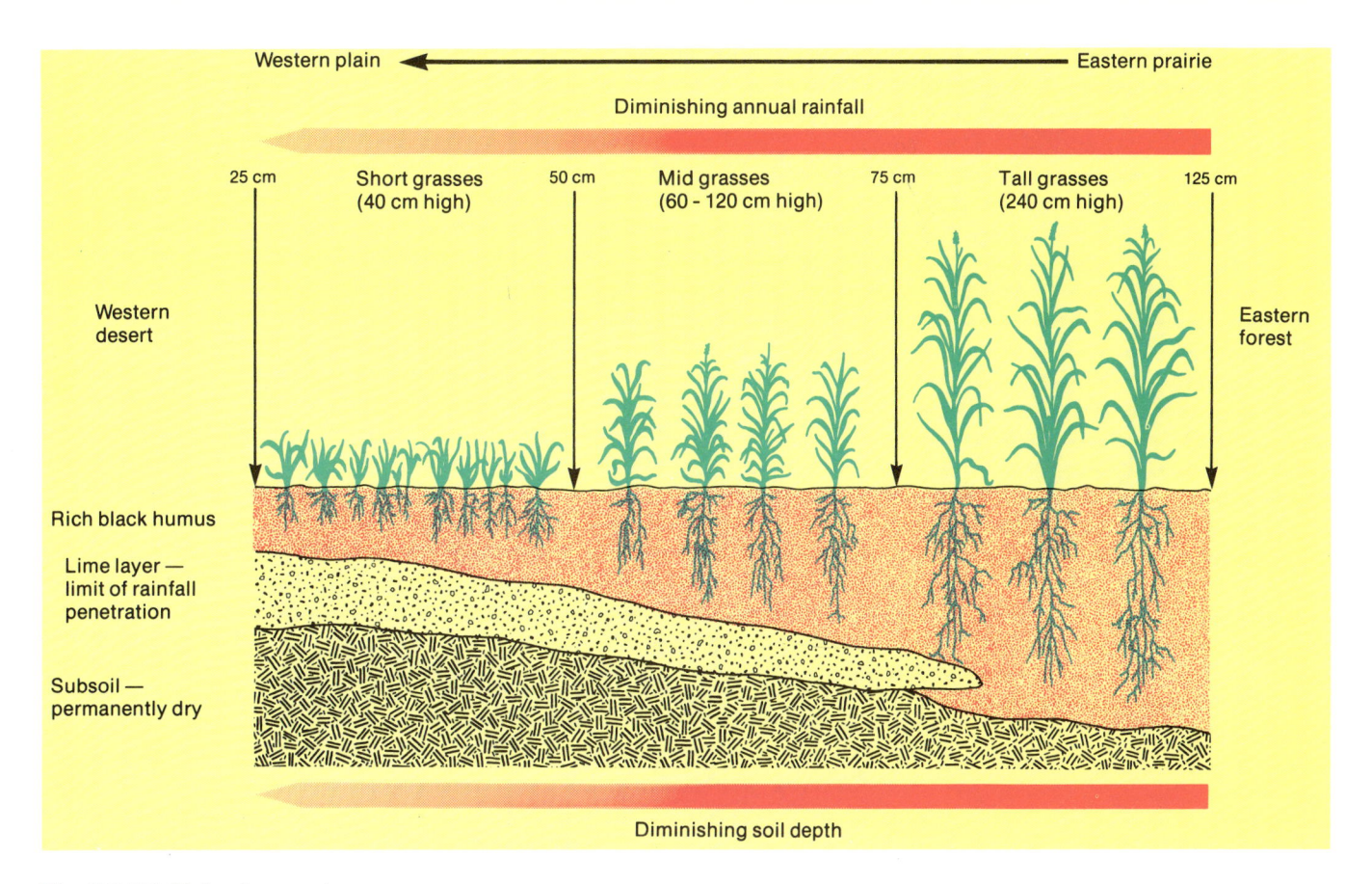

Fig. 36-27 This change in grass types is best observed in the United States' prairies. The tall grass prairies of the United States have become the famous corn belt.

Non-Living Factors

The grasslands biome is within the same latitudes as the deciduous forest. Therefore its seasons and energy supply (from the sun) are similar. However, the grasslands have a much lower precipitation. The annual rainfall is only 25-75 cm. This is enough to grow many grasses. But it is too low for tree growth.

Prairie soils are among the most fertile in the world. Grasses decay quickly. As a result, a deep layer of humus covers the prairies.

Living Factors

Vegetation The difference in rainfall produces three distinct types of grassland (Fig. 36-27). Moderate rainfall makes the eastern prairies a **tall grass zone**. Further west, the drier central grasslands support **mid grasses**. And, in the dry western plains, **short grasses** grow.

Tree growth is limited to stream valleys and low mountain ranges. Cottonwoods (poplars) are the common trees.

Animals Grasslands animals have many interesting adaptations to open country. Grazing animals like pronghorns have eyes located well above the snout. This enables them to watch for predators while grazing. Smaller mammals, like the ground squirrel, stand up on

Section 36.8

their haunches to see over the grass (Fig. 36-28). Others, like kangaroo rats, hop up and down to watch for enemies.

There are no trees to provide hiding places in the grasslands. Therefore animals rely on speed, burrows, and camouflage to escape enemies. Pronghorns, for example, can reach 100 km/h (Fig. 36-29)! Jack rabbits, using 8 m leaps, can go 70 km/h! Ground squirrels and prairie dogs escape into burrows. Other animals have learned to stand still in the grass and rely on camouflage for protection.

Insects are common in the grasslands. There are over 100 species of grasshoppers alone. The large number of insects and seeds attract a wide variety of birds.

Footnote

The grasslands are vitally important to Canada and the world. Much of the world's food is grown there. Yet no other biome has received greater abuse. Bison, the largest North American animal, were brought to the edge of extinction by hunters (Fig. 36-30). The badlands-grizzly and white-plains-wolf were hunted to extinction. Then careless farming techniques produced a dust bowl in the drier plains. Only a few small areas of true grasslands remain.

Farming practices are slowly improving. But, every year, thousands of hectares move closer to becoming a barren desert because we fail to respect the ecology of the grasslands.

Fig. 36-28 This position helps the ground squirrel see predators. Unfortunately (for the ground squirrel) it also helps the predators see the ground squirrel!

Fig. 36-29 Sturdy legs and strong lungs help the pronghorn move quickly. (Reproduction of art by Glen Loates courtesy of Setaol Incorporated.)

Fig. 36-30 Our government has introduced small herds into areas where bison once roamed by the millions.

Section Review

1. Where are the grasslands located?
2. Describe the environment of the grasslands.
3. Why can trees not grow in the grasslands?
4. Describe how three animals are adapted for survival in the grasslands.

36.9 Mountain Biomes

These biomes are small in area when compared to others. Yet they are an interesting and important part of the landscape in British Columbia, Alberta and the Yukon.

Non-Living Factors

A change in altitude affects the environment as much as a change in latitude. As altitude increases, the temperature drops about 1°C for every 150 m. Also, wind speed increases at higher altitudes. And,

Fig. 36-31 If you climb a tall mountain you will walk through ecosystems that resemble the biomes you would walk through if you walked from the Saskatchewan grasslands to the tundra in the Territories.

Fig. 36-32 The timberline marks the limit of tree growth. Beyond this is alpine tundra.

the soil is thinner near the tops of mountains. Erosion has carried the soil down.

These gradual changes in temperature, wind, and soil, create ecosystems on the mountain. These ecosystems resemble the biomes we have discussed so far in this chapter. Figure 36-31 compares the mountain ecosystems to the biomes.

Living Factors

Vegetation Forest creeps up most mountain slopes to the **timberline**. That is the uppermost limit at which trees can survive (Fig. 36-32). Above the timberline is the **alpine tundra**. In higher mountains there may be a permanent **snow belt** above this.

As in the Arctic, alpine tundra plants are small and stunted (Fig.

548 Chapter 36

Fig. 36-33 This willow is a mature tree. Yet it is less than 1 m tall!

36-33). To survive the cold and wind, alpine growth hugs the ground. Lichens, mosses, and a wide variety of flowering plants can be found in alpine meadows (Fig. 36-34). The species and adaptations are similar to those in the Arctic tundra.

Animals Alpine and Arctic wildlife share many of the same adaptations. Both have many species that change colour for camouflage. Both have species that hibernate. And both have species with body coverings that are well-insulated to conserve heat. Also, both have species that migrate to warmer regions in the winter. Arctic animals must migrate hundreds of kilometres. However, alpine animals like bighorn sheep need only walk down the mountain!

Section Review

1. What three factors help form bands of ecosystems on mountains?
2. What is the timberline?
3. What is alpine tundra?
4. In what ways are alpine plants and animals like those of the Arctic tundra?

Fig. 36-34 This alpine meadow, at the timberline, closely resembles the arctic tundra near the tree line.

Main Ideas

1. Five main environmental factors affect organisms. They are temperature, moisture, wind, light, and soil conditions.
2. Plants, animals, and other living things are adapted to their environments.
3. A biome is a large region of the earth with a characteristic climate and climax community.
4. Canada has five main biomes. They are the tundra, coniferous forest, temperate deciduous forest, grasslands, and mountains.

Main Ideas 549

Glossary

biome	a large region of the earth with a characteristic climate and climax community
conifer	a cone-bearing tree
ecotone	a zone between two different communities or biomes
muskeg	a marshy area in the tundra
permafrost	tundra soil that never thaws

Study Questions

A. True or False

Decide whether each of the following statements is true or false. If the sentence is false, rewrite it to make it true. (Do not write in this book.)
1. All terrestrial ecosystems are called biomes.
2. Spinach should be planted in July.
3. Permafrost is tundra soil that never thaws.
4. Muskeg is always frozen.
5. The largest coniferous trees in Canada are found in British Columbia.
6. All conifers are evergreens.
7. Deer feed mainly on grass.

B. Completion

Complete each of the following sentences with a word or phrase that will make the sentence correct. (Do not write in this book.)
1. Sphagnum moss is adapted to ▓▓▓ soil.
2. A biome is most easily recognized by its ▓▓▓ ▓▓▓.
3. A key link in many tundra food chains is a rodent called the ▓▓▓.
4. Trees that drop their leaves in the winter are called ▓▓▓ trees.
5. The rainfall in the grasslands is not enough to grow ▓▓▓.
6. The uppermost limit for trees in mountains is called the ▓▓▓.

C. Multiple Choice

Each of the following statements or questions is followed by four responses. Choose the correct response in each case. (Do not write in this book.)
1. One large area of Canada has about 2 cm of rain in July. It also has 24 h of daylight in midsummer. This area is in the
    a) tundra **b)** grasslands **c)** northern boreal forest **d)** mountains

2. Another large area of Canada is dominated by black spruce and tamarack trees. This area is likely in the
   a) tundra
   b) northern boreal forest
   c) temperate deciduous forest
   d) oceanic coniferous forest
3. The biome with the lowest annual precipitation is
   a) the grasslands
   b) the northern boreal forest
   c) the tundra
   d) the temperate deciduous forest
4. How many of Canada's five biomes have few, if any, trees?
   a) four
   b) three
   c) two
   d) one

D. Using Your Knowledge

1. Reptiles and amphibians are rare in the tundra. Why might this be so?
2. a) Compare the temperate deciduous forest to the northern boreal forest. How are they alike? How are they different?
   b) Where, in Canada, would you find an ecotone where these two forests blend?
3. Conditions in the alpine tundra are much like those in the Arctic tundra. However, there are some important differences. List as many as you can.
4. Look back to Figure 36-10. Which biomes are found in the United States that are not found in Canada? Which ones are found in Canada that are not found in the United States?

E. Investigations

1. Oil companies are searching Canada's north for oil deposits. Many conservationists fear permanent ecological damage will be done. This could be caused by oil spills, pipeline construction, and vehicles. Research this matter. Then write a paper of about 400 words giving your opinion.
2. Forestry is Canada's largest industry. It employs many thousands of Canadians. However, the spruce budworm threatens the forests, particularly balsam fir and black spruce. It kills vast tracts of forest and then people are put out of work. As a result, some provinces spray the forests with pesticides to kill the budworm.

   Environmentalists say the pesticides cause illnesses among people who live in sprayed areas. They also claim that the spraying won't work for long.

   Write a paper of 400-500 words on the spruce budworm problem. Does the spray work? Does it make people ill? Will it work as well in 10 a as it does now? Should spraying continue?
3. Find out what causes the heavy rains in the oceanic coniferous forest.
4. Four prairie species are on the verge of extinction. They are the prairie falcon, the tule elk, the black-footed ferret, and the kit fox. Select one of these. Find out why it faces extinction.
5. Find out why a cross-Canada network of National Parks is important to the survival of native plants and animals.

Unit 9

Building a Better Future

CHAPTER 37
Tobacco, Alcohol, and Other Drugs

CHAPTER 38
Human Population Growth: Problems and Solutions

CHAPTER 39
Conserving Natural Resources

The drug scene... disease... starvation... the energy crisis... wars... pollution... extinction of animals... loss of farmland...

It's easy to get discouraged these days, isn't it? The problems seem so great. And there are so many of them. However, you have learned enough in this course to help solve these problems. You can't do much alone. But neither can anyone else. So let's get our act together. Let's get involved and build a better future!

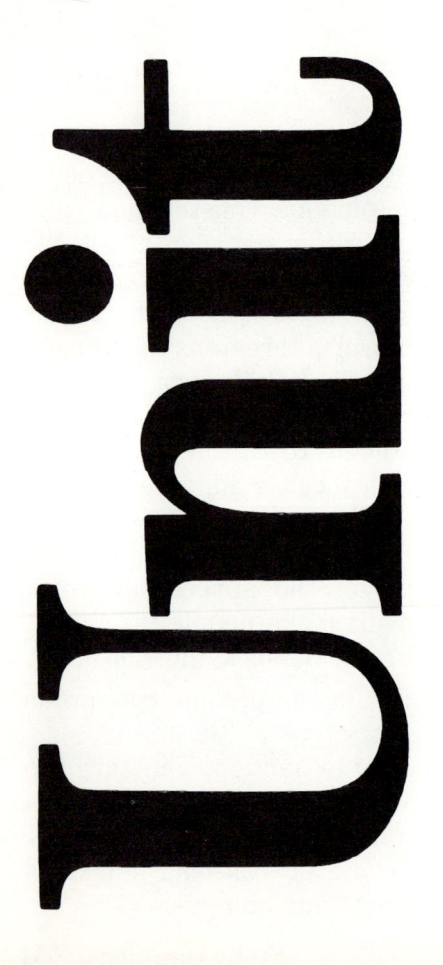

Fig. 37-0 Drugs... litter... pollution... waste... Who needs them? Do you?

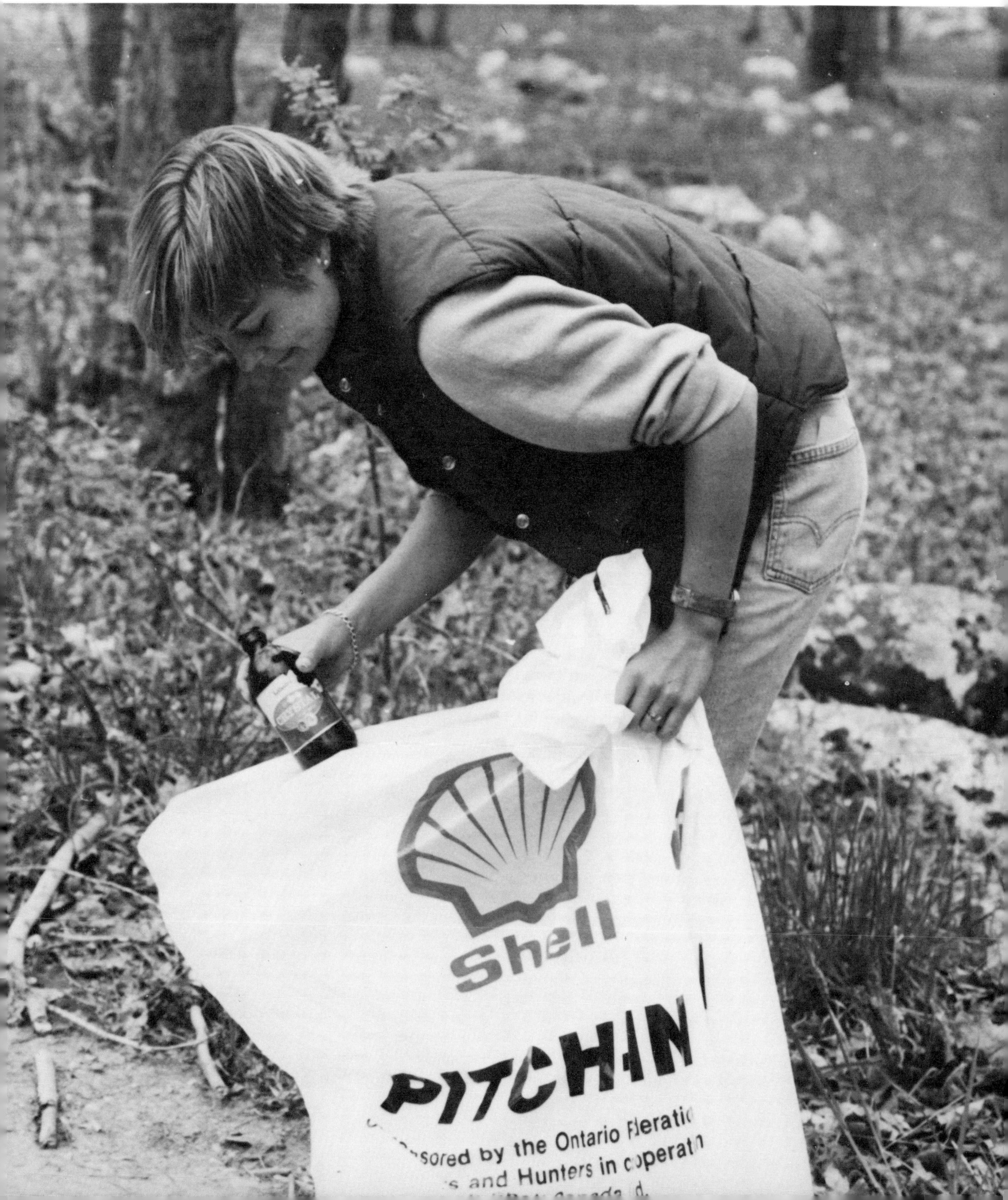
Shell
PITCH-IN

37 Tobacco, Alcohol, and Other Drugs

37.1   What Are Drugs?
37.2   Activity: What's in Tobacco?
37.3   Tobacco
37.4   Activity: What's in Tobacco Smoke?
37.5   Alcohol
37.6   Hallucinogens
37.7   Narcotics
37.8   Stimulants and Depressants

This is the first chapter in a unit called **"Building a Better Future"**. Perhaps you are wondering why we began this unit with a chapter on drugs. After all, there are many other problems in this world — an energy shortage, overpopulation, and starvation, to name a few. Well, here's our reason:

We feel that you cannot help solve these other problems if your life is messed up with drugs. Yet we know some of you are already into drugs. And we know that *all* of you are tempted in many ways to get into drugs. We know the pressures are great. We know the nice feelings some drugs produce. And we understand why you might like to try drugs. But we have also seen the results. We know what happens to the lives and bodies of drug-users. We thought you should know this, too.

37.1   What Are Drugs?

A Definition

The word "drug" has two meanings in our society. First, it is used as a catch-all term for all chemicals given to humans and other animals. This includes medicines prescribed by doctors. It also includes non-prescription chemicals like nose sprays, cough medicines, vitamins, aspirin, and antacids (Fig. 37-1). And it includes illegal drugs.

The second meaning is the one we will use in this chapter. **A drug is any chemical that affects the control systems of your body.** In

Fig. 37-1 A pharmacist sells both prescription and non-prescription drugs. Even non-prescription drugs like aspirin can be dangerous. Always read the directions carefully. Check with a pharmacist or doctor if you have any doubts.

other words, it is any chemical that affects your nervous system or endocrine system (see Chapter 28).

The nervous and endocrine systems work together to control all body functions. They control your heart beat, your breathing rate, and your mood. They control your reaction time and all aspects of your behaviour. Stop, for a moment, and make a list of things you think are drugs. Put a mark beside the ones you used yesterday.

Habit-Forming or Addictive?

Some drugs are **habit-forming**. If you use one of these for a long time, you begin to depend on it. You develop the habit of using it. Can you name some habit-forming drugs?

Other drugs are **addictive**. If you use one of these for a long time, you develop a physical need for it. Your control systems have adjusted to it. Therefore you will get **withdrawal symptoms** if you quit using it. These include illness and pains. A person in this state is an **addict**. This person cannot stop taking the drug without help. Can you name some addictive drugs?

Section Review

1. What is the broad meaning of "drug"?
2. What is the meaning of "drug" as used in this chapter?
3. What is a habit-forming drug?
4. **a)** What is an addictive drug?
   b) What are withdrawal symptoms?
   c) What is an addict?

37.2   ACTIVITY   What's in Tobacco?

Scientists often find out what is in a substance by distilling it. **Distillation** [dis-till-AY-shun] is the separation of a substance into its parts by heating it. The substance must *not be burned*, however. It must be heated in such a way that flames don't touch it.

In this activity you will distill tobacco. That is, you will heat it without burning it. Tobacco is a drug. It is a commonly used but dangerous drug. See what you can learn about this drug in this activity.

Problem

What kinds of substances are in tobacco?

Materials

test tube   cigarette
test tube clamp   Bunsen burner

Section 37.2   555

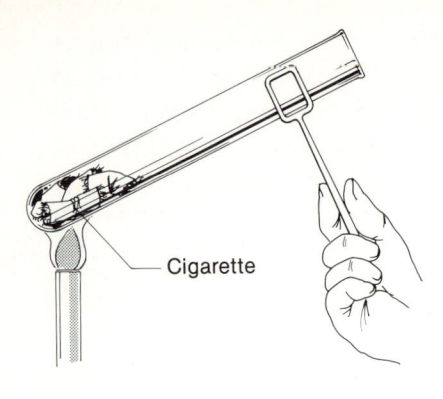

Fig. 37-2 Hold the test tube on an angle. Heat only the bottom 2-3 cm.

Procedure

a. Break a cigarette into 3 or 4 pieces. Put them in a test tube. Discard the filter.
b. Heat the test tube *gently* as shown in Figure 37-2. Hold the test tube on an angle as shown. Heat only the bottom 2-3 cm. Keep the top as cool as possible.
c. Continue to heat gently for 3-4 min. Record your observations.
d. Now heat strongly for 3-4 min. From time to time, bring the mouth of the test tube to the flame. Record your observations.
e. After the test tube has cooled, try to clean it out.

Discussion

1. What came out of the test tube?
2. What kinds of substances collected in the test tube?
3. Did you have trouble cleaning the test tube? Why?
4. Would it be better to say tobacco *is* a drug or that tobacco *contains* drugs? Why?

37.3 Tobacco

Tobacco is a plant. It is dried and shredded. Then it is sold for smoking and chewing. Thousands of millions of cigarettes are sold every year in Canada. And, every year, countless thousands of Canadians die from cancer, heart disease, accidents, and fires caused by smoking. Why does this happen? Why do people smoke? Why do people like having a piece of burning plant in their mouths?

Tobacco is a drug. Or, more accurately, tobacco contains drugs. When drugs and people are mixed, it is not easy to answer the "Whys". But let's look at some of the facts. Then, together, your class can explore the "Whys".

Some Facts

Everyone knows smoking is harmful. It says so right on the package (Fig. 37-3). Fortunately, more and more Canadians are paying attention to such warnings. Today, fewer than one third of Canadians smoke. Even so, over 30 000 Canadians will die of smoking-related diseases this year. This is more than all the deaths from accidents, murders, suicides, and other drugs put together. Smoking is the major single cause of death in Canada. It is by far our most serious health problem.

Lung cancer alone will kill about 9 000 people this year. At least 80% of lung cancer is caused by smoking. Also, 30% of all cancers are linked to smoking. This includes cancer of the mouth, larynx, and esophagus. It also includes cancer of the bladder, pancreas, and kidneys.

> Warning: Health and Welfare Canada advises that danger to health increases with amount smoked—avoid inhaling.
> "Tar"/"Goudron" 16 mg. Nic. 0.9 mg.
> Avis: Santé et Bien-être social Canada considère que le danger pour la santé croît avec l'usage—éviter d'inhaler.

Fig. 37-3 By law, cigarettes must carry this warning. Perhaps you have seen the same warning in the advertisements in magazines.

Smoking causes many other diseases besides cancer. Heart attacks, strokes, high blood pressure, emphesema, and chronic bronchitis are examples (see Section 24.7, page 350).

On the average, smokers die younger than non-smokers. Studies indicate that an average smoker can lose 8-14 years off his/her life. Non-smokers who live with smokers can expect to lose a few years, too. A recent study showed that non-smoking women living with smoking husbands lost 4 years. That's because sidestream smoke is laden with pollutants. In fact, it has more of some pollutants than mainstream smoke does. For example, it has 5 times as much carbon monoxide, 3 times more tar and nicotine, and 4 times more benzopyrene (a cancer-causing chemical).

Many car accidents are caused by smoking. Carbon monoxide in the smoke lowers reaction time. The smoke affects vision. And just handling the cigarette takes attention from the road.

Some studies show that at least half of home fires are caused by smoking. And thousands of hectares of forests are burned each year because of smoking.

Smokers are subject to poor health. They miss school and work much more often than non-smokers. They get colds easily. They cough a lot. They have a weakened sense of smell and taste. And they always smell like a dirty ashtray.

Why Do People Smoke?

It's easy to explain why people keep smoking once they start. Nicotine is a drug. It is certainly habit-forming and it is probably addictive. It has many effects on humans. It stimulates the nervous system. Then, later, it depresses it. It speeds up the heart and enlarges the pupils of the eyes. It increases the blood pressure. And, on top of all this, it makes the smoker want more.

Why do people start smoking? Most smokers blame peer pressure. Studies show that parental smoking habits are an important influence, too. And advertisements help — pretty girls and handsome guys beside a campfire on a scenic lake... with their cigarettes; a macho cowboy with a cigarette dangling from his mouth; you know the others.

How Do You Get Smokers to Quit?

Telling smokers the facts helps. But the best method is peer pressure. If you don't like your friends smoking, say so. Tell them you don't want to breathe pollutants. Tell them you don't want to smell of tobacco smoke. Tell them smoke bothers your eyes and chokes you. The government and cancer society are using such honest, forthright methods. Try them. They work.

Over 95% of smokers wish they had never started smoking. If you are one, quit now. If some of your friends are smokers, help them quit.

Fig. 37-4 A smoking machine at work

More and more people are speaking out these days. You have a right to expect clean air in public places. So speak up if you are forced to breathe polluted air. Remember, non-smokers are a large majority. Most people want to breathe clean air.

The Costs

Smoking costs lives and happiness. It also costs money. An average smoker spends $800 to $1000 a year on cigarettes. Even non-smokers pay to support the smoking habit. Health care costs alone are about $2 000 000 000 a year in Canada. That's about $100 a year for each Canadian. However, the cost is several times that when fires, accidents, and loss of work time due to illness are considered.

Smoking does make the country some money. Taxes bring in about $800 000 000 per year. But that won't even pay the health costs caused by smoking.

Section Review

1. Make a list of diseases that can be caused by smoking.
2. Why should non-smokers be concerned if they must live or work with smokers?
3. Why do smokers find it difficult to give up the habit?
4. What factors help start people smoking.
5. What is the best way to help someone quit smoking?

37.4 ACTIVITY What's in Tobacco Smoke?

In Activity 37.2 you distilled tobacco. This let you see some of the chemicals in tobacco. Some of these chemicals never get to your lungs. They get burned when the tobacco is lit. However, the burning makes new chemicals. And many of these are quite harmful. You will test for some of these chemicals in this activity.

Problem

What are some of the chemicals in tobacco smoke?

Materials

smoking machine
cigarettes (2)
air sampling pump

air sampling tubes for carbon monoxide, nitrogen dioxide, and sulfur dioxide

Procedure

a. "Smoke" a cigarette with the smoking machine (Fig. 37-4).
b. As the cigarette is being smoked, test the sidestream smoke for carbon monoxide, nitrogen dioxide, and sulfur dioxide (Fig. 37-5).

Fig. 37-5 Sidestream smoke contains many pollutants.

Follow the directions in the air sampling kit closely. (**Sidestream smoke** is the smoke that curls off the end of the cigarette.) Record your results.

c. Study the filter from the smoking machine closely. Record what you see.

Discussion

1. What gases were present in the sidestream smoke?
2. Are the concentrations high enough to be harmful?
3. a) What two substances could you see on the filter?
   b) What effects do they have on humans?

37.5 Alcohol

Like tobacco, alcohol is a legal drug. For as long as history has been recorded, alcohol has been consumed by humans. It is used at meals. It is widely accepted as a social drink. It is also used in some religious ceremonies. However, alcohol causes many problems. Let's look first at the effects of alcohol. Then we will look at some of those problems.

Effects

Alcohol is a **depressant**. That is, it *slows down* the activity of the nervous system. Many people think alcohol is a stimulant. But it isn't.

Alcohol enters the bloodstream quickly after being consumed. It can move directly from the stomach into the bloodstream. It is then carried to the liver. The liver breaks it down into harmless compounds. However, the liver can only break down about 10 to 30 mL of alcohol per hour. The glass in Figure 37-6 contains 30 mL of alcohol. It's not much, is it? If a person drinks more than this, the extra alcohol remains in the bloodstream.

This extra alcohol now moves through the bloodstream to the brain. In the brain it affects the medulla (see Section 28.4, page 409). The medulla controls breathing, heartbeat, blood pressure, and other automatic functions. Thus extra alcohol affects those functions.

Alcohol also affects a sorting centre in the brain. This centre sorts messages and sends them to the proper sections of the brain. Extra alcohol mixes up the messages. This causes reaction time to slow down. Just one glass of wine can do this in many people.

Alcohol also affects the part of the cerebrum that helps a person make judgments. Thus some drinkers often do things they are sorry about later.

Alcohol also affects mental processes. A drinker's thoughts can become disorganized. This person may stagger or have slurred speech. Too much alcohol can put a person to sleep. It can even cause coma and death. Prolonged heavy drinking can cause **cirrhosis**

Fig. 37-6 The liver can handle this much alcohol per hour.

[si-ROH-sis] of the liver. The alcohol damages the liver tissues. The liver hardens. Then it can no longer do its job. Cirrhosis will eventually kill the drinker. You can't live without a liver!

Alcoholism

About 1 of every 10 adult drinkers has an alcohol-related problem. Alcohol is a habit-forming drug for many people. But for some it is addictive. About 635 000 Canadians, or 1 of every 20 adult drinkers, is addicted to alcohol. In other words, they are **alcoholics**. These people must drink regularly. Their body demands alcohol. They may or may not drink until they are drunk. But alcoholics drink too much too often. And they cannot control their drinking. Table 37-1 gives the warning signs of alcoholism.

Table 37-1   Warning Signs of Alcoholism

- Lies about his/her drinking
- Drinks until he/she is dead drunk at times
- Hard to get along with when drinking
- Drinks to "calm his/her nerves"
- Neglects the family when drinking
- Can't remember some drinking bouts
- Doesn't eat properly when drinking
- Drinks "because he/she is depressed"
- Hides alcohol

There is no cure for alcoholism. Alcoholics must never drink again. However, alcoholics must be under medical care when alcohol is withdrawn. Sudden withdrawal can cause problems, even death. They may have a fit called **delirium tremens** [di-LIR-ee-um TREM-ens]. You may have heard of the "D.T's" or "the shakes". These people have hallucinations of creatures like insects crawling all over their bodies. They talk and act in a wild uncontrolled way. And the body shakes violently. About 15% of people who get the D.T's die.

Alcohol and the Car

Everyone agrees that alcoholism is a serious problem. But even if you are not an alcoholic, drinking too much can cause serious problems. The most serious of these is car accidents. Even a small amount of alcohol impairs judgment and reduces reaction time. It also affects the ability to judge distances and speed. Yet many people claim they drive better "with a few drinks under the belt". And many people have "one for the road" just before they leave a party.

Alcohol is involved in 40-50% of all fatal accidents. Thousands of people are killed and tens of thousands are hurt on our roads every year because of drinking drivers. *Never drink then drive.* And *never ride with a driver who has been drinking.* Your life and the lives of innocent people are at stake.

Section Review

1. How do you know alcohol is a depressant?
2. How fast can the liver break down alcohol?
3. Describe how alcohol in the bloodstream affects humans.
4. What is an alcoholic?
5. What is delirium tremens?
6. Describe a serious alcohol-related problem besides alcoholism.

37.6 Hallucinogens

Many drugs cause users to have **hallucinations**. For example, an alcoholic during withdrawal may have hallucinations. Your senses play tricks on you during hallucinations. You see, hear, smell, taste, and feel things that aren't there. Bugs crawl in your ears. Slimy creatures get in your bed. Strange noises come out of the walls of your room.

Drugs that cause hallucinations are called **hallucinogens** [ha-LOO-si-no-jens]. **Marijuana** [mar-i-WAHN-ah], **angel dust**, and **LSD** are hallucinogens. They can produce serious hallucinations.

Marijuana

Marijuana is the most commonly used illegal drug. It is made from the dried leaves of the hemp plant. **Hashish** is made from the hemp plant, too. But hashish is five to ten times stronger. Marijuana is also called pot, grass, reefers, joints, and tea.

Marijuana contains a chemical called **THC**. This chemical, in large doses, is a powerful hallucinogen. Smaller doses of THC act as a sedative. That is, the THC calms the person down. THC can remain in the body for at least 5-10 d after smoking marijuana.

THC produces a sensation of floating. The sense of time is distorted. Time seems longer than it is. The user "feels good". (Observers will probably say the user "acts silly".) Users do not hear well. Their judgment of speed and distances is not good. Short-term memory is poor. The pulse rate goes up and the blood pressure goes down. The eyes turn reddish.

Users would like to think that marijuana is harmless. But recent studies show that this is not so. Marijuana produces gradual changes in the personality. Regular users have lower coordination and ability to learn. The most dangerous effect, however, is its effect on driving. Many hours after a "high", the ability to drive safely is still impaired. The impairment is even more serious if alcohol is involved as well. *Never drive after using marijuana.*

Long-term effects include poorer performance on tests. Also, lung cancer can be caused by this drug. People who smoke 5 or more "joints" a day are called "burn-outs". They exist in a trance-like state. They live just to smoke marijuana.

Why do people use hallucinogens like marijuana? Evidence indicates that users tend to be immature and in conflict with society. In other words, they cannot cope with problems and things they don't like around them. If you are a marijuana user, does this description fit you? Does the "high" help you with your problems?

LSD

LSD is a powerful hallucinogen. Its hallucinations are sometimes pleasant. But sometimes they are terrifying. They cause some users to jump out of windows, leap in front of cars, or slash their own throats.

This illegal drug was originally obtained from a fungus called ergot. But "chemists" learned to make it. Fortunately it is not as available now as it was several years ago.

Another drug, **mescaline**, also comes from fungi. The fungi in this case are mushrooms. It too is a hallucinogen. Its effects are similar to those of LSD.

Angel Dust

Angel dust is also called **PCP**. It is an illegal drug and a hallucinogen. Some users swallow it. Others sprinkle it on marijuana and smoke it. Still others "snort" it. And the desperate user injects it into the veins. This is called **"mainlining"**.

Small doses create a floating effect. But large doses can make users violent. They may destroy property and even kill (themselves or others). This drug is stored in the brain for long periods of time. Weeks or even months later a **flashback** can occur. All the behaviours the user had weeks before can return without warning. This drug is extremely dangerous for this and other reasons.

Section Review

1. What is a hallucination?
2. What is a hallucinogen?
3. Make a summary of the effects of THC (the chemical in marijuana and hashish).
4. Why are the hallucinations caused by LSD serious?
5. Why is angel dust one of the most dangerous drugs?

37.7 Narcotics

A **narcotic** is a drug that induces sleep. Low doses relieve pain. But high doses cause stupor, comas, and convulsions. **Opium** and its derivatives — **morphine**, **codeine**, and **heroin** — are narcotics.

Opium

Opium is made from the juice of poppy pods that are grown mainly in Turkey and Mexico (Fig. 37-7). It is both eaten and smoked. It dulls the senses. It also causes a relaxed dreamy feeling. And it can make the user unconscious for hours. Opium is a powerful narcotic. And, it is highly addictive.

Morphine and Codeine

These opium derivatives are used by doctors. They are given to patients with severe pain that can't be controlled in other ways. "Morphine cocktails" are often given to dying cancer patients who are suffering from unbearable pain.

Both of these drugs, however, are addictive. A user develops a strong craving for the drug. If the craving is satisfied, the user needs more drug to satisfy it the next time. As a result, these drugs are usually prescription drugs. Doctors carefully control the amounts given to patients. Some codeine is found in certain brands of non-prescription pain-relieving pills.

Fig. 37-7 Poppy pods from certain kinds of poppies contain opium.

Heroin

Heroin is the most dangerous narcotic. It is a powerful addictive drug. It has no medical uses; it doesn't even relieve pain. It is also very expensive. As a result, the addict usually turns to crime to support an ever-increasing need for the drug.

Heroin is used three ways:
- Mainlining — injecting of the drug into the veins
- Skin-popping — injecting the drug under the skin
- Snorting — sniffing the drug into the nostrils

Section Review

1. What is a narcotic?
2. Name three opium derivatives.
3. Describe the symptoms of opium use.
4. Describe the medical uses of morphine and codeine.
5. Heroin is expensive. Why does this make heroin even more dangerous?

37.8 Stimulants and Depressants

Stimulants

Stimulants are drugs that excite users by activating the nervous system. **Caffeine** is the most commonly used stimulant. It is found in tea, coffee, cola soft drinks, chocolate, and some non-prescription pain-relieving pills. It is a mild stimulant. Nonetheless, doctors recommend that you don't use too much of it. Four cups of coffee a day is a maximum for most people.

Cocaine is a powerful stimulant. It is also a hallucinogen. This white powder is called "snow" by pushers. Users are called "snowbirds". This drug is very dangerous. At first it causes excitement and happy feelings. Then it causes depression and a split personality. A user may become aggressive and violent. An overdose can cause death. Cocaine is either snorted or injected.

Amphetamines [am-FET-ah-meens] are also powerful stimulants. Common amphetamines are **benzedrine**, **dexedrine**, and **methedrine**. These are sometimes prescribed by doctors to reduce appetite in overweight patients. However, they are also taken illegally to get "nice feelings". Methedrine is also called "**speed**". Users sometimes mainline this drug to get an extra high. Doctors say that the average life expectancy of a mainliner is just 5 years after mainlining begins.

Depressants

Depressants are drugs that have a calming effect on users. They are also called sedatives. Among the depressants are **barbiturates** and **tranquilizers**.

Most of these drugs are made for legal purposes. They help relieve tension and anxiety. They calm a person and help bring on sleep. Unfortunately, these drugs are being abused.

Barbiturates [bar-BICH-u-rits] are used in sleeping pills. They induce a feeling of relaxation. Then sleep usually follows. They are sometimes used to help a person through periods of unusual stress. They are, however, habit-forming. And addiction can result from continued use. Addictive effects can begin after 2 weeks of regular use.

Tranquilizers [TRANG-kwi-li-zers] are drugs that have a calming effect without inducing sleep. Canadians use over 6 000 000 tranquilizer tablets a week! Valium is the most commonly used tranquilizer. Again, 2 weeks of continued use can cause addiction.

Since tranquilizers don't induce sleep, they can be taken during the day. One serious problem results from this fact. The tranquilizer user may also use alcohol. Alcohol is also a depressant. And the two together create a dangerous situation. The most common cause of overdose seen in hospital emergency wards comes from the combined use of tranquilizers and drugs.

Section Review

1. What is a stimulant?
2. Name the most commonly used stimulant.
3. Describe the symptoms of cocaine use.
4. a) Give a legal use of amphetamines.
   b) How dangerous can the use of "speed" be?
5. What is a depressant?
6. What is the difference between barbiturates and tranquilizers?

Main Ideas

1. A drug is any chemical that affects the body's control systems.
2. Some drugs are habit-forming; others are addictive.
3. Tobacco smoking is our most serious health problem.
4. Alcohol is a habit-forming drug for most people and addictive for some.
5. Alcohol and marijuana users should NEVER drive while these drugs are in their bodies.
6. Marijuana, angel dust, and LSD are hallucinogens.
7. Opium, morphine, codeine, and heroine are narcotics.
8. Cocaine and amphetamines are stimulants.
9. Tranquilizers and barbiturates are depressants.

Glossary

addict	a person who cannot stop taking a drug without help
alcoholic	a person who must drink regularly
depressant	a drug that calms the user
hallucinogen ha-LOO-si-no-jen	a drug that causes hallucinations
narcotic	a drug that induces sleep and relieves pain
stimulant	a drug that excites the user

Study Questions

A. True or False

Decide whether each of the following statements is true or false. If the sentence is false, rewrite it to make it true. (Do not write in this book.)

1. All drugs are illegal.
2. The main cause of lung cancer is smoking.
3. Sidestream smoke is harmless.
4. Marijuana can affect driving ability.
5. Methedrine, or "speed", is a depressant.
6. Barbiturates are used in sleeping pills.

B. Completion

Complete each of the following sentences with a word or phrase that will make the sentence correct. (Do not write in this book.)

1. A person who suffers withdrawal symptoms is called an ▓▓▓ .

2. THC can remain in the body for ▓▓▓ days after smoking marijuana.
3. Many weeks after angel dust is used, ▓▓▓ can occur.
4. The most dangerous narcotic is ▓▓▓.
5. ▓▓▓ are drugs that have a calming effect without inducing sleep.

C. Multiple Choice

Each of the following statements or questions is followed by four responses. Choose the correct response in each case. (Do not write in this book.)

1. The main cause of death in Canada is
    a) car accidents
    b) smoking
    c) alcohol consumption
    d) murders and suicides
2. Alcohol is a
    a) depressant   b) stimulant   c) narcotic   d) barbiturate
3. Angel dust and LSD are
    a) narcotics   b) sedatives   c) stimulants   d) hallucinogens
4. To help people lose weight, doctors may prescribe
    a) tranquilizers   b) barbiturates   c) amphetamines   d) narcotics

D. Using Your Knowledge

1. a) Explain the difference between a habit-forming and an addictive drug.
    b) Which do you think caffeine is? Why?
2. Why has the government not made the sale of tobacco illegal? Make a list of possible reasons.
3. Make a list of methods we could use to keep drinking drivers off the roads.
4. Some people defend marijuana use with statements like these: "It's no worse than tobacco or alcohol." "What I do to my body is my own business." "I like it and it doesn't do me any harm." Write a one page reaction to these statements. Do you agree or disagree? Why?

E. Investigations

1. Collect ten cigarette ads from magazines. Write a report on the methods used by the advertisers to get people to smoke.
2. Find out what the legal blood alcohol level is for motorists in your province. Also, find out the drinking patterns that can lead to this level.
3. Research the laws that govern alcohol and driving. What are the penalties for exceeding the legal blood alcohol limit? for impaired driving? for drunk driving?

4. Prepare for and participate in a debate on drug use. The debate should deal with these areas:
   - Drugs and the law
   - The availability of drugs in your community
   - Dangers from drugs
   - Why people use drugs
   - Treatment of drug users
   - Alternatives to drug use

38 Human Population Growth: Problems and Solutions

38.1 Population Growth in Natural Populations
38.2 Human Population Growth
38.3 Activity: The Rate of Human Population Growth
38.4 The Future?

Hunger...starvation...sickness...wars. Are these symptoms of an overcrowded earth? Is the situation hopeless? Is there anything we can do?

In this chapter you will study population growth. Then you will look at the future and your role in **"Building a Better Future"**.

38.1 Population Growth in Natural Populations

A **population** is a group of individuals of the same species living in the same area. Thus the muskrats in a pond are a population. The deer in a meadow are a population. The poplar trees in a woods are a population. And the humans in a city are a population.

The first three populations — muskrats, deer, and poplar — are natural. Their numbers are controlled by the laws of nature. What are those laws? And how do these laws control population growth?

Population Growth Curves

Many scientists have studied population growth. One of these did her experiments with yeast cells (Fig. 38-1). She started a yeast

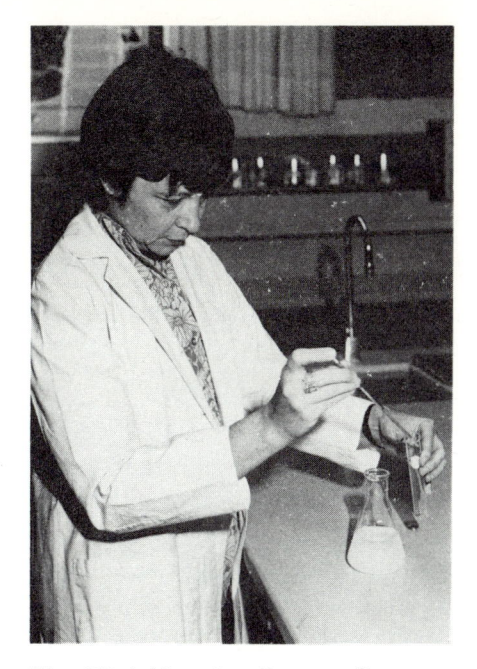

Fig. 38-1 Yeast cells are often used for population studies. These cells multiply rapidly. Therefore results are obtained quickly.

culture by putting a few yeast cells in a suitable medium (see Section 12.7, page 158). Then she counted the yeast cells in a certain volume of medium every 5 h for a total of 25 h. Then she graphed the results. This gave the **population growth curve** shown in Figure 38-2. Note that this graph has three main regions:

- Region A — The population size increases slowly.
- Region B — The population size increases more rapidly.
- Region C — The population size levels off and stays constant.

Most natural populations follow a growth curve like this one. Now let's look at some explanations.

Factors Affecting Population Size

Biotic Potential and Environmental Resistance In Region A the population size increases slowly. This is because the number of "parent" yeast cells is very low.

In Region B the population size increases more rapidly. This is because there are more and more "parent" yeast cells to make new cells.

In Region C the population size levels off and stays constant. Why is this so? All species have a tendency to reproduce. This tendency is called the **biotic potential**. The biotic potential tends to increase the

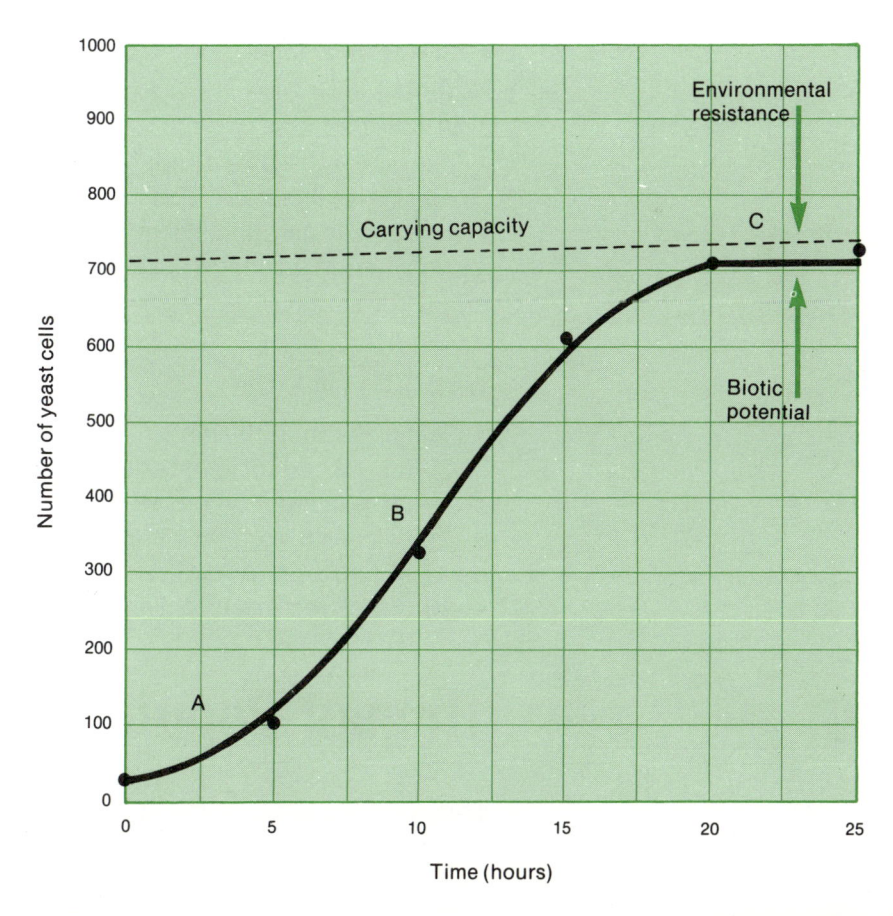

Fig. 38-2 Population growth curve for a yeast population

Section 38.1 569

population size. However, other factors work against that tendency. These factors are called the **environmental resistance**. As the graph shows, population size levels off when the environmental resistance balances the biotic potential. The ecosystem is now said to be at its **carrying capacity**. What factors do you think make up the environmental resistance?

Birth Rate Death Rate and Migration Here's another way of looking at factors that affect population size. There are three main factors that affect population size in an area:

- birth rate.
- death rate.
- migration.

The **birth rate** tells us how fast new individuals are being added to the population. The **death rate** tells us how fast individuals are dying in the population. And **migration** tells us how many **immigrate** (move into) or **emigrate** (move out of) the area. When the three are combined, we get an idea of how fast the population is growing.

Several factors affect the death rate:

- overcrowding.
- lack of shelter.
- disease.
- fighting.
- environmental pollution.
- shortage of food.
- human activities.
- predators.

The same factors tend to affect the birth rate. They make the parents less healthy. Thus fewer healthy young are born. Some of the same factors also affect migration. Animals tend to immigrate to less crowded areas where there is food. Or, if you like, animals emigrate from crowded areas where there is fighting and little food.

Section Review

1. What is a population?
2. What is the biotic potential of a population?
3. What is meant by environmental resistance?
4. Why does a population growth curve level off?
5. **a)** What 3 main factors affect population size?
   b) What factors affect those 3 factors?

38.2 Human Population Growth

Human population growth is controlled by most of the same factors that control other populations. These are disease, overcrowding, shortages of food, and fighting. Centuries ago these factors kept population numbers in check. Life expectancy in Europe was only 9 or 10 years in the 1500s. The death rate was high. Disease and wars killed many. As a result, few people were alive to reproduce.

In recent years, however, the death rate has decreased. Better health care, technology, and modern agriculture are the main reasons. These factors have lowered the environmental resistance. And population numbers are increasing at an alarming rate. Table 38-1 shows the population numbers of the world since the year 0. These are plotted in Figure 38-3. What forms of environmental resistance might stop this rapid growth?

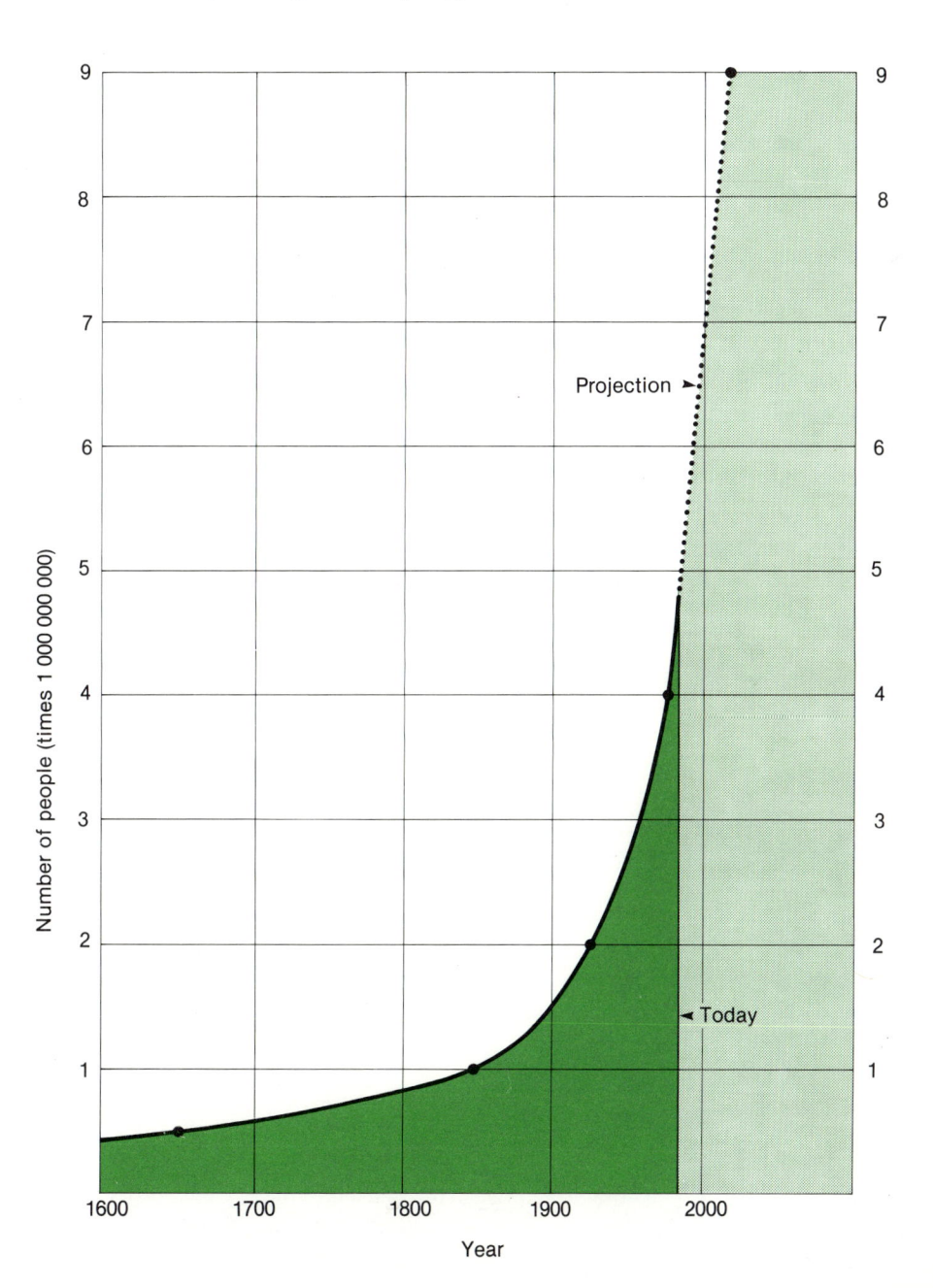

Fig. 38-3 Human population growth curve

Section 38.2   571

Table 38-1 Human Population Growth

Year	Number of people	Years for doubling to occur
0	250 000 000	
1650	500 000 000	1650
1850	1 000 000 000	200
1930	2 000 000 000	80
1976	4 000 000 000	46
⋮	⋮ Projection ⋮	⋮
2010	8 000 000 000	34

Section Review

1. What factors have lowered the environmental resistance in the human population since 1500?
2. Approximately how many people are on earth today?
3. In 1850 the earth's population was 1 000 000 000 people. Another 1 000 000 000 were added in the next 80 years. Copy Table 38-2 into your notebook. Then complete it using Figure 38-3.

Table 38-2

Number of people	Years needed to add another 1 000 000 000 people
1 000 000 000	—
2 000 000 000	80
3 000 000 000	
4 000 000 000	
5 000 000 000	
6 000 000 000	

38.3 ACTIVITY The Rate of Human Population Growth

Figure 38-3 gives you a good idea of how fast the earth's human population is growing. This activity should give you an even better idea. In this activity the jar represents the earth. You are to simulate the growth rate of the human population using these rules:

1. Each bean represents 250 000 000 people.
2. Each second of time represents 10 years.

Problem

Can you simulate the growth rate of the human population?

Materials

beans (about 40)   watch (with seconds)   jar

Procedure

a. Copy Table 38-3 into your notebook. Study the table closely. Then complete the last two columns.
b. Place 1 bean in the jar. This bean represents 250 000 000. That's the number of people on earth in the year 0 A.D.
c. Wait 165 s then add 1 bean. The 165 s represents 1650 years. That's the time taken for the population to double to 500 000 000 (represented by 2 beans).
d. Wait 20 s then add 2 beans. How many years does this represent? The jar now contains 4 beans. How many people does this represent?
e. Continue this process using the numbers you calculated for the table.

Table 38-3   Human Population Growth Rate

Year the population doubled	Number of people	Doubling time	Number of beans in jar	Time you have to put beans in jar
0	250 000 000	—	1	
1650	500 000 000	1650 years	2	165 s
1850	1 000 000 000	200 years	4	20 s
1930	2 000 000 000	80 years		
1976	4 000 000 000	46 years		
2010	8 000 000 000	34 years		

Discussion

1. What is meant by **doubling time**?
2. What happens to the doubling time as the years go by?
3. Describe in your own words the growth rate of the human population.

38.4   The Future?

The population of the earth is now increasing at the rate of about 250 000 people/d. That's about 90 000 000 people per year, or over three times the population of Canada! The doubling time is now about 34 years. Suppose it stayed at that level. Then in about 400 years every bit of land on earth would have densities of people as great as in our largest cities (Fig. 38-4). And in about 850 years there would be about 120 people for each square metre of land and water! It is clear that this cannot happen.

What Will Happen?

The human population growth must level off soon. Otherwise there will be little room for people and no room to grow their food. How might this levelling occur?

The growth rate can only be lowered by changes in the birth rate, death rate, or both. History shows that three major factors have increased the death rate. These are

- starvation
- war
- disease

Some experts feel that these factors will affect the death rate more and more as the years go by. That is, they tell us that we can expect more and more people to die from starvation, wars, and disease.

The same factors, of course, affect the birth rate. The birth rate drops during famines. It also drops during wars because families are separated. And diseases can lower the birth rate. So, if we do nothing, these natural factors may level off our population growth. This is not a pleasant thought, is it? Who wants to grow up in a world where the population is governed by famine, disease, and wars? But there is an alternative. Let's look at it.

Zero Population Growth

Suppose each family had only two children. Then, in the long run, population growth would level off. That is, there would be no population growth. This is called **zero population growth**.

Suppose every one on earth agreed today to aim for zero population growth. That doesn't mean that the earth's population would stay at its present 4 500 000 000. About half the people on earth haven't had their children yet. Also, these people would have their children while they and perhaps even their parents are still alive. Thus the earth's population would still continue to grow for 30 or 40 years. It would level off at about 8 - 10 000 000 000 people. Most scientists agree that, with good management, the earth can support that many people. In other words, that is the carrying capacity of the earth.

Food for the Future

How can we feed more people, even if we stay below the carrying capacity? Even now millions starve to death every year. And about three quarters of the people in under-developed countries don't get enough to eat (Fig. 38-5). They get enough to stay alive. But their bodies are weak and they get diseases easily.

Fig. 38-4 Imagine people this dense on all the land on earth. Where would we grow food?

Fig. 38-5 Hunger and disease are part of everyday life for millions of humans.

Most experts agree that there will be major famines in some countries in this decade. There have already been some smaller ones. These countries are mainly in Africa and Asia. Their crops are being grown on marginal farmland. Even in a normal crop year the people barely get enough to eat. A drought, flood, hot spell, or cold spell can reduce the harvest. If this happens, many people will not get enough to eat.

History shows that people will starve to death during famines, even though other countries like ours have surplus food. Today farmers in Canada and the United States cannot sell all the food they can grow. Yet millions in other countries are dying of starvation. And hundreds of millions go to bed hungry every single day.

Why does this happen? Most of the reasons seem to be political. It costs money to help other countries. Therefore helping others would raise our taxes. And governments that raise taxes too much may not stay in power very long. Some governments at the receiving end cause problems, too. They insist on distributing the food. But their distributing systems are poor. Therefore the starving often do not get the food.

Canada does have an active foreign aid program. But the amount

Section 38.4

we spend is small compared to what we spend on any one of highways, weapons, holidays, cigarettes, or alcohol.

In 1982 Britain and Argentina fought a war over the Falkland Islands. It was a short war. But it cost over $3 000 000 000. That same money would have fed, clothed, and educated 1 000 000 starving children for 10 years. Or it would have educated 100 000 000 people enough that they could become self-sufficient in agriculture, in education, and in medicine. Money can always be found for wars. But we never seem to be able to find enough to help others. We will, though, find enough when we really want to.

What Should We Do?
- Help provide food NOW for starving people.
- Help other countries become self-sufficient in agriculture.
- Continue to develop new strains of crop plants that grow well in other countries.
- Help other countries with irrigation projects.
- Help other countries with education projects. People must be able to read and write to work together to solve population problems.

Disease and War

Disease Disease and war are two more factors that could limit human population growth. A giant plague could sweep across the earth. Also diseases could kill millions of poorly fed people. But experts don't think diseases will play a big role in limiting human population growth. Modern medicine can control most diseases.

War We all know what a nuclear war could do. And the threat is always there. But no one wants this as a method of population control.

The world spends $1 500 000 000 a day on armaments. Poverty, starvation, and disease could be wiped from the surface of the earth if that money were used for health care, food, and education. People across the earth are speaking out more and more against wars (Fig. 38-6). Have you made your voice heard? Wars can disappear from this earth. But only if we want them to. Let's not allow wars to limit population growth.

A Last Word

Four main factors can limit human population growth. These are starvation, disease, war, and birth control. The first three are cruel and undesirable. The last one would be effective. However, there are problems. For example, it will be difficult to get people to limit families to just two children. What, then, do we do? We know that families in well-educated and well-fed populations tend to be small. Perhaps our best solution, then, is to help other countries with food and education.

Fig. 38-6 Thousands gather in cities across the earth to ask for an end to wars.

Section Review

1. What three factors can increase the human death rate?
2. What is zero population growth?
3. Why are people starving in other countries while we have surplus food?
4. Is disease likely to play a big role in limiting human population growth? Why?
5. Figure out how much the world spends every second on armaments.

Main Ideas

1. An ecosystem reaches its carrying capacity when the environmental resistance balances the biotic potential.
2. Three factors affect population size in an area: birth rate, death rate, and migration.
3. The human population is growing at an alarming rate.
4. The doubling time for the human population is now 34 years.
5. Starvation, war, and disease have increased human death rates in the past and present.
6. Zero population growth could solve many of the world's population problems.

Glossary

biotic potential	the tendency of a species to reproduce
doubling time	the time taken for a population to double its numbers
environmental resistance	factors in an ecosystem that work against the biotic potential
population	a group of individuals of the same species living in the same area
zero population growth	a situation in which each family has only two children

Study Questions

A. True or False

Decide whether each of the following statements is true or false. If the sentence is false, rewrite it to make it true. (Do not write in this book.)

1. Most populations grow very quickly near the start.
2. The earth's population in 1850 was 1 000 000 000 people.

3. The present population of the earth is about 4 500 000 000 people.
4. The doubling time is the time needed for a population to double its numbers.
5. Zero population growth means that everyone will have to stop having children.

B. Completion

Complete each of the following sentences with a word or phrase that will make the sentence correct. (Do not write in this book.)
1. The tendency for a species to reproduce is called the _____ _____.
2. Migration into an area is called _____.
3. The doubling time of the human population is now _____ years.
4. Three factors have increased the death rate in past years. They are _____, _____, and _____.

C. Multiple Choice

Each of the following statements or questions is followed by four responses. Choose the correct response in each case. (Do not write in this book.)
1. According to Figure 38-3, the number of people on earth in 1900 was about
   a) 2 000 000 000   b) 1 500 000 000   c) 1 000 000 000   d) 2 500 000 000
2. The earth's population increased from 1 000 000 000 to 2 000 000 000 in
   a) 200 years   b) 34 years   c) 46 years   d) 80 years
3. About how much does the world spend on armaments every second of the day?
   a) $173   b) $1730   c) $17 360   d) $173 600
4. About $250 a year will feed, clothe, and educate a child in an underdeveloped country. How many children could be helped for a year with just one day of the world's military budget?
   a) 6 000 000   b) 6 000   c) 600   d) 60

D. Using Your Knowledge

1. Canada has a fairly low population growth rate. What factors are responsible for this?
2. Some people have suggested that we can solve our population problems by colonizing the planets. What do you think of this? Why?
3. Some people have suggested that it is a mistake to feed starving people in under-developed countries. They say that if some of these people don't die, they will overpopulate their countries. What do you feel about this?

E. Investigations

1. Find out which five countries have the most serious food shortage.
2. Find out which five countries have the most rapidly growing populations.
3. Interview five people about zero population growth. Write a report on your findings.
4. Conduct a survey of 20 people to find out if they would like to see nuclear weapons abolished in all countries.

39 Conserving Natural Resources

39.1 Conserving Soil
39.2 Activity: Percolation Rate and Soil Erosion
39.3 Conserving Forests
39.4 Conserving Wildlife
39.5 Conserving Water
39.6 Conserving Minerals and Energy
39.7 Activity: Getting Methanol from Wood
39.8 Activity: Coal as an Energy Source

Fig. 39-1 The water, air, soil, rocks, and trees are natural resources. Use them wisely. Others need them, too.

Our environment contains **natural resources** (Fig. 39-1). These resources supply the needs of humans and other living things. Forests and wildlife are natural resources. Soil, water, and minerals are also natural resources.

Fig. 39-2 Soil can be washed away in heavy rainstorms. This is called water erosion.

Some natural resources are **renewable**. Trees can be grown, for example, to replace those we use. But other resources are **non-renewable**. Minerals like oil and uranium are examples. When you use them up, they are gone forever.

All resources, renewable and non-renewable, must be used wisely. Otherwise there may not be enough to meet future needs. We call the wise use of natural resources **conservation** [con-sur-VAY-shun]. Conservation is an important step in **"Building a Better Future"**.

39.1 Conserving Soil

The soil you see in fields took thousands of years to form. In fact, it can take up to 1000 years to form just 2-3 cm of soil. Yet, in some parts of Canada, soil is being blown or washed away at the rate of 1-2 cm a year (Fig. 39-2). This process is called **erosion** [ee-ROH-shun].

Causes of Erosion

Erosion is caused by
- leaving the soil bare so wind and water can move it away.
- growing the same crop on the same land for many years.
- removing windbreaks along fence lines.
- cutting down forests, particularly on hilly land.
- improper cultivation of fields (Fig. 39-3).
- failure to put organic matter into the soil.

Fig. 39-3 This field has been ploughed and planted with corn right into the grassy drainage area beside the road. Now nothing remains to slow down soil erosion during heavy storms.

Ways to Decrease Erosion

Many farmers realize that their farming methods are destroying the soil. Here are some of the things they are doing to lessen erosion.
- New types of ploughs are used that leave trash on top of the soil. This helps stop soil from blowing or washing away.

Section 39.1 581

Fig. 39-4 A windbreak. These spruce trees slow down the wind. Then it cannot blow the soil away.

Fig. 39-5 These trees took 10 years to grow this size. But now they help hold soil in place.

- Windbreaks are being planted between fields. These trees slow down the wind (Fig. 39-4).
- Trees are planted on hillsides. They stop the rain from washing the soil away (Fig. 39-5).
- Organic matter such as manure is put on the land. Or a crop such as clover is ploughed under to add organic matter to the soil. Organic matter soaks up water. Then the water cannot wash the soil away as easily.
- Crops are rotated. Growing just corn on land year after year can cause erosion. Now many farmers grow a different crop on the land every second year.
- Special cultivation methods are used. The farmer in Figure 39-6 is **contour ploughing**. He is following the contours of the land instead of ploughing up and down the hill. How will this lessen erosion?

Fig. 39-6 Contour ploughing is ploughing across the slope instead of up and down.

582   Chapter 39

Fig. 39-7 This large equipment gets the work done faster. It can also cause soil erosion if the farmer is not careful.

Section Review

1. Make a summary of the causes of erosion.
2. Make a summary of things that are being done to lessen erosion.

39.2   ACTIVITY   Percolation Rate and Soil Erosion

Many farmers use large equipment like that shown in Figure 39-7. It gets the job done faster. But, because of its weight, it can pack the soil. This can speed up soil erosion, particularly on hilly land. Try this activity and see if you can figure out why.

Percolation rate is how fast water soaks into soil.

Problem

How are percolation rate and soil erosion related?

Materials

watch (with seconds)
large juice can with both
　ends removed

small can (about 500 mL)
　with one end removed
small bucket of water

Fig. 39-8 This student found soil eroding on a hill in a park.

Procedure

a. Find a place where soil is eroding (Fig. 39-8).
b. Work the large can into the soil to a depth of about 2 cm (Fig. 39-9).
c. Pour a small can full of water into the large can.
d. Record the time required for all the water to soak into the soil.

Section 39.2   583

Fig. 39-9 Finding the percolation rate of soil

Fig. 39-10 Forests provide places for camping and hiking.

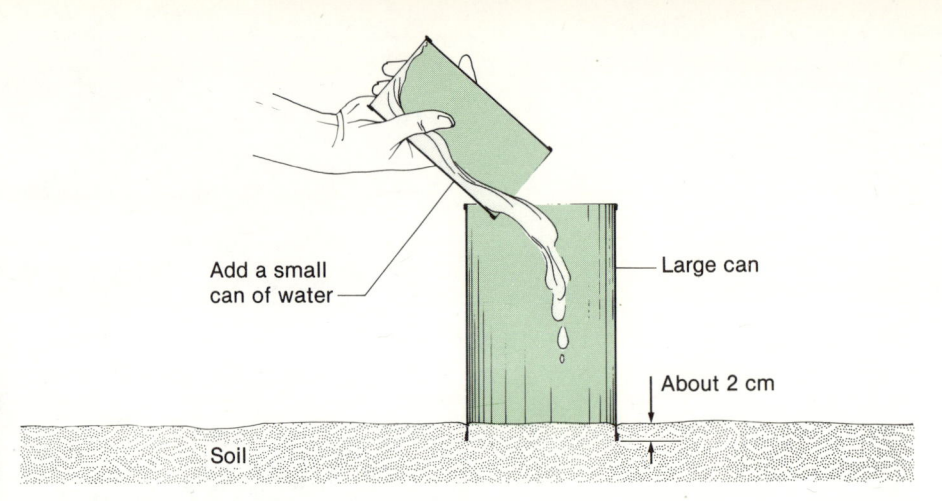

e. Repeat steps (b) to (d) in two more places in the eroding area. Average your results.
f. Now find a place that is not eroding. Pick a place as close as possible to the eroding area.
g. Repeat steps (b) to (e) at this place.

Discussion

1. Which area had the higher percolation rate, the eroding or the non-eroding area? (The one with the higher percolation rate is the one that took the longer time for the water to soak in.)
2. Think about the area with the high percolation rate. How easily can water soak into it? What will happen to rain that lands on it during a heavy rainstorm?
3. How are percolation rate and soil erosion related?
4. What factors could give soil a high percolation rate?
5. What factors could cause soil erosion?

39.3 Conserving Forests

Forests are one of Canada's most important natural resources. They support the lumber and pulp industry. They provide wood to build homes and other buildings. They provide pulp for making paper. They are the habitat for many animals. They prevent erosion and flooding. And they are a source of recreation for many (Fig. 39-10).

Generally speaking, we Canadians cannot be proud of our forest conservation practices. We have not, in many cases, used this resource wisely. Here are some of the things we have done:

- We have used poor practices when cutting forests. Little thought was given to the future.
- We have not replanted as many trees as we cut.
- We carelessly burn thousands of hectares of forest every year. Campfires and cigarettes are the main causes.

Fig. 39-11 Two or three trees are usually planted for every one that is removed. Why is this done?

In recent years, however, forestry practices have been improving.
- Forest fire control methods are much more effective. Helicopter crews move in quickly to stop fires started by lightning or thoughtless people.
- Forest management practices have improved. Stands of trees that are being managed for lumber are **thinned**. This lets the remaining trees grow faster. Also diseased and crooked trees are removed. This is called **improvement cutting**.
- Both governments and private companies realize the need for **reforestation** (Fig. 39-11). This is the practice of renewing a forest by planting seeds or small trees.
- Harvesting methods are being used that take the future into account. For example, **selective cutting** is used. That is, only the mature trees are cut. Also, **block cutting** is used. One area, or block, of trees is completely cut. But the trees around it are not cut. Then seeds blow into the block and reforest it.

Finally, **clear cutting** is used. All the trees in an area are cut. With certain species of pine, this results in a faster natural reforestation. With other species, however, reforestation must be done by humans. Clear cutting can cause problems. Soil erosion is one of these. Also, the area often grows over with unwanted species. Finally, some companies and governments don't always ensure that reforestation occurs.

Section Review

1. In what three ways have our forest conservation practices been poor in the past?
2. Why are forests sometimes thinned?
3. What is improvement cutting?
4. What is reforestation?
5. What is selective cutting?
6. What is the difference between block cutting and clear cutting?

39.4 Conserving Wildlife

The term **wildlife** means all the wild animals in an area. However, biologists who manage this resource often use the term to mean just mammals.

Food, Furs, and Fun

This resource is, too often, managed just for those species that are worth money to us. These tend to be species we want to kill for one of three reasons. We call these the three F's.
- food
- furs
- fun

Fig. 39-12 The black bear belongs to all the 3 F's. It can be eaten. Its fur has some uses. And killing a large mammal like this is the lifetime ambition of many sport hunters.

Because of this, many wildlife biologists classify mammals as
- food species,
- fur bearers, and
- game species (Fig. 39-12).

People in parts of our country kill animals for food. The deer and moose they shoot may be their only meat for the year. A few people in our country kill animals to get furs for their own clothing. Most furs, however, are shipped to foreign markets to be used by the fashion industry. And some people kill animals for pleasure. This is called **sport hunting**.

A great deal of money can be made out of killing animals. Hunting permits and hunters bring in millions of dollars. Therefore our provincial governments tend to manage the countryside to create more of the species we want to kill.

Recently, though, more attention has been given to other species. People are beginning to realize that other species have a right to live on this earth. They also realize that every time a species disappears, our ecosystem becomes more fragile. We are part of the ecosystem. Therefore our existence depends on the existence of the other species.

Habitat Destruction

Conservation is not simply a matter of protecting animals from hunters and trappers. It is true that hunting and trapping have harmed many species. Indeed, they have made some species **extinct**. These will never again be seen on earth. However, hunting and trapping are now carefully controlled by our governments. Thus these activities should no longer endanger animal species.

More harm is probably being done today by **habitat destruction**. We have replaced natural habitats with cities and farms. Cottages

Fig. 39-13 This racoon lives in a large city. Food (garbage) is easy to get. Ravines, garages, and attics provide homes.

586   Chapter 39

dot the countryside. And roads crisscross all parts of the country. Human activities are everywhere. Some animals adapt well to the changed environment. Deer, for example, are much more numerous since we have cleared the land of trees. And racoons love the city (Fig. 39-13). But most wildlife need their natural habitat.

Conserving Fish and Birds

Like wildlife, the fish and birds that get the most attention by wildlife biologists are those we want to kill. Streams and lakes are managed for trout and other sports fish. Other species are often deliberately killed. And marshes are managed for ducks and other game birds.

Fortunately, the habitat that is good for game species is often good for a host of other species. However, habitat destruction is taking place faster than management can replace it. Water pollution kills many species of fish. And the draining of wetlands has forced birds into smaller and smaller areas. Action is needed now to prevent further habitat destruction.

Section Review

1. What does the term "wildlife" mean?
2. How are mammals classified by many wildlife biologists?
3. How is animal habitat being destroyed?

39.5　Conserving Water

The Need for Water

All living things need water. Some of these are aquatic. They must live entirely in water. Fish are examples. Others, such as dogs, most birds, humans, and grass are terrestrial. They live on land. Yet they too need water. They would die if water were not present in the air and soil.

A great deal of water is needed to grow food. Growing just 1 kg of wheat requires 600 L of water; 1 kg of rice 2 000 L; 1 kg of meat 25 000 - 60 000 L; 1 L of milk 9 000 L. A single corn plant absorbs over 200 L of water from the soil in one growing season (Fig. 39-14).

Industry also uses large amounts of water. At least 500 000 L are needed to make one car!

Each one of us uses, directly or indirectly, about 9 000 L of water a day. We drink about 2 L of that. About 50 L are used in bathing. And the rest is our share of that used by agriculture and industry.

The Need to Conserve

Only 3% of the earth's water is fresh. Therefore care must be taken not to use large amounts needlessly. Global studies show that

Fig. 39-14 This one corn plant used all the water in the containers beside it. Imagine the amount of water needed by a large field of corn.

Section 39.5　587

Fig. 39-15 Do you recycle bottles, cans, and paper? If you do, you are helping to conserve our natural resources.

humans are removing fresh water from the land faster than the water cycle can replace it. In Canada we use about twice as much water as the water cycle returns. Every time we need more water, we pump it from a lake or river. Or, we drill a new well. Sooner or later a limit will be reached. Some scientists feel that, in the next decade, we will no longer be able to find all the fresh water we need.

If this is a problem in Canada, imagine what it is in India! India is short of food, has a large population, and a low average rainfall. India must have more water to grow food. As a result, India drilled 80 000 new wells in just one year. In the same year it installed 250 000 pumps to bring water from lakes and rivers. How long can this continue?

The purpose of water conservation is to make sure clean fresh water is always available. There's lots of water in the oceans. But it cannot be used to grow food unless the salt is taken out. This is a slow and very expensive process. Each one of us, then, must avoid wasting water. We must also speak out against water pollution. And we must speak out against the drainage of wetlands. These areas help hold water on the land. They also help purify it by taking nutrients out of it.

Section Review

1. Describe the need for water to grow food.
2. How much water do you use, directly or indirectly, in a day?
3. Canada has many lakes and rivers. Why should we conserve water?
4. What is the purpose of water conservation?
5. Why can't we use the water in the oceans to irrigate farmland?

39.6 Conserving Minerals and Energy

Minerals provide us with glass, iron, aluminum, uranium, nickel, copper, and many other things. Oil, natural gas, and coal are also considered to be minerals. We can grow more trees. We can replace animals that have been killed. But minerals are a non-renewable resource. Once used, they cannot be replaced in nature. Some of them can, however, be **recycled**. For example, we could save a great deal of our iron resources by recycling cans (Fig. 39-15). The cans are melted down and used over again. Recycling has another advantage. It reduces garbage and litter.

Oil, natural gas, and coal are energy sources. We are running out of oil. We have several decades of natural gas left. And we have enough coal for a few hundred years.

Oil is used to make gasoline, heating oil, diesel fuel, and other fuels. Our country would grind to a halt without oil. Therefore we are looking in the Arctic and Atlantic Oceans for more of it. This scarce and important resource must not be wasted. What are you doing to help conserve it?

Section Review

1. Name several important minerals.
2. What is meant by recycling?
3. Why is oil called a non-renewable resource?
4. Make a list of things you do to help conserve oil.

39.7 ACTIVITY Getting Methanol from Wood

We are running out of oil. Some day there won't be enough to run our cars and trucks. Therefore scientists are looking for other fuels. One possibility is **wood alcohol**. It is also called **methyl alcohol** or **methanol**. In this activity you will get methanol from wood and see if it is a fuel. Heat will drive methanol out of wood.

Problem

Is the methanol in wood a fuel?

Materials

test tube   several small pieces of dry wood
test tube clamp   Bunsen burner

CAUTION: Wear safety goggles during this activity.

Procedure

a. Put the pieces of wood in the test tube.
b. Heat the test tube gently as shown in Figure 39-16. Hold the test tube on an angle as shown. Heat only the bottom 2-3 cm. Keep the top as cool as possible.
c. Continue to heat gently for 3-4 min. Record your observations.
d. Now heat strongly for 3-4 min. From time to time, bring the mouth of the test tube to the flame. Record your observations.
e. Let the test tube cool. Pour out what is left of the wood. Examine it closely. What is it called?

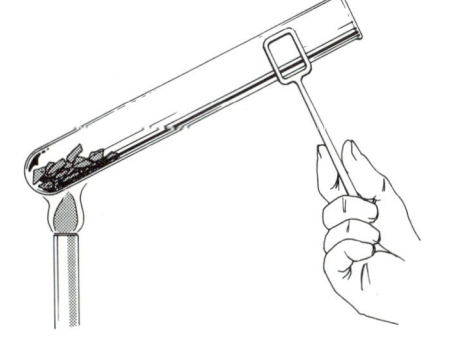

Fig. 39-16 Hold the test tube on the angle shown. Heat only the bottom 2-3 cm.

Discussion

1. What evidence do you have that methanol is a fuel?
2. Scientists are always looking for "clean" fuels. That is, they want fuels that will burn without leaving a black deposit. Is methanol a clean fuel?

3. What problems do you see in the future if we try to make a few million litres of methanol a day?
4. What is the black solid left in the test tube called? What is it used for?

39.8 ACTIVITY Coal as an Energy Source

Coal is abundant in Canada. It has been used and is still used as an energy source. It is also a source of many important chemicals. However, there are some problems associated with coal as an energy source. In this activity you will find out what some of them are.

Problem

What kinds of problems arise when coal is used as an energy source?

Materials

test tube   several small pieces of coal
test tube clamp   Bunsen burner

CAUTION: Wear safety goggles during this activity.

Procedure

Follow the procedure outlined in Activity 39.7. Simply replace the wood with coal.

Discussion

1. What evidence do you have that coal is a fuel?
2. Is it a clean fuel? Explain.
3. What problems do you see in the future if we start burning coal in many of our homes and industries?
4. What is the black solid left in the bottom of the test tube? What is it used for?

Main Ideas

1. Conservation is the wise use of our natural resources.
2. Soil erosion is a serious environmental problem.
3. Soil erosion can be reduced a great deal by proper management.
4. Soil with a high percolation rate tends to erode easily.
5. Forestry practices are improving in Canada.
6. Habitat destruction probably endangers more species than sport hunting does.
7. Canada must start conserving water.

8. Many minerals can be conserved by recycling.
9. Energy cannot be recycled. Therefore it must be conserved to the best of our ability.
10. Methanol is a clean energy source. However, considerable energy is needed to get it out of wood.
11. Coal produces pollution when it is used as an energy source.

Glossary

conservation	con-sur-VAY-shun	the wise use of natural resources
erosion	ee-ROH-shun	the washing or blowing away of soil
percolation rate	pur-ko-LAY-shun	the rate with which water sinks into soil
recycle		to use again
reforestation		the renewing of a forest by planting seeds or seedlings
wildlife		all the wild animals in an area

Study Questions

A. True or False

Decide whether each of the following statements is true or false. If the sentence is false, rewrite it to make it true. (Do not write in this book.)

1. Oil is a renewable resource.
2. Soil with a low percolation rate tends to erode easily.
3. During selective cutting of trees, all the trees in an area are cut.
4. Habitat destruction may harm species more than hunters do.
5. Canadians don't need to worry about water conservation.
6. Recycling can save some minerals.

B. Completion

Complete each of the following sentences with a word or phrase that will make the sentence correct. (Do not write in this book.)

1. The wise use of resources is called _____.
2. Bare soil tends to _____ more easily than soil with vegetation on it.
3. Wildlife biologists often classify mammals as _____, _____, and _____.
4. Wood contains a liquid fuel called _____.

C. Multiple Choice

Each of the following statements or questions is followed by four responses. Choose the correct response in each case. (Do not write in this book.)

1. Which one of the following is a renewable resource?
   a) trees   **b)** oil   **c)** coal   **d)** uranium
2. Which soil will likely have the highest percolation rate?
   a) woodland soil
   b) clay
   c) the soil of a meadow with a thick growth of grass
   d) sand
3. Removing diseased trees from a woodlot is called
   a) selective cutting
   b) block cutting
   c) thinning
   d) improvement cutting

D. Using Your Knowledge

1. Sport hunters are often called "the best conservationists" by defenders of hunting. What do you think of this?
2. City people demand cheap food. As a result, they may be partly responsible for soil erosion. Why is this so?
3. Make a list of examples of habitat destruction that you have seen recently.

E. Investigations

1. Visit a recycling depot. What things can be recycled there? Recycle at least one of these for a month. Write a report on the problems you encountered. Also include in your report the benefits to society of your recycling.
2. Get copies of the hunting and fishing regulations from your government. Study them carefully. Do you feel all species are being adequately protected?
3. Interview a sport hunter about his/her sport. Write a report on your findings.
4. Find out what is meant by an endangered species.
5. Find out what poaching is. What problems are associated with it? Why do people poach? How can it be stopped?
6. Find out about careers in forest management. Write a report on your findings.

Index

NOTE: The pages in **boldface** indicate illustrations.

A

Absorption, of nutrients 258-59, **259**
Acid rain 516
Acid stomach 338
Acne 402
Addict 555, 560
Adrenal gland 414
Adrenalin 368, 414
Air sac 344, **345**, 359, **359**
Alcohol 559-61
   see also Drugs
Alcoholic 560
Algae
   Blue-green, see Blue-green algae
   Brown, 167, **167**
   Green, 166, **166**, 171, **171**
   Protists, see Protists
   Red, 166, **167**
Algal bloom 106, **106**, 141, 515
Amino acids 314, 315
Ammonia, in nitrogen cycle 506
Amniocentesis 463-64, **464**
Amoeba
   feeding of, **135**, 246, **246**
   gas exchange in **264**
   reproduction of 135, **136**
   response to stimuli, 136
Amphetamines 564
   see also Stimulants, Drugs
Ampicillin 127
Amylase 331
   effect on starch 333-34
Angel Dust 562
   see also Hallucinogens, Drugs
Animal Kingdom 232 ff.
   characteristics of, 234-35
   main phyla 234-41
Animals
   breathing systems of, 262-71
   characteristics of, 234-35
   feeding and digestion of, 244-59
   reproductive systems of, 289-305
   transport systems of, 274-88
Annual rings 204, **205**
Antennae, of crayfish, 252
Antennules, of crayfish, 252
Anther 180
Antibiotics 127
Antibody 126
   in plasma 369
Antigens 126
Antiseptics 126
Antitoxin 126
Anus
   of earthworm, 254
   of human, 337
Aorta 360, **360**, 363
Appendicitis 338
Apple scab 157, **157**
Aquatic ecosystems 500
   chemical factors of, 514-17
   freshwater 511-13
Artery 279, 283, **283**, 355-57, **356**
Artificial respiration 348-49, **349**
Aristotle, 22
Arterioles 357, 359
Arthritis 397
Arthropods 238-39, **239**
Asexual reproduction 289-91
   definition of, 289
   see also Reproduction
Atrium 279, 359, 363
Autosome 443
Autonomic nervous system
   see Nervous system
Axil 203

B

Bacilli (bacillus) 107, **107**
Bacteria 107-29, **109**
   as causes of disease 123-25
   characteristics of, 108
   as coliform 337
   in decomposition 117-20
   and food spoilage, 121
   movement of, 109
   reproduction and growth of, 110
   in nitrogen cycle 506
   and nitrogen-fixing 118
   occurrence of 108
Barbiturates 564
   see also Depressants, Drugs
Bark 206
Behaviour 406
Bile 336
Binary fission **104**, 290, **290**
   in Amoeba 135, **136**
   in bacteria 110
   in Paramecium 137, **138**

Biodegradable 120
Biological movement 16
Biology,
   definition of, 4
   importance of, 5
Biome 482, **483**, 528, 534-51, **535**
   coniferous forest 539-42
   definition of 484, 534
   grasslands 544-47
   mountain 547-49
   temperate deciduous forest 542-44
   tundra 536-38, **536**, **537**
Biosphere 482, **483**
   definition of, 484
Biotic potential 569
   see also Population
Bird
   breathing system of, **273**
   heart structure of, **281**
   reproduction of, 295
Birth rate 570
Bladder 379, **379**
Blood 264, 355
   circuit of, 358-60, **358**
   parts of, 367-70
   types of, 370-72
Blood cells 367-70
   types of, 367 (table)
Blood circulatory systems 276
   see also Circulatory systems
   in earthworm 277, **277**
   in grasshopper 278, **278**
   in humans 355-73, **358**
   in vertebrates 279
Blood types 370-71, 370 (table)
   finding type 371-72
Blood vessel 277
Blue-green algae 104-107, **105**
   characteristics of, 104
   importance of, 105
   occurrence of, 104
Blue-green moulds 157, **157**, 158
Bogs 511
Bones
   see Skeletal system
   structure of, 394, **394**
Bone marrow 368
Bottom fauna 523
Botulism 121-22
Brain 240, 409, **409**
   see also Nervous system
Bread mould 154-55, **153**

593

Breathing
    how it occurs 344-46, **346**
    rate of, 347
Breathing systems
    of bird, **273**
    care of, 350-52, **351**
    see also Gas exchange
    how it occurs 344-46, **346**
    lung capacity 347-48, **348**
    parts of, 342-44, **343**, **344**
    rate of breathing 347
Breed
    definition of, 461
Breeding
    of animals 462, **462**
    cross 462
    of plants 461, **461**
    selective 461
Bronchioles 344, **345**
Bronchus 344, **344**
Brown, Robert 43
Bryophyllum **190**, 191
Bud 203
    terminal 203
Budding 290, **290**
    in hydra **296**, 297
    in yeast 156, **157**
Bulb 192, **193**

C

Cabomba 486, 508
Caffeine 415, 563
    see also Stimulants, Drugs
Calyx 180, **180**
Cambium 197, 204, **204**
Cancer
    of digestive system 338-39
    of lung 351, **351**
    of skin 402
Capillaries 259, **259**, 277, 283
    in human circulatory system 355-60, **357**, **359**, 359
    in human lung 344, **345**
    in human lymphatic system 361
Carapace 265, 268, **268**
Carbohydrates 310
    in food groups 310
    testing for, 311
Carbon
    in atmosphere as carbon dioxide, 505
    in carbon cycle 505, **506**
Carbon dioxide
    in gas exchange in animals 262-71
    percentage of air 344
    in photosynthesis 84
    in respiration in plants 87-89
    source of in flowing waters 515

Carnivore 235, 489, 491
Carrier
    of disease 125
Cartilage 394-95
Cells 18, **53**, **57**, **58**, **59**, **67**
    of animals 56, **56**
    cytoplasm of, see Cytoplasm
    definition of, 63
    diffusion of, 67-72
    discovery of, 42
    organization of, 62
    in osmosis 73-76
    in plants 52, **56**
    reproduction of, see Mitosis, 93-98
    theory of, 43
Cell division 92-97
    see also Meiosis, Mitosis
Cell membrane 54, 59, 61
    in bacteria 108
Cell mouth, as in Paramecium 137, **137**
Cell walls
    of bacteria 108
    in mitosis 95
    of plants 52-53
    in yeast 156
Cellulose 142, 164
    in food 311
Central nervous system
    see Nervous system
Centrioles 59
    in mitosis 94-95
Cerebellum 409, **409**
Cerebrum 409, **409**
Cervix 384, **384**
Chemical factors
    in aquatic ecosystems 514-17
    analyzing of in water, 519-20
Chloride, in water, 516
Chlorophyll, 60
    in blue-green algae 104
    in coloured flagellates 141
    in leaves 207
    in photosynthesis 82-83
    in plants 164
Chloroplast 52-53, 60, 62, 164, 171
Cholesterol 373
Chordates 240, **241**
Chromatid, in mitosis, 94-95
Chromatin
    definition of, 57
    function of, 61
Chromosome
    see also Genetics
    in humans 443-44, **444**
    in meiosis 436-40
    in mitosis 93-98, 435
    sex 442-48
    structure of, 61

Chronic bronchitis 352
Cilia 126, 133, 136, 344, 349, **350**
Circulatory system 264
    see also Blood
    blood circuits of 358-60, **358**
    care of, 372-73
    closed system 277, 279, 355
    double system 279, **279**
    open system 278
    parts of 355-57, **356**
    single system 279, **279**
    see also Transport systems
Class
    as in Classification 26-27
Classification
    definition of, 22
    history of, 22
    keys 34-37
    system of Aristotle, 22
    system of Linnaeus, 23
Climate 535
Clitellum 298, **298**
Cloaca 255, **255**
Cloning 464
Clostridium botulinum 121
Club fungi 149-53, **150**
    see also Fungi, club
Cocaine 564
    see also Stimulants, Drugs
Cocci (*coccus*) 107, **108**
Cocoon 298
Codeine 563
    see also Narcotics, Drugs
Coelenterates 236, **236**
Coliform 337
Colon 337
Colour blindness 446-47
Commensalism 496, **496**
Community 482, **483**, 484
    definition of, 484
Composting 119
    compost heap **119**
Conclusion, definition of, 8
Cones 176, **176**
Coniferous forest biome 539-42, **540**
    see also Biomes
Conifers 160, **170**, 175, 176, 540-41
Consumer
    in an ecosystem 488-89
    in food chains 491
Contractile vacuole 135, 137, **137**
Conjugation 171, **171**
Conservation 581-92
    see also Natural resources
Corm 192, **193**
Coloured flagellates **141**
    characteristics of, 141
    occurrence and importance of, 141

594

Corolla 180, **180**
Coronary circuit **358**, 360
Coronary heart attack 360
Cortex 199, **200**, 203, **203**
Cotyledons 177, **177**, 189, **189**
Cowper's gland 382, **383**
Crayfish
 breathing parts and movements of, 268, **268**
 circulatory system of, 278, **278**
 feeding and digestion of, 247-48, 252, 254
 gas exchange in, 264-65
 locomotion of, 252
 reproduction of, 298
 reproduction system of, **301**, 302, **302**
Crop, of earthworm, 247, **247**, 254
Crop rotation 118
Crustaceans **239**
Cuticle, of leaf, 208, **209**
Cuttings, of plants, 192, **192**
Cystic fibrosis 453-54, **454**
Cytoplasm 52, 59
 in Amoeba 135
 in bacteria 108
 definition of, 61
 streaming of, 276

D

Daphnia 251
Darwin, Charles
 theory of, 471-72
Data, definition of, 7
 see also Scientific method
Death rate 570
Deciduous 208, 542
Decomposer 111, 117, 148, 489
Delirium tremens 560
Depressant 417, 559, 564
Diabetes 414
Diaphragm 344-46, **346**
Dialysis tubing 381, **380**
Diatomaceus 143
Diatoms **143**
 characteristics of, 142
 occurrence and importance of, 142-43
Dicot 177, **177**, 204, **204**
Diet 323, 373
Diffusion
 definition of, 69
 experiment of, 68
 through membranes 69-72
 as a transport system, 275, **275**
Digestion 17, 245
 absorption of nutrients 258-59, **259**
 cavities and systems 253

definition of, 253
of earthworm 254
human system 327-41, **327**, **328**
of invertebrates 254, **254**
of vertebrates 255, **255**, 261
Digestive system (human) 327-38
 care of, 338
 chemical digestion 328-30
 function of, 327-28
 parts of, 327, **327**, **328**
Digestive tube 253, **253**
Digestive tract 327
Dihybrid cross 442
Dinoflagellates
 characteristics of, 142, **142**
 occurrence and importance, 142
Diplococcus 108
Disease
 of breathing system 350
 carriers of 125, **125**
 of circulatory system 372-73
 defence against 125-27
 of digestive system 338
 of reproductive system 387-88
 spread of, 124
 of urinary system 380-81
Division of labour 275
DNA 448, **449**
 definition of, 61
 and genetic engineering 464
 see also Genetics
 in mitosis 94
 in viruses 113
Dormancy 219
Doubling time 572
 see also Population
Down's syndrome 454-55
Drugs 417
 alcohol 559-61
 definition of, 554
 hallucinogens 561-62
 narcotics 562-63
 nicotine 557
 stimulants and depressants 564-65
Dujardin, Félix 43
Duodenum 334, 336-37
Dutrochet, Henri 43

E

Earthworm
 digestive system of, 247, **247**, 254, 256-58, **256**, **257**, **258**
 gas exchange in, 269, **269**
 reproduction of, 297-98, **298**
 reproductive system, **300**, 301
 transport system of, 277, **277**
Echinoderms 239-40, **239**

Ecology
 definition of,
 see also Ecosystem
 levels of 482-83
Ecosystem
 aquatic 500, 511-27
 definition of, 484
 energy flow in, 503, **504**
 feeding levels of, 488-98
 model of, 485-87, **486**
 non-living factors in, 499-500
 nutrient cycle in 504-507, **504**
 terrestrial, 500-10, 528-51
Ecotone **544**
Eczema 402
Egg 383
Egg cell 291, 293-300
 in meiosis 436
Egg nucleus 183, 185, **185**
Egg tube 301-303
Elodea **60**, 86
Embryo 186, 292, 294-300, **295**, **296**, 386, **386**
Embryo sac 183, **183**, 189, **189**
Emigrate 570
Emphysema 351-52, **351**
Endocrine glands 412, 413
 adrenal 414
 in pancreas 414
 pituitary 413
 sex 414
 thyroid 414
Endocrine system 412-14, **413**
 care of 411-12
 effect of caffeine on 415
 glands and hormones of 412-14
Endoplasmic reticulum 62
Endoskeleton 240
Endosperm 186, 189
Energy
 and cells 309
 conserving of, 588-90
 content in food, 309, 310, 321-22
 see also Natural resources
 need for 17, 244-45, 262-63
 in plants 215
 use of 322
Energy flow
 in food chain 503, **503**
Environment 482
Environmental resistance 570
 see also Population
Enzymes 253, 315, 317, 328
 amylase 331
 of pancreas 336-37
 pepsin 335
Epidermis 196, 200, **200**, 203, 208, 400
Epiglottis 332

595

Esophagus
  of earthworm 247, **247**, 254
  of human 330, **330**, 331-32
Evaporation 505
Evergreens 208
Evolution
  Darwin-Wallace theory 471-72
  Lamarck's theory 469
  natural selection 471-73
  theory of organic evolution 471
  protective colouration 475-76
Euglena 141, 143-44
Excretion, definition of, 377
Excretory system
  care of, 380-81
  organs of, 377-79, **379**
Exercise 322-23, **323**, 373, 399
Exhaling 346, **346**
Exoskeleton 238, 265
Experiment
  see Scientific method

F
Fallopian tubes 383, **384**
Family
  as in classification, 25-27
Fats
  food groups, 313
  role of, 313
  testing for, 314
Feces 255, 337
Fermentation
  definition of, 111
  in yeast, 156
Ferns 169, **169**, 172
  life cycle of, **173**, 174
Fertilization
  external 292, 303
  internal 292-303
  in plants, see Flowering plants
  problems of, 293-96
  in sexual reproduction 291
Fetus 386
Fever 126
Fibrous root **198**, 199
Fiddleheads 174
Filaments 171, 180
Fish
  breathing parts 265-67, **266**
  circulatory system of, **280**
  reproduction behaviour of, 295, **295**
Fitness level 367 (table)
Flagella (flagellum) 108-9, **110**, 133, 141, 142, 276, 293
Flagellates 141-44
Flatworms 236, **236**
Flower 176, **176**, 180-86, **180**

life cycle of 183, **183**
parts of 180, **180**
pollination and fertilization of, 182
Flowering plants 170, **170**, 175, **215**
  growing of, 215-27
  life cycle of, 183, **183**
  parts of, 180, **180**
  pollination and fertilization 182-86 **183**, **184**, **185**
  reproduction without seeds 191-92, **191**, **192**, 193
  seeds of, 189-90
  vegetative tissues of, 196-98, **196**, 197 (table)
Food 308
  energy of, 309, 310 (table)
  source of in flowing waters 512
Food chains 105, 490-91, **491**, 493-94
  energy flow in, 503-504, **503**
  pyramid of biomass 494, **495**
  pyramid of numbers 493-94, **494**
Food groups
  Canada's food guide 324
  carbohydrates 310
  fats and oils 313
  minerals 320
  proteins 315
  vitamins 320
Food poisoning 121-23, 338
Food vacuole 135, 137, 246, **246**
Food web 491, **491**
Forests
  conserving of 584-85
  see also Natural resources
Fossils 468, **468**
Fragmentation 290, **291**
Freshwater ecosystems 511-27
  flowing waters 512-13, **513**
  standing waters 511, 512 (table)
Frog
  breathing system of, 270-71, **271**
  circulatory system of, 279, **280**
  digestive system of, 255, **255**
  feeding of, 250, **250**
  life cycle of, 299
  reproduction of 294, 298-99
  reproductive system 302-303, **303**
Fronds 172, **173**
Fruits 176, **176**, 187, **187**
Fungi 100, 147-59
  black mould 153-55, **153**
  characteristics of, 147
  classification of, 149
  club, 149-53, **150**, see also Mushrooms
  importance of, 148

occurrence of, 148
sac 156-59, **156**, **157**, **158**, **159**
Fungicide 157

G
Gall bladder 336
Gamete 174, 183, 291, 382-83, **382**, 436-37
Gas exchange
  in animals 262-73
  in human lungs 359, **359**
  on land, 269-73
  problems of, 263
  requirements for, 263
  in water 264-69, **264**
Gastric pouches 254, **254**
Genes
  and chromosomes 439-47
  dominant and recessive 426
  and inheritance 455-57
  see also Genetics, Chromosomes
  in meiosis 436-40
Genetics 422-67
  in animal breeding 462
  in counselling 463
  disorders in humans 452-54
  DNA 448-49
  and environment 457
  Law of independent assortment 440-42
  linkage 440, 445-47
  in meiosis 436-40
  Mendel's experiments and laws 422-29
  mutations 447
  in plant breeding 461
  Punnett Square 426
  sex chromosomes 442-46
  symbols, use of, 428-29, **429**
  and twins, 458, **458**
Genital herpes 124-25, 388
Genotype 428
Genus, definition of, 24
  as in classification, 25-33
Germination 219-20, **219**
  affect of temperature on, 501-502
Glucose
  in photosynthesis 81, 87
  simple sugar 310
Gills
  in gas exchange 264-68, **265**, **266**, **268**
  of mushrooms 150, **151**
Gill arch, of fish, 266, **266**
Gill bailers, in crayfish, 265
Gill covers, of fish, 266, **266**
Gill filaments 265, 266, **266**
Gill rakers 266, **266**

Gizzard, of earthworm, 247, **247**, 254
Golgi bodies 59
Gonorrhea 124-25, 387-88
Grasshopper
  digestive system of, 254, **254**
  feeding of, 248
Grasslands biome 544-47
  see also Biomes
Growth rate
  affect of temperature on, 501-502
Guard cells 209, **209**, 210, **210**
Gullet
  of coloured flagellate 141, **141**
  of Paramecium 137, **137**, 246, **246**

H
Habitat, definition of, 487
Hallucinogens 561-62
  see also Drugs
Hard palate 331
Hardness, of water, 516
Heart
  of earthworm 277, **277**
  of grasshopper 278, **278**
  of human 355-57, 362-67, **362**
  of vertebrates 279, **280**, **281**, **282**
Heart attack 360
Heart Performance Score
  (HPS) 365-67, **366**
Heartburn 338
Hemoglobin 368
Herbivore 235, 489
Heredity 422
  see also Genetics
Hermaphrodites 292
Heroin 563
  see also Narcotics, Drugs
High blood pressure 373
Homologous pair 436
  see also Meiosis
Hooke, Robert 42
Hormones 275, 294, 368, 382, 384,
  412-14
  see also Endocrine system
Horsetails 168, **169**
Horticulture 215
Host 111, 495
Hybrids 423, **423**, 462
Hydra 246, **246**
  feeding of 247, **247**, 251, **251**
  gas exchange in, 264
  reproduction of 297, **297**
  transport system in, 276
Hydrochloric acid, in stomach, 335
Hydrogen, in water cycle, 504-505
Hydroponics 231
Hypertension 373
Hyphae (hypha) 147, 150, **151**, **153**, 154

Hypothesis, definition of, 6
  see also Scientific method
Humus 217

I
Immigrate 570
Index species 522
Ingestion 245
Inhaling 346, **346**
Inheritance 422-31
  blended 431-32
  multiple factor of, 455-57
Insects
  feeding and digestion of, 248-50
Insulin 414
Internode, of stem, 202, **203**
Intestine
  of earthworm, 247, **247**, 254
  glands of, 336
  of human, 336-37, **337**
Intravenous feeding 310

J
Jellyfish
  transport system of, 276, **276**
Joints 393, 395, **396**

K
Karyotype 443-44, **444**
Kidneys 379, **379**
  care of, 379-80
  machines 380, **380**
  transplants 381, **381**
Kilojoules (kJ) 309
Kingdom
  as in classification system, 26-27
  overview of, 27-33
Kwashiorkor 315, **316**

L
Labium
  of grasshopper 248, **249**
  of mosquito 249, **249**
Labrum
  of grasshopper 248, **249**
  of mosquito 249, **249**
Lacteal 259, **259**
Lakes 511
Lamarck, Jean, theory of, 469-70, **470**
Larynx **343**, 344
Law, definition of, 8
  of dominance, see Mendel, 426-27
  of independent assortment, see
  Genetics, 440
  see also Scientific method
  of segregation, see Mendel, 427-28

Leaves 196, 207, **207**, **208**
  deciduous and evergreen 208
  organization of 208
  simple and compound 208, **208**
  tissues of, 208, **209**
Leeuwenhoek, Antony van, 43
Lifestyle 127, 339, 350, 372-73, 380,
  416-17, 552-93
Ligaments 395
Light, effect on plants, 530
Linkage 440
  see also Genetics
Linnaeus, Carolus, 23
Litter 120
Liver 336, 378
Liverworts 167, **168**
Locomotion, definition of, 15
  of a Paramecium **137**
LSD 562
  see also Hallucinogens, Drugs
Lungs
  cancer of, 351, **351**
  capacity of human lung, 347-48, **348**
  of frog, 270, **271**
  of human, **343**, 344
Lung circuit 358-59, **358**
Lymph 355, 361-62
Lymph ducts 355, 361, **361**
Lymph nodes **361**, 362
Lymph vessel 259
Lymphatic system 259, 361, **361**
Lysosomes 59

M
Macronucleus 137
Mandibles 248, **248**, 249
Marijuana 561-62
  see also Hallucinogens, Drugs
Marrow 394
  see also Bones
Marshes 511
Matter, need for, 245
Maxillae 248, **248**, 249
Maxillipeds 248, **248**
Medulla 409, **409**
Meiosis 436-40, **437**, **438**
Membrane cell
  see also Cytoplasm
  in osmosis 76
  selectively permeable 70-72
  types of, 70
Mendel, Gregor 422-29
Menstrual cycle 384-85, **385**
Meristems 196-97, 199, 203
Mesophyll 197, 208
  palisade 208, **209**
  spongy 208, **209**
Metabolism, definition of, 17

597

Metamorphosis, of frog, 298-99, **299**
Micronucleus 137
Microscope **45**
  compound 44-45, **45**
  dissecting 44-45, **45**
  electron **56**
  history of, 42
  parts of, 44-45
  use of, 46-49
Microvilli 337, **337**
Midrib, of leaf, 207
Migration 475, 570
Minerals
  conserving of, 588-90
    see also Natural resources
  in food groups 320
  function and sources of, 320
  in plasma 368
  testing for 320-21
Mitochondria (mitochondrion) 59, **62**
Mitosis 93-98, 435
  definition of 93
  phases in **94**
Moisture, effect on organisms 529
Molluscs 237, **237**
Monera, kingdom of, 102
  characteristics of, 103
Monerans 100-111, 103
  see also Bacteria, Blue-green algae
  characteristics of, 103
Monocot 176-77, **177**, 203, **203**
Monohybrid cross 424
Morels 158, **159**
Morgan, Thomas S. 442
Morphine 563
  see also Narcotics, Drugs
Mosquito, feeding of, 248-50, **250**
Mosses 167-68, **168**
Mountain biome 547-49
  see also Biomes
Mouth 253, 330-31, **330**
Mouth breathing, in frog, 270, **271**
Mouth breeders 295
Mouth-to-mouth resuscitation 349, **349**
Mucin 331, 335
Mucus 126, 269
Muscles 397-99, **398**, **399**
  care of, 399
  how they work, 398-99
  types of, 398
Muscular system 397-99
  care of, 399
Mushroom **150**, **151**
  structure of 150, 152-53
Muskeg 536
Mutualism 495, **495**
Mutations 447
  see also Genetics

N
Narcotics 562-63
Nasal cavity 342, **343**
Natural resources 570-93, **570**
  conserving forests 584-85
  conserving minerals and energy 588-90
  conserving water 587-88
  conserving wildlife 585-87
  soil erosion 581-84
Natural selection 471-73
Nephrons 379, **379**
Nerve 407
  see also Nervous system
Nerve cell, types of, 406-407, **406**, **407**
Nerve cord 240
Nervous system 405-11, **406**
  autonomic system 411-12
  care of, 416-17
  central system 408-10
  function of, 405-406
  nerve cells 406
  reflex action 408, **408**
Nest breeders 295, **295**
Niche, definition of, 487
Nitrogen
  fixation of, 118
  in nitrogen cycle 506
  in water 516
Nitrogen cycle 506, **507**
Nodes, of stems, 202, **203**
Northern boreal forest 539-40, **540**
Nostrils 342, **343**
Nuclear membrane 57, **58**
Nucleoli (nucleolus) 57, **58**, 61
Nucleus, of cell, 52, 54, **57**
  of Amoeba 135
  function of, 61
  role in cell division, see Mitosis
  structure of, 57
Nutrients 245, 308
  absorption of, 258-59, 337
  basic groups of, **309**
  testing in soil 531
  in water 516
Nutrient cycle 504, **504**
  carbon cycle 505, **506**
  nitrogen cycle 506, **507**
  water cycle 504, **505**

O
Observations, 8
  see also Scientific method
Oceanic coniferous forest 539
Oil droplets 54
Oils 313
  in food groups 313
  testing for 314

Oken, Lorenza 43
Omnivore 235, 489
Opium
  see also Narcotics, Drugs
Oral groove 136, **137**, 246, **246**
Order, as in classification, 26-27
Organ, definition of, 63
Organ system
  digestive system of humans **327**, **328**
  examples of, 63
Organelles 59
Organic evolution, theory of, 471
Organic matter, affect on water quality, 521-22
Organisms
  adaptations to environment 513, 524, 528-30
  classification of, 23
  definition of, 13
  factors affecting 528-31
  indicators of water quality 522-23
Osmosis 73-76
Ovary 291
  of female reproductive system 383, **384**
  of Hydra 297, **297**
Oviducts 383, **384**, 385
  see also Fallopian tubes
Ovulation 384
Ovules 181
Oxygen
  in gas exchange in animals 262-71
  needs of bottom fauna 523
  percentage in air 344
  in photosynthesis 85-86
  solubility of, 514
  source of in flowing waters 512, 514
  in water cycle 504-505

P
PABA 402
Palate 331
Palps, of grasshopper, 248, 249
Pancreas 336
Paramecium 136-40, **137**
  feeding of, 246, **246**
  locomotion of **137**
  response to stimuli **139**
Paraplegic 416
Parasite 111, 147, 148, 154, 495
Parasitism 495
Particle theory, definition of, 69
Pathogens 111, 123, 124
Peat 539
Pellicle 136, **137**, 141
Penicillin 127

Penicillium 157
Penis 382, **383**, 385
Percolation rate 583-84
Peristalsis 332, **332**, 334-35
Permafrost 536
Petals 180, **180**
Petiole 207
pH 515
    measuring for in water 518-19
    scale for acidity 515
    testing for in soil 531-32
Phages 113, **114**
Pharynx
    of earthworm, 247, **247**, 254
    of human 331, 342, **343**
Phenotype 428
Phloem 196, **197**, 199, **200**, 203, **203**, 209
Phosphorous, in water, 516
Photosynthesis 17, 80-87, **87**
    in bacteria 111
    in blue-green algae 105
    and carbon dioxide 84
    and chlorophyll 82-83
    in leaves **206**, 207
    and light 83-84
    and oxygen 85-86
    in plants 165, 217
    summary and equation 87
Phylum, as in classification system, 26-27
Pincers, of crayfish, 247, **248**
Pistil 180, **180**
Pith 203, **203**
Pituitary gland 413
Placenta 295, **296**, 386
Planarian
    gas exchange in, 264, **264**
    transport system of, 276, **276**
Plankton 171, **171**
Plants
    characteristics of, 164-65
    coniferous, see Conifers
    effect of environmental factors on, 529-31
    flowering, see Flowering plants
    as food, 215-16, **216**
    growing of, 215-27
    requirements for growth 216-18
    uses of, 216
Plant kingdom 164
    main phyla 166
Plasma 361, 367, 368
Plastids 59, 62
    see also Chloroplast
Platelet 367-68, **367**
Pollen grains 180, **181**, 182, 184
Pollen tube 185, **185**

Pollination 182-84, **184**
Pollution
    chloride as a pollutant 516
    nitrogen as a pollutant 516
    organisms as indicators of, 523-24
    phosphate as a pollutant 516
    sewage and algal blooms 515
    thermal 515
Polyploidy 461
Ponds 511
Population 482, **483**
    definition of, 484, 568
    growth curve 569
    human growth rate 570-72
    future of, 573-76
Powdery mildews 157, **158**
Prairies 544
Precipitation 529
Predator 489
Pregnancy 385
Prey 489
Producers
    blue-green algae 105
    in food chains 490
    in an ecosystem 488
Prostrate gland 382, **383**
Protective colouration, theory of, 475-76
Protein
    breakdown of in stomach 335-36
    deficiency of 315, **316**
    in food groups 315
    nature of, 314
    in plasma 368
    role of, 315
    sources of, 315
    test for, 316-17
Protista, kingdom 132, **132**
Protists 132-46
    algal 141-44
    see also Protozoans
Protoplasm 18
    discovery of, 43
    function of, 61
    structure of, 60
Protozoans 132, **133**
    Amoeba 135-36
    characteristics of, 133
    Paramecium 136-38
Pruning 203
Pseudopods 133, 135, **135**, 246, **246**
PTC tasting 430-31
Pulse 364-65, **365**, 366
Punnett Square 426, 430, **430**
Pus 126, 368
Pyramid
    of biomass 494, **495**
    of numbers 493-94, **494**

Q
Quadriplegic 416

R
Range of tolerance 500, **500**
Reaction time 410-11, **411**
Recycling 588
Red blood cells 367-72, **367**
Reflex action 408, **408**, 410
Reforestation 585
    see also Natural resources
Relative humidity 529
Reproduction
    asexual 171, 180, 191, 289-91, **290**, **291**, **292**, **296**, **297**
    of Amoeba 135, **136**
    of animals 296-305
    of bacteria 110
    of cell (mitosis) 93-98
    of human 382-88, **383**, **384**
    of Paramecium 157, **158**
    sexual 171, **171**, 174, 180, 291-96, **292**, **297**, 382
    of viruses 113
    of yeasts 156, 157
Reptile, heart structure of, 281
Respiration 17, 81
    in animals 262-71
    in bacteria 111
    and carbon dioxide 87-89
    in humans 342
    summary and equation 88-89
Response, definition of, 16
    to stimuli, as in Amoeba **136**
    to stimuli, as in Paramecium 139
    to stimuli, as in human 405
Rhizobium 495, 506
Rhizomes 172, 191, **192**
Ribosomes 59, 62
Ribs 344-46, **346**
Rivularia 105
RNA 448
    see also Genetics
Root 196, **198**, 199-201, **199**, **200**
Root cap 199, **199**
Root hairs 200, **200**, 201
Rooting hormone 226
Roughage 311
Roundworms 237, **237**
Runners, in plants, 191, **191**

S
Sac fungi 156-59
    see also Fungi
Saliva 254, 331
Salivary glands 331
Salmonella food poisoning 221, **221**

Saprophytes 111, 147, 154
Scavenger 492
Schleiden, Matthias, 43
Schwann, Theodor 43
Scientific method 5-8
Scrotum 382, **383**
Slime layer 108
Slough 512
Seaweed 166, 167, **167**
Seed 175, **176**, 186, 189-90, **189**
   germination of, 219-24, **219**
Semen 382, 385
Segmented worms 238, **238**
Seminal vesicles 382, **383**
Sensitive plant 15, **15**
Sex, determination of, 443-44
   linkage 445-47
Sex glands 414
Sexual reproduction 171, **171**, 174,
   180, 291-96, **292**, **297**, 382
Shell 295, **295**
Sickle cell anemia 452-53, **453**
Skeletal system 391-97, **392**
Skin
   care of, 402
   use in excretion, 378, **379**
   of frog, 270
   structure and function of human 400-401
Smoking 350, 373, 556-58
Soft palate 331, **331**
Soil 217-18
   effect of conditions on organisms 530
Soil erosion 516, 581-84, **581**
Sori (sorus) 174, **175**
Species, definition of, 24
   in classification system, 24-33
Sperm cell 291, 293-304
   in human 382, **382**
   in meiosis 436
Sperm chambers 301-303
Sperm ducts 382, **383**
Sperm tube 301-303
Spinal cord 240, 409
Spiracles, in grasshopper, 270, **270**
Spirilla (spirillum) 107, **107**
Spirogyra 171, **171**, 172
Spleen 368, **368**
Sponges 235-36, **235**
Spores
   of black mould 154
   of ferns 174
   of flowers 183-84
   of mushroom 153
Sport hunting 586
Stamens 180, **180**

Starch
   in foods 311
   effect of amylase on, 333-34
   in photosynthesis 81-85
   testing for 312, **312**
Stem 196, 202-207, **202**, **204**
Stigma 181
Stimulant 417, 563-64
Stimulus, definition of, 16, 405
Stomach
   of crayfish 254
   of human 334-36, **335**
Stomata (stoma) 208-9, **209**, 210-11
Stoneworts 167, **167**
Staphylococcus 108, **108**
Staphylococcus aureus 121
Streptococcus 108, **108**
Streptomycin 127
Style 181
Suckers 191, **191**
Sucrose 310
Sugars, 310
   testing for 311
Sulpha 127
Sun Protection Factor 402
Sutton, Walter S. 439
Swallowing 331
Swamps 511
Sweat 378
   glands 378, **378**, 400
   pores 378, **378**, 400
Swimmerets 265, 298, **299**
Swollen glands 362
Syphilis 124, 387
Systemic circuit 359, **358**

T

Tadpole 299, **299**
Tap root **198**, 199
Taste buds 331, **331**
Tay-Sachs disease 453, 463
Teeth 331, **331**
Temperate deciduous forest 542-44
   see also Biomes
Temperature, effect on,
   breathing rate of fish, 265-67
   circulation in earthworm, 284-85
   germination and seedling growth rate, 501-502
   plant growth, 218
   organisms in environment 528-30
Tendons 397-99, **398**
Tension 373
Theory, see Scientific method, 8
Terramycin 127
Terrestrial ecosystem 500, 528-51

coniferous forest biome 539-42
factors affecting organisms 528-32
grasslands biome 547-49
non-living factors 500
temperate deciduous forest biome 542-44
tundra biome 536-38
Testes 291, 297, **297**, 382, **383**
Testicles 382, **383**
Tetanus 123, 125
Throat 342, **343**
Thyroid gland 413
Tissue, definition of, 63
   of flowering plants 197 (table)
Tobacco 555-58
   see also Drugs
Tongue
   of grasshopper 248, **249**
   of frog 250, **250**
Toxin 106, 121, 123
Trachea 269-70, **270**, 342, **343**
Tracheal breathing systems 269-70, **270**
Tracheoles 270, **270**
Traits
   dominant 424
   human, investigating of, 459-60
   recessive 424
Tranquilizers 564
   see also Depressants, Drugs
Transpiration 505
Transplanting 221, **222**, **225**
Transport systems
   in animals 276
   in earthworm 277, **277**
   in grasshopper 278, **278**
   necessity of, 274-75
   simple 275-76, **276**
   in vertebrates 279-88
Tree line 539
Truffles 158
Tuber 192, **193**
Tundra 536-38, **536**, **537**
   see also Biomes
Tubules, in kidney, 379
Turgid 76
Twins 458, **458**

U

Ulcer 335
Ureter 303, **303**, 379, **379**
Urethra 382, **383**, 379, **379**
Urinary system 379, **379**
   care of 380
Urine 379
Uterus 383, **384**, 385

V

Vaccination 126
Vacuole 52-53, **54**, 59
Vagina 384, **384** 385
Variable, definition of, 7
  see Scientific method
Variations 472
  see Natural selection, Evolution
Varicose veins 357
Vascular bundles 203, **203**, 207
Vascular tissue 196-97, **197**
Vegetative propagation 191, **191**, 226, **227**
Veins
  of humans, 279, 282, **282**, 356, **357**
  of leaves, 207
Venereal disease 124, 387-88
Ventricle 279, 359, 363
Venules 357, 359
Venus' flytrap **14**, 15
Vertebrae (vertebra) 240, 409
Vertebrates 240
Viability, of seeds, 219
Villi (villus) 259, **259**, 337, **337**
Viruses 112-13, **112**, **113**
  characteristics of, 113
  reproduction of, 113
  structure of, 113
Vitamins
  deficiency diseases of, 317
  in food groups, 318
  need for, 317
  sources and functions of, 317-18
  testing for Vitamin C
Vocal cords 344
Voice box **343**, 344

W

Wallace, Alfred 471-72
Water
  conservation of 587-88
    see also Natural resources
  need for in humans 379
  need for in plants 217
Water cycle 504, **505**
Water quality
  analyzing 519-20
  chemical factors of, 514-18
  measuring of, 518-19
  effect of organic matter on, 521-22
  organisms as indicators 522-23
Water vapour, in water cycle, 505, **505**
Wetlands 511
White blood cells 367-72, **367**

Wildlife, conserving of, 585-87
  see also Natural resources
Wilt 77
Wind, effect on plants, 530, **530**
Windpipe 342, **343**

X

Xylem 196, **197**, 199, **200**, 203, **203**, 209

Y

Yeasts 156-59, **156**
  importance of, 156-57
  reproduction of, 156, **157**
  structure of, 156
Yolk sac 295, **295**

Z

Zero population growth 574
  see also Population
Zygospore 171
Zygote 185, 291, 293-300, 385, **386**
  in meiosis 436